VICTORIAN SCIENCE IN CONTEXT

VICTORIAN SCIENCE IN CONTEXT

EDITED BY

BERNARD LIGHTMAN

The University of Chicago Press / Chicago and London

Bernard Lightman is professor of humanities at York University. He is the author of *The Origins of Agnosticism: Victorian Unbelief and the Limits to Knowledge* (1987) and coeditor of *Victorian Faith in Crisis: Essays on Continuity and Change in Nineteenth-Century Religious Belief* (1990).

The University of Chicago Press, Chicago 60637
The University of Chicago Press, Ltd., London
© 1997 by The University of Chicago
All rights reserved. Published 1997
Printed in the United States of America

06 05 04 03 02 01 00 99 98 97   1 2 3 4 5
ISBN 0-226-48111-5 (cloth)
    0-226-48112-3 (paper)

Library of Congress Cataloging-in-Publication Data

Victorian science in context / edited by Bernard Lightman.
    p.   cm.
  ISBN 0-226-48111-5 (cloth : alk. paper). — ISBN 0-226-48112-3 (pbk. : alk. paper)
    1. Science—Great Britain—History—19th century. 2. Great Britain—Social conditions—19th century. I. Lightman, Bernard.
Q127.G4V45 1997
306.4′5′094109034—dc21   97-20789
                                                         CIP

♾The paper used in this publication meets the minimum requirements of the American National Standard for Information Sciences—Permanence of Paper for Printed Library Materials, ANSI Z39.48-1984.

Contents

	Acknowledgments	vii
	Introduction by Bernard Lightman	1

PART ONE: DEFINING KNOWLEDGE

1. *Defining Knowledge: An Introduction* — George Levine ... 15
2. *The Construction of Orthodoxies and Heterodoxies in the Early Victorian Life Sciences* — Alison Winter ... 24
3. *The Probable and the Possible in Early Victorian England* — Joan L. Richards ... 51
4. *Victorian Economics and the Science of Mind* — Margaret Schabas ... 72
5. *Biology and Politics: Defining the Boundaries* — Martin Fichman ... 94
6. *Redrawing the Boundaries: Darwinian Science and Victorian Women Intellectuals* — Evelleen Richards ... 119
7. *Satire and Science in Victorian Culture* — James G. Paradis ... 143

PART TWO: ORDERING NATURE

8. *Ordering Nature: Revisioning Victorian Science Culture* — Barbara T. Gates ... 179

9	*"The Voices of Nature": Popularizing Victorian Science* Bernard Lightman	187
10	*Science and the Secularization of Victorian Images of Race* Douglas A. Lorimer	212
11	*Elegant Recreations? Configuring Science Writing for Women* Ann B. Shteir	236
12	*Strange New Worlds of Space and Time: Late Victorian Science and Science Fiction* Paul Fayter	256

PART THREE PRACTICING SCIENCE

13	*Practicing Science: An Introduction* Frank M. Turner	283
14	*Wallace's Malthusian Moment: The Common Context Revisited* James Moore	290
15	*Doing Science in a Global Empire: Cable Telegraphy and Electrical Physics in Victorian Britain* Bruce J. Hunt	312
16	*Zoological Nomenclature and the Empire of Victorian Science* Harriet Ritvo	334
17	*Remains of the Day: Early Victorians in the Field* Jane Camerini	354
18	*Photography as Witness, Detective, and Impostor: Visual Representation in Victorian Science* Jennifer Tucker	378
19	*Instrumentation and Interpretation: Managing and Representing the Working Environments of Victorian Experimental Science* Graeme J. N. Gooday	409
20	*Metrology, Metrication, and Victorian Values* Simon Schaffer	438

Contributors	475
Index	477

Acknowledgments

Without the support and encouragement of funding agencies, colleagues, friends, and family this book would never have appeared. All of the chapters were originally presented at a conference that took place in May 1995 at Bethune College, York University, with financial support from the Social Sciences and Humanities Research Council of Canada, as well as a number of York sources, including the Office of the President, the Office of the Vice-President (Academic), the Division of Humanities (Arts), Bethune College, the Department of Science Studies (Atkinson College), the Office of the Dean of Arts, and the Programme in Science, Technology, Culture and Society (Arts). York administrators to whom I am particularly indebted for their generous support of this project are Susan Mann, Michael Stevenson, George Fallis, Paul Delaney, and Margo Gewurtz. Thanks to William Whitla, Norman Feltes, Michael Collie, Suzanne Zeller, Sydney Eisen, Leslie Howsam, and especially Richard Jarrell for participating in the conference activities. The success of the conference was due in large part to the hard work of my graduate assistant Erin McLaughlin-Jenkins. The basic structure of the volume crystallized for me one afternoon during a delightful visit with the always thought-provoking Jim and Anne Secord in their home in Cambridge, England. Susan Abrams of the University of Chicago Press has been a tower of strength and a fountain of wisdom. Anna Foshay worked weekends and long into the wee hours of several nights to prepare the manuscript. Part of the funding to pay for the preparation of the manuscript came from a Faculty of Arts research grant.

The contributors were a joy to work with, reaffirming my belief that we are blessed with congenial as well as very talented colleagues in the field of the history of Victorian science. I owe a special debt of gratitude to some of those colleagues: to Jim Moore, for feeding me contextualist studies of science since the early eighties; to Martin Fichman, for the many hours of plotting and scheming to put the history of science on the map at York

University; to Frank Turner, for his friendship and guidance over the years; and to Syd Eisen, for inspiring me to pursue graduate work in the Victorian field in the first place and for acting as a sounding board for all my wild ideas ever since.

As usual, my family has provided a supportive environment essential for me to be productive. My brother-in-law, the astute but tragically capitalist Arthur Steinberg, supplied sound advice and more; in-laws Molly and Aaron Feldman stocked my wardrobe with jazzy shirts from Florida and kept my strength up by feeding me roast beef Friday nights; my sister E.J., my brother Jon and his wife Ryla, and my parents constantly gave me their love and encouragement; and finally, my wife Merle, son Matthew, and daughter Ilana create a happy home which centers my being.

Introduction

BERNARD LIGHTMAN

In the past, the phrase "Victorian Britain" was associated with the works of the literary giants of the period—*In Memoriam, Hard Times, Middlemarch* (the list of classics could easily be expanded)—and with key political events of the nineteenth century, such as the Reform Act of 1832, the Crimean War, and the various exploits associated with England's imperialistic ambitions. Why, then, a book drawing attention to science in Victorian Britain? The Victorians were fascinated by the strange new worlds that science opened to them. Exotic flora and fauna from across the empire poured into London daily, many later to be displayed in the British Museum (Natural History) or Kew Gardens to a public hungry for science. Although London was the center of British science—a place where the public could consume natural knowledge in lectures, clubs, museums, and theaters and where scientists could pursue their research with the aid of incomparable resources—there were some who preferred not to encounter nature through a gaslit smog (Morus, Schaffer, and Secord 1992). Charles Darwin perceived the irony in 1837. "It is a sorrowful, but I fear too certain truth," he remarked, "that no place is at all equal, for aiding one in Natural History pursuits, to this odious dirty smokey town, where one can never get a glimpse, at all, that is best worth seeing in nature" (Burkhardt and Smith 1986, 11). Darwin later retired to the rustic charm of Downe to work on his evolutionary theories in peace. Members of the public who shared Darwin's sentiments rushed off to the coast in search of sea anemones or combed the countryside for rare insects or ferns. They met to share their expertise in local clubs and pubs (Secord 1994). Victorians of every rank, at many sites, in many ways, defined knowledge, ordered nature, and practiced science.

The author is indebted to Martin Fichman, George Levine, Barbara Gates, Bruce Hunt, Simon Schaffer, Doug Lorimer, Harriet Ritvo, Susan Abrams, and particularly Jim Moore for suggestions in revising this piece.

Their science was central to their culture. Sometimes sensational (the anonymous *Vestiges of the Natural History of Creation* [1844] taught a bestial evolution), it could be spectacular (the awesome Crystal Palace exhibits of 1851 come to mind) or even ceremonial (as in the funerals of Charles Darwin and Lord Kelvin in Westminster Abbey). Always, Victorian science was political, as prime ministers knew well. The Tory Benjamin Disraeli declared himself "on the side of the angels" after evolutionists made human beings into apes (Davis 1976, 144). William Gladstone, having split the Liberal Party, tried to recover himself by backing Genesis against geology. Arthur Balfour, a future prime minister, pitched his Conservatism philosophically, attacking scientific naturalism head-on in his *Foundations of Belief* (1895).

The towering literary figures of the day also took a strong interest in science and in their works reinforced its close relationship with Victorian culture. Victorian novels abound with characters caught up in scientific pursuits, such as the astronomical protagonist of Thomas Hardy's *Two on a Tower* and the physician Lydgate in George Eliot's *Middlemarch*. Some novelists began to view the human condition through the entangled eye of the evolutionist. Hardy's *Jude the Obscure* depicted a bleak and brutal social world where those unable to adapt to the changing environment do not survive the struggle for existence. Some Victorian novelists who were friendly towards science quite actively studied various aspects of nature. Eliot, for example, accompanied by the philosopher and literary critic George Henry Lewes, set off for a seaside holiday in Illfracombe in May of 1856. As collectors, Eliot and Lewes were bumbling amateurs. The "deep well-like jars" they had dragged with them all the way from London for collecting specimens were not suited to the task. "When we put our anemones into our glass wells, they floated topsy-turvey in the water and looked utterly uncomfortable," Eliot reported in her journal. She was repeatedly obliged to stick her arm to the elbow into the salty water "to set things right." This did not prevent Eliot from experiencing the same delight expressed by so many Victorians when they sallied out into the field on their collecting expeditions. Every day presented her with "some little bit of naturalistic experience," whether she was looking through a microscope or hunting for interesting catches on the rocks (Byatt and Warren 1990, 220–21). For Eliot, this intense encounter with nature sparked a growing desire within her to pursue the scientific quest for clarity—to "escape from all vagueness and inaccuracy into the daylight of distinct, vivid ideas" (Byatt and Warren 1990, 228).

Even those Victorian novelists and poets who were critical of science acknowledged its centrality to Victorian culture or expressed an avid interest in scientific subjects. Alfred, Lord Tennyson's *In Memoriam,* elevated by the Victorians into a national hymn, recorded the painful religious doubts experienced by those who confronted the savage nature, "red in tooth and claw," depicted in midcentury geological and biological theory. Charles Dickens's *Hard Times* exposed the narrowness of the pervasive scientific, utilitarian perspective. Samuel Butler campaigned against Darwinian theory but nevertheless wrote books on evolution, arguing for a Lamarckian view. Despite his hostility towards certain aspects of modern science, John Ruskin collected together in his *Deucalion* all of his geological and botanical essays, claiming that had it not been for a freak of fortune, the gift of a book of poems from a friend, his "natural disposition for these sciences would certainly long ago have made me a leading member of the British Association for the Advancement of Science." Ruskin even admitted that becoming president of the Geological Society had always been the "summit of my earthly ambition" (Ruskin 1875, 6).

Just as the signs of interest in science are evident in all realms of Victorian culture, British scientists were deeply involved with general culture. Throughout the nineteenth century the simultaneous transformation of both British society and natural knowledge placed the scientific elite in a special position "to mediate the conflicts generated by changing conceptions of the sources and bases of the social order" (Schweber 1981, 2). The aristocratic gentlemen of science, those Oxbridge-educated Anglicans who dominated the scientific scene in the first half of the century, provided Victorians with a vision of culture and social order based on natural theology. The middle-class Young Turks of science like Thomas Henry Huxley and John Tyndall, who came from outside the Oxbridge environment, began at the middle of the century to vie with the gentlemen of science for the leadership of the British scientific world and the accompanying cultural authority. They presented an alternative view of culture and society that drew its inspiration from evolutionary modes of thought. Victorian science and culture were inextricably linked in the eyes of the Victorians themselves, scientists and nonscientists alike.

Victorianists have come to realize that the science of the period is central to an understanding of Victorian culture. In the past, Victorian science was singled out by historians of science as noteworthy for new developments in the life sciences associated with Darwin and crucial breakthroughs in the physical sciences linked to the contributions of Kelvin and James Clerk Maxwell. However, in recent years, historians of science have come to recognize that the Victorian era is a particularly important period, when significant features of the relationship between contemporary science and culture first assumed form. During this period many Western nations were transformed by the forces of industrialization, secularization, and urbanization, and they were increasingly dominated by a growing middle class.

Historians of Victorian science learned how to think about the interaction of Victorian science with these powerful social and cultural forces from scholars working in the 1970s and 1980s who were unhappy with an approach that paid too much attention to the intellectual dimension of science. By way of introduction to this volume I will trace the overall trajectory of the history of Victorian science from intellectual history to contextualism and indicate how the essays in this collection contribute to the ongoing task of refining our perception of the subtle interplay of Victorian science with its social and cultural context. Perhaps, had this volume been published ten, or even five, years ago, it would have been the duty of the editor to present a detailed theoretical justification of contextualism in the introduction. However, several fine essays undertaking this task have already appeared, and it is unlikely that those scholars who remain hostile toward contextualism will be convinced by yet another theoretical defense (Shapin 1982; Golinski 1990; Forman 1991). Contextualist historians no longer feel the need to defend an approach to doing history of science that has proven to be such a useful heuristic guide to research and that continues to inspire them to write some of the most exciting publications in the discipline.

Previous to the 1960s, scholarship in the history of science was dominated by approaches drawn from intellectual history in the tradition of A. O. J. Lovejoy. Historians tended to focus on the story of those scientific ideas perceived to be at the root of contemporary science. Naturally, the main plot of the story was the success of Western science and the scientist's intellectual mastery of nature. The struggles of the great scientific heroes were celebrated, while their opponents were either cast in the role of villain or virtually passed over in silence. The social and cultural context within which these heroes labored, when regarded as more than background setting, often was seen as an obstacle to the acceptance of correct scientific theories. What really interested historians of science was the development of ideas internal to science—that apparently autonomous realm of scientific discourse wherein scientific geniuses made their contributions to an ever-growing body of knowledge and fact. Of course this whole approach to the history of science, deeply imbued with the positivist spirit, conferred a privileged status upon scientific ideas that were correct by the standards of the historian's day.

From the 1960s to the late 1980s the discipline was shaped by a crucial debate that took place among historians and sociologists of science, the so-called externalism/internalism debate (Shapin 1992). Departing from the internalist's emphasis on intellectual history prior to the 1960s, historians of science in this period nevertheless strove to avoid the opposite extreme of "externalism," or the stress on external, extrinsic, and nonscientific factors as causing change in science. Whatever camp a scholar seemed to belong to, whether it was externalist or internalist, historians of science promoted an "eclectic selection of the respective 'factors' and a judicious admixture of elements from both orientations towards scientific change" (Shapin 1992, 342-43). However, no adequately informed and systematic debate over externalism and internalism ever took place. As a result, work in the history of science continued to be plagued by the false science/society dualism.

Throughout the sixties and seventies, when the central problematic of the history of science was the externalism/internalism debate, scholars investigating Victorian science contributed to the growing body of work on science, culture, and society. Walter (Susan) Cannon wrote a series of important essays leading up to the publication of *Science and Culture: The Early Victorian Period* (1978), which announced the need for historians of science who treated science "as an integral part of culture and society" (Cannon 1978, 256). Cannon argued that the publication of Darwin's *Origin of Species* in 1859 led to the fragmentation of modern culture, for it resulted in the demise of the truth complex, a universal norm for truth built on the corpuscular theories of Robert Boyle and the philosophic triumphs of Isaac Newton (Cannon 1978, 3, 268). During the same period John Greene focused on the history of evolutionary thought. In his *Science, Ideology and World View* (1981), Greene published a number of elegant essays dealing with Darwin, Herbert Spencer, Huxley, and Darwinism, designed to demonstrate that "the lines between science, ideology, and world view are seldom tightly drawn" (Greene 1981, 2).[1]

While Cannon and Greene tended to concentrate on overlapping intellectual contexts and the "transitions from the dominance of one world view to the dominance of another" (as Greene put it), Frank Turner analyzed the social dimension of the conflict between science and religion (Moore 1989, 4). In a series of important articles published during the seventies and later republished in *Contesting Cultural Authority* (1993) and in his *Between Science and Religion* (1974), Turner drew the attention of scholars to the clash between the scientists and amateurs who were part of the Anglican-Tory establishment, which controlled Oxbridge, the major scientific societies, and government resources for science right up until the middle of the century, and the middle-class professional scientists, bent on secularizing both British science and society. Scientific naturalists like Huxley, Spencer, Tyndall, William Clifford, Lewes, Edward Tylor, John Lubbock, Edwin Lankester, Edward Clodd, and Henry Maudsley put forward new interpretations of humanity, nature, and society derived from the theories, methods, and categories of empirical science, in particular evolutionary science. This cluster of ideas and attitudes was naturalistic in the sense that it would permit no recourse to causes not empirically observable in nature. The ideas of scientific naturalism provided the main weapons for middle-class members of the intellectual elite who were attempting to wrest control of English society from the Anglican clergy.

---

1. On Greene's influence see the "Introductory Conversation" (Moore 1989, 1–38).

Like Turner, the neo-Marxist historian Robert Young eschewed the idealist interpretation that had been pursued by previous scholars and adopted a social-intellectual history approach. In a set of provocative essays written and published in the late sixties and seventies, then later collected together in his *Darwin's Metaphor* (1985), Young attempted to overcome the science/society dualism through the presentation of an ambitious program of research for historians of Victorian science. The resolution of the externalism/internalism debate, in Young's mind, was to be found in a return to Marx and the construction of a more sophisticated version of the "base-superstructure" model of interpretation, which emphasized that "all intellectual and cultural phenomena [the superstructure] are ultimately determined by socioeconomic conditions [the base]" (Young 1985, 166). Young argued that the base-superstructure model could be made serviceable again if a richer and more subtle "theory of mediations and interactions between socioeconomic factors and intellectual life" could be developed (Young 1985, 208). This would allow the inclusion of many cultural factors—philosophical theories, religious views, and political thought—in the analysis. Many of Young's students brought to their work his historiographic approach and produced important contributions in the history of science not just limited to the nineteenth century. Other scholars came to view Victorian science after Young's manner independently of his Marxist assumptions.[2]

Throughout the corpus of his works, Young attempted to show that controversies between nineteenth-century biologists were part of a broader debate concerning humanity's place in nature, and that discussions on scientific, economic, philosophical, political, social, and religious issues took place within a common context, no part of which was isolated from others. Indeed, Young tried to set up an all-encompassing schematism that related pre- and post-Darwinian science to major transitions in the economic, political, social, and intellectual context of the nineteenth century. Whereas Young connected early-nineteenth-century Anglican natural theology to a pastoral, agrarian, and aristocratic world, the evolutionary theories of scientific naturalists, in which the deity was "identified with the self-acting laws of nature," reflected a competitive, urban, and industrialized world (Young 1985, 240). While Turner's interest in the sociology of intellectual change led him to emphasize the conflict accompanying the transitions, Young accentuated the subtle continuity from one period to the next that underlies the more obvious change (Moore 1981, 36–41). Young perceived a continuum running from Thomas Malthus and William Paley to

2. Bohlin has evaluated "the credibility of Young's general contextualist thesis as applied to the historical case of Darwin" (Bohlin 1991, 603). While Bohlin raises some serious objections to Young's handling of the Malthus-Darwin link, which concerns the impact of the context on Darwin, he deals in a cursory fashion with Young's views on Darwin's relationship to the broader social and cultural context (Bohlin 1991, 618–20).

Robert Chambers, Darwin, Social Darwinism, and even beyond to contemporary writings on biology and society. In what seems to be a flagrant contradiction in light of our understanding of the so-called Darwinian revolution, Young argued that natural theologians and scientific naturalists were "fighting over the best ways of rationalizing the same set of assumptions about the existing order. An explicitly theological theodicy was challenged by a secular one based on biological conceptions and the fundamental assumption of the uniformity of nature" (Young 1985, 191).

Though Young's social-intellectual history of Victorian science was vitally important, it really represented a halfway house between the Lovejoy intellectual history approach and the development of contextualism. Young's focus on how the middle class and aristocratic male intelligentsia sought to maintain cultural hegemony, a reading still very reliant on the Marxist notion of superstructure, neglected vast realms of fruitful research. Full-blown contextualist works began to appear in the eighties, as new ideas streamed into the history of science from cultural studies and from areas of study cognate to the field, such as science and gender and the sociology of science. The hallmark of contextualist studies is their emphasis on the way scientific ideas are embedded in material culture such that there are no insides or outsides of science. A rich interdisciplinary approach to examining Victorian science allowed historians to avoid the false analytical distinction between science and society (or base and superstructure), dissolve the categories external and internal, and begin to transcend the science/society dualism (Shapin 1992, 354–56).

Though a body of scholarship has arisen in the history of science in the eighties that shares an emphasis on science in its social and cultural context, contextualists can still find themselves in disagreement. There are, after all, many different kinds of contexts. Should the contextualist accentuate class, imperial, gender, or linguistic contexts? Are these accounts of context in conflict with each other or should the historian strive to synthesize them into a coherent whole?

Contextualist studies that appeared in the early to mid-eighties often continued to focus on the cultural elite. Morrell and Thackray's *Gentlemen of Science* (1981) and Rudwick's *The Great Devonian Controversy* (1985) both centered on the making of knowledge by aristocratic men of science who dominated British scientific institutions and societies in the first half of the nineteenth century. But in his *The Politics of Evolution* (1989), a history of science "from below," Desmond explored the world of radical, lower-class evolutionists that existed in the secular anatomy schools and Nonconformist colleges of London in the 1830s and demonstrated that a thriving scientific culture existed outside, and in opposition to, the elite establishment. By shedding light on a group of previously ignored scientists, Desmond encouraged historians to look beyond both establishment science offered by the scientific gentry and Oxbridge clergy and middle-class scientific naturalism, which challenged the authority of the Tory-Anglican establishment in the latter half of the century.

Other contextualist historians of Victorian science have tried to shift attention away from the class context to what they see as the more important imperial context. Stafford's *Scientist of Empire* (1989), a study of Roderick Murchison, focused on the way in which colonial data influenced important geological debates in Britain. In the same year Smith and Wise published their *Energy and Empire,* which explored how Lord Kelvin's vision of empire led him to a methodology and telegraph theory in opposition to Maxwell's electromagnetic theory. Both books shared an appreciation of how Britain's possession of a vast and expanding colonial empire shaped the development of science, just as it affected every dimension of Victorian life.

Yet another context of increasing interest to scholars centers on gender. Historians have examined how scientific thought provided a naturalistic basis to the sexual divisions of Victorian society and how scientific theory itself was shaped by notions of gender. In her *Sexual Science* (1989), Russett argued that the Victorian period is especially significant for an understanding of the relationship between science and gender, for it was during this time that the scientific conception of female nature first became widely influential, even though scientific interest in the topic dated back to Aristotle. The sexual science of the late nineteenth century possessed unprecedented power, because its practitioners attempted to be far more precise and empirical than researchers had been hitherto, could draw upon new developments in the life sciences as well as on the new social sciences of anthropology, psychology, and sociology, and spoke with the imperious tone of a discipline granted decisive authority in matters social and scientific. The essays of Evelleen Richards have shown that the ideas of the supposedly progressive scientific naturalists were no less gendered than those of the natural theologians of the early nineteenth century. In her piece on T. H. Huxley, Richards demolished the usual depiction of Darwin's bulldog as an enlightened defender of women's rights (Richards 1989). An earlier article by Richards investigated how Darwin deduced the natural and innate inferiority of women from his theory of evolution by natural and sexual selection (Richards 1983). Russett's and Richards's work on the sexist nature of male Victorian scientists opened the door for detailed studies on the struggle of Victorian women who wished to be a part of the scientific scene as well as the reaction of Victorian feminists and female intellectuals to the gendering of science.

Still another area that has received considerable attention in the eighties and nineties revolves around the linguistic context of Victorian science. In her *Darwin's Plots* ([1983] 1985), Beer examined the way in which evolutionary theory was assimilated and resisted by novelists like Eliot and Hardy and concluded that during the Victorian period there existed a shared discourse, which allowed ideas, metaphors, myths and narrative patterns to move "rapidly and freely to and fro between scientists and non-scientists" (Beer 1985, 7). Levine pursued a similar approach in his *Darwin and the Novelists* (1988), where he discussed Dickens, Anthony Trollope, Hardy, Joseph Conrad, and other writers indirectly influenced by Darwin. For Levine, the Victorian novel is the "cultural twin" to the project of Victorian science (Levine 1988, vii). The ideals of Victorian science—truth, detachment and self-abnegation—were echoed in the great aesthetic ideals of Victorian writers. "The Victorian novel," Levine declared, "clearly joins with science in the pervasive secularizing of nature and society, and in the exploration of the consequences of secularization that characterized mid-Victorian England" (Levine 1988, viii). Beer's and Levine's work has stimulated historians of science to look for the two-way traffic that exists between Victorian science and literature, in particular in the language, ideas, and even structure of scientific texts.

The field of the contextualist history of Victorian science, then, is marked by tremendous diversity. Scholars working in the field can disagree on the emphasis to be placed on different contexts, whether they be class, imperial, gender, or linguistic contexts. The recent interest in scientific practice and the audience for science, which will be explored at length in this volume, offer substantially different and sometimes conflicting accounts of context as well. However, the excitement of exploring new contexts of science may eventually give way to the need to see how they all fit together. After all, the historical actors who lived within these contexts may not have seen any conflict in living simultaneously within the multiple, overlapping contexts of one culture. Whereas modern scholars find it necessary to isolate a particular context in order to study the complex interaction with science, Victorian scientists, and those intellectuals and members of the popular reading audience who were influenced by science, may have seen all of these contexts as part of a single, seamless web.

The chapters of this volume reflect the diversity in the field. The authors examine the varied contexts of Victorian science, including its imperial, industrial, political, gendered, ideological, racist, literary, and religious nature. They also explore many areas within Victorian science, including biological thought, astronomy, field theory in physics, probability theory in mathematics, political economy, scientific nomenclature, instruments, laboratories, measurement, fieldwork, and the popularization of science. This book differs in many ways from eminent collective works on Victorian science which appeared in the past. Appleman, Madden, and Wolff's *1859: Entering an Age of Crisis* (1959) limited its focus to one context—religion—and many of the essays were informed by the now outdated notion of warfare between science and religion. Likewise, Inkster and Morrell's *Metropolis and Province* (1983) dealt with one particular context of British science, the social, and actually concentrated on an earlier period.

Though more interdisciplinary than the other collections, Knoepflmacher and Tennyson's *Nature and the Victorian Imagination* (1977) and Paradis and Postlewait's *Victorian Science and Victorian Values* (1981) both emphasized literary perspectives. Moore's *History, Humanity and Evolution* (1989) is closest to this collection in its contextualist spirit, but it is intended to zero in on the history of evolutionary thought.

The book is divided into three sections, which reflect the areas of research currently among the most important in the field. The chapters in part 1, "Defining Knowledge," examine how Victorians answered the question, What is science? As the authority of science grew during the nineteenth century, it became increasingly important for intellectuals, scientists, and social groups to fix the boundary between legitimate and illegitimate scientific knowledge and to locate themselves firmly within the domain of scientific orthodoxy. In this section the authors will address such topics as demarcation disputes involving mesmerism, phrenology, and spiritualism, biology and politics, literary challenges to scientific models of truth, the defining of knowledge so as to exclude women from science, the use of mathematics as a standard of knowledge, and the evolution of the principles of political economy. As these essays reveal, the protean quality of Victorian science persisted even as the professional characteristics with which we are familiar today became slowly established. But a sharply demarcated community of scientists working within specialized disciplines like biology and physics did not exist until the 1870s at the earliest. Although a number of the chapters refer to "biologists," "professionals," and "scientists" in the period before the final decades of the century, these terms are to be taken as legitimate anachronisms that convey to a contemporary reader features of Victorian science in the process of coming into being.

Chapters in part 2, "Ordering Nature," address the issue of audience, or the question, For whom is science written? How does the concern to reach a particular audience—whether it be women, the popular reader, the working class, or supporters of racism—lead scientists and intellectuals to read into nature a variety of messages charged with ideological significance? The title of the section signifies both the Victorian notion of nature as embodying an orderly system of necessary laws that have crucial implications for understanding the social order and the contextualist historian's perception of scientists as imposing this order on nature in accordance with their own vision of society.

In the final part, "Practicing Science," the authors examine the impact of various contexts on the way science was actually practiced during the Victorian period. How did precision instruments, systems of measurement, the use of the camera to represent reality, the development of scientific nomenclature, and the conventions of doing fieldwork discipline the senses and configure the ideas of Victorian scientists? How did practical problems posed by new technological projects interact with contextual factors to mold the direction and content of Victorian science?

Young once wrote that at the "heart of its science we find a culture's values" (Young 1985, 125). We have tried in this volume to explore the heart of Victorian science and have found ourselves coming face to face with the soul of Victorian culture. For scientific, theological, philosophical, and ideological issues are all a part of a common culture. In defining knowledge, human cultures often define themselves; by ordering nature to conform to a particular pattern, scientists and intellectuals frequently reveal the social order for which they yearn; and in the process of practicing science, of measuring, experimenting, and controlling phenomena, we not only find nature but also encounter ourselves as inquisitive, social, and political beings.

References

Appleman, Philip, William A. Madden, and Michael Wolff, eds. 1959. *1859: Entering an Age of Crisis.* Bloomington: Indiana University Press.

Beer, Gillian. [1983] 1985. *Darwin's Plots: Evolutionary Narrative in Darwin, George Eliot and Nineteenth-Century Fiction.* London: Ark Paperbacks.

Bohlin, Ingemar. 1991. "Robert M. Young and Darwin Historiography." *Social Studies of Science* 21:597–648.

Burkhardt, Frederick, and Sydney Smith, eds. 1986. *The Correspondence of Charles Darwin.* Vol. 2. Cambridge: Cambridge University Press.

Byatt, A. S., and Nicholas Warren, eds. 1990. *Selected Essays, Poems and Other Writings.* London: Penguin Books.

Cannon, Susan Faye. 1978. *Science in Culture: The Early Victorian Period.* New York: Dawson Science History Publications.

Davis, Richard W. 1976. *Disraeli.* Boston: Little, Brown and Company.

Desmond, Adrian. 1989. *The Politics of Evolution: Morphology, Medicine, and Reform in Radical London.* Chicago: University of Chicago Press.

Forman, Paul. 1991. "Independence, Not Transcendence, for the Historian of Science." *Isis* 82:71–86.

Golinski, Jan. 1990. "The Theory of Practice and the Practice of Theory: Sociological Approaches in the History of Science." *Isis* 81:492–505.

Greene, John C. 1981. *Science, Ideology and World View: Essays in the History of Evolutionary Ideas.* Berkeley: University of California Press.

Inkster, Ian, and Jack Morrell, eds. 1983. *Metropolis and Province: Science in British Culture, 1780–1850.* Philadelphia: University of Pennsylvania Press.

Knoepflmacher, U. C., and G. B. Tennyson, eds. 1977. *Nature and the Victorian Imagination.* Berkeley: University of California Press.

Levine, George. 1988. *Darwin and the Novelists: Patterns of Science in Victorian Fiction.* Chicago: University of Chicago Press.

Moore, James R. 1981. *Beliefs in Science: An Introduction.* Milton Keynes: Open University Press.

———, ed. 1989. *History, Humanity and Evolution: Essays for John C. Greene.* Cambridge: Cambridge University Press.

Morrell, Jack, and Arnold Thackray. 1981. *Gentlemen of Science: Early Years of the British Association for the Advancement of Science.* Oxford: Clarendon Press.

Morus, Iwan, Simon Schaffer, and Jim Secord. 1992. "Scientific London." In *London—World City 1800-1840,* edited by Celina Fox. New Haven, Conn.: Yale University Press, 129-42.

Paradis, James, and Thomas Postlewait, eds. 1981. *Victorian Science and Values: Literary Perspectives.* New York: New York Academy of Sciences.

Richards, Evelleen. 1983. "Darwin and the Descent of Woman." In *The Wider Domain of Evolutionary Thought,* edited by D. Oldroyd and I. Langham. Dordrecht: Reidel, 57-111.

———. 1989. "Huxley and Woman's Place in Science: The 'Woman Question' and the Control of Victorian Anthropology." In *History, Humanity and Evolution: Essays for John C. Greene,* edited by James R. Moore. Cambridge: Cambridge University Press, 253-84.

Rudwick, Martin. 1985. *The Great Devonian Controversy: The Shaping of Scientific Knowledge among Gentlemanly Specialists.* Chicago: University of Chicago Press.

Ruskin, John. 1875. *Deucalion. King of the Golden River. Dame Wiggins of Lee. The Eagle's Nest.* Boston: Dana Estes and Company.

Russett, Cynthia. 1989. *Sexual Science: The Victorian Construction of Womanhood.* Cambridge, Mass.: Harvard University Press.

Schweber, S. S. 1981. "Scientists as Intellectuals: The Early Victorians." In *Victorian Science and Values: Literary Perspectives,* edited by James Paradis and Thomas Postlewait. New York: New York Academy of Sciences, 1-37.

Secord, Anne. 1994. "Science in the Pub: Artisan Botany in Early Nineteenth-Century Lancashire." *History of Science* 32:237-67.

Shapin, Stephen. 1982. "History of Science and Its Sociological Reconstructions." *History of Science* 20:157-211.

———. 1992. "Discipline and Bounding: The History and Sociology of Science as Seen through the Externalism-Internalism Debate." *History of Science* 30:333-69.

Smith, Crosbie, and M. Norton Wise. 1989. *Energy and Empire: A Biographical Study of Lord Kelvin.* Cambridge: Cambridge University Press.

Stafford, Robert. 1989. *Scientist of Empire: Sir Roderick Murchison, Scientific Exploration, and Victorian Imperialism.* Cambridge: Cambridge University Press.

Turner, Frank M. 1974. *Between Science and Religion: The Reaction to Scientific Naturalism in Late Victorian England.* New Haven, Conn.: Yale University Press.

———. 1993. *Contesting Cultural Authority: Essays in Victorian Intellectual Life.* Cambridge: Cambridge University Press.

Young, Robert. 1985. *Darwin's Metaphor: Nature's Place in Victorian Culture.* Cambridge: Cambridge University Press.

PART ONE

Defining Knowledge

1

Defining Knowledge: An Introduction

GEORGE LEVINE

The chapters gathered in this section all see science in culture. They try to understand early Victorian science not only as sets of procedures for finding out what the natural world is really like but as human interventions in continuing political, social and religious struggles. On these accounts, science is no monolithic entity: always in process of becoming, its boundaries are never absolute, its definition never certain. Its contents and methods are never "innocent," never without influence from other nonscientific enterprises, and never to be understood without reference to particular historical perspectives and contexts. We are concerned here not with the standard histories of the period, with their important but by now all too thoroughly worked discussions of such important thinkers as John Herschel, his good friend William Whewell, and Whewell's not-so-good friend John Stuart Mill. This is not a section primarily given to the history of ideas, although ideas figure importantly, but with the history of ideas in cultural context.

While each chapter makes its argument with a particularity that forcefully demonstrates the connections between science and culture, each tends to suggest in a different way that the boundaries are always problematic. Whereas many scientists and even some philosophers of science hold out for the idea that science works primarily through the internal constraints of the discipline, these chapters all assume the inadequacy of this view. Science, they suggest, is always involved in the largest issues engaging the people and societies from which it emerges, and even the most individual of its achievements can be understood fully only through a wider understanding of a common cultural context. What emerges from these views of science is, therefore, rather less neat and well ordered than conventional histories of science or of particular disciplines: here mathematics, probability theory, evolutionary biology, and political economy are, each in its own way, embedded in the context of their cultures and societies.

The importance of boundaries to these chapters is manifest: between science and politics, science and religion, science and pseudoscience, expert and nonexpert, orthodox and unorthodox, the material and the transcendent, the material and the psychological. That importance is almost invariably shown to be complicated by the arbitrariness and inadequacy of border categories. While for the scientists involved, the establishment of borders was of enormous practical consequence (Margaret Schabas points out how in mathematizing economics, for example, Alfred Marshall in effect barred the gate of the discipline to amateurs), the analyses in these chapters, following much other contemporary work in the history and sociology of science, shows that they do not hold.

As a consequence of this contextual approach to science, it seems particularly useful to think about these studies in the light of a nonscientific, perhaps even antiscientific text of the first third of the nineteenth century, the opening of Thomas Carlyle's crucial *Sartor Resartus* (1831). The first pages of that book are full of science; indeed, in its fictional self-construction, it affirms itself as a scientific enterprise—an enterprise that, as Teufelsdröckh and the Editor suggest, incorporates and transcends all scientific subdisciplines. That enterprise is the study of humanity not from the perspective of natural history, or anatomy, or biology, or political economy, or even psychology, but from the perspective of "clothes." Working out of this large irony, Carlyle's book is an almost perfect medium for raising, testing, confirming, or complicating the questions and arguments that these chapters engage in their attempts to consider the relations of Victorian science to culture.

The fictitious Editor of *Sartor Resartus* begins by praising—perhaps—his culture for the way in which it has borne "the Torch of Science," to illuminate every nook, cranny, and "doghole in Nature." Only because science has been so pervasive can he then go on to wonder why, nevertheless, there has been so little written on "Clothes." It is significant that Carlyle here virtually opens his career as independent writer and thinker by responding to the omnipresence of science, fast becoming a dominant mode of knowledge and a practical influence on what people thought, what they did, how they lived. Carlyle cannot imagine science divorced from moral, social, and even political issues.

Sartor's significance for this volume has to do with its centrality to the intellectual and moral developments of Victorian culture and, in particular, with its timing: it is roughly contemporary with Herschel's influential *Preliminary Discourse on the Study of Natural Philosophy* ([1830] 1987), with the founding of the British Association for the Advancement of Science and Whewell's coining of the word "scientist," with the very struggles over intellectual authority traced here by Alison Winter and Martin Fichman, and with Whewell's voluminous and enormously important works *History of the Inductive Sciences* (1837) and *Philosophy of the Inductive*

Sciences (1840). Carlyle's voice seems to enter on the side of the "humanist" opposed to the "natural philosopher." But what matters most for this volume is that he sees science as the necessary starting point even for spiritual regeneration. *Sartor* also belongs to a tradition of irony both about and from the perspective of science (a tradition James Paradis very usefully explores in his essay). "Science" emerges from *Sartor* laden with ironies, so laden, indeed, that the question of its importance and validity remains unresolved. While Herschel and Whewell—followed by Mill in his *System of Logic: Ratiocinative and Inductive* (1843)—were affirming the importance of science as intellectual, moral, and religious endeavor and wrestling over how it could best be done, Carlyle was taking its pervasiveness for granted and putting its coherence and morality to question. In its countermovement to these foundational texts of the 1830s and 1840s, *Sartor Resartus* suggests that early Victorian science was triumphant and omnipresent, yet also troubling, and under the gun:

> Our Theory of Gravitation is as good as perfect; Lagrange, it is well known, has proved that the Planetary System, on this scheme, will endure forever; Laplace, still more cunningly, even guesses that it could not have been made on any other scheme. Whereby, at least, our nautical Logbooks can be better kept; and water-transport of all kinds has grown more commodious. Of Geology and Geognosy we know enough, what with the labours of our Werners and Huttons, what with the ardent genius of their disciples, it has come about that now, to many a Royal Society, the Creation of a World is little more mysterious than the cooking of a dumpling; concerning which last, indeed, there have been minds to whom the question, *How the apples were got in,* presented difficulties. Why mention our disquisitions on the Social Contract, on the Standard of Taste, on the Migrations of the Herring? Then, have we not a doctrine of Rent, a Theory of Value; Philosophies of Language, of History, of Pottery, of Apparitions, of Intoxicating Liquors? Man's whole life and environment have been laid open and elucidated; scarcely a fragment or fibre of his Soul, Body, and Possessions, but has been probed, dissected, distilled, desiccated, and scientifically decomposed: our spiritual Faculties, of which it appears there are not a few, have their Stewarts, Cousins, Royal Collards: every cellular, vascular, muscular Tissue glories in its Lawrences, Majendies, Bichâts (Carlyle [1831] 1937, 3-5).

The triumph of science and the traditional humanist critique are both immediately visible. Yet beyond those alternatives the ironies finally leave everything where, it seems, our own advanced intellectual culture likes to find them, undecidable.

It would be difficult to summarize the ways in which the questions raised in the chapters included in this section are shadowed forth here. The passage is about establishing boundaries (and about the virtual impossibility of doing so)—boundaries disciplinary, moral, scientific; it is about the degree to which the scientific (read also here "analytic" *and* "empirical") can provide an adequate description of the world and address the most important issues engaging human minds, hearts, and pocketbooks; it is about the question of, or the absence of, hierarchy, and about what constitutes scientific—that is, intellectual—authority; it is about the way scientific study was implicated in political and indeed imperial concerns; it is about what science may be leaving out or where it is not relevant (consider the debate about religious authority discussed by Joan Richards, for example). And it sets up that very tension between appearance and reality that Paradis discusses as central to the whole ironic enterprise of nineteenth-century literature and science. Whereas the great apologists for science— Herschel, Whewell, William Hamilton, in particular—see science as reflecting and discovering the divine, Carlyle seems to wonder whether science can possibly reveal God or, in its radical inconsistencies, shadow him forth. There is no "system of logic" here. As the passage plays over the various fields of science rooting out fact, it implies that fact is determined by consciousness, by the way the Editor and his readers can be understood to value the materials of science. And of course, because the passage is unrelentingly ironic, it opens far more questions than could be summarized here, while it entirely refuses to allow readers a point of stability from which to view the scientific panorama. Everything is satirized and literal at the same time. Self-evidently, *Sartor Resartus* belongs in the tradition of Swift's critique of science. Or does it?

The word "science" here still implicitly carries the pretechnical meaning of any systematic study, but technical meanings surface early: many of the investigations listed are concerned with technological and biological applications, yet they also carry over into philosophy and theory, into aesthetics, economics, and political science. In the early 1830s the power of science is contemporary with its multiplicity and pervasiveness. Could it, as John Stuart Mill, following Auguste Comte, argued, apply to the study of humanity as well as of nature? Whewell thought not; Carlyle thought not, too.

To the oddly innocent soon-to-be Editor of Teufelsdröckh's papers, all this scientific activity, which he if not Carlyle takes as wonderful, suggests no coherent notion of science as a culturally unified discourse or material project. Carlyle's ironic flattening out of the different levels of scientific discourse, the work of Newton and Laplace, for example, given no more emphasis than work on herring migration and apparitions, raises questions about the status of science by failing to acknowledge boundaries, definitions, distinctions, by refusing to recognize the differences among quackery, trivia, and "natural law." Implicitly, the claim is that the culture cannot make the distinctions. As Alison Winter points out, the very identification of a subject for scientific investigation becomes a question about the nature of disciplines themselves. As Carlyle raises such questions, he questions all of science as well.

Paradis's discussion of satire and irony points to the way such questioning implies a strong distinction between the activities of science and the ultimate nonmaterial realities that, implicitly, we are to value more. (In the light of Joan Richards's chapter, it is interesting to note that Laplace becomes an object of satire, but Herschel does not.) Carlyle's irony, even as it derives from a strong sense of the limitations of science, corrosively, like deconstruction itself, works against boundaries. Here it seems to affirm the superiority of humane culture over scientific, but there is no space in the language to treat the humane unironically. How can that extramaterial essence — if that is what it is — be embodied in clothing any more than in herring? The very absurdity of talking about clothing as though it were the "tissue of all tissues," while it may leave open some space for a symbolic reading, puts both the idea and its articulator into question. The double edge of this kind of writing, emphasizing continually but unstably the disparity between appearance and reality, is, as Paradis asserts, consistent with the effects, though not of course with the projects, of science. As Dwight Culler many years ago brilliantly associated Darwin's world view with a totally counterintuitive and ironic vision, and thus in particular with the characteristics of late-nineteenth-century writing, so Carlyle's passage, like science, exposes disjunctions and makes irony the dominant means by which scientific activity might be understood — if "understood" is the word to describe what happens here (Culler 1968).

In any case, the passage, its juxtapositions implying disjunction, confirms Alison Winter's point that by the 1830s there was "a dizzying variety of . . . arenas in which science was practiced and communicated." If one wants to complicate the picture of science as homogeneous, here is a place to start. The ironies depend on the recognition that there is indeed a dizzying variety of activities thought of as scientific; this variety is part of the problem, part of the culture's failure to make sense of itself in the very act of making sense of everything. Implicitly, I would argue, the passage suggests a project that unites those committed to a traditional religious view of the world and, oddly enough, those who, from varying scientific perspectives of the kinds discussed in the chapters gathered here, have displaced that view with science and scientific method: the project of turning multiplicity into unity, of making sense of the world by locating something like a "tissue of all tissues" — god or natural law or probability.

Carlyle's work is propelled by precisely that longing for coherence and meaning that the critical strategies of current cultural studies reject or deny, while at the same time his writing itself fairly glimmers with recognition of its possible delusiveness. There are ironies here at the expense of the quest for law that dominates all of these scientific enterprises. It is not only that the "laws" of science cannot begin to adumbrate the ultimate coherence that a living God could provide (Carlyle's own ambivalence about God is manifest in his ironic mode). It is that the laws themselves are so obviously not unified and coherent, whatever philosophers might want to make them.

But Carlyle's effort at coherence seems to reflect a fundamental disenchantment with the traditional notion of design. That Paleyesque notion, adopted by Herschel and tightly allied to strict rational, analytic argument, was no longer satisfying even to Whewell, as Richards demonstrates. Certainly, Carlyle, with his Romantic and irrationalist views of the way the world worked, could find no spiritual satisfaction in it. So the "order" of science becomes in *Sartor Resartus* a mad jumble of incompatible activities indistinguishable from quackery, from "bad science." Carlyle's project might then be thought of as parallel to the Laplacian project: the attempt to achieve certainty in a world of multiple perspectives, a world apparently driven by chance. Reliance on rational coherence leads to Carlyle's mocking description of Laplacian determinism as the view that the world "could not have been made on any other scheme." Faith in rational coherence seems to drive probabilism as well as mysticism, and does so too in Joan Richards's chapter.

Carlyle's strategy, through his somewhat ingenuous Editor, is to use that dizzying diversity of science to demarcate a world to which the activities of science are in fact irrelevant—irrelevant because science as it is imagined here is attempting through theory and method to account for the material world in ways publicly and universally applicable, and in so doing is missing what Joan Richards recurs to and concludes with, "the mystery."

The mystery keeps reappearing in these chapters, as the problems of perspectivism and subjectivity become unavoidable. Out of sympathy as Carlyle was with utilitarianism and the sort of economic thinking Margaret Schabas finds in Mill, and unhappy as he was with Laplace's sense of the complications of probability, and, certainly, disgusted as he would be with the way all of these chapters emphasize that matters other than fact significantly determine science, his work allows the mystery all of these unlikely colleagues share. The personal, the subjective, becomes the condition, too, for the ironic. Things are not what they seem. They can not be reduced to order and logic and analytic precision. The boundary between the analytic and the "psychological" is also the boundary between the hard scientific and the mysterious. That boundary breaks down everywhere in these chapters and in Carlyle's prose. In accounting for everything, science ends by accounting for nothing that Carlyle thinks really important—certainly not for what the Editor calls "the grand Tissue of all Tissues, the only real Tissue."

Carlyle belongs with Whewell to the British movement that rejects one fundamental element of Paleyan natural theology: the assumption that a scientific look at the material world will reveal its divine sources. At the same time he holds firmly to the reality of the divine. It is not that Carlyle denies the centrality of the material in issues of deep human import; rather, he rejects the idea that strictly rational and systematic study can get at the moral and spiritual. (There is virtually no literal connection between Carlyle and Whewell, but their mutual interest in German Romantic philosophy might account for their few intellectual similarities.) These issues produce strange bedfellows. Even John Henry Newman's probabilist thought, to which Joan Richards refers, connects significantly with Carlyle's Romantic organicism. Carlylean satire of science produces something akin to the world evoked by Newman (whom Carlyle accused of having the brain of a moderate-size rabbit) when he looks at it with the eyes of mere scientific reason: it is a vision to "dizzy and appal" (Newman [1864] 1967, 217).

Seeing Carlyle's boundary work as a reverse image of the natural philosophers' own helps illuminate what was culturally at stake in these boundary wars, not only for the most famous of the promoters of science—Herschel, Whewell, and Mill—but for the developing army of quasi-professionals and for the literate public at large. Sustaining the mystery, Mill's antirational anathema, for example, made possible justification of certain kinds of social and spiritual hierarchies that were threatened by Laplacian (French Revolutionary) rationalism. In British eyes rationalism and materialism and the potential violence that led to the French Revolution were connected. This kind of connection, central to Adrian Desmond and James Moore's discussion of evolution as it was taken up and, as it were, bourgeoisified by Darwin, is important both to Joan Richards's discussion of probability theory in Britain and to Martin Fichman's discussion of the relation between evolutionary theory and politics (ending with Wallace's reversion to a kind of socialism) (Desmond and Moore 1991). At the same time, the "revolutionary" aspects of materialist, or quasi-materialist, science contained their own rather nonrevolutionary implications within the established institutions of science. So Evelleen Richards shows that despite the attractiveness of evolutionary science in many respects to important Victorian women intellectuals, science too had its politics of exclusion and hierarchy. There was no recognized important place for women, either as practitioners or as active subjects.

Carlyle's mystifications, we know, while they made possible important critiques of "rational" political and economic reforms, ultimately confirmed a kind of intellectual anti-intellectualism, an ethic of industrious anti-industrialism, and deference to political authority. These helped mark the lines of resistance to the intellectual imperialism of science, but at the same time they deeply influenced the direction that British science would follow, not only into Darwin's evolutionary theory but into Augustus De Morgan and Robert Ellis's taming of probability. It is no small point that, as Frank Turner long ago showed, John Tyndall found Carlyle's ostensible hostility to science no obstacle to his inspirational significance for scientists (Turner 1975). The critique need not be seen as incompatible with the science so long as the boundaries of authority continue to be drawn between the public and the private, the material and the mysterious. Allowing for the significance and power of the mystery while at the same time claiming imperial sway over all of "nature," scientists could remain fairly comfortably within traditional social and spiritual organizations and at the same time employ the rationalist methods of revolutionaries in dealing with stars or the ether or bacteria.

Looking at these chapters collectively, one finds a whole set of Carlylean possibilities dramatized. It is not only that they assume that the move toward disinterest is a move toward self-authentication and professionalization rather than toward some universally detectable truth, but consistently these essays show that on the one hand the margins established to make distinctions are arbitrary—so Martin Fichman argues that the borders between politics and biology are permeable—and on the other that uniformity yields everywhere to multiplicity.

Here, then, the arguments for social construction are made most forcefully by close historical examination of several developing disciplines. Alison Winter, breaking down the distinction between real and fake science, makes a strong case that orthodoxy got established not by being right about nature but, as in the case of William Benjamin Carpenter, by careful marshaling of experts who ultimately determined what being right could mean. Here as throughout these studies, the particularities of contingent social and personal structures become more important to scientific definition than epistemological correctness. Messiness is part of the picture our science studies produce with some consistency as they almost invariably undermine or revise orthodox readings of the history of science. It is a messiness that Carlyle would have recognized, although, unlike the essayists here, he would have continued to insist that somewhere we should be able, mysteriously, to locate that tissue of all tissues that unites all apparent incoherences in organic unity. This section, it should be clear, does not attempt to produce that kind of tissue.

References

Carlyle, Thomas. [1831] 1937. *Sartor Resartus: The Life and Opinions of Herr Teufelsdröckh*. Edited by C. F. Harrold. New York: Odyssey Press.
Culler, A. Dwight. 1968. "The Darwinian Revolution and Literary Form." In *The Art of Victorian Prose*, edited by George Levine and William Madden. New York: Oxford University Press, 224–46.
Desmond, Adrian, and James Moore. 1991. *Darwin*. New York: Warner Books.
Eliot, George. [1872] 1965. *Middlemarch*. Harmondsworth: Penguin Books.
Herschel, John. [1830] 1987. *Preliminary Discourse on the Study of Natural Philosophy*. Chicago: University of Chicago Press.

Newman, John Henry. [1864] 1967. *Apologia pro vita sua.* Edited by Martin Svaglic. Oxford: Clarendon Press.

Turner, Frank M. 1975. "Victorian Scientific Naturalism and Thomas Carlyle." *Victorian Studies* 18:325–43.

2

The Construction of Orthodoxies and Heterodoxies in the Early Victorian Life Sciences

ALISON WINTER

Over the past fifteen years, a number of important studies have developed an account of early Victorian scientific leadership by the "gentlemen of science" (Morrell and Thackray 1981; Rudwick 1985; J. A. Secord 1986). Individuals such as Charles Lyell, Adam Sedgwick, William Whewell, and John Herschel were members of a well-defined group. They endorsed the holy alliance of the established church with natural philosophy, were suspicious of French materialism, and regarded plebeian and "amateur" science, when not carried out under their supervision, as dangerous. In these important respects they were united. Moreover, they were clearly demarcated from their subordinates and from the general public. This account is still fundamental to our understanding of Victorian science, but the picture is becoming increasingly complex.

It is no longer possible to regard the gentlemanly community as homogeneous, or even as necessarily typifying what science and nature meant to other early Victorians. Over the past decade we have come to appreciate that scientific communities were less defined than was hitherto assumed, and that definitions of science itself were very fluid during these years. Early Victorian science was volatile and underdetermined. People could not agree about what one could safely claim about natural law, nor was it obvious when, where, and to whom such claims could be made. What were the implications of this more fluid, chaotic state of affairs for the formation of a public scientific identity?

The launching of the physiologist William Benjamin Carpenter's scientific career is a particularly revealing example of how the status of "orthodox" or "heterodox" came to be accorded to individuals and their work in

I would like to thank the following for useful discussions relating to this chapter: Katherine Anderson, Kevin Gilmartin, Adrian Johns, George Levine, Bernie Lightman, James Moore, Mac Pigman, Robert Richards, Simon Schaffer, Anne Secord, and James Secord. Research for this chapter was made possible by the generous support of the California Institute of Technology.

this period. As a young researcher in the late 1830s, Carpenter entered the field of physiology at a particularly tumultuous time and wished to stake claims in a particularly sensitive area. He used a variety of strategies to try to construct a prestigious and stable context for his work. Eventually, he succeeded in gaining a reputation for himself as legitimate whereas many of his colleagues did not. His story reveals some of the specific resources scientists could seize upon in their efforts to influence how their work would be received and understood. It also illustrates the significant degree of ambiguity that could surround the status of a new claim or practitioner.

I

The 1840s were characterized by a heady optimism about the powers and achievements of scientific inquiry. Britons seemed to be annexing a huge range of new terrains, not only in the exploration of new territory like the Arctic and Africa, but also in other new "realms." They mastered the history and future of living things, for example, and the invisible interplay of natural forces (S. F. Cannon 1978, chap. 3). These advances, and the confidence they inspired, suggest a characterization of an assured scientific elite engaged in ever-increasing mastery of nature. This portrayal would seem to be borne out by the fact that many of the fields, institutions, and structures of scientific organization we now retrospectively regard as modern appeared for the first time during these years or shortly thereafter. Examples would include the term "scientist" itself, the disciplines of physics and biology (among others), the founding of the British Association for the Advancement of Science, and the reformed Royal Society. However, there now exists a substantial body of work on the social history of Victorian science that makes it impossible to accept such a picture at face value.

We now know that the practices, practitioners, contexts, and audiences that existed for early Victorian science were extremely diverse—far more so than their eighteenth-century predecessors. By the late 1830s and 1840s there was a far wider range of specialist journals and societies, and a dizzying variety of other arenas in which science was practiced and communicated (Altick 1978; Morrell 1976; Morus 1991; Porter 1978). Indeed, the very phenomena that might seem like signs of the consolidation of orthodoxy in science, such as the increase in specialization in various fields, may be understood differently. Specialization caused intense concern that scientific communities, far from becoming united, well defined, and authoritative, were actually growing more diverse and even chaotic (Yeo 1984; Porter 1978). Moreover, this period is well marked by movements such as phrenology and mesmerism, which have been polemically termed "pseudo," "alternative," or "heterodox" science and, in medicine, "quackery." This diversity indicates that there may have been as many ways of defining proper science as there were constituencies for science. Moreover, sciences we now retrospectively regard as heterodox or marginal cannot be considered unambiguously to have held that status at a time when no clear orthodoxy existed that could confer that status upon them.

There are two areas that have been particularly revealing of the varieties and status of supposedly heterodox scientific projects: the study of "radical" sciences and of the so-called alternative medical practices. Take first the radical evolutionary campaigns that flourished during the 1830s. They present some of the most fruitful areas for the study of the politics of Victorian natural law because of the explicit political claims they could be used to support, and because of their variety. Radical evolutionary projects have been portrayed as alternative options for individuals hostile to more conservative lines of scientific explanation. Long before Darwin's *Origin of Species* (1859), and even before the publication of Chambers's *Vestiges of the Natural History of Creation* (1844), radical artisans adapted evolutionary thought to give a blueprint in natural law for their socialist and cooperative projects.

Jean-Baptiste Lamarck's doctrine of the inheritance of acquired traits helped radical evolutionists argue that the environment surrounding an organism induced individual changes that could eventually transform a whole population. When this argument was applied by radicals to the development of human society it provided a useful piece of ammunition against a variety of liberal arguments. For instance, radical evolutionary projects could be used against representations of nature and society that were indebted to Thomas Malthus's claims (about the likelihood that uncontrolled population growth would outstrip food supply). Such representations underpinned the items of 1830s government policy that were most detested by working-class radicals, such as the New Poor Law of 1834 (Desmond 1987). There was also a middle-class radicalism in the evolutionary projects of a number of individuals. The most prominent of them was Robert Grant, whose comparative anatomy put forward the notion of a single law structuring all of nature, in self-conscious opposition to more conservative and traditional portrayals of a Creation the constituent parts of which had been individually designed by God (Desmond 1987, 1989*b*).

Historical reconstructions of these endeavors, then, portray the natural history of the 1830s and 1840s as underpinning, and underpinned by, rival visions of a healthy polity. Related historical accounts have developed a picture of how competing projects in the life sciences supplied pedigrees for the conservative, liberal, and radical agendas of their advocates. One classic account has suggested that the political fault lines in early Victorian life sciences can be traced according to whether individuals stressed "immanence" or "transcendence" in explaining vital phenomena (Jacyna 1983). The former notion—that life and activity were an essential, inherent part of the organic world—lent itself to stances which were vulnerable to charges of materialism. Conversely, portrayals of "transcendence," which claimed that life was something superadded to nature, breathed into it by God, could be used to defend a traditional account of the moral and natural order. These alternatives had been the subject of famous debates by the physiologist William Lawrence and the surgeon John Abernethy in the 1810s. The two men disputed whether the faculties of mind were no more than the results of organic processes (as Lawrence argued) or there was some transcendent substance or property superadded to organic matter that was the locus of vital power (as Abernethy claimed) (Jacyna 1983, 312-16). More generally, historians both of the life sciences (Jacyna 1987; Desmond 1987, 1989*a*, 1989*b*) and of religion (Hilton 1988) have shown that scientific projects associated with these rival perspectives were integrated in rival concepts of the moral order. According to this literature, Victorians could choose between two broad conceptions of the theological and moral order: a monistic or materialistic representation of the world, on the one hand, and, on the other, a world in which spirit and matter are separate (Hilton 1995).

The various studies described above complement each other in describing various facets of a world in which politics mapped onto nature and vice versa. Evolutionary radicals, utilitarians, and evangelicals were far apart from each other ideologically, but these accounts place them in the same intellectual framework. They differed strikingly in the claims they wished to make about the natural world, but they operated within the same logical parameters. In effect, they took part in the same debate, since their claims could confront each other head-on. If one combined such claims to produce a more general account of what kinds of knowledge counted as legitimate and valuable among different constituencies of Victorian society (something that the authors of these works refrain from doing), a map of natural knowledge in some scientific projects could look like a map of political positions in Victorian culture.

While these studies have overwhelmingly demonstrated the political significance that could be attached to claims about nature, it would be a mistake to conclude from any part of this literature that any single framework that opposes rival ideologies or class orientations (such as radical versus conservative sciences, or gentlemanly versus artisanal projects) would suffice as a means of categorizing the heterodox and orthodox constituencies of early Victorian science. For example, we have come to appreciate— partly as a result of these studies themselves—that conservative thinkers felt themselves to be under threat from *within* the community of gentlemanly natural philosophers as well as from without by widespread materialism. In view of these and other factors, it is clear that there were greater and more significant differences among the "gentlemen of science" than was once supposed (Desmond 1989*b:* Bloor 1983, 612-19; J. A. Secord 1991; Desmond and Moore 1992). Moreover, even texts that have traditionally been portrayed within histories of evolutionary theory as having been received as materialist by early Victorian audiences—such as Robert Chambers's anonymous *Vestiges*—had, it is argued, any number of possible political associations. This is indicated for *Vestiges* by the diversity of individuals who were speculated to have written it. For instance, the political commitments of putative authors of *Vestiges* ranged from radical Tory to Benthamite (J. A. Secord forthcoming, chap. 5; Chambers 1994, xxviii–xxix). This fluidity is significant, of course, because one's religious and political orientation was crucial to how one defined oneself and how others defined one as a natural philosopher. Such diversity would therefore suggest that there flourished a range of conflicting definitions of proper science. It is therefore dangerous to use specific views about the natural order as a straightforward test to tell the "insiders" from the "outsiders" of science, to tell the "orthodox" from the "heterodox."

A second literature that reveals the ambiguities surrounding the definitions of legitimate knowledge and scientific practice is the social historical study of the so-called alternative sciences and medical therapies. Much of this literature complements the studies of radical science mentioned above, but there are some important differences. We now know that a number of Chartists and Owenites were at one time or another mesmerists, phrenologists, spiritualists, herbalists, and homeopaths and that they imbued their projects with political significance (Harrison 1987; Barrow 1991; Shapin 1979; Winter 1991, 1994*a*). The itinerant artisan mesmerist Spencer Timothy Hall, for instance, argued that his ability to create the mesmeric trance (and thereby both to heal the sick and to reveal new truths about the mind) demonstrated that knowledge was not to be considered the property of the "professional" classes but rather accessible to the "common man" (Winter 1994*a*). Alternative knowledge and practices could serve as a validation of ongoing political projects. During the decline of Chartism and related campaigns after the late 1840s, they could also provide a repository for frustrated political ambitions (Barrow 1980, 1986; Harrison 1979).

But the literature on alternative medical therapies also reveals that explicitly ideological concerns did not necessarily have to be the dominant factor in the appeal and character of many supposedly heterodox sciences. The claims such sciences made about nature did not always lend themselves to metaphorical or literal political lessons. Moreover, individual alternative projects facilitated a range of conflicting political interpretations. For instance, a crucial component of these sciences was often their accessibility to new practitioners. Their popularity might be influenced by a perception that they were not already identified as the domain of skilled experts or elite communities. It was particularly important that their fundamentals could be quickly learned (Cooter 1984; Winter 1994*a*). The crucial attribute of certain sciences, therefore, could be characteristics that made

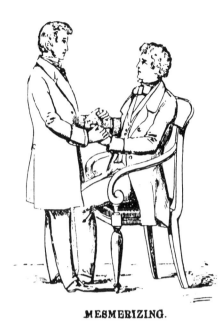

MESMERIZING.

Figure 2.1 A standard posture for producing the trance. (From Davey 1854.)

them less the property of professionals and more accessible to anyone. While these attributes could complement a politically radical agenda, they did not necessarily have to accompany it.

Mesmerism is again a good example. Radicals found useful the notion that mental phenomena could be provoked and managed by a physical force emanating from an individual. For them, it substantiated claims about the materiality of the mind and the broader radical platform this materialism helped to support. Moreover, because mesmerism was an *experimental* science, the sight of one person placing another in an altered state of mind could add force and immediacy to these claims (figure 2.1). But mesmerism was popular not only with radical lecturers, but also with Tory Evangelicals, Whig aristocrats, middle-class utilitarians, and other disparate groups. These different constituencies offered conflicting explanations for the phenomena, each of which had different implications for how one portrayed the nature of human relations. Consequently, when a mixed group viewed a mesmeric display, its members gave rival explanations for what they saw. Mesmerism could be the leveling force that, once establishing the materiality of mind, would supply the epistemological foundation for a democratic society. Alternatively, it could provide a pedigree in natural law for traditional relations of rank or spiritual guidance when an aristocratic mesmerist subdued a servant or a preacher a member of his flock (Winter 1994*a*, 335–36; 1994*b*, 81).

Mesmerism was debated not only in aristocratic homes and provincial lecture halls, but also in those institutions and forums that were designed to be places where the future state of science was to be determined. For instance, in 1837-38 an intensive series of increasingly public experiments in animal magnetism took shape at the hospital of the recently founded University College London. University College was intended to be a place where faculty and students could develop innovative understandings of natural law unimpeded by the constraints of tradition. It was, therefore, an appropriate place for the investigation of an exciting but highly controversial new science.

These experiments, led by John Elliotson, professor of practical medicine, raised fundamental issues regarding what kinds of questions could be asked and answered regarding the relationship between physical forces, including electricity and magnetism, and physiological phenomena, and regarding *where* they should be answered. Ambiguities regarding the relations between physical and living phenomena, the nature of the variously forming scientific disciplines, and the question of where and how proper scientific research should be carried out made these experiments both intensely interesting to the scientific and medical community and to the public and extremely controversial. In the same year, and in fact the same week in which the mesmeric experiments began to be carried out publicly, the Royal Society sought to resolve its members' sense of uncertainty regarding the parameters of various sciences by forming a set of committees to monitor and guide the development of work in discrete subject areas of scientific research (M. B. Hall 1984). Many of their members attended the mesmeric experiments taking place at University College throughout the rest of the year (Winter 1991, 1994*a*).

The significance of the debates over mesmerism is underlined by comparison with other sciences with which historians are more familiar. The differences between geology and electricity provide another example of the patchy state of the sciences. One could make a case for the existence of a clear orthodoxy in the example of the geological community. The Geological Society was a gentleman's club dominated by a discrete coterie. It provided a relatively disciplined set of contexts for scientific communication and a specific site at which controversies could be resolved (Rudwick 1985; J. A. Secord 1986).

If one compares this to the state of research in electricity, the picture looks very different. Contrast the electrical worlds of William Sturgeon and Michael Faraday (Morus 1992*a*). Sturgeon prepared his apparatus and phenomena so as to produce the widest and most spectacular effects and to show off the piece of technology on display to his paying audiences. In contrast, Faraday designed his experiments to conceal the work that had gone into them and to encourage his audience to look past the piece of apparatus at the laws of nature he wanted it to reveal. If one turns to the electrical production of life by Andrew Crosse in 1836 the picture is even messier (J. A. Secord 1989a). When Crosse found that insects of the genus *Acari* seemed to have been produced by his electrochemical apparatus, he concluded that the phenomenon demonstrated the electrical production of life. As the controversy over his phenomenon developed, it was not clear where and by whom it would be authoritatively resolved. It was not even agreed to which science the experiments belonged.

In each of the above scientific controversies, issues of place, practice, and audience have been central to the construction of scientific authority and orthodoxy. In the most rudimentary way, attention to these issues — which were often related to explicitly ideological concerns but as frequently independent of them — has come to be central to the social history of science in this period. This literature has broadened historical appreciation of the extent to which science was undefined in the 1830s and 1840s by refining our appreciation of what counted as orthodoxies and heterodoxies in these years and by documenting the extent to which rival conceptions of natural law flourished. While the so-called alternative sciences have long been portrayed as vehicles of protest for individuals outside cultural establishments of one kind or another, it has become clear that they had far more adherents among the so-called scientifically orthodox than we might have once supposed. For instance, most individuals who encountered mesmerism — whether or not they approved of it — found that its phenomena forced them to confront fundamental issues about the nature of scientific inquiry.

Controversial projects and bodies of theory, then, did not exist in any straightforward relationship with "real" or "orthodox" sciences as their "others." Fine-grained social histories of these projects have, instead, revealed two surprising attributes. It was often impossible for Victorians to agree on what counted as illicit or pseudoscience or medical quackery in specific instances (Morus 1992b; Winter 1991, 1995). And within those undefined areas, researchers used their scientific work itself to develop the basic principles that would underpin that practice. In the variously forming disciplines, disputes about the nature of their objects were played out as disputes about how to define those disciplines. Sciences of mind offered guides (in the case of phrenology, a literal map) to which mental and physiological characteristics gave one the ability to understand the mind and human behavior (figure 2.2). In physiology, attempts to define the mind were attempts to define procedures of analysis and experiment; in mathematics, debates explored the nature of proof and the laws of reason (see Joan Richards's chapter in this volume); controversy about problems of forecasting in meteorology prompted debate about the nature of prediction (Anderson 1994); and models of the relations between the forces of nature coincided

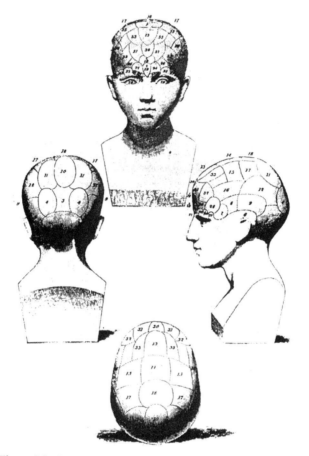

Figure 2.2 Frontispiece of *Elements of Phrenology* (Combe 1824).

with proposals for how a scientific community should be organized (Morus 1991). Such issues did not preoccupy everyone to the same extent, but they were present in varying degrees throughout the sciences.

II

Given the degree of uncertainty regarding the parameters of legitimate knowledge, it should not be surprising that immense uncertainty surrounded the communication of scientific claims. It was not to be assumed that a set of assertions would be interpreted in the manner in which its author intended. A particular statement could be taken to have a variety of readings depending on the context in which it was heard or read. This slipperiness was particularly true of claims that were to any extent open to charges of determinism or materialism. Scientific writers dealing with the relationship between physical and living or mental phenomena, or positing the consistent action of natural laws in natural history and throughout creation, had special vulnerabilities. They needed to be careful in stage-managing the publication of their claims in order to enculturate readers to take in what they said in the way they had intended.

Recent studies of scientific correspondence, publishing, reading, and conversation have shown how much is to be learned by examining the change in meaning a scientific claim could undergo depending on the particular context in which it was made or communicated. More generally, they have revealed the existence and importance of treating particular social relationships, or conventions of correspondence, as "contexts" of scientific communication in the same way that one might historically examine the context provided by a scientific institution. Consider, for instance, the function of correspondence in the work of Victorian mathematicians in Cambridge and Dublin who, at hundreds of miles distance, could not form a research school together in the material contexts in which they worked (Warwick 1995). Their correspondence formed a space for collaboration that gave specific meaning to the sorts of scientific claims they wished to make.

An example of how correspondence could restructure a very different sort of space concerns the communications between artisan naturalists and elite botanical specialists. Correspondence allowed intellectual exchanges between gentlemen and artisans to take place in a manner that their class differences would have made extremely difficult in face-to-face interaction. The essential role of correspondence in constructing a space in which gentlemen and artisans could collaborate is underlined by cases in which face-to-face confrontation destroyed relationships that had been carefully nurtured through long years of collaboration by post (A. Secord 1994, 396–97). Similar work is revealing the significance of conversation, soirée culture, and other forms of interaction (J. A. Secord forthcoming, chap. 5). More generally, such work broadens our notion of a historical context for scientific work and sensitizes us to the importance of very specific social conventions in structuring the meaning and reception of particular scientific claims.

The importance of such conventions indicates the need for closer attention to the role they played in establishing a particular assertion as orthodox or heterodox. In particular, it is clear that the way in which one publicized a scientific statement or the publisher one chose, for instance, strongly influenced how it would be received. But there is much more work to be done in excavating how the interpretation of a scientific claim could be orchestrated through the careful use of such conventions. There were surely significant (and at present insufficiently understood) opportunities and pitfalls involved in moving from one forum into another, particularly given the volatile and combative nature of many early Victorian public forums. For instance, we know that the Geological Society, the X Club (a group of elite members of the Royal Society who set themselves the task of defining the relations of the sciences and scientific communities) (Barton 1976; Jensen 1971–72; Morus 1991), and many of the small specialist societies functioned as safe places in which scientists could try out new theories. While this much is clear, we could profit from knowing much more about how the course from these spaces into more chaotic public ones was navigated.

In particular we could explore how some places, ostensibly secure, became dangerous despite the best efforts of practitioners who worked within them. An extreme example is University College and its medical school. University College was intended not only to allow practitioners of a single mind to work in concert, but also to make it possible for people openly and safely to disagree about the nature of the new natural and medical sciences emerging during these years. But the difficulty many faculty had in maintaining even a veneer of respectful interaction with one another demonstrates how difficult it was to establish a forum "for all the talents." For instance, the surgeon Robert Liston and the physician John Elliotson were "at daggers drawn" with each other, and students formed two "poles" around them. Each of them, for different reasons, hated their colleague, the physiologist William "the Serpent" Sharpey; he returned the sentiments (Clarke 1874, 146). University College may be a particularly extreme example. However, it does illustrate the explosive possibilities of a space in which scientists could interact and disagree in pursuing their different projects without the constant danger of destroying itself (Merrington 1976; Desmond 1989*b*).

There are very good examples of how individuals failed spectacularly to manage their publics in such a way that their work looked legitimate. One of the most famous is, perhaps, the case of William Lawrence. His *Introduction to Comparative Anatomy and Physiology* (1816) was an attempt to make physiology more lawlike. The work was intended for a small gentry audience—not a wider and more heterogeneous readership. Later, after the work was declared blasphemous and Lawrence lost his copyright, the radical agitator Richard Carlile reprinted it in a cheap paper edition. Carlile's imprint and the lower price dramatically changed the meaning of the words Lawrence had written, making them definitively materialistic (Butler 1993; Desmond 1989*b*, 217–21; Goodfield-Toulmin 1969).

Other examples include the case of John Elliotson, whose attempts to stage-manage his research program in animal magnetism in the late 1830s were a notorious failure (Kaplan 1974, 1982; Winter 1991). During Elliotson's early experiments, and before they became highly public, individuals such as Michael Faraday, Charles Wheatstone, Dionysius Lardner, and Peter Mark Roget attended them. To varying degrees, they became involved in the experimental program and concurred with his conclusions as to the validity of the notion that a physical force could induce an altered state of mind. However, Elliotson's program unraveled as it became more public. Some of his most powerful potential supporters backed away over the following months, partly because they found the experiments hard to stage-manage and their phenomena difficult to validate. The experiments became more public, highly prominent in the weekly and medical press, and harder to control (Winter 1991, 1994*a*). By midsummer of 1838 Elliotson's claims about the relations between physical forces and the mind were widely represented by the medical press as both materialistic and foolish. His enemies among the medical faculty made increasingly forceful complaints that the publicity was undermining order in the hospital and harming the reputation of the medical school. Within a year of beginning his experiments Elliotson had resigned from University College.

But in the very same year as Elliotson's fall, a younger and ultimately far more prestigious scientist, William Benjamin Carpenter, was more successfully negotiating his debut on the London scientific stage. Carpenter's difficulties exemplify some very common dangers that early Victorian scientists encountered. They also illustrate a very important series of maneuvers that individuals could make to establish themselves. Moreover, this story can be used to make some more general reflections about both the dangers and the opportunities inherent in the ambiguities surrounding scientific claims in this volatile period.

III

William Carpenter established himself through his work in comparative physiology and the physiology of the mind and by the many textbooks he produced on a wide range of subjects (see bibliography in W. B. Carpenter 1888). He had become established by the 1850s as an exemplary scientific figure. Centrally concerned to assert the nature of orthodox and heterodox forms of knowledge and scientific research, he followed William Robert Grove in asserting the "correlation" of different forces.[1] He was also associated with Grove's project in appropriately different branches of science and of scientific communities. In debates over mesmerism, spiritualism, and psychical research, he sought to demarcate the legitimate from the illegitimate experiments and phenomena (W. B. Carpenter 1877). Finally, he became interested in developing a physiological basis for sound judgment and reason. He wished to relate this to a model of the proper relations between an authoritative scientific community and the general public (W. B. Carpenter 1852, 1874, 1877). Carpenter was not only someone who could

---

1. The term "correlation" was used to refer to the mutual relations of different forces in nature, for instance, how one force influenced or could be converted into another. "Correlation" was a useful term because it resisted the reductionist or materialist interpretations that could easily be attached to such dynamics. On Grove, and for a full explanation of the meanings of "correlation," see Morus 1991; on Carpenter and correlation see Hall 1979 and [Carpenter] 1851.

claim, by the 1850s at least, a fairly secure status within various scientific communities, but also someone who was active in asserting the very nature of orthodox and heterodox knowledge.

Yet in 1839, when he launched his career with the publication of a substantial work, *Principles of General and Comparative Physiology,* his eminence seemed anything but inevitable. This work promoted a controversial claim (though hardly an unusual one at this time) about natural law: that physiology should become as lawlike as the physical sciences (Jacyna 1984, 59-60). Specifically, he wished to assert that the same kinds of laws that governed living phenomena governed physical ones. Carpenter desired to reduce physiology to a set of naturalistic laws with himself as the systematizer of those laws. His enterprise was reflexive: his claims about the lack of boundaries between different phenomena in creation also applied to his definition of science. He wished to redefine physiology along the lines of the physical sciences and to break down as fully as he could the borders between different scientific disciplines. He described the progress of physiology as an increasingly "natural" disciplinary state in which "man-made" boundaries would disappear and creation would be increasingly shown to operate via general unchanging laws. As physiological research facilitated the articulation of general laws, he wrote, so would scientists "find the boundaries which at present divide the sciences disappear; just as the aeronaut, in enlarging his horizon, successively loses sight of the divisions which the art of man or the hand of nature has interposed to separate from each other, estates, provinces and kingdoms" (W. B. Carpenter 1838*a*, 318).

Seeming exceptions to natural law—such as supposed miracles and monsters—were also included in the natural terrain of progressive science. Miracles were evidence of a higher law, as yet not discovered. The fundamental reason for the convergence of the sciences was the fact that all of nature was the ultimate creation of the "Almighty *fiat* which created matter out of nothing," which "impressed upon it one simple law, which should regulate the association of its masses into systems." This was the law that "should harmonize and blend together all the innumerable multitude of these actions, making their very perturbations sources of new powers" (W. B. Carpenter 1839*a*, 463; Jacyna 1984). Thus, Carpenter's physiology facilitated a view of a creation run by constantly acting laws, and of the scientist's role in observing natural "experiments" produced by the "perturbations" of the actions of matter (see figure 2.3) (see also W. B. Carpenter 1838*a*, 342; and compare Babbage 1837, 48-49).

Carpenter's claims were similar to the ones that other individuals such as Southwood Smith, Robert Grant, Marshall Hall, and John Elliotson were making, and that had provoked extreme and sustained controversy. Carpenter's connection to these individuals was extensive. He was a supporter and publicist of Geoffroy St. Hilaire, had almost certainly attended Grant's

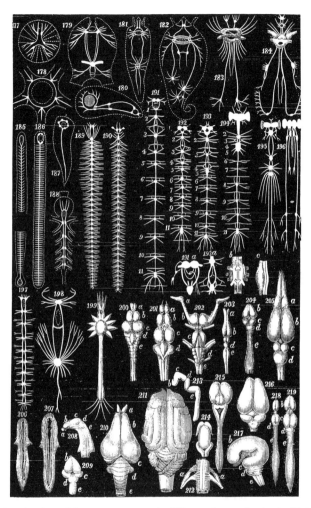

Figure 2.3 Sketches of the nervous system in different classes of animals. (Plate 6 of W. B. Carpenter 1839a.)

lectures on comparative anatomy, and was a friend of Hall (Desmond 1989b, 213-20). He was also interested in animal magnetism. If one were to place Carpenter on the sort of political map of nature described earlier in relation to comparative anatomy and physiology in the 1830s, he might look as heterodox as these others (Desmond 1989a; 1989b, 213-20). Yet after a brief skirmish in the medical press Carpenter had far fewer difficulties than the rest of this cohort. Why?

One of the clearest reasons for Carpenter's success was the rigor with which he solicited a large number of specific elite scientists who could be represented as constituting scientific and religious orthodoxy. This set him apart from other radicals like Grant and Elliotson, who tried less hard to solicit patronage for their work and were certainly less successful (Desmond 1989b, 114; Kaplan 1974, 1982). His success in this respect helped his physiology survive and look increasingly orthodox during a period when the views of his London teachers were not only highly controversial, but increasingly marginal (Desmond 1989b, 236-75).

Carpenter consulted a number of the elite figures in science and religion before the publication of his *Principles* in 1839 and sent presentation copies to many others with flattery and solicitations of support. This careful move helped to protect him from the attacks his work received from a small number of the more conservative reviewers. One of them, for instance, described his writings as materialist, reductionist, and "detrimental to the best interests of mankind" ("Carpenter's Principles" 1840, 228). This anonymous review in the *Edinburgh Medical and Surgical Journal* accused him of creating a soulless, Godless world in which "the visible creation was at first made so perfect that the machine of nature runs its allotted course without requiring the superintendence of the Creator" and in which mind had no separate existence from inanimate matter. Carpenter's *Principles* tended "to lead the mind to the doctrines of materialism" ("Carpenter's Principles" 1840, 228). The *Medico-Chirurgical Review* ran a similar, though less vitriolic, review, claiming that Carpenter's "flights" of theorizing as to the mode of God's action in organizing animate nature reached "heights too high, or depths too low, for our timid philosophy." His researches had "begun in fancy or in skepticism" and ended in "mysticism, dogmatism or nonsense" ("Principles of General and Comparative Physiology" 1839, 170). The reviewer presumably associated Carpenter with the controversial lecturers who marketed such fodder as clockwork universes and the transmutation of species.

In this period such an attack could be sufficient, if unanswered, seriously to compromise one's career. It placed at risk not only Carpenter's future career as a physiologist, but even his present livelihood as a medical practitioner and scientific tutor. He therefore took immediate and vigorous action to vindicate himself. He published as an appendix to one of the moderate progressive medical periodicals a personal defense of the spiritual respectability of his work. Carpenter's defense involved systematically identifying his characterization of natural law in relation to living phenomena with the works of prestigious writers in the field and associating his current work with well-liked previous writings. First, he argued that his *Principles* were merely an amplification of an essay entitled "On the Laws Regulating Vital and Physical Phenomena," which won the Students Prize in 1838 and was subsequently published in Robert Jameson's *Edinburgh New Philosophical Journal*. Carpenter's *Principles* was intended to introduce students to the field, whereas the essay was for a far more restricted audience. One might therefore have expected that the way natural law was represented in the *Principles* would bear closer scrutiny since, as the *Edinburgh Medical and Surgical Journal* reviewer had emphasized, it was offered as "a safe guide" in the study of the life sciences. A research essay might have enjoyed a more tolerant reception, since its audience was assumed to be more restricted and more specialist. Carpenter's association of the two publications was consequently a powerful strategy because it suppressed the issue of audience. In doing so it allowed him to claim that to criticize the *Principles* was "virtually" to charge his Edinburgh professors "with having sanctioned opinions which are 'detrimental to the best interests of mankind' " (W. B. Carpenter 1840, 2).

The most serious charge was that Carpenter's definition of natural law represented a clockwork, deterministic universe—a charge that would imply, Carpenter said, his "disbelief in Revelation" (W. B. Carpenter 1840, 2). Carpenter described, both in his *Principles* and in his defense, a world run by laws that had themselves been ushered into existence by a single divine act. His depiction of natural law was reminiscent of Babbage's recently-published *Ninth Bridgewater Treatise* (1837) and of contemporary Unitarian conceptions of natural law. However, he emphasized his view—stated in the *Principles* itself—that "when a law of Physics or of vitality is mentioned, nothing more is really implied than a simple expression of the mode in which the Creator is constantly operating on inorganic matter, or on organized structures" (W. B. Carpenter 1840). This formulation was intended to mediate the reception of Carpenter's investigative plans into the "laws" governing physiological phenomena and to protect him from the charge of removing God from nature.

Carpenter also carefully allied his specific characterization of vital phenomena with prestigious individual researchers. He made two assumptions regarding these phenomena. First, they were the result of properties of organized tissues called into action by regular laws. Second, these properties were not "superadded to matter in the process of organisation; but . . . this act calls out or developes [*sic*] the properties which previously existed in the particles subjected to it, but which are not manifested except under the peculiar circumstances which this new disposition of them produces" (W. B. Carpenter 1840, 3). The first assumption, Carpenter maintained, was commonly held by all physiologists; the second was more difficult. Carpenter asserted that it was similar to claims made by James Cowles Prichard and James Fletcher, each of whom, he argued, had made similar or complementary arguments.

Finally, there was the matter of suggesting that the characterization of natural law in the life sciences should be modeled on the physical sciences. To defend himself against the charge of reducing the life sciences to the physical sciences, he quoted a passage from Peter Mark Roget's article on physiology in the *Encyclopaedia Britannica,* which referred to Carpenter's views on the life sciences and suggested that there were fewer differences between living and physical phenomena than might be supposed (W. B. Carpenter 1840, 3).

Carpenter therefore took care to rebut objections to his work in some detail. But more important than his own defense were the letters of reference themselves—letters solicited, as Carpenter put it in the request he sent to John Henslow, because the individuals he chose could be represented as embodying orthodoxy in science and religion (Carpenter to Henslow 1840, Botany School, Cambridge). They included the Dissenting theologian John Pye Smith, the Reverend William Daniel Conybeare, the Reverend Baden Powell, the Reverend John Stevens Henslow, the Reverend William Clark, John Herschel, Peter Mark Roget, James Cowles Prichard, William Pulteney Alison, and Henry Holland. We can plausibly assume that these individuals had already corresponded with Carpenter about his *Principles.* Correspondence with Henslow and Herschel still exists; Smith was a family friend, and Alison was Carpenter's teacher.

The religious figures Carpenter enlisted were not only divines, but professors at Oxford and Cambridge (aside from Smith, theological tutor at Homerton Academy). At Oxford, Conybeare was Bampton Lecturer and Powell Savilian Professor; at Cambridge, Clarke was Professor of Anatomy and Henslow Professor of Botany. The other scientific and medical figures were well chosen to make up a powerful body of authorities: Herschel's *Preliminary Discourse* had already come to be regarded as stating the highest ideal for scientific method; Roget was the secretary of the Royal Society during the 1830s and 1840s; Prichard, Alison, and Holland were among the most powerful physiologists of the 1830s.

They were also notable for their known interest in "cosmologies" and their interest in the construction of general laws governing the creation of life. Smith had just published a work on the formation of general laws of divine action in geology, and his well-known antipathy to Unitarianism helped distance Carpenter's physiological work from his family's religion, which six months before had led to a controversy over the dedication of Lant Carpenter's *Apostolic Harmony of the Gospels* to the queen (J. P. Smith 1839; Corsi 1988, 252; Brooke 1979; Chadwick 1966–70, 1:395; R. L. Carpenter 1842). Powell's 1839 *Tradition Unveiled* stated bluntly that modern science in no way "collided" with the authority of the church, and his *Connection of Natural and Divine Truth* (1838) had dismissed objections to "physical inquiries" into the mode of action of the Creator (Powell 1839, 64–6; 1838, 67–70; W. B. Carpenter 1838*b*, 548–49).

The testimonials provided support on two specific fronts. They defended as theologically sound both Carpenter's use of "natural law" and his use of the theoretical and experimental apparatus of the physical sciences to investigate physiological phenomena. With respect to the use of the term "natural law," Powell and Conybeare were most helpful. Conybeare claimed that he had himself "repeatedly expressed the same opinions" as had Carpenter's *Principles,* though "never half so well or eloquently"; both he and Powell saw no "dangerous tendency" in what Carpenter had said regarding laws of nature (W. B. Carpenter 1840, 7). Herschel would not comment as a scientist on the content of the *Principles* but argued that it was "common sense" that Carpenter had not put forward the notion of a machine wound up at Creation and thereafter running itself. Regarding Carpenter's experimental work, Henslow argued that only the "narrow-minded" could fail to realize "how possible it is for a man to be duly impressed with the truths of revelation, though he is equally satisfied that they were never intended to interfere with the freedom of his researches into those great natural laws by which God frames and governs the Universe" (W. B. Carpenter 1840, 7-8). That is, the fullest piety did not conflict with researches into the relationship between physical and vital forces. Holland agreed that Carpenter had "never exceeded the authorized bounds of physical research, as pursued by the most eminent physiologists" (W. B. Carpenter 1840, 8).

These statements were fairly successful, in that Carpenter was no longer troubled by outright accusations of materialism. In the wake of the controversy over his *Principles,* however, his reputation as a writer did not help his medical practice and scientific tutoring, and "the struggle to maintain his position was severe" (Estlin Carpenter "Memorial Sketch," 32 in W. B. Carpenter 1888). But his finances and his professional standing improved steadily with the success of his publications during the 1840s. His *Principles of General and Comparative Physiology* had gone through four editions by 1854; his *Principles of Human Physiology,* first published in 1842, reached a fourth edition in 1853; and the several-volume series of the *Cyclopaedia of Natural Science* was published between 1841 and 1844. Along with these works he published a steady stream of articles and reviews in the *British and Foreign Medical Review,* which he began to edit in 1847. Carpenter's works became the standard medical textbooks of his time, the texts which embodied orthodox medical knowledge for the medical students of the 1840s and 1850s.

Thus, through careful canvassing of his elite colleagues, Carpenter saved himself from the fate which John Elliotson had suffered for similar claims, which had been made without the deferential solicitation of individual patrons and without the concern for careful phrasing that had helped Carpenter claim that his depictions of natural law were not materialist. This is an illustration of how social networks supported their members and of the patronage tactics that individuals could deploy in creating a hospitable space for themselves in various intellectual communities. But Carpenter's case also suggests further reflections on how a group of eminent individuals could be represented as an orthodox community by someone in his position, as well as how authors could influence the way a particular work was read and interpreted in this period.

For authors wishing to ensure that a potentially controversial claim be given a desired reading, there were specific opportunities and dangers involved in moving from a more bounded context for scientific communication, such as the patronage correspondence Carpenter undertook in the late 1830s, to that of the various publications in which his work appeared. As indicated earlier, Carpenter consulted several prestigious natural philosophers and medical writers before his *Principles* came out in print, and then before publishing the rebuttal to his medical critics. There are significant differences between private correspondence and public press as contexts for the interpretation of scientific claims. This has been documented with respect to the very issues that Carpenter got into trouble for addressing (W. Cannon 1960; Hyman 1982). The kind of claims being made by Charles Babbage were far less controversial when they were being discussed in private correspondence—for instance, in correspondence between John Herschel and Charles Lyell—than when they came out in print. One might argue that the very meaning, or, should one wish to make a distinction, the significance of the claims changed when they were made in the more volatile medium of the printed word. This change was something that was explicitly recognized in early-nineteenth-century society. The debates over "useful knowledge" and newspaper taxes, for instance, were related to such concerns. The libel laws in effect at this time give some perspective on the way this issue was perceived. Libel laws focused on effects rather than meanings, so the context of what was said was the only determining factor in the decision as to whether it was illicit. There were debates in Parliament in the 1810s and 1820s about what kinds of claims could be made in which context, and particularly about the difference in significance of a piece of information communicated round a middle-class dining table as opposed to a radical artisan broadsheet (Gilmartin 1996, chap. 2; Vincent 1989, 235).

For early Victorian scientists, this dynamic made for significant dangers in moving private statements into print, since in consequence they could look more controversial and heterodox to readers. Historians are well aware that printing potentially controversial claims was more dangerous than communicating them via correspondence. However, it is worth considering that the move from correspondence into print may have provided opportunities as well as obstacles to early Victorian scientists. Carpenter's move to secure consent to his various claims via epistolary correspondence —both the claims printed in the *Principles* and in his later statement of self-defense—suggests the possibility that the private solicitation of support by people in Carpenter's position not only helped to secure patronage, but also could help to fix a particular interpretation of a potentially controversial scientific claim, at least in terms of how a work might be understood in the short term and by specific constituencies. That is, if one could secure an important potential supporter's consent to a claim one wished to make in the more bounded forum of private correspondence, one could transfer that support, if necessary, into the more volatile arena of print and thereby help to stabilize the claim's meaning. Once a potential supporter had expressed approbation in a private letter, it would be difficult for him to refuse to state this approbation publicly, even though both parties would be aware of the change in significance both of the scientific claim and of the statement of support. In Carpenter's case, it is certainly true that the appearance of the letters would have tended to have this effect, though it is not clear whether Carpenter self-consciously intended to produce it.

IV

This chapter has sought to fulfill two related agendas. It began with the assertion that no secure, stable, bounded community of definitive authorities or set of rules governing scientific work existed as such during the early Victorian period. A vast array of different scientific and medical projects flourished, and along with them rival portrayals of what kind of enterprises should be considered legitimate. I have been particularly concerned to show that if proper science could be defined differently in different contexts, then scientific claims could have radically different status and even, perhaps, different meanings depending on where they were read or heard and by whom.

The account of William Benjamin Carpenter's early career has explored what implications this messy state of affairs had for how individuals could negotiate the status of controversial scientific claims. The uncertain status of his depiction of natural law, and its ultimate characterization as not "exceeding the authorized bounds" (Holland as quoted in W. B. Carpenter 1840, 8), shows that the significance of a scientific assertion could be profoundly influenced by being provided with a particular context in which to be read. This account of early Victorian science, and Carpenter in particular, also has implications for understanding how individuals played a role in defining the communities which counted as orthodox for them in particular situations.

The story of Carpenter's success in securing scientific respectability points to a picture of scientific orthodoxies and heterodoxies as emerging together and as being constantly subject to redefinition. While the figures who came to Carpenter's aid were individually eminent, they did not (outside the context of this debate) constitute a body of orthodoxy that shared and policed certain assumptions about the nature and bounds of proper science. It was Carpenter's act of juxtaposing the names and statements of individually eminent personages that constructed them as an authoritative and definitive community. He assembled the group of eminent individuals who supported him; he selected the particular passages of their letters he wished to be printed; he chose which credential, of the various positions and honors held by each individual, would follow their name in the printed statement; and he chose the order in which their names and statements would appear. This work cannot, then, be regarded as merely aesthetic. It had the effect of constructing on paper and for a specific publication and reading event authorities whose individual significance was carefully specified by Carpenter and who were presented as a group whose assembled authority surrounded his work. The specific work that was necessary to secure the status of orthodoxy for himself was the assertion of what counted as an authoritative community for him. That is, the list of individuals marshaled to Carpenter's cause should not be understood as constituting a set of scientific leaders recognized by contemporaries as a bounded elite. Rather, it was Carpenter's maneuver that asserted the existence and membership of a community of definitive experts—those individuals juxtaposed against one another as leaders in the field—and, simultaneously, the status of orthodoxy for his own claims. Carpenter surrounded himself with people who formed, as a composite, a body of authority tailored to accomplish local, transient goals, but they existed as a unified group only for that purpose and for the temporary period in which their services were required.

Carpenter's actions suggest that one way of building on the literature on Victorian heterodoxy that has characterized early Victorian science in terms of indeterminacy and chaos would be to examine how authoritative communities are constructed temporarily and for local purposes. This would have the advantage of offering a perspective on scientific authority that has the potential to learn from the literature discussed in the first half of this chapter without denying that prestigious and authoritative individuals did exist during these years. It can take into account the power and significance of individual scientific luminaries without assuming an overly homogeneous and artificially unified picture of a scientific elite.

Bibliographical Note

The strongest area of secondary literature in the study of the construction of orthodoxy and heterodoxy in Victorian scientific practice reconstructs the projects of so-called alternative sciences and medical practitioners. Wallis (1979) and Bynum and Porter (1987) provide good collections of essays that include studies of Victorian Britain. There is a substantial literature on mesmerism and phrenology. Cooter (1984) provides the definitive study of the latter and includes references to several other studies; see also Parssinen (1974) and Shapin (1975); and for a particularly striking study of phrenology that documents how different communities saw nature in different ways, see Shapin (1979). There exists, as yet, no monograph on Victorian mesmerism, but Gauld's more general, and staggeringly well-researched general history and bibliography of hypnotism (1992) and Ellenberger's classic study (1970) provide a rich source of information. Useful social histories of mesmerism include Kaplan (1974) and Parssinen (1977). Individual articles on mesmerism that give a sense of how its heterodox status was defined and contested include Parssinen (1979), Palfreman (1977), Cooter (1985), and Winter (1991, 1994*a*). Harrison (1987) is particularly useful, as it brings together a wide range of "radical" and "fringe" medical practices, from mesmerism to homeopathy and herbalism. The vast literature on Victorian spiritualism ranges from richly researched overviews of psychical research (Oppenheim 1989) to studies of how different groups used spiritualism as a vehicle for constructing authority for themselves or their political projects. Among the best examples of these are Barrow (1986) and Owen (1989). Barrow's study of plebeian spiritualism argues that socialist artisans constructed a "democratic epistemology" in relation to their spiritualist projects. Owen documents how Victorian women mediums used spiritualism to subvert Victorian conventions of femininity. Finally, there were several incidents and publications that became the focus for intense contests over the nature of orthodox and heterodox knowledge, one of the most widely debated of which was Robert Chambers's *Vestiges of the Natural History of Creation*. On the debates over this work see Yeo (1984) and A. J. Secord (1989*b*, forthcoming).

References

Altick, R. D. 1978. *The Shows of London*. Cambridge, Mass.: Belknap Press.
Anderson, Katherine. 1994. "Practical Science: Meteorology and the Forecasting Controversy in Mid-Victorian Britain." Ph.D. thesis, Northwestern University.
Babbage, Charles. 1830. *Reflections on the Decline of Science in England and on Some of Its Causes*. London: B. Fellowes.
———. 1837. *The Ninth Bridgewater Treatise*. London: John Murray.
Barnes, Barry, and Steven Shapin. 1976. "Head and Hand: Rhetorical Resources in British Pedagogical Writing 1770-1850." *Oxford Review of Education* 2:231-54.
Barrow, Logie. 1980. "Socialism in Eternity: The Ideology of Plebeian Spiritualists 1853-1913." *History Workshop* 9:37-69.
———. 1986. *Independent Spirits: Spiritualism and English Plebeians 1850-1910*. London: Routledge and Kegan Paul.
———. 1991. "Why Were Most Medical Heretics at Their Most Confident around the 1840s? (The other side of mid-Victorian Medicine)." In *British Medicine in an Age of Reform*, edited by Roger French and Andrew Wear. Cambridge: Cambridge University Press, 165-85.
Barton, Ruth. 1976. "The X Club: Science, Religion, and Social Change in Victorian England." Ph.D. thesis, University of Pennsylvania.
Bloor, David. 1983. "Coleridge's Moral Copula." *Social Studies of Science* 13:605-19.

Brooke, J. H. 1979. "The Natural Theology of the Geologists: Some Theological Strata." In *Images of the Earth: Essays in the History of the Environmental Sciences,* edited by L. J. Jordanova and R. S. Porter. Chalfont St. Giles, Buckinghamshire: British Society for the History of Science, 39-64.

Butler, Marilyn. 1993. "The First Frankenstein and Radical Science." *Times Literary Supplement,* 9 April, 12-14.

Bynum, W. F., and R.S. Porter, eds. 1987. *Medical Fringe & Medical Orthodoxy.* London: Croom Helm.

Cannon, S. F. 1978. *Science in Culture: The Early Victorian Period.* New York: Dawson Science History Publications.

Cannon, Walter. 1960. "The Problem of Miracles in the 1830s." *Victorian Studies* 4:5-32.

Carpenter, R. L. 1842. *Memoirs of the Life of Lant Carpenter.* Bristol: Philp and Sons.

Carpenter, W. B. 1838*a*. "Physiology an Inductive Science." *British and Foreign Medical Review* 9:317-42.

———. 1838*b*. "Prof. Powell's Natural and Divine Truth." *British and Foreign Medical Review* 5:548-49.

———. 1839*a*. *Principles of General and Comparative Physiology.* London: John Churchill.

———. 1839*b*. *Prize Thesis: Inaugural dissertation on the physiological inferences to be deduced from the structures of the nervous system in the invertebrated classes of animals.* Edinburgh: Edinburgh University.

———. 1840. "Remarks on some Passages of the 'Review of Principles of General and Comparative Physiology' in the Edinburgh Medical and Surgical Journal, January, 1840, by William B. Carpenter." *British and Foreign Medical Review* 9, appendix (separately paginated).

[———]. 1851. "The Correlation of the Physical and Vital Forces." *British and Foreign Medico-Chirurgical Review* 8:206-38.

———. 1852. "On the Influence of Suggestion in Modifying and Directing Muscular Movement, Independently of Volition." *Proceedings Royal Institution of Great Britain,* 1851-54, 1:147-53.

———. 1874. *Principles of Mental Physiology.* London: H. S. King and Company.

———. 1877. *Mesmerism, Spiritualism, &c., Historically and Scientifically Considered: Being Two Lectures Delivered at the London Institution.* New York: D. Appleton.

———. 1888. *Nature and Man: Essays Scientific and Philosophical, With an Introductory Memoir by J. Estlin Carpenter.* Edited by J. Estlin Carpenter. London: K. Paul, Trench and Company.

"Carpenter's Principles of General and Comparative Physiology." 1840. *Edinburgh Medical and Surgical Journal* 53:213-28.

Chadwick, Owen. 1966-70. *The Victorian Church.* 2 vols. New York: Oxford University Press.

[Chambers, Robert]. 1844. *Vestiges of the Natural History of Creation.* London: John Churchill.

———. 1994. *Vestiges of the Natural History of Creation and Other Evolutionary Writings.* Edited by J. A. Secord. Chicago: University of Chicago Press.

Clarke, J. F. 1874. *Autobiographical Recollections of the Medical Profession.* London: J. and A. Churchill.

Cooter, R. J. 1984. *The Cultural Meaning of Popular Science: Phrenology and the Organization of Consent in Nineteenth Century Britain.* Cambridge: Cambridge University Press.

———. 1985. "The History of Mesmerism in Britain: Poverty and Promise." In *Franz Anton Mesmer und die Geschichte des Mesmerismus: Beiträge zum internationalen wissenschaftlichen Symposion anlässlich des 250. Geburtstages von Mesmer, 10. bis 13. Mai 1984 in Meersburg,* edited by H. Schott. Stuttgart: F. Steiner, 152–62.

Corsi, Pietro. 1988. *Science and Religion: Baden Powell and the Anglican Debate, 1800–1860.* Cambridge: Cambridge University Press.

Darwin, Charles. 1859. *On the Origin of Species.* London: John Murray.

Davey, William. 1854. *Illustrated Practical Mesmerist: Curative and Scientific.* Edinburgh: W. H. Lizars.

Desmond, Adrian. 1987. "Artisanal Resistance and Evolution in Britain, 1819–1848." *Osiris* 3:77–110.

———. 1989a. "Lamarckism and Democracy: Corporations, Corruption, and Comparative Anatomy in the 1830s." In *History, Humanity and Evolution: Essays in Honor of John C. Greene,* edited by J. R. Moore. Cambridge: Cambridge University Press, 99–130.

———. 1989b. *The Politics of Evolution: Medicine, Morphology and Reform in Radical London.* Chicago: University of Chicago Press.

Desmond, Adrian, and J. R. Moore. 1992. *Darwin.* London: Michael Joseph.

Ellenberger, H. F. 1970. *The Discovery of the Unconscious: The History and Evolution of Dynamic Psychiatry.* New York: Basic Books.

Gauld, Alan. 1992. *A History of Hypnotism.* Cambridge: Cambridge University Press.

Gilmartin, Kevin. 1996. *Print Politics: The Press and Radical Opposition in Early Nineteenth-Century England.* Cambridge: Cambridge University Press.

Goodfield-Toulmin, June. 1969. "Some Aspects of English Physiology 1780–1840." *Journal of the History of Biology* 2:283–320.

Hall, Marie Boas. 1984. *All Scientists Now: The Royal Society in the Nineteenth Century.* Cambridge: Cambridge University Press.

Hall, Vance. 1979. "The Contribution of the Physiologist, William Benjamin Carpenter (1813–1885) to the Development of the Correlation of Forces and the Conservation of Energy." *Medical History* 23:129–55.

Harrison, J. F. C. 1979. *The Second Coming: Popular Millenarianism 1780–1850.* New Brunswick, New Jersey: Rutgers University Press.

———. 1987. "Early Victorian Radicals and the Medical Fringe." In *Medical Fringe and Medical Orthodoxy,* edited by W. F. Bynum and Roy Porter. London: Croom Helm, 198–215.

Hilton, Boyd. 1988. *The Age of Atonement: The Influence of Evangelicalism on Social and Economic Thought 1795–1865.* Oxford: Clarendon Press.

———. 1995. "The Politics of Nature and the Nature of Politics." Ford Lecture delivered at University of Oxford, November 1995.

Hyman, Anthony. 1982. *Charles Babbage: Pioneer of the Computer.* Princeton, N.J.: Princeton University Press.

Jacyna, L. S. 1983. "Immanence or Transcendence: Theories of Life and Organization in Britain 1790-1835." *Isis* 74:311-29.

———. 1984. "Principles of General and Comparative Physiology: The Comparative Dimension to British Neuroscience in the 1830s and 1840s." *Studies in the History of Biology* 8:47-93.

Jensen, J. Vernon. 1971-72. "Interrelationships within the Victorian 'X Club.'" *Dalhousie Review* 51:539-52.

Kaplan, Fred. 1974. "'The Mesmeric Mania': The Early Victorians and Animal Magnetism." *Journal of the History of Ideas* 35:691-702.

———. 1982. *John Elliotson on Mesmerism*. Princeton, N.J.: Princeton University Press.

Lawrence, William. 1816. *An Introduction to Comparative Anatomy and Physiology: Being Two Introductory Lectures Delivered at the Royal College of Surgeons, on the 21st and 25th of March, 1816*. London: J. Callow.

Mandler, Peter. 1990. *Aristocratic Government in the Age of Reform: Whigs and Liberals 1830-1852*. Oxford: Clarendon Press.

Merrington, W. R. 1976. *University College Hospital and Its Medical School: A History*. London: Heinemann.

Morrell, Jack. 1976. "London Institutions and Lyell's Career, 1820-41." *British Journal for the History of Science* 9:132-46.

Morrell, Jack, and Arnold Thackray. 1981. *Gentlemen of Science: The Early Years of the British Association for the Advancement of Science*. Oxford: Oxford University Press.

Morus, I. R. 1991. "Correlation and Control: William Robert Grove and the Construction of a New Philosophy of Scientific Reform." *Studies in the History and Philosophy of Science* 22:598-621.

———. 1992a. "Different Experimental Lives: Michael Faraday and William Sturgeon." *History of Science* 30:1-28.

———. 1992b. "Marketing the Machine: The Construction of Electro-Therapeutics as Viable Medicine in Early Victorian England." *Medical History* 36:34-52.

"Notices of Some New Works." 1839. *Medico-Chirurgical Review* 31:497-98.

Oppenheim, Janet. 1989. *The Other World: Spiritualism and Psychical Research in England, 1850-1914*. Cambridge: Cambridge University Press.

Owen, Alex. 1989. *The Darkened Room: Women, Power and Spiritualism in Late Nineteenth Century England*. London: Virago.

Palfreman, J. 1977. "Mesmerism and the English Medical Profession: A Study of a Conflict." *Ethics in Science and Medicine* 4:51-66.

Parssinen, T. M. 1974. "Popular Science and Society: The Phrenological Movement in Early Victorian Britain." *Journal of Social History* 8:1-20.

———. 1977. "Mesmeric Performers." *Victorian Studies* 21:87-104.

———. 1979. "Professional Deviants and the History of Medicine: Medical Mesmerists in Victorian Britain." In *On the Margins of Science: The Social Construction of Rejected Knowledge*, edited by Roy Wallis. Sociological Review Monograph 27. Keele: University of Keele, 103-20.

Porter, Roy. 1978. "Gentlemen and Geology: The Emergence of a Scientific Career, 1660-1920." *Historical Journal* 21:809-36.

Powell, Baden. 1838. *The Connection of Natural and Divine Truth; Or, the Study of the Inductive Philosophy Considered as Subservient to Theology.* London: J. W. Parker.

———. 1839. *Tradition Unveiled; Or, an Exposition of the Pretensions and Tendency of Authoritative Teaching in the Church.* London: J. W. Parker.

"Principles of General and Comparative Physiology." 1839. *Medico-Chirurgical Review* 31:165–70.

Rudwick, Martin. 1985. *The Great Devonian Controversy: The Shaping of Scientific Knowledge among Gentlemanly Specialists.* Chicago: University of Chicago Press.

Secord, Anne. 1994. "Corresponding Interests: Artisans and Gentlemen in Nineteenth-Century Natural History." *British Journal for the History of Science* 27:383–408.

Secord, J. A. 1986. *Controversy in Early Victorian Geology: The Cambrian-Silurian Dispute.* Princeton, N.J.: Princeton University Press.

———. 1989a. "Extraordinary Experiment: Electricity and the Creation of Life in Early Victorian England." In *The Uses of Experiment: Studies in the Natural Sciences,* edited by David Gooding, Trevor Pinch, and Simon Schaffer. Cambridge: Cambridge University Press, 337–83.

———. 1989b. "Behind the Veil: Robert Chambers and *Vestiges.*" In *History, Humanity and Evolution: Essays for John C. Greene,* edited by J. R. Moore. Cambridge: Cambridge University Press, 165–94.

———. 1991. "Edinburgh Lamarckians: Robert Jameson and Robert E. Grant." *Journal of the History of Biology* 24:1–18.

———. Forthcoming. *Evolution for the People* (provisional title).

Shapin, Steven. 1975. "Phrenological Knowledge and the Social Structure of Early Nineteenth-Century Edinburgh." *Annals of Science* 32:219–43.

———. 1979. "The Politics of Observation: Cerebral Anatomy and Social Interests in the Edinburgh Phrenology Disputes." In *On the Margins of Science: The Social Construction of Rejected Knowledge,* edited by Roy Wallis. Sociological Review Monograph 27. Keele: University of Keele, 139–78.

Smith, John Pye. 1839. *On the Relations Between the Holy Scriptures and Some Parts of Geological Science.* London: Jackson and Walford.

Smith, Roger. 1973. "The Background of Physiological Psychology in Natural Philosophy." *History of Science* 21:75–123.

———. 1977. "The Human Significance of Biology: Carpenter, Darwin and the Vera Causa." In *Nature and the Victorian Imagination,* edited by U. C. Knoepflmacher and G. B. Tennyson. Berkeley: University of California Press, 216–30.

Vincent, David. 1989. *Literacy and Popular Culture: England 1750–1914.* Cambridge: Cambridge University Press.

Wallis, Roy, ed. 1979. *On the Margins of Science. The Social Construction of Rejected Knowledge.* Sociological Review Monograph 27. Keele: University of Keele.

Warwick, A. C. 1995. "The Sturdy Protestants of Science: Larmor, Trouton, and the Earth's Motion through the Ether." In *Scientific Practice: Theories and Stories of Doing Physics,* edited by Jed Z. Buchwald. Chicago: University of Chicago Press, 300–343.

Winter, Alison. 1991. "Ethereal Epidemic: Mesmerism and the Introduction of Inhalation Anaesthesia to Early Victorian Britain." *Social History of Medicine* 4:1–27.

———. 1994a. "Mesmerism and Popular Culture in Early Victorian England." *History of Science* 32:317-43.

———. 1994b. "'Compasses all awry': The Iron Ship and the Ambiguities of Cultural Authority in Victorian Britain." *Victorian Studies* 38:69-98.

———. 1995. "Harriet Martineau and the Reform of the Invalid in Victorian Britain." *Historical Journal* 38:597-616.

Yeo, Richard. 1984. "Science and Intellectual Authority in Mid-Nineteenth Century Britain: Robert Chambers and *Vestiges of the Natural History of Creation*," *Victorian Studies* 28:5-31.

3

The Probable and the Possible in Early Victorian England

JOAN L. RICHARDS

In 1830, the young John Herschel published his *Preliminary Discourse on the Study of Natural Philosophy* as the first volume of Dionysius Lardner's *Cabinet Cyclopedia*. In the first part, entitled "Of the General Nature and Advantages of the Study of the Physical Sciences," the up-and-coming scientist explained the value of the physical sciences. The external world is so multifarious, Herschel there asserted,

> that as the study of one [subject] prepares him [the scientist] to understand and appreciate another, refinement follows on refinement, wonder on wonder, till his faculties become bewildered in admiration, and his intellect falls back on itself in utter helplessness of arriving at an end. (Herschel [1830] 1966, 4-5)

Being thus overwhelmed is a positive first step in the scientist's pilgrimage. It turns his gaze inward, where again he

> feels himself capable of entering only very imperfectly into these recesses of his own bosom, and analysing the operations of his mind,— in this as in all other things, in short, "a being darkly wise;" seeing that all the longest life and the most vigorous intellect can give him power to discover . . . serves only to place him on the very frontier of knowledge, and afford a distant glimpse of boundless realms beyond. (Herschel 1966, 6)

"Is it wonderful," Herschel continued,

> that a being so constituted should first encourage a hope, and by degrees acknowledge an assurance, that his intellectual existence will

---

Research for this chapter was supported by a grant from the Dibner Institute for the History of Science and Technology, Massachusetts Institute of Technology, Room E56-100, 38 Memorial Drive, Cambridge, Massachusetts 02139.

not terminate with the dissolution of his corporeal frame, but rather that in a future state of being . . . he shall drink deep at that fountain of beneficent wisdom for which the slight taste obtained on earth has given him so keen a relish? (Herschel 1966, 7)

In these passages Herschel paints a picture in which science leads to the very borders of human knowledge, from which we glimpse a reality that is much larger than our knowing. This reality lies beyond scientific understanding, but we do have indications of it. The personal experiences of wonder, bewilderment, relish, and hope are signposts marking the route to the understanding Herschel described.

The central value Herschel assigned to these personal experiences reflects an essential aspect of the culture of which he was a part. The institutional locus of this kind of personal knowledge was religion, which was an ever-present part of life in his culture: as Joseph Altholtz remarked, "the most important thing to remember about religion in Victorian England is that there was an awful lot of it" (Altholz 1988, 150). The essential point, for the purposes of this chapter, is that religious and scientific knowing were neither separate nor separable categories. It was not clear whether there were boundaries between them or, if there were, where they should be drawn.

As the early Victorians came in contact with the science being developed on the Continent they were forced to examine this unclear boundary. This chapter focuses on a particular aspect of the discussion that revolved around probability theory. Herschel's typically English, personally weighted formulation of the nature and purpose of knowledge stands in stark contrast to the rationalist assumptions of Continental probabilists. For them scientific thinking was construed as dispassionate, grounded in an epistemological realm far from the religious one of human affect. It was a significant challenge for English thinkers in the 1830s and 1840s to assimilate probability theory into their culture, where the boundary between scientific and religious knowing, between rational and affective knowledge, was not clearly drawn. The process took decades and left neither the science nor the culture unaffected.

English attempts to assimilate Continental mathematics in the first half of the nineteenth century have long provided historians with a rich case study of the interaction between mathematics and views of knowledge. The central narrative revolves around the Analytic Society, whose members vowed to bring French analysis to England in the second decade of the century. The young analytics, as well as much of their posterity, presented this as a relatively straightforward question of translation. They were attracted by the raw power of Continental symbolism and simply wanted to introduce that symbolism into England, in particular into the Tripos examination at

Cambridge, so that students educated there could follow Continental work.

Bolstered by impressive archival resources, recent historians have begun to construct a more historically nuanced picture of the analytics' project (Enros 1983). Several historians have followed the group into the 1820s, when many, notably Herschel, Charles Babbage, and, somewhat later, Augustus De Morgan, moved out of undergraduate Cambridge into cosmopolitan London. There their fascination with symbolical power became entwined with commercial, industrial, and political issues and also spilled into analogies between the human mind and machines, epitomized by Babbage's calculating engines (Miller 1986; Ashworth 1994; Durand-Richards forthcoming; Schaffer 1994).

Such thinking was less comfortable in the pastoral parochial world of Victorian Cambridge, where William Whewell remained for all of his life. Whewell's early relations to the Analytic Society work are debated (Fisch 1991, 1994; Becher 1992), but it is clear that by the 1830s he was deeply disturbed by the mechanical implications of French analysis. In 1840 Whewell countered the epistemological implications of French analysis with a philosophy that dismissed symbolical manipulation as empty. Although it was not widely accepted, Whewell's work was a milestone that defined the terms of discussion for the next generation (Fisch and Schaffer 1991; Fisch 1991; Yeo 1994).

The historical school that is embedding the analytics and their mathematics in the larger world of English culture has been supported by a number of more mathematically focused studies. It is as algebraists that mathematicians remember these Englishmen, and there has been considerable interest in the epistemological complexities of their enterprise. A number of studies have charted the ways that often hidden epistemological assumptions shaped English mathematical interests and insights into algebra and analysis (Richards 1980, 1991, 1992; Pycior 1981, 1982, 1983; Fisch 1994).

In mathematics, probability theory is less central than algebra. However, in the last two decades a number of historians and philosophers have recognized the central importance of this theory to Western concepts of knowledge, in both the physical and the social sciences. For those interested in the ways that scientific knowledge has affected larger cultural issues, the mathematization of chance and its application to social thinking have been centrally important (Hacking 1975; Gigerenziger et al. 1989; Krüger, Daston, and Heidelberger 1987; Krüger, Daston, and Morgan 1987).

The early Victorians play a relatively small part in this tale. However, as the analytics spilled out of Cambridge in the 1820s it became clear to them and their contemporaries that probability theory had important practical applications, particularly for astronomy and life insurance. They were again in the position to import a Continental theory onto their island. As with analysis, this was a complicated process. An important strand of seventeenth- and eighteenth-century probability theory recognized it as an exemplar of rationality, which meant that many of the epistemological issues that were implicit in analysis were explicit in probability (Daston 1980, 1988).

The discussion was initially based in natural theology, which is the subject of the first section of this chapter. Developed throughout the eighteenth century primarily as an attempt to ground religion in the new science, this genre received an unexpected boost in the 1830s when the Earl of Bridgewater left the considerable sum of eight thousand pounds for the support of works devoted to "the Power, Wisdom and Goodness of God, as manifested in the Creation" (Whewell [1833] 1836, "Notice"). Several of the eight treatises that resulted are distinguished from the rationalism of eighteenth-century works, because they approach the discussion more personally. Notable in this regard is William Whewell's *Astronomy and General Physics considered with reference to natural theology*, which developed a view of science in explicit opposition to the rationalism exemplified by Continental probabilists. The theological parameters of Whewell's position are suggested by comparing his views with those of his contemporary, the Oxford theologian John Henry Newman.

Thus conjoining Whewell and Newman may seem highly artificial from a historical perspective that recognizes the sharp disagreements that divided the Cambridge scientific aficionado from the leader of the Oxford movement. However, the commonalities of their epistemological outlook were compelling enough that Charles Babbage lumped them together and responded with a spirited defense of rationalism and the probability theory that mathematized it. Babbage's *Ninth Bridgewater Treatise* provoked considerable discussion. In the period immediately following its publication Whewell and Herschel both objected, while the somewhat younger Augustus De Morgan tried to understand what probability theory said about knowledge. This discussion is the subject of the second section of this chapter.

The third section will follow English considerations of probability theory into the next decade. Whewell's immediate response to Babbage was laconic. In 1840, however, he published *The Philosophy of the Inductive Sciences*, which effectively moved English considerations of the nature of knowing out of natural theology and into philosophy. Whewell did not directly consider probability theory in this work. However, the young Robert Leslie Ellis turned his attention to reconciling probability theory with Whewell's philosophy. By 1850 a new interpretation had emerged that allowed the mathematics and applications of probability to stand but challenged the tight rationalism of its classical devotees. Although short-lived, Ellis's treatment can be seen as the culmination of a long attempt to assimilate the French import to early Victorian culture.

I. Knowledge and Natural Theology

The 1830s were turbulent years in British history; the Reform Bill of 1832, which greatly increased suffrage, was a central event that entailed major political change. Centrally important in the swirling scene were changing relations between church and state, spearheaded by two previous bills: the Test and Corporation Acts that in 1828 allowed Protestant Dissenters to become full citizens, followed swiftly by the Catholic Emancipation Bill of 1829. With these bills began a process that eroded the bonds of church and state, a particularly pressing issue for the Universities of Cambridge and Oxford, which were landed Anglican establishments. At issue in both places, though played out in rather different ways, was the relationship between religious and intellectual life. At Cambridge, the heir to Newton's science, the issues developed around natural theology; at Oxford they were framed by the Oxford movement. The two traditions were often in conflict with each other, but it is also true that they were both firmly rooted in the same Anglican Church.

Whewell's *Astronomy and General Physics* was the first of the *Bridgewater Treatises*. On the surface it is devoted simply to constructing a design argument for the existence of God around the lawlike motions of the heavens. However, a closer reading reveals that for Whewell, design suggests more than it proves; to quote a characteristically tentative statement, "Many persons, . . . especially those who are already in the habit of referring the world to its Creator, will probably see something admirable in itself in this vast variety of created things" (Whewell 1836, 74). Knowledge in Whewell's natural theology was recognized by the individual beholder rather than established by the structure of the argument; it was indicated rather than proved.

Whewell expounded his orientation in the third section of his book, entitled "Religious Views." His major thesis is captured in the subtitle of the final chapter: "On the Impossibility of the Progress of our knowledge ever enabling us to comprehend the Nature of the Deity." Whewell builds to this conclusion with a consideration of the roles induction and deduction play in finding knowledge.

In Whewell's construction, induction describes the tortuous process of trial and error by which great scientific discoverers—Newton and Kepler are his favorite examples—came to their discoveries. Their investigations clearly established laws, but these were not their deepest insights; those were engendered by the humbling anterior process. As he put it,

> The effort and struggle by which he [the scientist] endeavors to extend his view, makes him feel that there is a region of truth not included in his present physical knowledge; the very imperfection of the light in which he works his way, suggests to him that there must

> be a source of clearer illumination at a distance from him. (Whewell 1836, 334)

For Whewell, the great scientific discoverers practiced this kind of inductive science.

Another tier of investigators devoted themselves to "deductive reasoning, exhibiting the consequences and applications of the laws which have been discovered" (Whewell 1836, 326). Rather than standing on the brink of the unknown, the attention of deductive thinkers is focused on the few

> general principles, which form the basis of their explanations and applications. . . . they make these their ultimate grounds of truth. . . . Their thoughts dwell little upon the possibility of the laws of nature being other than we find them to be, . . . and still less on those facts and phenomena which philosophers have not yet reduced to any rule. (Whewell 1836, 331)

This orientation, which Whewell attributed to Jean d'Alembert, Alexis Clairault, Leonhard Euler, Joseph-Louis Lagrange, and Pierre-Simon Laplace, produces no real insight. It does not force the same humbling recognition that much that is real is unknown to us; a related weakness is that, as his exemplars indicate, it does not conduce to religious conviction.

It was difficult to analyze the personally enriching learning process Whewell illustrated by historical example in the terms of the new science, and Whewell did not try. However, the process he described was amenable to religious characterization. This can be seen in the work of Newman. In the 1830s, the Oxford theologian was arguably the most articulate theological voice in the same Anglican Church that housed Whewell throughout his life. The terms in which Newman described religious knowledge suggest the underlying assumptions behind Whewell's views.

In 1837 Newman devoted the seventy-third of the *Tracts for the Times* to a defense of religion against "rationalism." "To Rationalize," he explained,

> is to ask for reasons out of place; to ask improperly how we are to account for certain things, to be unwilling to believe them unless they can be accounted for. ([Newman] 1836, 2)

This approach suffers from hubris,

> measuring the credibility of things, not by the power and other attributes of God, but by our own knowledge. . . . Nothing is considered to have an existence except so far forth as our minds discern it. . . . Mystery is discarded. ([Newman] 1836, 2)

The specific people Newman cited as rationalists, "Mr. Erskine and Mr. Jacob Abbott," are a far cry from the "continental mathematicians" against whom Whewell railed. However, they share the conviction that their knowledge is adequate to grasp the world's realities, and the consequence is the same. For both groups "mystery is discarded" and religious understanding is not attained.

Whereas Whewell countered the atheism of deductive science with examples illustrating the religious power of inductive pursuits, Newman insisted on the primacy of faith. He defined faith as an "agent" that "may be supposed as acting in unknown ways" ([Newman] 1836, 2): "the reaching forth after and embracing what is beyond the Mind" ([Newman] 1836, 5). Newman supported this view of faith with St. Paul: "Those all died in faith, *not having received* the promises, but *having seen them afar off,* and were persuaded of them, and embraced them" ([Newman] 1836, 5). Except that they lacked the authority of the apostle, Newman might equally have used Herschel's "distant glimpse of boundless realms beyond" or Whewell's feeling "that there is a region of truth not included in . . . present physical knowledge" to illustrate his dynamic concept of faith. Their descriptions of the inductive process incorporated the essential aspects of Newman's active faith into the very heart of science.

This was a highly charged position in an intellectual world assessing the value of the new science, and in some ways it can been seen as having pleased no one. It threatened the traditional church by claiming for science insights that were traditionally located in religion; certainly Newman granted little value to Herschel's and Whewell's science or to the natural theology in which it was embedded. Equally upset were those who valued the kind of science Whewell dismissed as deductive. From this side, the issue was joined by Charles Babbage, who wrote an uncommissioned *Ninth Bridgewater Treatise* to defend deductive science from the strictures Whewell's interpretation placed on it. Probability theory played a central role in Babbage's arguments for the central importance of the knowledge and insights to be gained from deductive science.

II. Knowledge and Probability Theory

Probability theory was the epitome of the kind of deterministic, mechanical thinking Whewell labeled deductive and attributed to French analysts. Its epistemological claims are clear from the first paragraph of Laplace's 1814 *Essai Philosophique sur les Probabilités:*

> Here I shall present . . . the principles and general results of the Theorie, applying them to the most important questions of life, which are indeed, for the most part, only problems in probability. One may even say, strictly speaking, that almost all our knowledge is only probable; and in the small number of things that we are able to know with certainty, in the mathematical sciences themselves, the principal

> means of arriving at the truth—induction and analogy—are based on probabilities, so that the whole system of human knowledge is tied up with the theory set out in this essay. (Laplace 1995, 1)

Laplace's probability theory is generally classified as "subjectivist" because it locates probabilities squarely in the human mind. Thus, human minds make a subjective, probabilistic prediction of the way a thrown die will fall, but the outcome of the throw is objective, completely determined by the laws of physics. By assuming that chance is an epiphenomenon of the mind and an expression of its epistemological limitations, Laplace insured that his theory was not misunderstood to indicate that there were actually random, or chance, events in the world. Events happen in what appear to the human mind to be random ways, but in a greater reality these events are strictly determined; the experience of chance and the probability theory that mathematizes it are rooted in the gap between what humans can know and what is (Daston 1992).

The model of the knowing mind on which Laplace based this claim rests on the central metaphor of an urn filled with black and white balls. An "event" consists of drawing a ball of a certain color, the "probability" of an event is the ratio of "the number of cases favorable to the event whose probability is sought" to the number of "all possible cases" (Laplace 1995, 8). Thus, if there are sixty balls in an urn, of which twenty are black, the probability of picking a black ball is 20/60, or 1/3. In this way of thinking, certainty is attained when the urn is completely filled with balls of only one color.

In Laplace's view, probabilities of everyday events can be calculated, and rational decisions made on the outcomes of those calculations. In practice, however, seemingly rational people differ in their opinions. These individual differences are problems. They indicate that people are accepting different data on authority or that they are calculating probabilities differently; they could and should be eliminated by the determined application of a sophisticated probability theory. To reach the right conclusion requires

> great precision of mind, a nice judgement, and wide experience in worldly affairs. It is necessary to know how to guard oneself against prejudice, against illusions of fear and hope, and against those treacherous notions of success and happiness with which most men lull their *amour-propre*. (Laplace 1995, 12)

Laplace's theory places real knowledge squarely in the rational realm. His dispassionate gaze transforms Herschel's "hopes" into mere prejudice, his "assurance" of a future life into a self-serving form of spiritual gluttony. There is no place for Herschel's direct personal knowledge in Laplace's probabilistic outlook.

The implications of this kind of probabilistic thinking for traditional Christianity were well known to the early Victorians, having been spelled out by David Hume. In his *Essay on Miracles,* first published in 1748, Hume argued that the probability nature would follow its normal course was so huge that no amount of personal testimony could persuade a rational person that it had diverged, and a miracle had occurred. Personal conviction simply could never counterbalance probabilistic evidence; a religion that rested on events attested to by personal experience and conviction had no standing in probabilistic discourse.

Whewell's blast against deductive science drew the analytics into a discussion of the epistemological implications of Continental analysis, in particular probability theory. The discussion did not take place immediately, but in 1837, after the issues had been reformulated in Newman's seventy-third tract, Babbage responded in an uncommissioned *Ninth Bridgewater Treatise.* The targets Whewell had named were all Continental and dead at the time he wrote, but Babbage's *Treatise* and the response it engendered indicate that in the England of the 1830s, their ideas were not.

Babbage's *Treatise* answered the charge that analytic mathematics subverted religion by changing the subject. Whewell and Newman had maintained that knowledge of God could neither be attained nor sustained through rational argument. Babbage countered by constructing God in a rationally comprehensible world. This entailed refuting Hume's argument against miracles and thereby showing that even if one fully accepted the probabilistic restriction of legitimate knowledge, traditional Christianity could be rationally defended.

To this end, Babbage offered a contemporary twist on the classical design argument. In its traditional form the argument constructed a conception of divine intelligence by analogy with a human designer, which explained how purposive things like the eye have come to be and could equally allow for purposive events like miracles. It stands in marked contrast to the strictly deterministic world that lay behind Laplace's probability theory, where all events are determined by unyielding mechanical laws (Daston 1992).

Babbage tried to mediate the designing nature of classical natural theology and the grinding regularity of determinism with his calculating machine. He pointed out that a mechanical computer could be programmed to do one thing for the first hundred million terms and then to change for the next 2,762 terms only to change again for the next 1,430 terms. "It is more consistent," Babbage argued, "to look upon miracles not as deviations from the laws assigned by the Almighty for the government of matter and of mind; but as the exact fulfillment of much more extensive laws than those we suppose to exist" (Babbage 1838, 92; W. Cannon 1960). Under this model of a completely determined world, miracles were not as impossible as Hume had argued. They could be accepted as natural events.

By giving a naturalistic interpretation for miracles Babbage thought he had defended religion, but few agreed. Whewell responded in a measured open "Letter to Charles Babbage":

> It is only by recognizing the utter dissimilarity of moral and religious grounds of belief, from mathematical and physical reasonings upon established laws of nature;—that he [the mathematician] can make his way to the conviction of a moral constitution and providential government of the world. (Whewell 1838, 4-5)

Herschel wrote from the Cape of Good Hope: "I have objections in toto to any application of the calculus of probabilities to the case in question, as a ground for belief one way or other." Miracles, he asserted, are simply not comprehensible on the probabilistic model.

> It is precisely because we refuse in our hearts to admit that essential postulate without which the theory of probabilities cannot stir a step . . . because in short we cannot help a lurking sentiment that a subversion of the law of nature is in reality, in a certain sense, less possible than its continuance—that we regard it *as* a miracle and are affected by its occurrence . . . by other profounder emotions. Human testimony cannot prove a miracle. . . . The mind must be predisposed to its admission. (Herschel 1837)

Knowledge of a miracle requires a particular orientation; it cannot be rationally established or evaluated, and so it lies outside the purview of probabilistic reasoning.

Whewell and Herschel said little more in their letters. One can, however, trace the outlines of a more detailed argument in the work of another Cambridge-educated mathematician, Augustus De Morgan, professor of mathematics at the University of London. De Morgan did not need to respond to Babbage on paper because he could simply speak to him. However, he did interpret Laplacean probability theory for the English audience in two works published in 1838: a mathematical "Theory of Probabilities" published in the *Encyclopaedia Metropolitana* and a more practical *Essay on Probabilities: and on their application to life contingencies and insurance offices* published in Lardner's *Cabinet Cyclopedia*.

The University of London was a self-consciously secular institution and De Morgan was passionately committed to the separation of religious and public life; nonetheless, residues of early Victorian religious preoccupation can be found in his work. So, for example, he found the salient feature of the nonprobabilistic world hidden from our inquiring gaze to be not its mechanical determinism, be it modeled by a steam engine or computer, but rather its providence. This position was possible because De Morgan followed Laplace in locating the uncertainties probability theory was mathematizing in the mind rather than in the external world.

In an important way, however, De Morgan pushed beyond the Frenchman's position by acknowledging individual variations to be legitimate. As he put it,

> It is wrong to speak of any thing being probable or improbable in itself. The thing may be *really* probable to one person and improbable to another. And thus men may be justified in drawing different conclusions upon the same subject. (A. De Morgan [1838] 1845, 394)

In some cases, the consequences of adopting a poor value might become obvious rather quickly; poor gamblers could lose their shirts. In other areas, however, De Morgan did not believe that individual differences could be resolved by experience or by fine-tuning probabilistic calculations. What is more, he did not think they should be.

This is because, probabilist though he was, De Morgan was as unwilling to let the theory define rationality or epistemological legitimacy as were Whewell or Herschel. His motivations were different, though. In the relatively homogeneous context of natural theology, personal process guaranteed certainty and hence legitimacy; in the midst of the heady diversity of London, personal certainty remained but could not guarantee assent. De Morgan recognized the implications of this for probabilists as well as churchmen and insisted that all of their certainties had to be kept in check. As he explained in the discursive introduction to his *Essay*,

> Two spectators [standing by a probabilistic urn] . . . may be very differently affected with the notion of likelihood in respect to any ball being drawn. . . . And thus we see that the real probabilities may be different to different persons. The abomination called intolerance . . . arises from the inability to see this distinction. (A. De Morgan [1838] 1981, 7)

This conviction is reflected in De Morgan's practice. At the time he was writing he espoused a rational religion: he wrote his evangelizing mother in 1836, "Such matters are not with me matters of feeling, they are to be tried by reason and evidence." This was, however, a private conviction expressed in private, and immediately followed by the caveat: "That is *by me*, for I do not object to anyone who thinks he can find truth by another method trying what he can do" (quoted in S. De Morgan 1882, 144). In his published "Theory of Probabilities," De Morgan only felt "at liberty to say, that though a result of the theory of probabilities, upon a moral question, is not to be lightly or easily adopted, when it differs from usual notions, yet, on the other hand it is not therefore to be immediately rejected" (A. De Morgan 1845, 473).

De Morgan's interpretation, in which probability theory was valid only for the individual whose certainty it measured, led him to emphasize a distinction between two basic kinds of probability:

> 1. *Moral probability* is the impression existing with regard to the happening of an event depending upon the constitution of the individual, his knowledge of the circumstances, and the effect the event will produce.
> 2. *Mathematical probability* is the moral probability in that case, and in that case only, in which the mind is disposed to consider equal successive changes of favourable circumstances into unfavourable or *visa versa,* as of equal importance: not regarding certainty as possessing any peculiar value. (A. De Morgan 1845, 396-97)

De Morgan illustrated the division with the example of the man whose life depends on drawing a black ball from an urn. To him, any change in the ratio of black to white balls, from 5/10 to 6/10 for example, would be significant, but the mathematically identical difference between 9/10 and 10/10 would be immeasurably large. The mathematical regularity of probability theory was simply inadequate to model such a person's judgment.

The distinction between moral and mathematical probabilities was an eighteenth-century commonplace (Daston 1980). What marks De Morgan's characterization is, first, that he was so clear that most situations were relevant to moral, as opposed to mathematical, probabilities, and, second, that he made no attempt to fix them mathematically.

This stance greatly limited the scope of probabilistic implications, since it rendered the precision of the mathematical theory inapplicable to virtually any situation in which one had a personal stake. It certainly rendered probabilistic discussion of religious matters suspect: as De Morgan put it, Hume "would have been (had he understood his own assertion) of a morbid degree of faith, willing to believe a miracle the moment more than an even chance was made out in its favour" (A. De Morgan 1845, 472).

In the end, then, Babbage's attempt to cast all knowing in a rational mold and to limit the possible by the probable was not accepted. Whewell simply reiterated his position, but Herschel and De Morgan struggled to define a middle ground that protected the sanctity of personal conviction by distinguishing it from probabilistic rationalism. Structurally their positions were similar; all found religious issues too personally weighted to be decided by probabilistic argument.

Although it was not their intent, their divisive solutions threatened the validity of personal conviction by marginalizing it epistemologically and sociopolitically. When, in the 1840s, the English discussion of probabilities moved out from under the umbrella of natural theology into the newly emerging rational arena of philosophy, maintaining the validity of personal knowledge became a major challenge. Whewell's *Philosophy of the Induc-*

tive Sciences, first published in 1840, can be seen as an attempt to meet it. He did not there focus on probability theory, *per se,* but Robert Leslie Ellis took the solution he offered into the heart of probability theory.

III. Knowledge in Philosophy

In his 1833 *Astronomy,* Whewell had noted, "It is no easy matter, if it be possible, to analyze the process of thought by which laws of nature have thus been discovered. . . . We shall not here make any attempt at such an analysis" (Whewell 1836, 304). One could argue that the next decade of his life was devoted to just such an attempt, and that his *Philosophy of the Inductive Sciences* was a concentrated effort to define the process that led true discoverers to their insights.

However, "process" is a misleading term here. Its personal and religious overtones are appropriate to Herschel's scientific activity, to Whewell's scientists become wise, or to Newman's active faith, but not to the dispassionate context of philosophy. There one speaks of "method." With this terminological shift personal experience vanishes. Concurrently the emphasis shifts. A process may lead to understanding, but that understanding is so personal as to be indistinguishable from the process itself. Method, on the other hand, is a means to the end of attaining or establishing a truth that is external to the self, fixed and known rather than changing and elusive. It was a formidable challenge to defend the dynamic personal knowing of natural theology in the rational context of philosophy (Yeo 1979; Yeo 1994).

Whewell attempted it by developing a novel view of induction that emphasized the central importance of individual input into theory construction. Theories were not constructed from piles of observations but rather created when the investigator identified the Fundamental Idea that served to explain observed phenomena. This meant that truth was not discovered but recognized through a long process of engaged human interaction with the natural world; the establishment of inductive truth entailed an active interaction between the conceptual framework of the investigator and the external world with which he was engaged.

In important ways, Whewell's Fundamental Ideas reflect the religious values he had claimed for scientific investigation in his *Astronomy.* In their inception and characteristics they are quintessentially human, generated and identified by a process of conceptualization. What is more, even when pinned down with a system of axioms and definitions the Fundamental Idea is not defined or encompassed by them. As Whewell put it,

> The Idea is disclosed but not fully revealed, imparted but not transfused, by the use we make of it in science. When we have taken from the foundation so much as serves our purpose, there still remains behind a deep well of truth, which we have not exhausted, and which we may easily believe to be inexhaustible. (Whewell [1840] 1967, 73)

For Whewell knowing the truth about something meant thinking about it properly, bringing the mind in line with that of the God whose understanding suffused the world.

This kind of knowing was fundamentally different from that which grasped, manipulated or calculated the truth. It placed the personal process of discovery in the center of the inductive method. In a friendly but critical review, De Morgan objected to this violation of his public/private distinction: "Let induction mean, as it always has done, the generalization by collection of particulars: let the act of the discoverer by which he divines the general notion under which the properties can be brought, receive its own proper name" (A. De Morgan 1859, 44).

Certainly Whewell's philosophy was completely at odds with that on which probability theory was constructed. From the Cambridge man's perspective, the process of drawing balls of one color or another from an urn might spark insight in an observer, but it was emphatically not the basis for understanding. Perhaps for this reason he evinced little interest in it, and the problem of reconciling probability theory with his philosophy was taken up by a much younger man, Robert Leslie Ellis.

Ellis belonged to the generation subsequent to the one that had spawned Whewell, De Morgan, Herschel, and Newman. He emerged as first wrangler from the 1840 Tripos, very skilled in analytic mathematics and in a state of nervous and physical collapse from which he never truly recovered. His frailty was reflected in an inability to complete a work of more than an article's length to the end of his life. Among the snippets he did produce were two articles, in 1844 and 1854, on the foundations of probability theory.

Ellis's first paper, "On the Foundations of the Theory of Probabilities," was an attempt to reinterpret probability theory in such a way that it would "cease to be, what I cannot avoid thinking it now is, in opposition to a philosophy of science which recognizes ideal elements of knowledge, and which makes the process of induction depend on them" (Ellis 1863, 11). Toward this end he launched a frontal attack against what he called the "sensational philosophy" embodied by the probabilists' empirical urn.

In Ellis's view, the urn models a rational method that is amenable to mathematical calculation, but not adequate to the way people, including scientists, actually think. Ellis elaborated his position in an attack on one of De Morgan's examples. In his 1838 "Theory of Probabilities," the Londoner had calculated the probability that a vessel will have a flag on the basis of the previous ten vessels having had one. "Let us suppose the ten vessels to be Indiamen," Ellis objected.

> Is the passing up of any vessel whatever, from a wherry to a man of war, to be considered as constituting a next occasion? or will an Indiaman only satisfy the conditions of the question?
>
> It is clear that in the latter case, the presumption that the next *In-*

diaman would have a flag is much stronger, than that, as in the former case, the next *vessel* of any kind would have one. Yet the theory gives 11/12 as the presumption in both cases. (Ellis 1863, 7)

Ellis then elaborated on the formidable series of obstacles that would have to be negotiated for the argument to be valid. "The most perfect acquaintance with the nature of the case would not enable us to say what was the *a priori* probability of the event," he insisted, "for this depends, not only on the event, but also on the mind which contemplates it" (Ellis 1863, 9). Herschel had insisted that knowing a miracle required a mind predisposed to its admission, but for Ellis rational mathematical argument is too simplistic even to describe everyday ships on a river.

In 1844 Ellis complained that his countrymen had paid too little attention to the foundations of probability theory, but he did not remain a lone voice for long. In 1848 a number of his contemporaries took up the question of how much could be established by probabilistic argument. The occasion for the discussion was a disarmingly simple statement about double stars in Herschel's 1848 *Outline of Astronomy*. Double stars had attracted sporadic interest since William Herschel had first observed their rotation around a common center in the previous century. William's son marshaled probabilistic arguments to argue that their positioning was not merely an epiphenomenon of random distribution, but rather evidence that a physical cause grouped them together: "The conclusion of a physical connexion of some kind or other is therefore unavoidable," John Herschel wrote (quoted in Forbes 1849, 132).

The response came from the Scottish natural philosopher James D. Forbes, who was a friend of both Whewell and Ellis. "Though I am not trying to controvert the truth of the general result," he wrote in 1850,

> I hope clearly to prove, that it has no absolute and compulsory form addressing itself alike to all understandings and to all capacities, and to persons ill and well-informed alike. The grouping of stars is like any phaenomena occurring in physical investigations, which suggests further inquiry; which points at a result not improbable, but requiring to be inductively established by bringing together other considerations, whose accumulation may impel conviction. (Forbes 1850, 403)

Forbes was defending a Whewellian model of induction against the rigid mechanical overtones of a probabilistic one. He moved seamlessly from a personal interpretation of knowing to a physics wherein phenomena "suggest further inquiry," data "point at results not improbable," and their accumulation "may impel conviction." Under this construction knowledge is grounded in personal insight; real understanding lies tantalizingly beyond rational constructions, however powerful they might be.

Forbes's article was but the public expression of an extensive correspondence involving the Royal Astronomer, George Biddel Airy; the bishop of Edinburgh, George Terrot; the Irish mathematician Philip Kelland; and Ellis. The Scot even tried to draw De Morgan into the discussion by sending him a copy of the paper, but the Londoner merely responded crisply: "I am much obliged to you for your paper on the chances of distribution of stars—a subject it has not fallen my way to consider" (De Morgan to Forbes, 18 December 1850 [Forbes Archives]). There was a considerable range of opinion among Forbes's other correspondents about the proper purview of probabilistic argument. Airy was unsure: "I think that the force of induction admits of numerical expression, though I have not arrived at it yet" (Airy to Forbes, 12 November 1850 [Forbes Archives]). Bishop Terrot disagreed: "I think the *regularity* is a matter not subject to numerical expression" (Terrot to Forbes, 27 August 1850 [Forbes Archives]). Ellis, for his part, was incensed. "Between ourselves I am beginning to think the great Sir John Herschel is rather a charlatan: honourably distinguished no doubt . . . but neither clear nor deep" (Ellis to Forbes, 20 September 1850 [Forbes Archives]). In another letter he fumed, *"Avec des chiffres on peut tout démonstrer."* To turn "the theory of probabilities—which in it's own nature and according to the plain view of it, is only a developement of the theory of combinations," into "the philosophy of science, is in effect to destroy the philosophy of science altogether" (Ellis to Forbes, 3 September 1850 [Forbes Archives]).

Protecting the philosophy of science from the mechanical calculations of the probability calculus led Ellis, in 1850, to write a second paper: "Remarks on the Fundamental Principle of the Theory of Probabilities." That fundamental principle was "On a long run of similar trials, every possible event tends ultimately to recur in a definite ratio of frequency" (Ellis 1863, 49). It can be interpreted as an expression of the law of large numbers that Bernoulli had proved in 1704. In his first paper Ellis had challenged the sensationalist point of view from which this principle required proof. "Are we prepared to admit," he asked,

> that our confidence in the regularity of nature is merely a corollary from Bernoulli's theorem? That until this theorem was published, mankind could give no account of convictions they had always held, and on which they had always acted? (Ellis 1863, 1)

In 1850 he reiterated his conviction that it was not known empirically or mathematically, but intuitively

> the word being used, as in all similar cases, with reference to the intuitions of a mind, which has fully and clearly apprehended the subject before it, and to which therefore to have arrived at the truth and to perceive that it has done so are inseparable elements of the same act of thought. (Ellis 1863, 49)

For classical probabilists, the Herschel of the late 1840s included, such intuitions were, at best, personal and therefore should be private. For Ellis, however, it was simply impossible to confine knowledge recognized in this way to the private world of the individual mind:

> Man in relation to the universe is not *spectator ab extra,* but in some sort a part of that which he contemplates. . . . The *veritas essendi* is the fountain from whence the *veritas cognoscendi* is derived.

Applying his fundamental principle, he continued:

> It is only when in thought we remove the action of disturbing causes to an indefinite distance, that we can conceive the absolute verification of any *a priori* law. Only on the horizon of our mental prospect earth and sky, the fact and the idea, are seen to meet, though in reality the atmosphere is everywhere present. Everywhere it surrounds and interpenetrates the [black earth] on which we stand; making it put forth and sustain all the numberless forms of organization and of life. (Ellis 1863, 51)

For Ellis, any separation of the personal from the real was only apparent: in essence the two were always, everywhere conjoined. Despite his considerable mathematical prowess, personal knowing, rather than De Morgan's dispassionate mathematics, defined Ellis's reality.

The issues that divided these two mathematicians might be located in their personal circumstances, and from Ellis's perspective this would be appropriate. When he wrote his paper Ellis was entering his final decade. Of his short, sickly life he commented just before he died, "The curse of Moses 'thy life shall hang in doubt before you night and day' has been fulfilled here if anywhere" (Ellis to Walton, n.d. [Whewell Archives]). This characterization of his situation is eerily evocative of the example De Morgan had used to illustrate the distorting power of personal involvement in probabilistic situations. The man whose life depended on drawing from a probabilistic urn was an abstract example for De Morgan but all too real for Ellis; the personally infused knowing that De Morgan relegated to the sidelines was central to the sick younger man.

The story is larger than these two individuals, though; to encompass it the perspective must be broadened to include the larger circumstances that joined them with their contemporaries. Ellis, like Whewell, was harbored in Anglican Cambridge defending the conjunction of religion with knowing on which the university stood; in this homogeneous community personal certainties were routinely reflected back to him. De Morgan, like Babbage, was immersed in the cosmopolitan life of a large city, startling in its newly recognized diversity; here it was clear that personal convictions differed and could not be the ground for public consensus. Herschel, for his part, was institutionally free, and defined his interests according to concerns that arose from his research.

Whether one looks to the microlevel of personal biography or the larger one of institutional affiliation to understand the concerns that motivated these men, their ideas and the context that supported them were short-lived. The publication of Darwin's *Origin of Species* can be seen as marking the beginning of a new era in the understanding of scientific knowing; within a decade all of the principals treated in this chapter were dead.

IV. Conclusion

Historians of probability theory usually find the early nineteenth century to mark a transition between the subjectivism of the Enlightenment, including Laplace, which located probabilities in the rational mind, and the frequentism of statisticians, including Darwin, which located them in the external world. If mentioned at all, the English here considered are positioned in these categories: De Morgan becomes a latter-day subjectivist; Ellis, because of his opposition to those views, some strange kind of frequentist; Herschel, because he defended the Belgian statistician Adolph Quetelet, a herald for the new world to come (Daston 1988; Porter 1986).

The story line is neat, but it is achieved at the expense of the pre-Darwinian world of the early Victorians, who were neither frequentists nor subjectivists. For the most part this group was approaching science from a religious tradition wherein knowing was a transformative personal experience that moved one beyond one's human limitations. This vision could not be mapped simply onto a grid that separated the subjective from the objective and erected a probabilistic bridge between.

Their confrontation with Continental probability theory, which had been erected on this bifurcated interpretation, severely challenged their vision; in the long run, with a new generation, that vision was abandoned. But for several decades probability theory served as a challenge for English attempts to pursue "distant glimpses of boundless realms beyond" even as they tried to build a scientific view of the nearer world.

Bibliographical Note

There is not yet a monograph that focuses primarily on early Victorian mathematics within the larger picture of early Victorian science. Pycior (1981, 1982, 1983) and I (Richards 1980, 1991, 1992) considered the epistemological implications of algebra in this period. Though their focus is not on mathematics per se, Ashworth (1994) and Schaffer (1994) consider the ways that mathematical ideas were embedded in the culture at large, focusing primarily on developments in London. The picture they paint is balanced by a large literature on the Cambridge-based William Whewell, which again is not explicitly mathematical but bears directly on mathematical issues. Fisch (1991) and Yeo (1994) paint complementary pictures of the development of Whewell's philosophical ideas that are full of mathematical implications. Fisch (1994) suggests a set of further issues that await consideration. The literature on probability theory is large and sprawling but little focuses on the early Victorians. The best leads into the area are Gigerenziger et al. (1989), Krüger, Daston, and Heidelberger (1987), and Krüger, Daston, and Morgan (1987). For an overview of the historiography of natural theology, as well as a consideration of the role of the personal in that arena see Brooke in Fisch and Schaffer (1991). A larger monographic case study of an institutional and intellectual interaction of religion and science is Corsi (1988); Hilton (1988) provides an excellent introduction to the religious intellectual scene more broadly considered.

References

Altholz, J. L. 1988. "The Warfare of Conscience with Theology." In *Religion in Victorian Britain.* Vol. 4, *Interpretations,* edited by Gerald Parsons. New York: Manchester University Press in association with the Open University, 150-69.

Ashworth, William J. 1994. "The Calculating Eye: Baily, Herschel, Babbage and the Business of Astronomy." *British Journal for the History of Science* 27:409-41.

Babbage, Charles. [1837] 1838. *The Ninth Bridgewater Treatise: A Fragment.* 2d ed. London: John Murray.

Becher, H. W. 1980. "Woodhouse, Babbage, Peacock and Modern Algebra." *Historia Mathematica* 7:389-400.

——. 1992. "The Whewell Story." *Annals of Science* 49:377-84.

Cannon, Susan. 1978. *Science in Culture.* New York: Dawson and Science History Publications.

Cannon, Walter. 1960. "The Problem of Miracles in the 1830s." *Victorian Studies* 4:5-32.

Corsi, Pietro. 1988. *Science and Religion: Baden Powell and the Anglican Debate, 1800-1860.* Cambridge: Cambridge University Press.

Daston, Lorraine. 1980. "Probabilistic Expectation and Rationality in Classical Probability Theory." *Historia Mathematica* 7:234-60.

——. 1988. *Classical Probability in the Enlightenment.* Princeton, N.J.: Princeton University Press.

——. 1992. "The Doctrine of Chances without Chance: Determinism, Mathematical Probability, and Quantification in the Seventeenth Century." In *The Invention of Physical Science,* edited by Mary Jo Nye, Joan L. Richards, and Roger H. Stuewer. Boston: Kluwer Academic Publishers, 27-50.

De Morgan, Augustus. [1838] 1981. *An Essay on Probabilities: and their application to life contingencies and insurance offices.* New York: Arno Press.

——. [1838] 1845. "Theory of Probabilities." In *Encyclopaedia Metropolitana,* vol. 2. London: B. Fellowes, 393-473.

——. 1859. "Review of *Novum Organon Renovatum.*" *Athenaeum* 1682:44.

De Morgan, Sophia Elizabeth. 1882. *Memoir of Augustus De Morgan.* London: Longmans, Green, and Company.

Durand-Richards, Marie-José. Forthcoming. "L'École Algébrique Anglaise: les conditions conceptuelles et institutionnelles d'un calcul symbolique comme fondement de la connaissance." In *Myths et Réalités de l'Europe Mathématique*, edited by J. Gray, C. Goldstein, J. Ritter.

Ellis, Robert Leslie. 1863. *The Mathematical and Other Writings of Robert Leslie Ellis, M.A.* Edited by William Walton. Cambridge: Deighton, Bell and Company.

Enros, Philip. 1983. "The Analytical Society (1812-1813): Precursor of the Renewal of Cambridge Mathematics." *Historia Mathematica* 10:24-47.

Fisch, Menachem. 1991. *William Whewell: Philosopher of Science.* Oxford: Clarendon Press.

———. 1994. "'The Emergency Which Has Arrived': The Problematic History of Nineteenth-Century British Algebra—a Programmatic Outline." *British Journal for the History of Science* 27:247-76.

Fisch, Menachem, and Simon Schaffer. 1991. *William Whewell: A Composite Portrait.* Oxford: Clarendon Press.

Forbes, James D. Archives. University of St. Andrews.

———. 1849. "On the Alleged Evidence for a Physical Connexion between Stars forming Binary or Multiple Groups, arising from their Proximity Alone." *London, Edinburgh and Dublin Philosophical Magazine and Journal of Science*, 3d ser., 35:132-33.

———. 1850. "On the Alleged Evidence for a Physical Connexion between Stars forming Binary or Multiple Groups, deduced from the Doctrine of Chances." *London, Edinburgh and Dublin Philosophical Magazine and Journal of Science*, 3d ser., 37:401-27.

Gigerenzer, Gerd, Zeno Swijtink, Theodore Porter, Lorraine Daston, John Beatty, and Lorenz Krüger. 1989. *The Empire of Chance.* Cambridge: Cambridge University Press.

Hacking, Ian. 1975. *The Emergence of Probability.* Cambridge: Cambridge University Press.

Herschel, John Frederick William. [1830] 1966. *A Preliminary Discourse on the Study of Natural Philosophy.* New York: Johnson Reprint Corporation.

———. 1837. Letter to Charles Babbage. Herschel Archive of the Royal Astronomical Society, London.

Hilton, Boyd. 1988. *The Age of Atonement: The Influence of Evangelicalism on Social and Economic Thought, 1795-1865.* Oxford: Clarendon Press.

Krüger, Lorenz, L. Daston, and M. Heidelberger, eds. 1987. *The Probabilistic Revolution.* Vol. 1, *Ideas in History.* Cambridge, Mass: MIT Press.

Krüger, Lorenz, L. Daston, and M. Morgan, eds. 1987. *The Probabilistic Revolution.* Vol. 2, *Ideas in the Sciences.* Cambridge, Mass: MIT Press.

Laplace, Pierre-Simon. 1995. *Philosophical Essay on Probabilities.* Translated by Andrew I. Dale. New York: Springer-Verlag.

Miller, David Philip. 1986. "The Revival of the Physical Sciences in Britain: 1815-1830." *Osiris*, 2d ser., 2:107-34.

Newman, John Henry Cardinal. 1949. *Sermons and Discourses (1825-39).* New York: Longmans, Green and Company.

[———]. 1836. "On the Introduction of Rationalistic Principles into Religion." No. 73 of *Tracts for the Times,* by Members of the University of Oxford, vol. 3 for 1835-36. London: J. G. and F. Rivington.

Porter, Theodore. 1986. *The Rise of Statistical Thinking*. Princeton, N.J.: Princeton University Press.

Pycior, Helena. 1981. "George Peacock and the British Origins of Symbolical Algebra." *Historia Mathematica* 8:23-45.

———. 1982. "Early Criticisms of the Symbolical Approach to Algebra." *Historia Mathematica* 9:413-40.

———. 1983. "The Three Stages of Augustus De Morgan's Algebraic Work." *Isis* 74:211-22.

Richards, Joan L. 1980. "The Art and the Science of British Algebra: A Study in the Perception of Mathematical Truth." *Historia Mathematica* 7:343-65.

———. 1991. "Rigor and Clarity: Foundations of Mathematics in France and England, 1800-1840." *Science in Context* 4:297-319.

———. 1992. "God, Truth and Mathematics in Nineteenth Century England." In *The Invention of Physical Science,* edited by Mary Jo Nye, Joan L. Richards, and Roger H. Stuewer. Boston: Kluwer Academic Publishers, 57-78.

Schaffer, Simon. 1994. "Babbage's Intelligence: Calculating Engines and the Factory System." *Critical Inquiry* 21:203-27.

Todhunter, Isaac. 1876. *William Whewell: An Account of his Writings with Selections from his Literary and Scientific Correspondence.* London: Macmillan and Company.

Whewell, William. Archives. Trinity College Library.

———. [1833] 1836. *Astronomy and General Physics considered with reference to natural theology.* 5th ed. London: William Pickering.

———. 1838. "Letter to Charles Babbage, ESQ. A. M." Trinity College Library.

———. [1840] 1967. *The Philosophy of the Inductive Sciences.* 3d ed. London: Frank Cass and Company.

Yeo, Richard. 1979, "William Whewell, Natural Theology and the Philosophy of Science in Mid-nineteenth Century Britain." *Annals of Science* 36:493-516.

———. 1994. *Defining Science.* Oxford: Oxford University Press.

4

Victorian Economics and the Science of the Mind

MARGARET SCHABAS

> The sceptre of psychology has decidedly returned to this island. The scientific study of mind, which for two generations, in many other respects distinguished for intellectual activity, had, while brilliantly cultivated elsewhere, been neglected by our countrymen, is now nowhere prosecuted with so much vigour and success as in Great Britain.
>
> JOHN STUART MILL ([1859] 1978, 341)

Apart from Thomas Tooke's first volume of his *History of Prices* (1838) the year of Queen Victoria's coronation was an uneventful one in the history of economics. Much the same could be said of the year of her death in 1901. Yet political economy dominated intellectual discourse throughout that century and was particularly ascendant in Britain. As John Maynard Keynes later remarked, "Ricardo conquered England as completely as the Holy Inquisition conquered Spain" (Keynes [1936] 1964, 32). And while there were a number of excellent Continental economists, the "Age of Capital" clearly belonged to the English economists (Hobsbawm 1975, 316). However, it was not just the presence of Ricardo, Mill, and Marshall that made England famous for the subject. As Joseph Schumpeter correctly noted, the strength and quality of the second tier of nineteenth-century economists greatly contributed to "the unrivaled prestige that English economists then enjoyed" (Schumpeter 1954, 382-83, 757).

David Ricardo's *Principles of Political Economy and Taxation* (1817) remains one of the great classics in the history of economics, and in terms of pure theoretical analysis surpassed even Adam Smith's *Wealth of Nations* (1776). Ricardo, however, confined his policy recommendations, and broader social philosophy, to his correspondence and parliamentary ad-

I wish to thank Bernard Lightman, A. W. Coats, Anthony Brewer, Myles Jackson, David Millet, George Stocking, and Stephen Stigler for specific comments and suggestions.

dresses. Both John Stuart Mill's *Principles of Political Economy* (1848) and Alfred Marshall's *Principles of Economics* (1890) aimed at clarifying and tempering the central principles laid down by the immortal Ricardo. But they also addressed a much broader set of concerns, ranging from behavioral assumptions and methodological heuristics to the question of economic well-being. Ricardo may have conquered England, but it was Mill and Marshall who ruled the land. Each book served as the classic text for some forty years. Put together, they span, more or less, the Victorian era, and thereby conveniently provide us with a tidy historical chapter.

One need only take note of the slight variation in the titles of Mill's and Marshall's tomes, however, to see that something had changed. The discipline was no longer overtly tied to political imperatives. When Marshall helped to found the British Economic Association in 1891 (later renamed the Royal Economic Society), he took great pains to insure that the charter membership spanned the political spectrum and thus tolerated political pluralism (Coats 1968). Unlike their counterparts in the United States, who were deeply divided by political allegiances, economists as Marshall portrayed them had matured past the point of political dogma (Coats 1968; Haskell 1977). Certainly he put to rest any remaining controversies that Ricardo and the Ricardian socialists had stirred up in the 1820s and 1830s. As John Maloney has argued, Marshall coated economic theory and its professional trappings with a veneer of ideological neutrality, and he did such an excellent job that the task has not been repeated (Maloney 1985).

It is uncommon to package Mill and Marshall together because a more important watershed in Victorian economics—the Marginal Revolution of the 1870s—rent them asunder. The leading instigator in Britain was William Stanley Jevons, whose *Theory of Political Economy* (1871) called for a radical transformation of the conceptual foundations and methodological principles of the classical theory of Ricardo and Mill. And the changes that ensued were profound and permanent. Value was determined by utility, not labor. The distribution of goods and services was the result of individual deliberations at the margin, not the incessant struggle between laborers, landlords, and owners of stock. Jevons also campaigned for the adoption of mathematics, particularly the calculus, and envisioned the time when probability and statistics would make sense of the abundant data compiled by every office clerk. He thus set in motion the program for a unified mathematical theory derived from a limited set of behavioral axioms, purportedly verified by econometric testing, that has been much more fully developed in this century (Mirowski 1989; Morgan 1990). Certainly Marshall, in his advanced courses at Cambridge, erected a mathematical barrier to entry that has served to this day to demarcate the professional from the amateur (Maloney 1985, 233-34; Schabas 1990*a*, 126-34).

My task here will be to develop a preliminary characterization of Victorian economics, the period from Mill to Marshall. Most scholarly efforts thus far have highlighted the methodological transformation, but there also transpired a new conceptualization of the economic order, with appeals to psychology as a key source of inspiration. This can be only a preliminary characterization, since there are virtually no works that explicitly feature Victorian economics, perhaps because of the methodological rift stirred up by Jevons and Léon Walras. Up until the present, the majority of scholars have chosen to focus on the contributions of one or, at best, two leading figures in the field and to leave synthetic treatments to the assorted textbooks that contemporary economists so dearly love to use. One of the reasons why there is no work specifically on Victorian economics may be a preference among historians of economics to trace analytical progress across national boundaries. That is to say, they are much more inclined to take a specific theoretical issue and trace its development than to consider the possibility of a distinct national school or style. Even in the case of recognizable national groupings—the Austrian school initiated by Carl Menger in the 1870s, the Swedish school of economists that flourished with Knut Wicksell during the early decades of this century, and the American institutionalists made notorious by Thorstein Veblen circa 1900—the unifying force was theoretical and not the direct result of nation-specific policy debates.[1]

During the eighteenth century, professorships in political economy were established at various universities on the Continent (Austria, Sweden, Italy, Germany, and France), but not in Britain. The French group of self-proclaimed "économistes" were arguably at the forefront in the 1760s with their own journal and doctrine of Physiocracy. While matters proceeded apace in the next century, the academic status of economics did not grow by the same leaps and bounds as it did across the channel. The two most brilliant French economists alive during Victoria's reign, Antoine Augustin Cournot and Léon Walras, spent most of their working lives qua economists in relative obscurity, many efforts at promotion notwithstanding.

If English universities were tardy about recognizing political economy during the Enlightenment, they marched quickly ahead in the nineteenth century. Thomas Robert Malthus held the first academic post, at Haileybury College for the East India Company starting in 1805. Professorships in political economy at Cambridge, Oxford, and the newly founded London colleges, University College and King's College, were set up in the 1820s. By the mid-1880s, Oxford and Cambridge employed nine lecturers on political economy, including the recently appointed Alfred Marshall recruited from the University of Bristol. But many of the steps that enabled Marshall and his colleagues to entrench economics in academia—the founding of the *Eco-*

---

1. It seems that the more central force in giving rise to national schools in economics is philosophical traditions or related intellectual currents (Social Darwinism in the case of American institutionalism, for example).

nomic Journal (1891), the London School of Economics (1895), and the Economic Tripos at Cambridge (1903)—had been undertaken in the early Victorian period.

In the 1830s and 1840s, the leading promulgators of science, John Herschel, William Whewell, Charles Babbage, and John Stuart Mill, not only praised political economy as a respectable science, but also, with the exception of Herschel, made important contributions in their own right to the subject. Leading periodicals of the time, such as the *Quarterly Review* and the *Westminster Review,* were chock full of articles and commentaries on political economy. In a survey of a few journals for the years 1802–53, George Stigler found almost twelve hundred entries on the subject (Stigler 1965, 41). Most telling, perhaps, was the placement of Nassau Senior's article on political economy in the pure sciences section of the *Encyclopedia Metropolitana.* By the late 1840s, with the founding of Section F of the British Association for the Advancement of Science and the instigation of the Tripos in the moral sciences at Cambridge (1848), political economy had about as much status as chemistry or geology.[2] Dismal or not, the science of political economy was widely respected.

This might seem self-evident, given Britain's economic superiority at the time. Presumably the content and quality of economic discourse, or the esteem in which it was held by the learned community, had much to do with the extant economy. Yet, however straightforward such claims might sound, they are remarkably difficult to establish. Ricardian economics would be inconceivable without a system of capitalism, with its developed markets for land, labor, and capital, financial institutions, and nation states. But it is much harder to make the case that specific national features determine the content of the body of literature that one finds in a given place and time. As an economist—and all great economists have done this—one can help oneself to phenomena from across the globe and as far back as historical records permit. There is much to be learned about the economics of ancient Rome, sixteenth-century Spain, eighteenth-century China, or nineteenth-century India by reading the leading texts of classical political economy. In short, the content of the central theoretical core of economic discourse is underdetermined by the specific national features of the economy that shape and govern the life of a given theorist.[3] Political economy

---

2. Both in English- and German-speaking regions, all three subjects had traditionally been developed as adjuncts, chemistry to medicine, geology to mining or natural history, and political economy to law or moral philosophy. Around the same time, the 1830s and 1840s, those three subjects began to gain autonomy in the university curriculum.

3. The one exception to this claim might be the "machinery question," which, as Maxine Berg has so persuasively argued, was central to many of the debates among English political economists. But there was more than enough mechanization to be witnessed on the Continent to stimulate comparable debates. That the debate was most intensive in England had much more to do with Ricardo's celebrated chapter 31 on the subject than England's head start at industrialization.

has ideological components, to be sure, but possibly no more or less than any other branch of knowledge.

English economists were certainly oriented toward advancing their own national economy, but the principle of the mutual gains from trade made it clear that their own economic growth would be assisted rather than harmed by a concomitant growth from other nations. In the mid-eighteenth century, David Hume had declared that, "as a British subject, I pray for the flourishing commerce of Germany, Spain, Italy, and even France itself" (Hume [1752] 1985, 331). This cosmopolitan spirit, at least in the realm of intellectual trade, was sustained well into the next century. English economists were often full of praise and appreciation for the writings of those abroad: Ricardo of Jean-Baptiste Say, Mill of H. Saint-Simon and J. C. L. de Sismondi, and Marshall of Cournot and J. H. von Thünen. Even Jevons, who made little effort to join forces with Walras, conceded in 1879 that "the truth is with the French School" (Jevons [1871] 1957, preface to the 2d ed., xlv). One would be hard-pressed to find any other area of science at the time where the English were more open to foreign ideas.

Some economists, such as Senior, Babbage, and Herbert Somerton Foxwell, were eager to claim credit for economic theory itself as one of the reasons British industry was unrivaled at the time (Maloney 1985, 7-8; Foxwell 1887). There may well be some merit to the claim that economics as a science thrives in the climate of a strong economy, but since all economies wax and wane by decade or by century, and since the well-being of most learned individuals is subject to economic conditions, it would seem a gross generalization to endorse such a causal connection. The economic superiority of Britain may have been conducive to the growth of economic knowledge, but it underdetermined its theoretical content. A more likely explanation of the dominant position of Victorian political economy was the brilliance of Ricardo both in pen and in Parliament and the more widespread ascent of science in British universities and institutions.

Political economists have commonly been accused of serving the status quo, and the nineteenth century was no exception. Certainly a good case can be made for Malthus and Senior, who advocated harsh legislation on working conditions. But economic theory could also be embraced as a tool for extensive programs for reform, as it was by Mill, if not revolutionary manifestos, as in the case of Karl Marx. Yet all of these writers, from Malthus to Marx, took Ricardo as their main source of theoretical inspiration. And even in the case of the early neoclassical economists, who had moved well past the central tenets of Ricardian economics and were more likely to look to Jevons, one finds conservatives like Francis Ysidro Edgeworth engaging civilly with socialists like Philip Henry Wicksteed and Sidney Webb. Economics, or political economy, is inherently political, but it does not occupy only one place on the political spectrum.

Appeals to the scientific status of political economy were supported by lengthy epistemological arguments. John Stuart Mill's 1836 essay "On the Definition of Political Economy and on the Method of Investigation Proper to It" and more extensive *System of Logic* (1843) offered the most developed position, but many others—John Ramsay McCulloch, Nassau Senior, even William Whewell—made contributions as well during the 1830s and 1840s. The primary preoccupation was the extent to which political economy was or was not like natural science, particularly Newtonian physics. For Mill, it had the same axiomatic and deductive character, but was much more inexact at the stage of verification (Hollander 1985; Hausman 1981, 1992). For Whewell, induction was deemed more relevant, although in his own essays on the subject, he explored the use of mathematical economics and was steadfastly deductive. What is most striking about these methodological musings is the unanimity over the scientific standing of political economy. The detailed imprimaturs of Mill and Whewell were of considerable service in sustaining the prestige and respect bestowed on the subject at least until the "methodenstreit" of the 1860s. After that, English economists could turn to John Cairnes, Jevons, and John Neville Keynes for extensive arguments endorsing economics as a science (Schabas 1990*a*, chap. 6). In most cases, however, they preached to the converted.

Implicit in these assorted reflections on the scientific standing of political economy are ontological commitments as well. In what respects are the phenomena specific to political economy—prices, interest rates, trade, and so on—different from the phenomena covered by the natural sciences? And why might they be governed by laws analogous to natural laws? To what extent is there an autonomous entity known as an economy that is a social entity quite apart from the natural world? During the Victorian era, the conception of the economy and its salient features underwent a significant transformation. One critical factor in this process was an unprecedented but relatively short-lived enthusiasm by economists for the science of psychology, starting with Mill's declaration of 1859 (see the opening quote) and ending more or less with Marshall.[4]

The Jevonian theory, for example, depicted economic phenomena, prices and the like, as the product of individual choice. In a sense, the entire economy emanated from the mind, or rather the aggregate of independent minds. Prices were the product of a Benthamite calculus of pleasure and pain rather than the return to physical inputs such as labor and capital. And even though, in the long term, the cost of production matters, the factors

---

4. Two possible exceptions to the early end of this enthusiasm are Thorstein Veblen and John Maynard Keynes, both of whom made frequent appeals to psychological traits, instincts, and habits. But neither one explicitly sought out developed theories of psychology, as we find in the Victorian period.

responsible—labor and capital—were recast in terms of utility or mental wear and tear, so to speak.[5] In short, the economy was mind driven through and through. Humans do not produce any new matter. They merely reconfigure what is there, and order it according to their calculus of pleasure and pain. All that man can produce and alter is utility. As Alfred Marshall later remarked, "Man cannot create material things. In the mental and moral world indeed he may produce new ideas; but when he is said to produce material things, he really only produces utilities; or in other words, his efforts and sacrifices result in changing the form or arrangement of matter to adapt it better for the satisfaction of wants" (Marshall [1890] 1920, 53).

To a significant degree, Victorian economists repositioned their concept of the economy. They cut themselves free of the Enlightenment associations with physical nature that once saw the production and distribution of wealth as part of a providential order. The economy was now depicted in terms of man-made social institutions. To put it most emphatically, the economy went from a natural entity to a social one. This did nothing, however, to diminish the high esteem and confidence in the scientific standing of political economy among its practitioners. If anything, it suggested that economic theorists, by discovering the laws that governed the production and distribution of wealth, might also be in a position to change social arrangements.

In the mid-eighteenth century, when political economy arguably emerged as a distinct discipline, the various features of wealth, which formed the domain of discourse, were generally viewed as extensions of physical nature. David Hume, for example, treated the flow of money from one nation to the next as a natural process in terms of the ebb and flow of the tides. Gold, like water, always seeks its own level, regardless of legal restraints. And the Physiocrats maintained that economic wealth literally comes from the gifts of nature, the sun, rain, and soil that provide us with our daily bread. The activities of artisans and merchants were deemed sterile or unproductive, in that they merely transformed matter but created no net surplus. The most prominent member of the group, François Quesnay, represented the economy in terms of a circular flow or tableau with explicit analogies drawn to the circulation of the blood and the body politic (Christensen 1994). A central tenet of the group was the doctrine of laissez-faire, which literally meant to let nature take its course.

One reason for this privileging of the natural world stemmed from appeals to that "rude and early state" that preceded the rise of nations. Social activities and institutions were always derivative and thus a less intrinsic

---

5. Land, the third factor traditionally posited in classical political economy, was dropped from the analysis. No extensive explanations were given for this, but the root of it lay in Ricardo's analysis of rent as a derivative and hence dispensable cost.

part of the order of things. Adam Smith posited the existence of a "natural progression of opulence" from agriculture to manufacturing to foreign trade that transcended any specific institutional arrangements. Nevertheless, Smith accorded a larger role to human agency in the economy than the Physiocrats. Economic harmony is engendered by our "natural propensity" to "truck, barter, and exchange," as well as "the desire of bettering our condition." Clearly, the configuration of human labor was critical in determining the yield of the earth's crust, but this came about without any intention or overarching plan. Moreover, there was still a strong inclination to view institutions as something best dismantled such that physical nature could run its course.

With Ricardo we find a more developed conception of an autonomous and self-governing economy. While it was still subject to natural laws—Malthusian conditions and the principle of diminishing returns most notably—there was greater scope for institutional reform. One need only think of Malthus's own recommendations for overcoming the persistent problem of a burgeoning population: moral restraint under the guidance of the church. For Ricardo, taxation became the central means for supervising economic growth and assaulting the unproductive sectors of the economy. Nevertheless, there was still a strong conviction that economic development proceeded according to principles that no group could change at a fundamental level. The best that could be achieved was to accelerate or retard the rate by which the economy unfolded.

Joseph Schumpeter once described Mill's *Principles* as a halfway house (Schumpeter 1954, 603). There is a large grain of truth in this remark, although in a sense different from that intended by Schumpeter, who focused on Mill's analytical oscillations between Ricardian and neoclassical tenets. As I have argued elsewhere, Mill undertook numerous steps that recharacterized the economy vis-à-vis physical nature (Schabas 1995). Both in his posthumous essay "On Nature" (Mill [1874] 1969) and in his economic writings, Mill struggled with the question of human activity and came down firmly on the side of humans dominating rather than submitting to physical nature. Even human nature was malleable and thus perfectible. The economy was thus set apart from the natural order and seen as an instrument for the amelioration of humankind. Furthermore, political economy was no longer a material science. Although it presupposed the operation of the laws of physiology, chemistry, mechanics, and so forth, political economy could take mental phenomena as its proper domain of inquiry. He thus paved the way for the subsequent declarations by the early neoclassical economists who grounded the subject so firmly in the mind.

Mill's enthusiasm for psychology, or the science of the mind, was part of a larger movement at the time, one that he helped to spearhead. As we see in the opening quote to this chapter, he declared psychology to be a new immigrant to England, but one that had settled in successfully.[6] The key figure was Alexander Bain. His *The Senses and the Intellect* (1855) did much to wed physiology with the associationist psychology of David Hartley and James Mill. Introspection was thereby made respectable, in that it was correlated with physical states at the neural or muscular level, as well as facial expressions and bodily gestures. This inaugurated a period of psychological research that took seriously questions of emotion, consciousness, and volition. Some, such as Henry Maudsley and William Benjamin Carpenter, even collapsed the cherished dualism between mind and body, although subsequent investigators, James Sully most notably, felt compelled to restore the sacrosanct divide. Evolutionary biology, in the hands of Herbert Spencer, and the German experimental research of Gustav Fechner and Wilhelm Wundt also played a role in shaping a vibrant community of psychologists in Victorian Britain (Smith 1973; Jacyna 1981; Daston 1978, 1982). Although concrete knowledge of neurophysiology then as now was grossly inadequate to the task, the mere presence of appeals to physiology infused the discipline with an aura of scientific objectivity. This in turn served to dissipate some of the thorny religious and ethical debates that surrounded the question of free will.

A contemporary of Mill's, Richard Jennings, also drew a line between the "province of human nature" and the "external world." His *Natural Elements of Political Economy* (1855) is replete with remarks about the nature and scope of the subject, and most notably highlights the importance of psychology in the development of political economy (White 1994). In his view, "all the phenomena of Political-economy are of two kinds, caused severally by the action of matter on man, and of man on matter"; thus, "there occur simultaneously mental phenomena and physical phenomena, mutually connected by laws, to determine which is the chief object of abstract Political-economy." There are for Jennings laws of human nature that are as "fixed and invariable" as the laws of nature, and to some extent these have already been discerned by statisticians. But the main point he drives home is that the phenomena of political economy, such as exchange value, are mental in origin. It is imperative, therefore, to develop psychological inquiry. He even, quite perspicaciously, proposes that "from this law of the variation of sensations consequences will be found to ensue, affecting more or less all the problems of Price and of Production" (Jennings [1855] 1969, 22, 9–10, 140, 99–100).

John Elliott Cairnes is often classified as the last prominent economic

---

6. Recent scholarship indicates that psychology had already congealed into a coherent discourse in the eighteenth century, mostly in the German- and French-speaking regions. Gary Hatfield argues that "psychology as a natural science was not *invented* during the eighteenth century but *remade*" (Hatfield 1995, 188). Christopher Fox (1987) and Fernando Vidal (1993) also demonstrate the widespread appeal to the science of the mind, or psychology, in the Enlightenment.

theorist of the classical era. His *Character and Logical Method of Political Economy* (1875) confronts directly the issue of the epistemological status of the science of wealth. In his view, political economy is on a par with astronomy. "What Astronomy does for the phenomena of the heavenly bodies," he declared, "Political Economy does for the phenomena of wealth" (Cairnes [1875] 1965, 35). Notwithstanding the seeming messiness of the empirical record, everything in the economy is law governed, and the task of the political economist is to discover those laws. With an implicit debt to Darwin's entangled bank, Cairnes asserts that the phenomena of political economy, "the prices of commodities, the rent of land, the rates of wages, profits, and interest, differ in different countries; but here again, not at random. The particular forms which these phenomena assume are no more matters of chance than . . . the fauna or flora which flourish on the surface of those countries are matters of chance" (Cairnes 1965, 36).

Cairnes makes the interesting argument that political economy, while it draws on both the material and the mental, is in some sense neither a material nor a mental science. There is an equal dependence on the laws from both domains, but in some unspecified sense, wealth, the true subject matter of political economy, is a domain unto itself:

> Neither mental nor physical nature forms the *subject-matter* of the investigations of the political economist. . . . The subject-matter of that science is wealth; and though wealth consists in material objects, it is not wealth in virtue of those objects being material, but in virtue of their possessing value—that is to say, in virtue of their possessing a quality attributed to them by the mind. (Cairnes 1965, 48)

This equivocation, but clear recognition of the mental dimension of the subject, was to be resolved by the early marginalists, notably Jevons, Edgeworth, Wicksteed, and Marshall. With Jevons, economics was to be placed entirely in the domain of the mental: "The theory presumes to investigate the condition of a mind, and bases upon this investigation the whole of Economics" (Jevons 1957, 14–15). All prices were said to be reducible to the feelings of pleasure and pain at the margin, in terms of "the final degree of utility," as he coined it. Moreover, while prices changed as the result of the aggregate effect of individual deliberations, there was no need for a common measuring rod between minds, or even for one mind to directly sway another:

> Every mind is thus inscrutable to every other mind, and no common denominator of feeling seems to be possible. But even if we could compare the feelings of different minds, we should not need to do so; for one mind only affects another indirectly. Every event in the outward world is represented in the mind by a corresponding motive, and it is by the balance of these that the will is swayed. But the motive in one mind is weighed only against other motives in the same mind,

never against the motives in other minds. . . . Hence the weighing of motives must always be confined to the bosom of the individual. (Jevons 1959, 14)

Jevons has here granted considerable autonomy to individual minds as the source of all economic features of the world.

Even capital was defined in terms of mental attributes, as something fixed in objects through the passage of time and the intentions of the person who uses the object. What is critical is that an object be intended for the production of additional wealth. A loaf of bread can "feed the hardworking navvy, the idle beggar, the well-to-do annuitant" (Jevons 1957, 296). But only in the first case is the bread an object of capital. In sum, "There is nothing which marks off certain commodities as being by nature capital as compared with other commodities which are not capital. The very same bag of flour may have to change its character according to the mental changes of its owner" (Jevons 1957, 285).

Rent and wages were also recast in terms of the utility theory of value and were thus defined in terms of mental states. Labor was simply the production of utility. It creates nothing material. And rent, as Ricardo had already demonstrated, was essentially a function of the configuration of property relationships in a given region. All of economic development came from human wants and desires, which were taken to be sui generis, and it was in the act of deliberation that economic phenomena were formed and altered.

Francis Ysidro Edgeworth was even more emphatic about the psychological turn taken by economics. His only book, *Mathematical Psychics* (1881), called for a full mathematization of economic theory in terms of the utility calculus and drew direct inspiration from the psychophysiological work of the German experimental psychologists, such as Hermann von Helmholtz and Gustav Theodore Fechner. As Philip Mirowski has recently shown, Edgeworth was also closely allied with the British psychologists, especially James Sully, who in turn had studied with both Bain and Helmholtz (see Mirowski 1994, 7-15). For Edgeworth, the first principle of economics is that "every agent is actuated only by self-interest." Pleasure, as the Comte de Buffon had shown, was a property of human evolution and thus "an essential attribute of civilisation." Utility was viewed as a kind of energy and thus measurable. Humans were deemed "pleasure machines." Indeed, the day might come when there would be an instrument, "a psychophysical machine, continually registering the height of pleasure experienced by an individual" (Edgeworth [1881] 1967, 16, 77, 15, 101). For Edgeworth, economics was firmly rooted in psychology, and for him this was something to relish, not fear. It would bring greater rigor and objectivity to the subject and provide the proper ontological foundation for the newly developed mathematical theory.

Philip Henry Wicksteed came to economics by way of the Unitarian ministry and was so taken with the new ideas of Jevons that he hired a private tutor to reacquaint him with the calculus. He also endorsed the psychological turn. In his entry "Political Economy and Psychology" for *Palgrave's Dictionary of Political Economy* (1896) he remarked on the links between the two fields that had been recently forged: "The economist must from first to last realise that he is dealing with psychological phenomena, and must be guided throughout by psychological considerations" (Wicksteed [1910] 1933, 767). This is true not only of the analysis of consumption, which gives psychology a "conspicuous place" in economics, but also of all the other areas of the science, such as production, distribution and money. These latter categories are all governed by the law of supply and demand and thus by the psychological questions of satisfaction and motivation. In sum,

> The direction taken by economic study in recent years tends to a more express and generous recognition of the close connection between psychology and political economy, and the necessity of constantly keeping in touch with our psychological basis even when pursuing those branches of economic inquiry which appear to be remotest from it. (Wicksteed 1933, 769)

It was clear to Wicksteed that the character of economics had changed dramatically since the 1870s and that a central factor was the harnessing of psychological theories.

The final figure to be considered in this brief overview of Victorian economics is none other than Alfred Marshall (1842-1924), who dominated not only the last decade of Victoria's reign but, given his imprint on young Maynard Keynes, the first half of this century as well. Marshall was very interested in psychology in his formative years, and wrote several unpublished essays on the subject that have recently been made available in print. The editor of those papers, Tiziano Raffaelli, has suggested that Marshall took the human mind to be a machine of relative simplicity and thus still adhered to some of the basic tenets of the associationist school. Borrowing heavily from Alexander Bain, Marshall believed that most mental connections were grounded in contiguity and similarity. All actions stem from the mind, but within the mind there is room for the reassembly of sensations from the external world, and possibly some internal machinery as well. This latter belief may stem from Marshall's affinity for Kant, even though his psychological inquiries were more in keeping with the empiricist tradition of Hume and Bain. With little delving, he tried to draw a line between the mental and the physical: "My psychological facts are independent of my physical facts, although in any hypothesis or theory by which I attempt to connect my psychological facts I shall be indebted at every step to my corresponding physical theories" (Raffaelli 1994, 113). Deliberation comes about because we place different values on future actions than present ones.

All market phenomena thus come from the mind, and even the formation of capital is essentially the result of an investment of time, of forgoing immediate consumption. A person in an economic context makes very straightforward decisions that revolve around his preference for the present over the future: "Sometimes he is like the children who pick the plums out of their pudding to eat them at once, sometimes like those who put them aside to be eaten last" (Raffaelli 1991, 50). In sum, man produces nothing material, only utility. And it is investment in one's mind that matters most. Physical capital is taken to be subordinate to intellectual capital: "The most important machine is man, and the most important thing produced is thought" (Raffaelli 1991, 52).

This brief tour through the leading writers of Victorian economics has hopefully lent weight to the significant presence of psychology at the time. Needless to say, there were many other novel developments. Herbert Somerton Foxwell, a contemporary of Edgeworth, wrote a succinct overview entitled "The Economic Movement in England" (1887) that also acknowledged the emergence of the historical school or economic history as we now know it, of socialism and Marxism, and of a general shift away from crude laissez-faire reasoning by one and all. I have not discussed these trends here, in part because they have already been addressed by other historians (Dobb 1973; Kadish 1982; Maloney 1985) and in part because they do nothing to alter the theme highlighted here. Indeed, these three developments reinforce the general transformation of the economy as one that can be understood and managed, and not left to the laws of nature. Economic history tended to undercut the belief in laws altogether, by emphasizing the unique and ideographic features of the economic landscape. Socialism was predicated on reform and the refusal to commit the naturalistic fallacy. And the widespread dismissal of laissez-faire principles speaks for itself.

The advent of concentrated appeals to psychology in economic discourse has not been given proper recognition, possibly because most contemporary historians of economics are first trained in economics and acquire nothing but disdain for psychology. Its inherent subjectivism seems to cast a murky shadow over a science as solid and rigorous as economics. Such suspicions date back to the early 1900s. Irving Fisher, arguably the most prominent analytical economist in the first third of the twentieth century, was strongly opposed to the use of psychological findings in economics (Chaigneau 1995). But it was the legacy of positivism that nailed the coffin shut, particularly in the work of Paul Samuelson, who even purged economic discourse of the concept of utility because it was too subjective. Economists have since been content to speak of revealed preferences, and leave the inner workings of the mind to others (Mirowski 1989, 222–31, 378–86).

Why was there this fleeting fancy for psychology in Victorian Britain? It was not found on the Continent at the time, nor in the treatises of the seventeenth and eighteenth centuries. There is no single or simple answer. Certainly Mill's preeminence, and his own familial acquaintance with the subject, played an important role in bringing psychological reasoning into the foreground. His enthusiasm for Bain's efforts to join physiology with associationist psychology may have been the critical turning point. Perhaps this was enough to weaken any resistance that might have been posed by Auguste Comte and his refusal to concern himself with anything as inscrutable as the human mind. Fred Wilson has argued in great detail that Mill assuaged these doubts and convinced himself that there could be an empirical science of psychology. More important, it was Mill who pointed the way to a purely qualitative analysis of pleasure and thus the significant notion of an ordinal ranking of inner states (Wilson 1990, 220). Even when later economists such as Jevons repudiated Mill on doctrinal and methodological commitments, they waxed enthusiastic about psychology and upheld the view that one could infer mental states from manifest actions.

One factor that may have sustained this favorable attitude toward the science of the mind in the post-Millian period was a predilection for the individual as the point of departure for all economic theory. In a nutshell, the early neoclassical economists dissolved economic classes as the unit of analysis and built their model of the world from individuals. Whereas with Ricardo, the central question was how to divide the pie between the landlords, capitalists, and laborers, with Jevons and Marshall, the key question became one of maximizing individual utility and taking the aggregate to arrive at meaningful claims about social welfare. There were no more classes; indeed it was proposed that every laborer might also be an owner of capital and certainly partook in the cash and credit nexus. To put it another way, individuals rarely made an appearance on the stage of the classical economists; the forces of capital accumulation and the ongoing rise of commerce dwarfed individual differences. For the early neoclassical economists, individuals—albeit faceless and nameless individuals—were the *prima causa* of all economic phenomena, which, as we have seen, were fundamentally mental and not material. Moreover, the properties and motions of market phenomena were directly the result of the fact that human minds differed one from another, at least in terms of their evaluations of pleasure and pain. Introspection and studies of the mechanisms of the mind were thus just the license required by Victorian economists to reorient the discipline around individual agency.

Enthusiasm for psychology and the inner life of the mind was also to be found among natural scientists of the time, although these seemed to run parallel to economics rather than impinge directly on it. Evolutionary biology had profound implications for our understanding of the origins and nature of language, memory, and reasoning. Darwin and Spencer both rose to the challenge, though with little by way of concrete evidence to guide them. The full consequences were felt only much later. But of more immediate significance for political economy was the question of the origin of a moral sensibility. Darwin, as Robert Richards has argued, took care to develop an approach different from that of the utilitarians (Richards 1987, 218-19). Darwin looked much more to the group, and to the longevity of the species, than to individual states of pleasure and pain. The two ends were more likely to conflict than to coincide. Altruistic acts were aimed at the general good and were thus not likely to increase the happiness of the individuals concerned. This and other tenets of evolutionary biology suggest that political economy was on an orthogonal track. They both fed on a common interest in psychology, but one is hard-pressed to find points of intersection in their theoretical developments. Indeed, for all of Marshall's proclamations, his economic theory was remarkably impervious to evolutionary biology (Schabas 1994). As Philip Mirowski has argued at length, it was much more infused with the conceptual and methodological constituents of physics (Mirowski 1989, 262-65).

A striking feature of the community of physicists in the latter half of the Victorian period was their reverence for the world beneath the given of experience. Perhaps in reaction to Comtian positivism and its atheistic associations, Victorian physicists delighted in the spiritual dimensions of the "unseen universe," as it was dubbed in 1875 by Balfour Stewart and Peter Guthrie Tait. With the formulation of the law of the conservation of energy, the luminiferous ether became the seat of the electromagnetic field, of light, and of heat, but most of all of the deity (Heimann 1972; Wilson 1977; Wynne 1979). This was the period when British physicists such as Oliver Lodge and George Gabriel Stokes were smitten with spiritualism and psychics. Such predilections for the mental resonated well with the declarations of the psychologists of the time, such as Carpenter and Fechner, to wit, that the mental phenomena are part of the same unitary power that manifests itself in the various forms of energy.

As Jacyna has suggested, the Victorian science of the mind was part and parcel of a broader movement to unify nature and restore a moral foundation to scientific inquiry that had passed away with the demise of natural theology (Jacyna 1981, 129). This was probably less true for political economy, which seemed to ease into its secular state much sooner and without significant challenges. In many respects, it had been the voice of secular reason since the late Enlightenment and did not seem in need of filling a void.[7] But even if motivated by different factors, appeals to the mental

---

7. Peter Minkowitz (1993) has argued forcefully that Adam Smith had already emancipated political economy from theological concerns. While his case is somewhat overblown, there is a large grain of truth in it. Certainly with Ricardo and Mill the subject was fully secular. A small minority promoted Christian political economy throughout the nineteenth century, for example, Thomas Chalmers (1780-1847) and Charles Kingsley (1819-75).

realm were commonplace among both physicists and economists in the latter half of the nineteenth century.

My emphasis here on the significance of psychology in Victorian economics is not meant to imply that appeals to the mind were absent from previous or subsequent economic theories. If anything, all economic theory makes some implicit commitment to a model of human behavior. But rarely have economists been as explicit about embracing the scientific tenets of psychology as they were in the Victorian period. To a considerable degree, the eighteenth-century conception of economic phenomena was much more naturalistic, much more inclined to treat rationality as derivative of more fundamental natural instincts and propensities, than the Victorian conception, which granted so much efficacy to mental deliberation. In Victorian political economy, as we have seen, the presence of psychology was quite pronounced. This was in sharp contrast to Enlightenment political economy, in which reason was subordinate to the passions. As Joseph Cropsey noted, "Smith's formulation is that nature did not leave it to man's feeble reason to discover that and how he ought to preserve himself, but gave him sharp appetites for the means to his survival" (Cropsey 1975, 143). In the classical theory, the individual mind did not make choices that determined the pricing and distribution of economic goods. This was rather the result of the configuration of large groups of people, distinguished by their need to labor, and the cycle of the harvest. Moreover, Hume and Smith took humans to be much more like animals, suggesting in numerous passages that human intelligence was merely refined animal instinct (Schabas 1994). Even within the human species, they suggested, we are all more or less alike in terms of our rational faculties. It is education and happenstance that weeds out the philosopher from the street porter.

Robert Young among others has argued that "Darwin is Social." He means by this that the Darwinian movement served to bring man into nature and that an important source of inspiration for this was classical political economy, the economic and demographic analyses of Adam Smith and Thomas Robert Malthus. But where Young went astray was in assuming that the economy was conceived as a social entity at the time. For Smith and Malthus there was no separate social realm that had its own set of autonomous laws. Rather, there were economic properties of a single natural order which were all part of one grand design. Moreover, as I have suggested elsewhere, the story of Darwin and political economy is a complicated one (Schabas 1990*b*, 1994). If anything, Charles Lyell was much more instrumental in importing economic ideas into biology, and it was the economics of Ricardo, not Adam Smith. Young is correct to see an economic component to the Darwinian theory of evolution, but the economics that was incorporated into that theory was more natural than social. Political economy only became a full-blooded social science, that is to say, it only mapped onto a separate social realm, during the Victorian period. Ironically, at the very time it was said to be stimulating biological thinking, it was moving away from, not toward, natural history.

Enlightenment economists took the economy to be a natural entity and saw *homo economicus* as a creature of animal passions and instincts bent on outcomes such as excess population and the dreaded stationary state that were at odds with the dictates of reason. Subsequent economists, such as Mill, Jennings, Cairnes, and the early neoclassicists, took human beings out of nature. The economy was seen to be the result of rational agency and thus no longer subject to the forces of physical nature. Jevons openly repudiated Malthus, and even Mill from an early age believed that reason could thwart the passion between the sexes. Economic well-being was not like the ebb and flow of the oceans, as Hume had once suggested, but something that could be planned if not controlled. Humans need no longer struggle against nature. Once one is armed with a firm understanding of the principles of political economy, as Mill declared, "the ways of Nature are to be conquered, not obeyed" (Mill 1969, 380–81). Victorian economists thus tugged in a direction different from the one in which Darwin and the Social Darwinists such as Herbert Spencer pulled. From our own vantage point a century later, it appears that they are still holding the same end of the rope.

Bibliographical Note

Possibly because the corpus of economic texts in nineteenth-century Britain is so formidable in size, historians have resisted the temptation to lump them together under a single rubric. But I can point the reader to specific works on several prominent economists—John Stuart Mill, William Stanley Jevons, Francis Ysidro Edgeworth, Alfred Marshall—as well as the development of political economy at Oxford and Cambridge (see Hollander 1985; Schabas 1990*a;* Creedy 1986; Maloney 1985; and Kadish 1982). There are also articles and chapters of books that treat the rise of socialist economics in Britain, the Owenites, Ricardian socialists, or the later group of Fabians (see Berg 1980; King 1983; Henderson 1985; and Stigler 1965). A series of papers on Marshall's predecessor, Henry Fawcett, provides a colorful prism on the period (Goldman 1989). An excellent overview of the rise of political science, with its strong ties to economics, is offered by the collaboration of Stefan Collini, Donald Winch, and John Burrow (1983).

Historians of science have examined many points of intersection between economics and the natural sciences in the hands of such polymaths as George Poulett Scrope, Charles Babbage, and Fleeming Jenkin (see Rudwick 1974; Alborn 1994; Berg 1980; and Wise 1989–90). There are also a few celebrated cases of exchange between political economy and a given science among those who were more inclined to specialize: Darwin's purported assimilation of the subject via Malthus (see Young 1985*a,* 1985*b;* Schweber 1985; Schabas 1990*b*); the absorption of physics, both Newtonian and Maxwellian, by the early neoclassical economists (see Mirowski 1989); and Marshall's appeals to biology as the Mecca for economics (see Maloney 1985; and Schabas 1995). None of these claims are uncontroversial; even the one that Darwin was influenced by Malthusian political economy has been challenged (Gordon 1989). But these debates have at least begun the task of embedding political economy within the scientific culture of the Victorian period. Perhaps the best efforts in that direction are Theodore Porter's, although his main concern is with the questions of quantification and objectivity rather than political economy per se (Porter 1986, 1995).

References

Alborn, Timothy L. 1994. "Economic Man, Economic Machine: Images of Circulation in the Victorian Money Market." In *Natural Images in Economic Thought*, edited by Philip Mirowski. Cambridge: Cambridge University Press, 173-96.

Berg, Maxine. 1980. *The Machinery Question and the Making of Political Economy, 1815-1848.* Cambridge: Cambridge University Press.

Cairnes, John Elliott. [1875] 1965. *The Character and Logical Method of Political Economy.* New York: Augustus M. Kelley.

Chaigneau, Nicolas. 1995. "A Sketch in the History of Indifference Curves: Edgeworth, Fisher and the Role of Psychology." Unpublished manuscript, GRESE, Université de Paris I, Panthéon-Sorbonne.

Christensen, Paul. 1994. "Fire, Motion, and Productivity: The Proto-energetics of Nature and Economy in François Quesnay." In *Natural Images in Economic Thought*, edited by Philip Mirowski. Cambridge: Cambridge University Press, 249-88.

Coats, A. W. 1968. "The Origins and Early Development of the Royal Economic Society," *Economic Journal* 78:349-71.

———. 1972. "The Economic and Social Context of the Marginal Revolution of the 1870's." *History of Political Economy* 4:303-24.

Collini, Stefan, Donald Winch, and John Burrow. 1983. *That Noble Science of Politics.* Cambridge: Cambridge University Press.

Creedy, John. 1986. *Edgeworth and the Development of Neoclassical Economics.* Oxford: Basil Blackwell.

Cropsey, Joseph. 1975. "Adam Smith and Political Philosophy." In *Essays on Adam Smith,* edited by Andrew Skinner and Thomas Wilson. Oxford: Oxford University Press, 132-53.

Daston, Lorraine. 1978. "British Responses to Psycho-Physiology, 1860-1900." *Isis* 69:192-208.

———. 1982. "The Theory of Will versus Science of Mind." In *The Problematic Science: Psychology in Nineteenth-Century Thought,* edited by William R. Woodward and Mitchell G. Ash. New York: Praeger, 88-115.

De Marchi, Neil B. 1973. "The Noxious Influence of Authority: A Correction of Jevons's Charge." *Journal of Law and Economics* 16:179-89.

———. 1974. "The Success of Mill's *Principles.*" *History of Political Economy* 6:119-57.

Dobb, Maurice. 1973. *Theory of Value and Distribution since Adam Smith: Ideology and Economic Theory.* Cambridge: Cambridge University Press.

Edgeworth, Francis Ysidro. [1881] 1967. *Mathematical Psychics: An Essay on the Application of Mathematics to the Moral Sciences.* New York: Augustus M. Kelley.

Fox, Christopher. 1987. "Defining Eighteenth-Century Psychology: Some Problems and Perspectives." In *Psychology and Literature in the Eighteenth Century,* edited by Christopher Fox. New York: AMS Press.

Foxwell, Herbert Somerton. 1887. "The Economic Movement in England." *Quarterly Journal of Economics* 2:84-103.

Goldman, Lawrence, ed. 1989. *The Blind Victorian: Henry Fawcett and British Liberalism.* Cambridge: Cambridge University Press.

Gordon, H. Scott. 1989. "Darwin and Political Economy: The Connection Reconsidered." *Journal of the History of Biology* 22:437-59.

Groenewegen, Peter D. 1988. "Alfred Marshall and the Establishment of the Cambridge Economic Tripos." *History of Political Economy* 20:627-67.

Haskell, Thomas. 1977. *The Emergence of Professional Social Science.* Urbana: University of Illinois Press.

Hatfield, Gary. 1995. "Remaking the Science of Mind: Psychology as Natural Science." In *Inventing Human Science: Eighteenth-Century Domains,* edited by Christopher Fox, Roy Porter, and Robert Wokler. Berkeley: University of California Press.

Hausman, Daniel. 1981. "John Stuart Mill's Philosophy of Economics." *Philosophy of Science* 48:363-85.

———. 1992. *The Inexact and Separate Science of Economics.* Cambridge: Cambridge University Press.

Heimann, P. M. 1972. "The *Unseen Universe:* Physics and the Philosophy of Nature in Victorian Britain." *British Journal for the History of Science* 6:73-79.

Henderson, James P. 1985. "An English Communist: Mr. Bray and His Remarkable Work." *History of Political Economy* 17:73-95.

Hobsbawm, E. J. 1975. *The Age of Capital, 1848-1875.* London: Abacus.

Hollander, Samuel. 1985. *The Economics of John Stuart Mill.* 2 vols. Toronto: University of Toronto Press.

Hont, Istvan, and Michael Ignatieff, eds. 1983. *Wealth and Virtue: The Shaping of Political Economy in the Scottish Enlightenment.* Cambridge: Cambridge University Press.

Hume, David. [1752] 1985. *Political Discourses.* In *Essays, Moral, Political and Literary,* edited by Eugene F. Miller. Indianapolis: Liberty Classics Reprint.

Jacyna, L. S. 1981. "The Physiology of Mind, the Unity of Nature, and the Moral Order in Victorian Thought." *British Journal for the History of Science* 14:109-32.

Jennings, Richard. [1855] 1969. *Natural Elements of Political Economy.* New York: August M. Kelley.

Jevons, William Stanley. [1871] 1957. *The Theory of Political Economy.* New York: Augustus M. Kelley.

Kadish, Alon. 1982. *The Oxford Economists in the Late Nineteenth Century.* Oxford: Clarendon Press.

Keynes, John Maynard. [1936] 1964. *The General Theory of Employment, Interest and Money.* New York: Harcourt Brace.

King, J. E. 1983, "Utopian or Scientific? A Reconsideration of the Ricardian Socialists." *History of Political Economy* 15:345-73.
Maloney, John. 1985. *Marshall, Orthodoxy, and the Professionalisation of Economics.* Cambridge: Cambridge University Press.
Marshall, Alfred. [1890] 1920. *Principles of Economics.* London: Macmillan.
Mill, John Stuart. [1836] 1967. "On the Definition of Political Economy and on the Method of Investigation Proper to It." In *Essays on Economics and Society.* Vols. 4-5 of *Collected Works of John Stuart Mill*, edited by John M. Robson. Toronto: University of Toronto Press, 4:309-39.
———. [1843] 1973. *A System of Logic: Ratiocinative and Inductive.* Vols. 7-8 of *Collected Works of John Stuart Mill*, edited by John M. Robson. Toronto: University of Toronto Press.
———. [1848] 1965. *Principles of Political Economy.* Vols. 2-3 of *Collected Works of John Stuart Mill*, edited by John M. Robson. Toronto: University of Toronto Press.
———. [1859] 1978. "Bain's Psychology." In *Essays on Philosophy and the Classics.* Vol. 11 of *Collected Works of John Stuart Mill*, edited by John M. Robson. Toronto: University of Toronto Press, 341-73.
———. [1874] 1969. "On Nature." In *Essays on Ethics, Religion, and Society.* Vol. 10 of *Collected Works of John Stuart Mill*, edited by John M. Robson. Toronto: University of Toronto Press. 373-402.
———. 1962-90. *Collected Works of John Stuart Mill.* Edited by John M. Robson. Vols. 2-3, *Principles of Political Economy* (1965). Vols. 4-5, *Essays on Economics and Society* (1967). Vols. 7-8, *A System of Logic: Ratiocinative and Inductive* (1973). Vol. 10, *Essays on Ethics, Religion, and Society* (1969). Vol. 11, *Essays on Philosophy and the Classics* (1978). Toronto: University of Toronto Press.
Minowitz, Peter. 1993. *Profits, Priests, and Princes: Adam Smith's Emancipation of Economics from Politics and Religion.* Stanford, Calif.: Stanford University Press.
Mirowski, Philip. 1989. *More Heat than Light: Economics as Social Physics, Physics as Nature's Economics.* Cambridge: Cambridge University Press.
———. 1994. "Marshalling the Unruly Atoms: Understanding Edgeworth's Career." In *Edgeworth on Chance, Economic Hazard, and Statistics*, edited by Philip Mirowski. Lanham, Md.: Rowman and Littlefield, 1-79.
Morgan, Mary S. 1990. *The History of Econometric Ideas.* Cambridge: Cambridge University Press.
Paradis, James. 1989. "*Evolution and Ethics* in Its Victorian Context." In *Evolution and Ethics*, edited by James Paradis and George C. Williams. Princeton, N.J.: Princeton University Press, 3-55.
Porter, Theodore M. 1986. *The Rise of Statistical Thinking, 1820-1900.* Princeton, N.J.: Princeton University Press.
———. 1995. *Trust in Numbers.* Princeton, N.J.: Princeton University Press.
Raffaelli, Tiziano. 1991. "The Analysis of the Human Mind in the Early Marshallian Manuscripts." *Quaderni di Storia dell'Economia Politica* 9:29-58.
———. 1994. "The Early Philosophical Writings of Alfred Marshall." *Research in the History of Economic Thought and Methodology,* archival supplement, 4:53-159.

Rashid, Salim. 1981. "Political Economy and Geology in the Early Nineteenth Century: Similarities and Contrasts." *History of Political Economy* 8:726–44.

Richards, Robert J. 1987. *Darwin and the Emergence of Evolutionary Theories of Mind and Behavior.* Chicago: University of Chicago Press.

Robson, John M. 1976. "Rational Animals and Others." In *James and John Stuart Mill: Papers of the Centenary Conference,* edited by John M. Robson and Michael Laine. Toronto: University of Toronto Press, 143–60.

Romano, Richard M. 1982. "The Economic Ideas of Charles Babbage." *History of Political Economy* 14:385–405.

Rudwick, Martin J. S. 1974. "Poulett Scrope on the Volcanoes of Auvergne: Lyellian Time and Political Economy." *British Journal for the History of Science* 7:205–42.

Ryan, Alan. 1974. *J. S. Mill.* London: Routledge and Kegan Paul.

Samuels, Warren. 1966. *The Classical Theory of Economic Policy.* New York: World Publishing Company.

Schabas, Margaret. 1990a. *A World Ruled by Number: William Stanley Jevons and the Rise of Mathematical Economics.* Princeton, N.J.: Princeton University Press.

———. 1990b. "Ricardo Naturalized: Lyell and Darwin on the Economy of Nature." In *Perspectives on the History of Economic Thought,* edited by Donald Moggridge. London: Edward Elgar, 40–49.

———. 1994. "The Greyhound and the Mastiff: Darwinian Themes in Mill and Marshall." In *Natural Images in Economic Thought,* edited by Philip Mirowski. Cambridge: Cambridge University Press, 322–35.

———. 1995. "John Stuart Mill and Concepts of Nature." *Dialogue* 34:447–65.

Schumpeter, Joseph A. 1954. *A History of Economic Analysis.* New York: Oxford University Press.

Schweber, S. S. 1985. "The Wider British Context in Darwin's Theorizing." In *The Darwinian Heritage,* edited by David Kohn. Princeton, N.J.: Princeton University Press, 35–69.

Smith, Roger. 1973. "The Background of Physiological Psychology in Natural Philosophy." *History of Science* 11:75–123.

Stigler, George. 1965. *Essays in the History of Economics.* Chicago: University of Chicago Press.

Turner, Frank Miller. 1974. *Between Science and Religion.* New Haven, Conn.: Yale University Press.

Vidal, Fernando. 1993. "Psychology in the Eighteenth Century: A View from Encyclopaedias." *History of the Human Sciences* 6:89–119.

Whewell, William. [1829, 1831, 1850] 1971. *Mathematical Exposition of Some Doctrines of Political Economy.* New York: Augustus M. Kelly.

White, Michael V. 1994. "The Moment of Richard Jennings: The Production of Jevons's Marginalist Economic Agent." In *Natural Images in Economic Thought,* edited by Philip Mirowski. Cambridge: Cambridge University Press, 197–230.

Wicksteed, Philip Henry. [1910] 1933. *The Common Sense of Political Economy.* 2 vols. Edited by Lionel Robbins. London: Routledge and Kegan Paul.

Wilson, David B. 1977. "Concepts of Physical Nature: John Herschel to Karl Pearson." In *Nature and the Victorian Imagination,* edited by U. C. Knoepflmacher and G. B. Tennyson. Berkeley: University of California Press, 201–15.

Wilson, Fred. 1990. *Psychological Analysis and the Philosophy of John Stuart Mill.* Toronto: University of Toronto Press.

Wise, M. Norton. 1989-90. "Work and Waste: Political Economy and Natural Philosophy in Nineteenth Century Britain." *History of Science* 27:263-301, 391-449; 28:221-61.

Wynne, Brian. 1979. "Physics and Psychics: Science, Symbolic Action, and Social Control in Late Victorian England." In *The Natural Order,* edited by Barry Barnes and Steven Shapin. Beverly Hills, Calif.: Sage Publications, 167-86.

Young, Robert M. 1985*a.* "Darwinism *Is* Social." In *The Darwinian Heritage,* edited by David Kohn. Princeton, N.J.: Princeton University Press, 609-38.

———. 1985*b. Darwin's Metaphor: Nature's Place in Victorian Culture.* Cambridge: Cambridge University Press.

5

Biology and Politics: Defining the Boundaries

MARTIN FICHMAN

The search for a suitable—and viable—demarcation between scientific and nonscientific discourse was one of the more notable, if elusive, endeavors of the Victorian period. As the impetus to organize a more professionally oriented scientific community gathered strength during the late nineteenth century, so also did the need to specify more precisely what constituted the scientific aspect of the pronouncements of scientists on a broad range of issues, including politics, education, and social values. This subject involves not only the complex question of the ideological context of science, but also the methodological issues raised by the convoluted history of the interactions between various models of "natural science" and "social science" since the Scientific Revolution (Cohen 1994). Defining the territory of professional science posed a particular dilemma for evolutionary biologists. Charles Darwin's and Alfred Russel Wallace's theory of natural selection seemed to provide a powerful basis for establishing the scientific status of biology, and thus for improving the prospects of professional and cultural rewards for its practitioners. Yet precisely because evolutionary biology was at an interface between the natural and social sciences, it was notoriously susceptible to sociopolitical influences and deductions (Jones 1980; Greene 1981; Moore 1989; Bowler 1993b). This chapter examines why the efforts to construct appropriate professional boundaries for evolutionary biology proved so challenging and contentious. These efforts testify eloquently to the complexities inherent in the process by which any age defines or redefines the domain of science and fixes, for itself, the malleable border between scientific and nonscientific discourse (Oldroyd and Langham 1983; R. Young 1985).

The first section of this chapter briefly examines two of the major paradigms of evolutionary biology, Darwinism and Lamarckism, to demonstrate the fecundity of the evolutionary metaphor for political thought. Focusing on Herbert Spencer and Francis Galton, I indicate how, at one level, scientific constructs could be deployed explicitly to enunciate models of social evolutionism that assumed a nonproblematic, unidirectional transition from biological to political speculation. The second section broadens the scope of analysis by situating the Victorian debates concerning evolutionary biology within the wider context of the ideologically charged strategies, especially by the scientific naturalists, to construct a definition of value-neutral and hence objective professional science—and to demarcate this from the now pejorative depictions of value-laden and hence subjective nonscience or pseudoscience.

The scientific naturalists—such as Thomas Henry Huxley and John Tyndall—recognized that by proclaiming the ideological neutrality of science, they created a highly effective strategy for advancing the professional status of biologists. By divesting evolutionary biology of its manifold ideological accretions—as they appeared to be doing—Huxley and his camp could claim that they spoke as objective experts, not political or ideological partisans. This metascientific strategy, however, necessarily involved erecting a sharp demarcation between biology and politics—at least overtly—in order to claim that the objective study of nature scientifically supported certain specific political positions. The strategy was essentially enunciated by the end of the 1860s and served Darwin, Huxley, and their colleagues well for several decades (Moore 1991). Focusing on Huxley, this section illustrates how the scientific naturalists deployed the postulated neutrality of science to construct an "ideologically pure" biology that concealed its varied sociopolitical agendas behind the banner of a rigorous professionalism.

It is significant if somewhat ironic that Wallace, by the 1880s, emerged as one of the most outspoken critics of such a strategy of ideological neutrality. His refusal to recognize any objective demarcation between biology and politics was particularly irksome to the scientific naturalists because of his copaternity of natural selection. The final section of the chapter demonstrates how the defection of Wallace from the camp of the scientific naturalists elucidates both the initial potency of the politics of neutrality and its ultimately fatal flaw. By attempting to insulate biology from politics, evolutionary science became hostage to pervasive ideological manipulation by the scientific naturalists themselves. During the 1890s, the controversy over Wallace's biological socialism was marked by hostility and a patronizing marginalization accorded his evolutionary worldview by many of his scientific colleagues. In the end, however, Wallace's candid conflation of biology and politics—with his insistence upon their reciprocally constitutive dynamics of interaction—signaled the inadequacy of scientific naturalism to maintain the facade of objective neutrality.

I. Darwinism, Lamarckism, and Social Evolutionism

The goal of employing the scientific method to elucidate the problems of society—and ultimately to produce objective and demonstrably valid solutions—dates back, of course, at least to the period of the Scientific Revolution. Victorian biologists and their wide audience, therefore, were hardly novel in their efforts to educe political guidelines from evolutionary theory. As Porter has noted, however, "the urge to elevate politics above *mere* politics by achieving a consensus of experts" usually fails to "force a consensus, particularly when practical applications are at issue. In part, this is because conflicting political or social visions are often only masked by the ostensibly neutral language of science." Equally pertinent is the fact that the natural sciences, particularly evolutionary biology, "present neither a unified nor any single readily-applicable model for social science. Science envy, far from elevating political and social thought above politics, has provided instead a pervasive idiom of debate" (Porter 1990, 1024). Evolutionary biology conjured up variant, often conflicting readings within the British scientific community. When the debates are widened to include the German, French, American, and other scientific communities, the meaning of evolution becomes more complex still (Glick 1974). An additional factor complicating the analysis of the interaction between biological and political thought is the ambiguity surrounding the crucial, and value-laden, term "progress" (Desmond 1982; R. Richards 1992). Particularly in the English-speaking world, "progressive movements," "progressive thinkers," and "progressive political parties" all benefited from the wide, if confusing, scope afforded by the concept of evolutionary progress (Gascoigne 1991, 434-35; Bowler 1989).

Analysis of the political impacts of evolutionary theory must begin with an accurate conception of what were then considered the basic precepts of evolutionary science. Historians of biology have now successfully challenged the view that Darwinism was the dominant evolutionary hypothesis; in the later decades of the nineteenth century, many biologists became outspoken opponents of the theory of natural selection, erecting a variety of alternative mechanisms for evolution. Most influential was the Lamarckian concept of the inheritance of acquired characteristics, but alternatives also included nonadaptationist (orthogenetic) and discontinuous (nongradual) hypotheses (Bowler 1983). Political conclusions drawn from biology relied mainly upon analogies to the two main mechanisms proposed for evolutionary change—mechanisms often confounded or conjoined by the participants in the Victorian debates.

The idea of progressionism was central to Lamarckism; since traits acquired by the purposeful behavior of animals were inherently adaptive, evolution would thus be guided along beneficial lines, as organisms gradually became fitter as they responded to changing environmental demands

(Burkhardt 1981, 132, 206). In contrast, the theory of evolution by natural selection presents a more complex epistemological relation to progressionism (Greene 1994, 334; Nitecki 1988). It is crucial to distinguish at the outset between the general theory of evolution and the specific theory of natural selection. Both Darwin and Wallace, in contrast to Lamarck, convincingly demonstrated the *fact* of evolution to many (though not all) contemporaries. Their writings provided a vast body of evidence—drawn from embryology, comparative behavior, mimicry (the fact that certain species so closely resemble another unrelated species as to be mistaken for the latter), animal and plant breeding (domestic variation), biogeography, and paleontology—showing that organisms cannot have been separately created in their present forms, but must have evolved from earlier forms by gradual transformation (D. Young 1992). Moreover, for Darwin and Wallace, evolution was a two-step process: first, the appearance of (random) variations in nature, and second, the sorting of this variation by natural selection. They argued that the existence of *heritable* variations within a species, coupled with the production of more offspring than could possibly survive, constituted the conditions under which "favorable variations" tended to be preserved and "injurious variations" eliminated. Over many generations, and under the continued selective influence of the environment (the so-called struggle for existence), a group of organisms would eventually have accumulated sufficiently numerous variations to constitute a new taxonomic status: thus the "origin of species" (DeBeer 1963).

An ironic result of Darwinism's success, however, was that Lamarckism seemed more plausible after *The Origin of Species* had given evolutionism greater, and more widespread, credibility than it possessed in the first half of the century. Before the advent of August Weismann's theory of the germ plasm (the concept that the transmission of heritable traits cannot be affected by the environment) in the 1880s, many biologists probably found it difficult to distinguish between natural selection and Lamarckism. Thus, there were a variety of evolutionary theories upon which biologists—and other interested parties—could erect rival "scientifically sanctioned" political systems. Despite the social, professional, theological, philosophical, gender, and empirical constraints that conditioned the shaping of biological theories, there was still a wide range of maneuvering for biologists to draw political conclusions from their science (Bowler 1993*b*, 16, 61, 89-92). Darwinism and Lamarckism—and their variant readings—thus provided fertile ground for theorists of social evolutionism.

Social evolutionists emphasize directional change in societies with time. Although attempts to construct social theory predicated upon scientific analogies were common in the Enlightenment, it was the concepts of Lamarck, Darwin, and Wallace that provided the most striking opportunity to appropriate biological metaphors to sociopolitical thought (Jones 1980; R. Young 1985; Bowler 1993*b*). Despite the fact that evolutionary theories are complicated by the question of the empirical and logical connection between directional change with time and the concept of progress, social evolutionists—of whatever stripe—seek to justify their theoretical models as congruent with some version of natural—that is, "scientific"—evolution. It was the profound political, technological, industrial, and urban transformations of Europe and North America in the late nineteenth century that generated the greater urgency, authority, and popularity of evolutionary accounts of societal change—whether descriptive, prescriptive, or as rationalization—in Victorian culture (R. Smith 1981, 133-35). I shall focus on Spencer and Galton as influential examples of social evolutionists who presumed a nonproblematic transition from biology to the realm of political discourse.

Although often misunderstood or maligned, Spencer was one of the grandest systematizers of evolutionary thought. From his first book, *Social Statics* (1851), Spencer elaborated an enormous lifelong project—what he termed the "system of synthetic philosophy"—which incorporated the entire realm of human knowledge and experience, from biology to religion, from psychology to sociology, into the framework of evolutionism. He had become a Lamarckian through his reading of Charles Lyell in 1840. Although Spencer applied Malthusian principles to animal populations, deduced a struggle for survival, and coined the phrase "survival of the fittest"—thus incorporating natural selection partially into his grand system—his perspective remained Lamarckian. He continued throughout his career to maintain that the inheritance of acquired characteristics was the major mechanism of evolutionary change (Haines 1991, 416-22).

That Spencer explicitly defended Lamarckism against Weismann's critique of use inheritance suggests one powerful reason why he refused to abandon his Lamarckian perspective: he would have undermined what he wanted most to maintain, namely, the fundamental identity of biological evolution and of psychic and social evolution (Spencer 1887, 1893). Spencer envisioned higher forms of organic being emerging from earlier ones by a gradual process of adaptation to the environment. The mental development of man, he argued, lay from egotism to altruism. Correspondingly, society developed from a "militant" phase, in which rigid coercion was needed to hold men together, to an "industrial" phase. In this latter stage, altruism and a harmonious individualism permitted the decline of external state control and the emergence of a fully evolved and integrated social order, in complete and peaceful adaptation to its environment (Peel 1975, 570-71).

Needless to say, Spencer's vast evolutionary synthesis lent itself to the most diverse political readings. He has been interpreted as providing a biological rationale for society as a ruthless struggle for existence, in which relentless, individual competition provides the engine for social progress in accordance with nature's laws, which put all alike under trial (Hofstadter

1955, 31-50; Jones 1980, 56). Yet it has been persuasively claimed that it was his use of the social-organic analogy, not his championing of competition, that best expressed Spencer's political views—and that also provided a frequent opening for socialists to claim him as an intellectual comrade (Pittenger 1993, 20-22). His synthetic philosophy was permeated not by materialism, but by the notion that the ultimate goal of social development was a moral one (Bowler 1993*b*, 65-69; R. Richards 1987, 287, 303-9). Spencer is significant for his efforts to provide a synthesis of much of the accepted physical and biological science of his day, coupled with the nascent social sciences, in the integrating framework of an evolutionary secularization of ethics. His unified—if often technically problematic—vision contributed greatly to the acceptance of science as a major cultural force in Victorian society (Peel 1975, 570).

Francis Galton, Darwin's cousin, adopted a more draconian approach toward elucidating the relationship between biology and politics than did Spencer. Or, to be more precise, Galton simply subsumed politics under biology. Fascinated by what he perceived to be the biological transmission of talent—scholarly, artistic, and athletic—he provided in his 1865 essay "Hereditary Talent and Character" the first persuasive statistical evidence for the presumed inheritance of physical and mental traits in humans. In *Hereditary Genius* (1869), Galton further suggested that the races of mankind could be ranked according to the frequency with which each race produced individuals of high natural ability—which he defined as intellectual capacity, eagerness for work, and power of doing superior work. Races that did not produce such individuals would be swept away by their increasing contact with superior (read "advanced Western technological") races as global industrialization proceeded (Galton 1865, 1869; Mazumdar 1992, 39). Galton coined the term "eugenics" in 1883. His arguments for societal programs to foster talent, health, and other "fit" traits (positive eugenics) and to suppress feeblemindedness and other "unfit" traits (negative eugenics) became influential in the closing years of the nineteenth century and the decade following. It was his brilliant disciple Karl Pearson who developed Galton's insights into a science of biometry, that is, the application to biological phenomena, including human social evolution, of precise and sophisticated statistical correlation techniques (Searle 1976, 7). Although both Galton and Pearson were skeptical—at times hostile—towards the implications of the rediscovery of Mendelian genetics in 1900, biometry became an important tool in the development of statistical methods for modern biology (Kevles 1985, 16-17, 35-39).

For Galton, eugenics was preeminently a scientific repudiation of conservative, aristocratic privilege; politically, he reflected the middle-class outlook of much of the liberal intelligentsia (Jones 1980, 35-36). As with Spencer's system, however, a wide variety of political strategies could be educed from Galton's concept of eugenics predicated upon state intervention in human breeding. Socialist intellectuals, from Pearson to George Bernard Shaw, fashioned systems of "reform eugenics." Conversely, agitated conservatives, anxious to counter the emergence of social welfare politics in the years preceding the Great War, saw in eugenics a scientifically constructive alternative both to the prewar Liberal government's programs and to socialism (Searle 1976, 112-15). Galton also impressed those who began to take the threat of racial degeneration seriously in the 1890s and who were receptive to hereditarian theories that appeared to justify imperialism and racism (Bowler 1993b, 78, 90). The potency of eugenics as a political force is a significant feature of twentieth-century history (Kevles 1985; Adams 1990).

As part of the wider movement of scientific naturalism, Galton's eugenics was "a celebration of the work of the professional elite [that] was also a bold attempt to colonise intellectual territory previously occupied by science's rivals" (MacKenzie 1981, 51). Convinced of the obligatory ideological function of biological evolutionism, Galton committed five hundred pounds a year to University College (in 1904) for a research fellowship in national eugenics—which he defined as "the study of agencies under social control that may improve or impair the racial qualities of future generations either physically or mentally" (Galton 1909, 81). In the influential *Natural Inheritance* of 1889, Galton declared that the statistics of heredity and their eugenic imperatives "are the only tools by which an opening can be cut through the formidable thicket of difficulties that bars the path of those who pursue the Science of man" (Galton 1889, 62-63). It was, however, an evolutionary science constructed upon a political infrastructure.

II. Huxley's Metascientific Strategy: The Ideology of Neutrality

To be sure, debates about the nature of science—such as those involving the work of Spencer and Galton—were important features also of the early Victorian period. William Whewell, John Herschel, and David Brewster, among others, raised fundamental questions concerning the epistemological and cultural status of science. However, science in this earlier period did not enjoy the cultural and institutional security it acquired after midcentury. Accordingly, it is only after the 1860s that there emerged "a scientific culture, rather than science in culture" (Yeo 1993, 32). Numerous works appeared in the 1870s and after, attesting to the ascendancy and autonomy of science—what John Stuart Mill termed the "general property of the age"—and making methodological claims for the broad scope of scientific philosophy that would not have received a sympathetic hearing in the 1830s, nor even the 1850s (Butts 1993, 313-17). Evolution was a key ingredient in this potent cultural vision. However, since evolutionary biology was itself an ongoing, contentious discourse, it imposed no single set of conclusions on political thinkers and activists. As a set of resources, evolution offered defining questions, if not always definitive answers (Pittenger 1993, 8). The scientific naturalists, therefore, turned to the idea of science itself, not merely evolutionary theory in particular, to ground their campaign for ideological neutrality. It was Huxley who most clearly recognized that a metascientific strategy—the politics of neutrality—was necessary to render social evolutionism more objective to its broad audience.

As Frank Turner has noted, the spokesmen for scientific naturalism constituted one of the most vocal and visible groups on the Victorian intellectual landscape. With a combination of research achievements, polemic wit, and literary eloquence, this influential coterie—including Huxley, Tyndall, Leslie Stephen, and John Morley (as well as Spencer and Galton)—helped to create a largely secular climate of opinion in which the theories and metaphors of modern science penetrated the institutions of education, industry, and government. They preached a gospel of social and material progress allied to the advance of science and technology to enthusiastic audiences ranging from skilled mechanics to members of the aristocracy (Turner 1993, 131-32).

From the 1840s onwards, the Victorian scientific world was essentially transformed into a modern professional community (Turner 1993, 179). By the 1870s, in terms of editorships, professorships, and offices in the major scientific societies, such figures as Huxley, Tyndall, Joseph Dalton Hooker, John Lubbock, Galton, and Lyon Playfair emerged as spokesmen for the new scientific elite. A key ingredient in the rising cultural status of these professional scientists was their insistence upon—and a growing popular acquiescence in—the authority of a thoroughly naturalistic approach to science; this would presumably free contemporary science of metaphysical and theological residues and exclude the kinds of troublesome questions as well as answers that characterized traditional approaches to natural knowledge. Huxley's contemptuous, and enduring, caricature of Auguste Comte's own spiritual embellishment of his and Claude Henri Saint-Simon's system as "Catholicism *minus* Christianity" necessitated a sanitized positivism. In 1869, Huxley coined the term "agnostic," which permitted scientific naturalists to present their version of the scientific method not as a rival creed, but as the "unsectarian" method of inquiry of the professional scientist. This neutral stance was seductive—although important evolutionists such as Wallace and St. George Mivart would have none of it—and served to discredit the wider cultural influence of organized religion (Desmond 1994, 373-75; Turner 1993, 180-82).

The interaction between biology and politics during this period, therefore, assumes greater significance when it is recognized as part of the broader process of the fundamental redistribution of cultural authority in Victorian society. Even those biologists who fought fiercely among themselves—such as Huxley and Richard Owen in their celebrated "hippocampus minor" controversy on the significance of the comparative cerebral anatomy of humans, apes, and monkeys—joined forces to contest the influence of theologians and members of other groups, such as the judiciary, whose authority was traditionally recognized as extending over questions of natural history. Owen wrote regularly for nontechnical periodicals—such as Dickens's *Household Words* and *Blackwood's Edinburgh Magazine*—to enhance the reputation of the emerging scientific profession in major public controversies. In these "gladiatorial shows," including debates over the great sea serpent, the longevity of man, and vivisection, Owen emerged as an expert who exposed traditional ignorance and demonstrated the superiority, as well as the usefulness, of scientific knowledge (Rupke 1994, 287–352). Although Owen's expertise was often deployed on behalf of the Tory Anglican establishment, thus separating him from the politically liberal stance of many of his colleagues (including Huxley), his voice was a powerful one attesting to the growing cultural authority of science. Similarly, the often acrimonious disputes between the rival (anti-Darwinian) Anthropological Society and the (Darwinian) Ethnological Society during the 1860s must be seen in the context of their mutual aim of establishing the paradigm of the "scientific study of man." Indeed, Huxley finally succeeded in amalgamating the two groups in the newly formed Anthropological Institute of Great Britain and Ireland in 1871 (E. Richards 1989a, 431–32).

The broader issue, then, is to what degree the later Victorians accorded scientific authority to evolutionary naturalists in questions concerning human society and its politics. Precisely because biologists were becoming part of a more clearly defined professional scientific community, the cultural context of professional pronouncements is essential to an understanding of the political function of evolutionary biology. The issue is further complicated by the fact that the biological and physical sciences were not the only fields aspiring to the status of professions. As Collini has shown, there was a widespread, albeit exceedingly complex, movement toward professionalization of many intellectual disciplines in the late Victorian period. There were rival voices claiming the status of experts in the field of the emerging moral and political sciences. The National Association for the Promotion of Social Science, founded in 1857, was yet another Victorian institution that sought to promote the scientific investigation of fields ranging from legal reform to penal policy, education, public health, and "social economy"; both William Gladstone and Mill addressed at least three of the association's meetings. Such new professionals would (in Henry Sidgwick's phrasing) then be "able, to a certain extent, to pour the stream of pure science into the somewhat muddy channel of current [public] opinion" (Collini 1991, 200) on a wide range of issues. That these new professional experts would also speak authoritatively to the public from the lofty plateau of academic "neutrality" underscores the importance of Huxley's campaign to establish the objective neutrality of the emerging biology profession—so that the scientific naturalists could claim a privileged voice among the competing groups of experts vying for the ears and minds of that diverse audience which constituted Victorian culture, both high and low (Collini 1991, 199-200, 204-5, 210-11, 224).

From the early 1870s onward, Huxley's efforts to cleanse evolutionary naturalism of its accumulated, if diverse, political utilizations were effective. He and his allies (such as his "X Club" comrades) were successful in obtaining modest state funding for research, reorganizing and realigning scientific institutions, and implementing certain educational reforms—all of which, though they tended to legitimate certain aspects of the liberal political agenda, could be viewed as victories of scientific naturalism. Huxley was one of the first to see the polemical advantage of adopting agnosticism as a public philosophy—and Darwin, too, soon appropriated the term, in part to extricate himself from the religious controversies that had long hounded his theory (Lightman 1987, 1989; Moore 1991, 405-6). However, Huxley's stratagems, while undoubtedly effective, were not palatable to all evolutionists. For some, their "science" was inextricably and explicitly linked to ideological and cultural determinants.[1]

Since Huxley's career is emblematic of the professionalization of science, he is the most central of all the evolutionary naturalists to the critical analysis of the relationship of biology to politics (Paradis and Williams 1989; Desmond 1994). His Romanes Lecture of 1893, "Evolution and Ethics," is generally taken as a humanistic distancing of the authority of evolutionary science from sociopolitical and ethical policy disputes. In contrast to most of his contemporaries—and to his own earlier convictions—Huxley is deemed to have discredited the pervasive utilization of biological analogy (Paradis and Williams 1989, 53) by rejecting nature itself as a moral

---

1. During the 1880s and 1890s, for example, evolutionary biology was used in the popularizing agnostic press to buttress arguments against radicalism and socialism. The "new agnostics," in contrast to the elite, middle-class scientific naturalists such as Huxley, explicitly aimed their efforts at a broader, lower-middle-class audience (Lightman 1989, 294-300). In the same period, certain socialist thinkers, notably Wallace, Karl Pearson, the Fabian Annie Besant, and Edward Aveling, utilized evolutionary theory to substantiate their collectivist visions of a scientifically grounded social order (Pittenger 1993, 23). Historians are now examining such rival "scientific campaigns" in terms of their different embedding sociopolitical, professional, religious, and class contexts. This necessitates a more critical examination of the ways in which scientists, and their disciples, use language—in writing texts, giving lectures, preparing research reports, in conversations and correspondence, and in popularizing their concepts. New insights into the functioning of scientific metaphors and analogies in specific cultural circumstances permit fields as diverse as phrenology (Cooter 1984) and evolutionary biology to be interpreted, in part, as modes of discourse whose success stemmed from their manifest adaptability as ways of making sense of a wide range of human social relations. Discussion of the varied attempts at demarcation of biology and politics—or the denial that such demarcation is possible—must, therefore, rely upon these tools of semantic, rhetorical, and symbolic analysis (Golinski 1990, 110-23).

or sociopolitical norm: "Social progress means a checking of the cosmic process at every step and the substitution for it of another, which may be called the ethical process; the end of which is not the survival of those who may happen to be the fittest, in respect of the whole of the conditions which obtain, but of those who are ethically the best" (T. Huxley 1894, 9:81). However, a persuasive argument has been offered by Helfand that Huxley's celebrated essay, rather than limiting and depoliticizing the authority of evolutionary science, subtly invoked it to support his own political views. In particular, Huxley deployed a version of competitive biological selection to sanction the centralized and paternalistic sociopolitical agenda by which Liberals proposed to solve the problems of the decline in English prosperity which had occurred in the period from the mid-1870s to the mid-1890s (Helfand 1977, 159–77).

Huxley would have had two motives in doing so. First, he was personally and publicly opposed to what he regarded as the radical and utopian reform strategies of groups as diverse as eugenicists, anarchists, the Salvation Army, and socialists (Paradis and Williams 1989, 6, 24). Second, the bureaucratic sophistication Huxley had acquired in the corridors of power, as he became one of the foremost spokesmen for professionalized science, required a repudiation of his earlier support of the laissez-faire reading of Spencer's philosophy, which denied certain powers to the government (including the provision of public science education) that Huxley now felt were important to ensure the success of England in commercial competition with other European powers. The Romanes Lecture, which he described as an "egg-dance," was a rhetorical tour de force in which he apparently repudiated the ethical authority of evolutionary biology to deny scientific legitimacy to both Spencerian individualists and the radical land socialists, only to appropriate evolutionism for his own political stance. In dismissing as unrealistic the socialist theories of Henry George and Wallace, for example, because they challenged Malthusianism, Huxley was compelled to readmit a modified theory of biological struggle to political discourse. In a letter to Romanes, Huxley affirmed that "though there is no [direct] allusion to politics in my lecture," he would "never have taken the pains I have bestowed on these 36 pages" if his audience had failed to apply his ideas to draw appropriate "liberal" political conclusions (L. Huxley 1901, 2:375). In 1887, Huxley had written an influential letter to the *Times* urging Britain to strengthen its economic and military position by relying more substantially upon the political expertise and vision of its professional scientific and technical elite (Turner 1993, 206).

Huxley's Romanes Lecture, therefore, is paradoxical in one crucial sense. He and other advocates of the rising cultural potency of professional science had striven to publicly divorce science from ideology in their efforts to ensure more adequate government and public recognition and support of science as a self-governing entity. In his presidential address to the British Association for the Advancement of Science in 1874, Tyndall, for example, announced that all "theories, schemes and systems, which . . . reach into the domain of science must, *in so far as they do this,* submit to the control of science, and relinquish all thought of controlling it. Acting otherwise proved disastrous in the past, and it is simply fatuous to-day" (Tyndall 1874, 61; italics in original). Similarly, Huxley emphasized the presumed neutral and apolitical character of science in his influential efforts to promote acceptance of the view that science, including the "human sciences," should be regarded as definitive in its conclusions (Fichman 1984, 482). Although Tyndall's and Huxley's views were controversial, their strategy to foster the concept of an ideologically pure, and hence objective, scientific naturalism had become a crucial tool of the Victorian lobbyists for the "cult of science." The Romanes Lecture, however, exposed a critical weakness in such a strategy.

The dilemma of William Whewell, late in his career (he died in 1866), affords testimony to the tensions inherent in the growing cultural authority of professional science championed by Huxley. In sociological and institutional terms, the decade of the 1860s had been notable for the concerted strategy by members of the scientific community to enunciate a professional identity, often coupled with an attack on the unscientific attitudes and classical education of the political establishment. Their polemical activity was dictated by the reality that the social status of career scientists was, with some exceptions, still relatively low in late-nineteenth-century Britain—as were government levels of financial and institutional support for scientific research—particularly in comparison to the situation in Germany (Alter 1987, 72-74, 131-37, 214-45). This confrontational style differs sharply from the tactics of the early British Association for the Advancement of Science and its Cambridge and Oxford managers. Whewell and William Buckland, for instance, never treated their Tory patrons during the 1830s and 1840s as adversaries. Whewell could then afford to be candid, not defensive, about the presence of values and commitments in his own attempts to use science as a basis for action in other areas—such as the reform of moral philosophy or the support of the Anglican Church and its link with the state—because he did not proclaim that science was a neutral discourse or an autonomous cultural enterprise. By the 1860s, however, with the Darwinian debates in mind, he had become uncertain as to the precise implications of the cultural role of an increasingly specialized, professionalized community of scientists (Yeo 1993, 32, 254-55). Whewell's candor concerning his own explicitly ideological uses of science was a troublesome anachronism by the 1860s; it was precisely such candor that Huxley and his cohort spent the next three decades trying to suppress.

III. Wallace's Critique of Ideological Neutrality

In "Evolution and Ethics," Huxley was forced to confront, albeit ambiguously, the implications of the politics of neutrality raised earlier by Whewell. Scientific naturalism had never been ideologically neutral. Any further pretense to that effect could only be an invitation to exacerbate the emerging sociopolitical and environmental dilemmas posed by an increasingly self-confident and imperialist technoscientific culture (Paradis and Williams 1989, 34, 55). It was Wallace who most clearly refocused on the dilemma posed by Whewell. By the 1880s, Wallace attacked the central paradox of the politics of neutrality: the legerdemain by which Victorian evolutionists sought to erect a boundary between their biological theorization (and its empirical validation) and their sociopolitical views in order to then assert that the objective study of nature scientifically supported specific political agendas. Wallace's enunciation of an evolutionary worldview, his biological socialism, abandons any pretext of ideological neutrality and unveils the hollowness of the scientific naturalists' claim to objectively demarcate biology from politics. Specifically, Wallace's volte-face on the question of human sexual selection is not simply a further modification of his and Darwin's theory of natural selection. It is a manifesto of the necessary ideological context and texture of evolutionary biology. In this sense, Wallace retains Spencer and Galton's goal of linking biology and politics while shedding their comforting armature of objectivity.

In 1890 Wallace contributed an article entitled "Human Selection" to the *Fortnightly Review,* which, though short, he considered the "most important contribution I have made to the science of sociology and the cause of human progress" (Wallace [1905] 1969, 2:209). He began by noting that in one of his last conversations with Darwin, the latter "expressed himself very gloomily on the future of humanity" because natural selection no longer operated effectively: those who succeeded in the race for wealth were not necessarily the best or the most intelligent. Darwin further lamented that "it is notorious that our population is more largely renewed in each generation from the lower than from the middle and upper classes." Wallace dismissed as possible solutions to the apparent check on human progress any proposals based solely upon beneficial environmental influences, such as education and public hygiene. He felt that Galton and Weismann had demolished the theory of the inheritance of acquired characteristics; there remained thus "some form of selection as the only possible means of improving the race" (Wallace 1890, 325-26).

But Wallace also rejected what he termed artificial selection, under which he included such schemes as Galton's and Pearson's eugenics. Aside from its objectionable moral implications, artificial selection would be biologically ineffective, only slightly increasing "the number and rais[ing] the

"PROGRESS IS THE INTEREST OF BOTH."

Figure 5.1 A satirical cartoon from the *Clarion,* an influential socialist newspaper established in 1891, ridiculing the idea that capital benevolently guides labor as partners in progress. Wallace was a frequent contributor to the *Clarion* from the late 1890s until his death (1913). (From "Mild and Bitter," *Clarion,* 2 January 1892, p. 1.)

standard of our highest and best men," while at the same time leaving the bulk of the population unaffected (Wallace 1890, 328). Wallace's "fundamental objection" to eugenics schemes, however, betrays the motive — his commitment to socialism (figure 5.1) — behind the biological critique:

> They all attempt to deal . . . by direct legislative enactment, with the most important . . . of all human relations, regardless of the fact that our present phase of social development is . . . vicious and rotten at the core. . . . Let any one consider, on the one hand, the lives of the wealthy . . . with their almost inconceivable wastefulness and extravagance; and, on the other hand, the terrible condition of millions of workers. . . . Can any thoughtful person admit for a moment that, in a society so constituted that these overwhelming contrasts of luxury and privation are looked upon as necessities, and are treated by the Legislature as matters with which it has practically nothing to do, there is the smallest probability that we can deal successfully with such tremendous social problems? (Wallace 1890, 330)

Wallace argued that capitalism, aside from its immorality, precluded the operation of any effective selective agency. This article, along with one entitled "Human Progress: Past and Future" (in the Boston *Arena*) (Wallace 1892*a*), marks not only Wallace's first public declaration as a socialist but, significantly, the "first scientific application of my conviction," namely, sexual selection (Wallace 1969, 2:267). To be sure, Wallace had long embraced certain socialist ideas, dating from as early as his youthful attendance at Owenite lectures in the working-class Halls of Science and Mechanics' Institutes. Also, the ending of his 1864 essay entitled "The Origin of Human Races" echoed the Owenite social utopian vision (Wallace 1864, clxix-clxx). But Wallace changed the ending in an 1870 version, which emphasized spiritualist rather than socialist themes (Wallace 1870).

By 1890, however, the advent of socialism had become the explicit precondition for the operation of "beneficial" sexual selection in human society, which would bring about the improvement of civilized races that Wallace considered impossible under capitalism. Socialism, by removing inequities of wealth and rank, would free females from the obligation to marry solely on the grounds of financial necessity. Female choice, which Wallace considered to have been hitherto ineffectual—or distorted—in human evolution, would now result in selection of only the most "desirable" husbands, with the inevitable result that the race would be bettered. Wallace considered the principle of sexual selection under socialism to be "by far the most important of the new ideas I have given to the world" (Wallace 1969, 2:389). This principle, however, was not a new idea of Wallace's; he borrowed it almost verbatim from the American utopian thinker Edward Bellamy. But Wallace's particular use of sexual selection under socialism was crucial for the development of his own brand of polemicized political biology.

It is curious that it should be sexual selection that Wallace now advanced as an agency for human evolution. For one of the major theoretical differences between Wallace and Darwin derived from Wallace's refusal, initially, to accord any scientific status to sexual selection (Cronin 1991, 131-36, 155-64; Marchant 1975, 130). The publication of *The Descent of Man, and Selection in Relation to Sex* in 1871 had elicited a critical review from Wallace in which he controverted Darwin's assertion that sexual selection accounted for the racial differences and other characteristics of mankind (Wallace 1871, 177-83; Russett 1989, 80-81; E. Richards 1983, 70; Kottler 1985). In *Contributions to the Theory of Natural Selection* (1870) and *Tropical Nature* (1878), Wallace argued that natural selection alone produced the marked sexual and species differences among animals (e.g., protective coloration and recognition markings), including humans; he declared that natural selection sufficed to explain the striking racial divergences (e.g., selection by disease) and other external characteristics of the human species (Wallace 1870, 1878). Though Wallace maintained his position with respect to the absence—or relative unimportance—of sexual selection among animals, he reversed it by 1890 with respect to humans. Moreover, in contrast to Darwin, who employed a problematic dual mechanism of sexual selection, that is, male combat *and* female choice (Jann 1994), Wallace focused on the crucial (potential) role of female choice in human evolution. Why?

The year before (1889), Wallace had read Bellamy's *Looking Backward* (1888), a book that changed his social views "once for all." Previously, Wallace had been torn between the conflicting claims of Owenite socialism and Spencerian individualism. This tension was finally resolved by Bellamy's book, in which every "sneer, every objection, every argument I had ever read against socialism was . . . met and shown to be absolutely trivial or altogether baseless" (Wallace 1969, 2:266-67). But *Looking Backward* provided Wallace with more than a cogent defense of socialism. It yielded an explicit mechanism by which a progressive human mental and moral evolution could be effected. In Bellamy's egalitarian future state, "for the first time in human history the principle of sexual selection, with its tendency to preserve and transmit the better types of the race, and let the inferior types drop out, has unhindered operation." Freed from poverty and subjugation, women could choose as biological fathers of their children only those males who possess the admirable qualities of "wit, eloquence, kindness, generosity, geniality, and courage. . . . Every generation is sifted through a finer mesh than the last" (Bellamy [1888] 1960, 179-80).

Wallace's earlier reading of Henry George's *Progress and Poverty* (1879) had further reinforced his conviction that the Malthusian principle, though valid in the case of animals and plants, did not apply in the case of mankind, "still less that it has any bearing whatever on the vast social and political questions which have been supported by a reference to it" (Marchant 1975, 260). When Wallace urged Darwin to read it, the latter replied (in the last letter he sent Wallace) that he would "certainly order 'Progress and Poverty,' for the subject is a most interesting one. But I read many years ago some books on political economy, and they produced a disastrous effect on my mind, viz. utterly to distrust my own judgment on the subject and to doubt much everyone else's judgment!" (Marchant 1975, 261). The weary honesty of Darwin's response is testimony to the divergent outlooks of the cofounders of natural selection concerning the explicitness of the political context of evolutionary discourse.

What Bellamy and George provided for Wallace was a more critical appreciation of the complex relationship of evolutionary biology to sociopolitical ideas. Most important of all, Bellamy's work suggested a plausible mechanism for social advance. Although Wallace never abandoned his belief in the guidance of spiritual intelligences as agents in human evolution— nor his insistence that spiritualist claims could be verified empirically and thus constituted a body of demonstrable knowledge (Oppenheim 1985,

320)—he would clearly have appreciated the polemical advantages of a "naturalistic" sexual selection in his own biological argumentation (Durant 1979, 31-58). However, Wallace's social progressionism informed his biological progressionism and reinforced his position that science did not function as a neutral blueprint for political philosophy. George's thesis that material progress had engendered, rather than alleviated, human poverty and misery reinforced and reanimated Wallace's own claims. Responding to George's crusades for increased taxation of rural and urban landlords, Wallace had assumed a prominent role in the public debate on land reform, a debate that during the 1870s and 1880s provided a major focus for the broader question of social and political reform in Great Britain. Wallace's new role did not, as Darwin feared, force him to "turn renegade to natural history" (Marchant 1975, 262). Rather, Wallace was poised to analyze and reassess the use (or misuse) of his and Darwin's biological theories to buttress particular social and political ideologies and policies. Wallace's views on land reform, for example—culminating in the publication of *Land Nationalisation* in 1882 and his election as president (1881) of the newly formed Land Nationalisation Society—must be seen as integral elements in his biological philosophy, in which evolutionary and socialist arguments reacted upon one another. Moreover, Wallace's biological philosophy cannot be viewed as a modification of scientific naturalism. For Wallace, science does not "authorize" his political position: socialist convictions and biological insights are equally constitutive components of a broader evolutionary worldview. Science is but one element, albeit a crucial one, in his construction of a comprehensive cultural vision (Wallace 1882).

Wallace's combination of socialism and biology, specifically sexual selection, was not unique. Furthermore, his admixture of feminism, spiritualism, and reformist social evolutionism was not uncommon (Owen 1990, 26-27). Aside from Bellamy, a number of feminist writers, notably the American Charlotte Perkins Gilman, also utilized sexual selection within a socialist framework. In *Women and Economics* (1898), Gilman agreed that sexual selection was a force in human evolution, but one distorted by capitalism's exploitation of women—which nullified any genuine free female choice. Economic equality and independence would restore to women the evolutionary potential to make "their rightful contribution to the future of the race" (Russett 1989, 84-86). The striking similarity between aspects of Wallace's and Gilman's views—she was also a Bellamy enthusiast (Love 1983, 121)—suggests the possibility of an Anglo-American faction for socialism and sexual selection. Gilman had affinities with the Fabians (Pittenger 1993, 72-79). In a like vein, the socialist Eliza Burt Gamble declared that under capitalism, women have become "economic and sexual slaves . . . dependent upon men for their support" and dispossessed of their "fundamental prerogative" of aesthetic choice. Gamble envisioned a noncapitalist future, when women would regain their rightful power of sexual selection and, through the transmission of their "more refined instincts and ideas peculiar to the female organism" (such as altruism and sympathy) to their offspring, found a "new spiritual age" (E. Richards 1983, 110 n. 155). Wallace argued similarly (Wallace 1913*b*, 163-64); he also enthusiastically endorsed the more fully developed feminist socialism of *Equality* (1897), Bellamy's sequel to *Looking Backward* (Wallace 1969, 2:268-72; 1905), which eliminated some middle-class and patriarchal values that marked the earlier work (Strauss 1988, 80, 88; Bellamy 1897, 128-38).

Socialism, therefore, provided both the motive and the rationale for Wallace's espousal of sexual selection as an agency auxiliary to natural selection. The motive was to provide an alternative to eugenics schemes, which he feared would perpetuate class distinctions and postpone social reform. "Eugenics," he declared, "is simply the meddlesome interference of an arrogant scientific priestcraft" (Marchant 1975, 467). It is interesting that Huxley—although Wallace's socialist politics were anathema to him—also expressed concerns about harsh treatment of many individuals, arguing that eugenic intervention would destroy the bonds of social sympathy (Paradis and Williams 1989, 47-48). Of course, this was precisely what most eugenists considered the virtue of their schemes: scientific experts would manage societal evolution. Wallace's view of the incompatibility of socialism and eugenics was not shared by all of his contemporaries. Pearson, as did certain of the Fabians, saw eugenics as compatible with an "elitist socialism"—a planned socialism by middle-class experts and administrators (MacKenzie 1981, 75-79). This was not Wallace's, or George's, socialism, which was more cognizant of the necessary participation of the working classes in effecting social change.

Moreover, Wallace's previous theoretical objections to the efficacy of sexual selection—in the human realm—were no longer valid. He had never denied that females could exercise some degree of individual choice in mating. But mate choice is not equivalent to sexual selection, although it is a critical component of it. For sexual selection to occur, mate choice must bring about differential reproduction rates favoring those individuals who display the preferred traits and who vary genetically in this respect from others of their sex (Cronin 1991, 114, 168-74). Socialism, Wallace believed, was the only political system under which these conditions were satisfied (Wallace 1913*b*, 163-65). Male mortality (partly due to the males' more dangerous occupations under capitalism) would be decreased; the natural preponderance of males, as was demonstrated by birth statistics, would be maintained. In such a state, women would be the minority, and female choice could function as a dynamic evolutionary mechanism (Wallace 1890, 336-37). The fact that Wallace continued to deny or minimize the efficacy of sexual selection among nonhuman animal species (Wallace

1892b, 749-50) underscores the singularity of his reversal with respect to human sexual selection.

Wallace became increasingly adamant in his fundamental conviction that to divorce science from its sociopolitical and moral context is both logically indefensible and historically dangerous. In books and articles from the 1880s until the year of his death (1913), Wallace—in contrast to many of his liberal colleagues (including Spencer, Galton, Huxley, and Darwin), who were not so terribly discomfited with Victorian capitalist imperialism—argued forcefully, on political grounds, that capitalism had corrupted human evolution. In *The Revolt of Democracy* (1913), he condemned capitalist technoscientific advance, which had created a luxurious upper class while leaving one-fourth of the population in poverty:

> [Thus] the principle of competition—a life and death struggle for bare existence—has had more than a century's unbroken trial under conditions created by its upholders, *and it has absolutely failed.* The workers, now for the first time, know *why* it is that with ever-increasing production of wealth so many of them still suffer the most terrible extremes of want and of preventable disease. There must, therefore, be no further compromise, no mere talking. To allow the present state of things to continue is a crime against humanity. (Wallace 1913a, 2-3, 76-77; italics in original)

Wallace castigated establishment science, which, by naturalizing inequalities in industrial society, had along with religion "agreed in upholding the competitive and capitalistic system of society as being the only rational and possible one" (Wallace 1913a, 5). Little wonder that his political works, as James Marchant puts it, "produced feelings of regret amongst many of his scientific friends, [as] his advocacy of spiritualism caused them (as Tyndall said) 'feelings of deep disappointment'" (Wallace 1913a, xxxviii-xxxix). For the majority of them, who were busily utilizing science to expound a variety of their own political platforms, radicals like Wallace and Besant went too far; according to the agnostic Frederick Millar, an "Evolutionist Socialist" was a contradiction in terms (Lightman 1989, 296-97). Yet other contemporaries applauded Wallace's biological socialism. The magazine *The Social Democrat* emphasized (in 1910) that "Wallace shares the honours with Darwin for the discovery of the law of evolution—and we may proudly add, is a Socialist" (Jones 1980, 25). It is significant that Wallace also explicitly linked his spiritualist convictions to his advocacy of socialism (Wallace 1898).

Wallace had come a long way since the early 1860s. Then, still regarded as a member of the Darwinian clique advocating a nonteleological evolutionary naturalism (Moore 1991, 379), he had (in 1864) spoken against the Ethnological Society's admission of women to its meetings on the grounds that "consequently many important and interesting subjects cannot possi-

Figure 5.2 An 1879 cartoon from *Punch* demonstrating a recurrent Victorian theme that the incongruity of fragile and attractive women tackling "masculine" jobs rendered women doctors and dentists, for example, comic or amusing. It provides evidence that Wallace's own earlier ambivalence regarding female participation in the professions was not uncommon. (From *Punch* 77 [1 November 1879]: 203.)

bly be discussed there" (E. Richards 1989b, 264) (see figure 5.2.) By the early 1890s, Wallace had become a biological-socialist-feminist. His odyssey, intriguing as it is in its own right, symbolizes a fundamental shift in the cultural context of evolutionary biology. Because he was the ultimate insider become outsider—professionally, politically, and metascientifically—Wallace's own evolution signified that scientific naturalism could no longer contain biology within the confines that had enabled it to emerge as a potent professional science. In this respect, Wallace's conviction that science was not—and could never become—a uniquely privileged source of cultural authority is reflective of the broader European critique of, and reaction against, positivism at the very close of the nineteenth century (Turner 1974). The cultural and ecological impacts of science and technology at the dawn of the twentieth century demanded a more critical investigation of the politics of neutrality, with its attendant program for defining knowledge, than was deemed necessary in the optimistic climate of Victorian scientific naturalism. To be sure, Wallace's remained somewhat of a lonely voice among the professional scientific community, which continued to develop under the aegis of positivism and the politics of neutrality in the first decades of the new century. His deconstruction of scientific naturalism, nonetheless, prefigures in important respects the contemporary contextualist approach to science studies.

Bibliographical Note

There is a large and controversial literature on the interaction between nineteenth-century evolutionary biology and sociopolitical theory. A concise, reliable guide to recent scholarly assessments is Peter Bowler, *Biology and Social Thought: 1850-1914* (1993b). An earlier, but still valuable analysis, is Greta Jones, *Social Darwinism and English Thought* (1980). Robert Young's classic *Darwin's Metaphor: Nature's Place in Victorian Culture* (1985) should be consulted in conjunction with Ingemar Bohlin's sympathetic but cogent critique, "Robert M. Young and Darwin Historiography" (1991). Mark Pittenger's *American Socialists and Evolutionary Thought, 1870-1920* (1993) contains insights applicable to Britain. Finally, Frank Turner's *Contesting Cultural Authority* (1993) has much to offer on the professional, political, religious, and ideological dimensions of scientific naturalism.

References

Adams, Mark, ed. 1990. *The Well-Born Science: Eugenics in Germany, France, Brazil, and Russia.* Oxford: Oxford University Press.

Alter, Peter. 1987. *The Reluctant Patron: Science and the State in Britain, 1850-1920.* Oxford: Berg.

Bellamy, Edward. [1888] 1960. *Looking Backward.* New York: New American Library.

———. 1897. *Equality.* New York: Appleton and Company.

Bohlin, Ingemar. 1991. "Robert M. Young and Darwin Historiography." *Social Studies of Science* 21:597-648.

Bowler, Peter J. 1983. *The Eclipse of Darwinism: Anti-Darwinian Evolution Theories in the Decades around 1900.* Baltimore: Johns Hopkins University Press.

———. 1988. *The Non-Darwinian Revolution: Reinterpreting a Historical Myth.* Baltimore: Johns Hopkins University Press.

———. 1989. *The Invention of Progress: The Victorians and the Past.* Oxford: Blackwell.

———. 1993a. *Darwinism.* New York: Twayne Publishers.

———. 1993b. *Biology and Social Thought: 1850-1914.* Berkeley: University of California at Berkeley, Office for History of Science and Technology.

Burkhardt, Richard W. 1981. "Evolution" and "Inheritance of Acquired Characters."

In *Dictionary of the History of Science,* edited by W. F. Bynum, E. J. Browne, and Roy Porter. Princeton, N.J.: Princeton University Press, 131-33, 206-7.

Butts, Robert E. 1993. *Historical Pragmatics: Philosophical Essays.* Dordrecht: Kluwer Academic Publishers.

Cohen, I. Bernard. 1985. *Revolution in Science.* Cambridge, Mass.: Harvard University Press.

———, ed. 1994. *The Natural Sciences and the Social Sciences: Some Critical and Historical Perspectives.* Dordrecht: Kluwer Academic Publishers.

Collini, Stefan. 1991. *Public Moralists: Political Thought and Intellectual Life in Britain, 1850-1930.* Oxford: Clarendon Press.

Collini, Stefan, Donald Winch, and John Burrow. 1983. *That Noble Science of Politics: A Study in Nineteenth-Century Intellectual History.* Cambridge: Cambridge University Press.

Cooter, Roger. 1984. *The Cultural Meaning of Popular Science: Phrenology and the Organization of Consent in Nineteenth-Century Britain.* Cambridge: Cambridge University Press.

Cronin, Helena. 1991. *The Ant and the Peacock: Altruism and Sexual Selection from Darwin to Today.* Cambridge: Cambridge University Press.

De Beer, Gavin R. 1963. *Charles Darwin: Evolution by Natural Selection.* London: Nelson.

Desmond, Adrian. 1982. *Archetypes and Ancestors: Palaeontology in Victorian London, 1850-1875.* London: Blond and Briggs.

———. 1989. *The Politics of Evolution: Morphology, Medicine, and Reform in Radical London.* Chicago: University of Chicago Press.

———. 1994. *Huxley: The Devil's Disciple.* London: Michael Joseph.

Durant, John R. 1979. "Scientific Naturalism and Social Reform in the Thought of Alfred Russel Wallace." *British Journal for the History of Science* 12:31-58.

Fichman, Martin. 1984. "Ideological Factors in the Dissemination of Darwinism in England, 1860-1900." In *Transformation and Tradition in the Sciences: Essays in Honor of I. Bernard Cohen,* edited by Everett Mendelsohn. Cambridge: Cambridge University Press, 471-85.

Galton, Francis. 1865. "Hereditary Talent and Character." *Macmillan's Magazine* 12:157-66, 318-27.

———. 1869. *Hereditary Genius: An Inquiry into Its Laws and Consequences.* London: Macmillan.

———. 1889. *Natural Inheritance.* London: Macmillan.

———. 1909. *Essays in Eugenics.* London: The Eugenics Education Society.

Gascoigne, Robert M. 1991. "Julian Huxley and Biological Progress." *Journal of the History of Biology* 24:433-55.

George, Henry. 1879. *Progress and Poverty: an inquiry into the cause of industrial depressions, and of increase of want with increase of wealth; the remedy.* New York: D. Appleton.

Glick, Thomas F., ed. 1974. *The Comparative Reception of Darwinism.* Austin: University of Texas Press.

Golinski, J. V. 1990. "Language, Discourse and Science." In *Companion to the History of Modern Science,* edited by Robert C. Olby, G. N. Cantor, J. R. R. Christie, and M. J. S. Hodge. London: Routledge, 110-23.

Greene, John C. 1981. *Science, Ideology, and World View: Essays in the History of Evolutionary Ideas.* Berkeley: University of California Press.

———. 1994. "Science, Philosophy, and Metaphor in Ernst Mayr's Writings." *Journal of the History of Biology* 27:311-47.

Haines, Valerie A. 1991. "Spencer, Darwin, and the Question of Reciprocal Influence." *Journal of the History of Biology* 24:409-31.

Helfand, Michael S. 1977. "T. H. Huxley's 'Evolution and Ethics': The Politics of Evolution and the Evolution of Politics." *Victorian Studies* 20:159-77.

Hofstadter, Richard. 1955. *Social Darwinism in American Thought.* Boston: Beacon Press.

Huxley, Leonard, ed. 1901. *Life and Letters of Thomas H. Huxley.* New York: D. Appleton.

Huxley, Thomas Henry. 1894. "Evolution and Ethics." In *Collected Essays,* vol. 9. London: Macmillan.

Jann, Rosemary. 1994. "Darwin and the Anthropologists: Sexual Selection and Its Discontents." *Victorian Studies* 37:287-306.

Jones, Greta. 1980. *Social Darwinism and English Thought: The Interaction between Biological and Social Theory.* Sussex: Harvester Press.

Kevles, Daniel. 1985. *In the Name of Eugenics: Genetics and the Uses of Human Heredity.* New York: Knopf.

Kitcher, Philip. 1993. *The Advancement of Science: Science without Legend, Objectivity without Illusions.* New York: Oxford University Press.

Kottler, Malcolm Jay. 1985. "Charles Darwin and Alfred Russel Wallace: Two Decades of Debate over Natural Selection." In *The Darwinian Heritage,* edited by David Kohn. Princeton, N.J.: Princeton University Press, 367-432.

Lightman, Bernard. 1987. *The Origins of Agnosticism: Victorian Unbelief and the Limits of Knowledge.* Baltimore: Johns Hopkins University Press.

———. 1989. "Ideology, Evolution and Late-Victorian Agnostic Popularizers." In *History, Humanity and Evolution: Essays for John C. Greene,* edited by James Moore. Cambridge: Cambridge University Press, 285-309.

Love, Rosaleen. 1983. "Darwinism and Feminism: The 'Woman Question' in the Life and Work of Olive Schreiner and Charlotte Perkins Gilman." In *The Wider Domain of Evolutionary Thought,* edited by David Oldroyd and Ian Langham. Dordrecht: D. Reidel Publishing Company, 113-31.

MacKenzie, Donald A. 1981. *Statistics in Britain, 1865-1930: The Social Construction of Scientific Knowledge.* Edinburgh: Edinburgh University Press.

Marchant, James. 1975. *Alfred Russel Wallace: Letters and Reminiscences.* New York: Harper and Brothers Publishers, 1916. Reprint, New York: Arno Press.

Mazumdar, Pauline M. H. 1992. *Eugenics, Human Genetics and Human Failings: The Eugenics Society, Its Source and Its Critics in Britain.* London: Routledge.

Moore, James, ed. 1989. *History, Humanity and Evolution: Essays for John C. Greene.* Cambridge: Cambridge University Press.

———. 1991. "Deconstructing Darwinism: The Politics of Evolution in the 1860s." *Journal of the History of Biology* 24:353-408.

Nitecki, Matthew H., ed. 1988. *Evolutionary Progress.* Chicago: University of Chicago Press.

Oldroyd, David, and Ian Langham, eds. 1983. *The Wider Domain of Evolutionary Thought.* Dordrecht: D. Reidel Publishing Company.

Oppenheim, Janet. 1985. *The Other World: Spiritualism and Psychical Research in England, 1850-1914.* Cambridge: Cambridge University Press.

Owen, Alex. 1990. *The Darkened Room: Women, Power and Spiritualism in Late Victorian England.* Philadelphia: University of Pennsylvania Press.

Paradis, James, and George C. Williams. 1989. *Evolution and Ethics: T. H. Huxley's 'Evolution and Ethics,' With New Essays on Its Victorian and Sociobiological Context.* Princeton, N.J.: Princeton University Press.

Peel, J. D. Y. 1975. "Herbert Spencer." In *Dictionary of Scientific Biography,* edited by Charles C. Gillispie, vol. 12. New York: Charles Scribner's Sons, 569-72.

Pittenger, Mark. 1993. *American Socialists and Evolutionary Thought, 1870-1920.* Madison: University of Wisconsin Press.

Porter, Theodore M. 1990. "Natural Science and Social Theory." In *Companion to the History of Modern Science,* edited by Robert C. Olby, G. N. Cantor, J. R. R. Christie, and M. J. S. Hodge. London: Routledge, 1024-43.

Richards, Evelleen. 1983. "Darwin and the Descent of Woman." In *The Wider Domain of Evolutionary Thought,* edited by David Oldroyd and Ian Langham. Dordrecht: D. Reidel Publishing Company, 57-111.

———. 1989a. "The 'Moral Anatomy' of Robert Knox: The Interplay between Biological and Social Thought in Victorian Scientific Naturalism." *Journal of the History of Biology* 22:373-436.

———. 1989b. "Huxley and Woman's Place in Science: The 'Woman Question' and the Control of Victorian Anthropology." In *History, Humanity and Evolution: Essays for John C. Greene,* edited by James Moore. Cambridge: Cambridge University Press, 253-84.

Richards, Robert J. 1987. *Darwin and the Emergence of Evolutionary Theories of Mind and Behavior.* Chicago: University of Chicago Press.

———. 1992. *The Meaning of Evolution: The Morphological Construction and Ideological Reconstruction of Darwin's Theory.* Chicago: University of Chicago Press.

Rupke, Nicolaas A. 1994. *Richard Owen: Victorian Naturalist.* New Haven, Conn.: Yale University Press.

Russett, Cynthia Eagle. 1989. *Sexual Science: The Victorian Construction of Womanhood.* Cambridge, Mass.: Harvard University Press.

Searle, G. R. 1976. *Eugenics and Politics in Britain, 1900-1914.* Leiden: Noordhoff International Publishing.

Smith, Roger. 1981. "Evolutionism in Mind and Society." In *Dictionary of the History of Science,* edited by W. F. Bynum, E. J. Browne, and Roy Porter. Princeton, N.J.: Princeton University Press, 133-35.

Spencer, Herbert. 1887. *The Factors of Organic Evolution.* London: Williams and Norgate.

———. 1893. "The Inadequacy of Natural Selection." *Contemporary Review* 63:153-66, 439-56.

Strauss, Sylvia. 1988. "Gender, Class, and Race in Utopia." In *Looking Backward: 1988-1888: Essays on Edward Bellamy,* ed. Daphne Patai. Amherst: University of Massachusetts Press.

Turner, Frank M. 1974. *Between Science and Religion: The Reaction to Scientific Naturalism in Late Victorian England.* New Haven, Conn.: Yale University Press.

———. 1993. *Contesting Cultural Authority: Essays in Victorian Intellectual Life.* Cambridge: Cambridge University Press.

Tyndall, John. 1874. *Address Delivered before the British Association, assembled at Belfast.* London: Longmans, Green.

Wallace, Alfred Russel. 1864. "The Origin of Human Races and the Antiquity of Man Deduced from the Theory of 'Natural Selection.'" *Journal of the Anthropological Society of London* 2:clviii–clxx.

———. 1870. "The Development of Human Races Under the Law of Natural Selection." In *Contributions to the Theory of Natural Selection: A Series of Essays.* London: Macmillan, 303–31.

———. 1871. "Rev. of The Descent of Man, and Selection in Relation to Sex, by Charles Darwin." *Academy* 2:177–83.

———. 1878. *Tropical Nature and Other Essays.* London: Macmillan and Company.

———. 1882. *Land Nationalisation: Its Necessity and its Aims.* London: Trubner and Company.

———. 1890. "Human Selection." *Fortnightly Review*, n.s., 48:325–37.

———. 1892a. "Human Progress: Past and Future." *Arena* 5:145–59.

———. 1892b. "Note on Natural Selection." *Natural Science* 1:749–50.

———. 1898. "Spiritualism and Social Duty." *Light* 18:334–36.

———. [1905] 1969. *My Life: A Record of Events and Opinions.* 2 vols. Westmead, England: Gregg International Publishers.

———. 1905. "If there were a Socialist Government—How should It Begin?" *Clarion* (London) 715:5a–f.

———. 1913a. *The Revolt of Democracy: With the Life Story of the Author by James Marchant.* London: Cassell and Company.

———. 1913b. *Social Environment and Moral Progress.* New York: Cassell and Company.

Yeo, Richard. 1993. *Defining Science: William Whewell, Natural Knowledge, and Public Debate in Early Victorian Britain.* Cambridge: Cambridge University Press.

Young, David. 1992. *The Discovery of Evolution.* Cambridge: Cambridge University Press.

Young, Robert M. 1971. "Evolutionary Biology and Ideology: Then and Now." *Science Studies* 1:177–206.

———. 1985. *Darwin's Metaphor: Nature's Place in Victorian Culture.* Cambridge: Cambridge University Press.

6

Redrawing the Boundaries: Darwinian Science and Victorian Women Intellectuals

EVELLEEN RICHARDS

In *The Descent of Man, and Selection in Relation to Sex,* his long-awaited work on human evolution of 1871, Charles Darwin wrote, "The chief distinction in the intellectual powers of the two sexes is shown by man's attaining to a higher eminence in whatever he takes up, than can woman— whether requiring deep thought, reason, or imagination, or merely the use of the senses or hands." Those aspects of intelligence conventionally attributed to women, such as intuition, rapid perception and imitation, Darwin dismissed as "characteristic of the lower races, and therefore of a past and lower state of civilization." For Darwin, the intellectual differences between the sexes, like their physical differences, were entirely predictable on the basis of a consideration of the long-continued action of natural and sexual selection aided by use inheritance. Male intelligence, he argued, would have been consistently sharpened through the struggle for possession of the females (sexual selection) and through hunting and other male activities such as the defense of the females and young (natural selection). "Thus," he concluded, "man has ultimately become superior to woman" (Darwin 1871, 2:326-29).

By 1871, Darwinism, in the capable hands of its leading popularizers and propagandists, the scientist Thomas Henry Huxley and the social theorist Herbert Spencer, was well on its way to becoming the new orthodoxy in Victorian science and society. Darwin's theory of evolution (first publicly presented in *The Origin of Species* in 1859) was accepted into the body of scientific knowledge in a period of extraordinary social and economic transformation, in which preindustrial modes of legitimation, religion in particular, were giving way to a secular, naturalistic redefinition of the world. In the process, the natural sciences increasingly took over from reli-

---

Extracts from the Huxley Papers are given by permission of the Archives, Imperial College, London. I would also like to thank the Syndics of Cambridge University Library for permission to reproduce quotations from the Charles Darwin Papers, and Bernie Lightman for his encouragement, sage advice, and patience.

gion the task of defining and upholding the moral and social order. Darwinism was central to this transition.

Darwin took the biological "struggle for existence," the basis of his theory of natural selection, from Malthusian social theory, and it has been compellingly argued that the image of nature presented in Darwin's work was contingent upon his own social context of mid-Victorian capitalist enterprise (Young 1985; Desmond and Moore 1991). Earlier evolutionary doctrines had been closely associated with political radicalism (Desmond 1989). Natural selection, with its emphasis on progress through competition and the elimination of the less well adapted, dissociated evolution from revolution and, at the same time, brought it into line with the competitive, free-trading ideals of the newly powerful industrialists and reform-oriented professionals who constituted a ready-made receptive audience for Darwin's views.

Huxley, in particular, capitalized on the opportunity thus provided to promote his claims for social progress through scientific advance. Darwinism was his lever for shifting power from an old, privileged, ecclesiastical elite to a new, technocratic elite of professional scientists whose authority to guide the conduct and organization of society rested in right reasoning and reliable natural knowledge, not mythical "truths." Throughout the sixties, he actively popularized and institutionalized a form of evolutionary naturalism that recruited support from a wide spectrum of society. In an increasingly secular and scientifically minded age, "progressives" of all kinds, including many feminists, rallied to a scientifically credentialed creed that its leading advocate overtly opposed to outmoded theological modes of explanation and linked with social and technological progress (Desmond 1994, 310–63).

But, in certain respects, the new Darwinism represented less a revolutionary break than an underlying continuity with the natural theology tradition it displaced. For all their differences, both doctrines were concerned to justify much the same set of underlying assumptions about economic and social relations, to preserve the status quo (Young 1985). In a context of imperial expansionism, economic uncertainty, urban and industrial unrest, the emergence of mass socialist working-class movements all over Europe, and the increasing urgency of the demands by women for the suffrage, higher education, and entrance to middle-class professions, the origin of "man" by natural law rather than divine creation was made more palatable for its Victorian audience by Darwinian concepts of "natural" and inevitable white, middle-class male supremacy.

Huxley led the way with his widely read "Emancipation—Black and White" of 1865. Here he steered a carefully calculated middle course, advocating votes and education for both women and blacks but invoking the Darwinian natural law of fair competition and no favors to reassure their oppressors and his threatened fellow professionals that "Nature's old salique law will not be repealed, and no change of dynasty will be effected."

Women, like blacks, were the natural inferiors of white men and would remain so. Not "even the most skilfully conducted process of educational selection," Huxley asserted, could remove the "physical disabilities under which women have hitherto laboured in the struggle for existence with men" (Huxley [1865] 1968, 74).

With the publication of *The Descent of Man*, Darwin put his imprimatur on such evolutionary ratification of Victorian values. His reconstruction of human evolution is pervaded by Victorian racial and sexual stereotypes and assumptions of the inevitability and rightness of the sexual division of labor. By asserting the instinctively maternal and inherently modest traits of the human female and the male's innate aggressive and competitive characteristics, Darwin provided naturalistic corroboration of woman's narrow domestic role and contemporary social inequalities (Richards 1983; Rosser and Hogsett 1984; Jann 1994). Following this, there was scarcely an evolutionist who did not take up and pronounce upon the woman question.

The more conservative, even reactionary, position was hammered out by Spencer, chief architect of "Social Darwinism." Spencer opposed the extension of the franchise and higher education to women primarily on the grounds that they were less highly evolved than men and constitutionally less fit to handle political or social and professional responsibilities. Other prominent Darwinians such as George John Romanes, Francis Galton, and Patrick Geddes joined forces with anthropologists, psychologists, and gynecologists to forge a formidable body of biological determinist theory that purported to show that women were inherently different from men in their anatomy, physiology, temperament, and intellect—that women, like the "lower" races, could never expect to match the intellectual or cultural achievements of men or obtain an equal share of power and authority. Victorian science (and evolutionary science in particular), as feminist scholars have documented, was strongly gendered (Conway 1970; Fee 1974; Rosenberg 1975; Mosedale 1978; Russett 1989).

The response by the women concerned to counter the concerted Darwinian reinforcement of traditional views of their social and cultural roles has also begun to be charted (Alaya 1977; Love 1983; Tedesco 1984; Egan 1989; Erskine 1995). But few studies that locate individual Victorian women in precise relation to the institutional and wider sociopolitical contexts of Darwinian science and its practitioners have been undertaken (Richards 1989). By unpacking specific instances of the engagements of particular representative women with Darwinism and its institutions, we may better our understanding of the ways in which the discursive categories of gender, sexuality, and science were constructed and contested by some of the scientists and women concerned and how that discourse was rooted in a historically specific set of ideas and practices about gender, sexuality, and science (Outram 1987; Haraway 1989; Hall 1992).

In keeping with such contextual, comparative approaches, I want here to examine the contrasting responses of two Victorian women intellectuals, Eliza Lynn Linton and Frances Power Cobbe, to Darwinian science and its institutions. My two case studies are intended to uncover something of the diversity and complexity of Victorian feminism and the contradictions inherent in its general reliance on Victorian stereotypes of femininity upon which Victorian science was also contingent. They also offer a means of exploring the various strategies adopted by the dominant Darwinians in redrawing the boundaries of organized science against the incursions by women into this most masculine profession.

The women I have chosen were born in the same year—1822. Both became self-supporting writers with a keen interest in science and the related social and political issues of their time. Both were political conservatives who fully endorsed the Victorian conventions of womanhood. But there the similarities end.

Eliza Lynn Linton (1822-98) was a Victorian paradox, an "emancipated woman opposed to women's emancipation" (Anderson 1986, x). Successful journalist and ardent Darwinian, Lynn Linton represents the extreme pole of the biological determinist position on the woman question as advocated by the Darwinians. She campaigned vehemently against women's higher education, birth control, suffrage, and entry into the professions, largely on Darwinian grounds. But she also confronted Huxley over the exclusion of women from the Ethnological Society—a confrontation that brings to the fore all the contradictions of her position as a woman and an evolutionist in Victorian society and exposes Huxley's own manipulations of the woman question in pursuit of his interrelated goals of the professionalization of science and the Darwinian control of anthropology.

Frances Power Cobbe (1822-1904) represents another response to Darwinism—the theistic alternative that accepted the evolution of the body but not of the mind. Cobbe was well known in middle-class circles as a leading advocate of women's rights and an antivivisectionist. Her antivivisection crusade brought her into conflict with the professional and social aspirations of the Darwinians, notably Huxley and Darwin, who strongly defended the right of the scientist to animal experimentation. Their conflict also illustrates the ways in which women like Cobbe tested and extended the limits of the sphere of femininity and constructed political identities for themselves on a terrain different from that of the scientists.

I. Eliza Lynn Linton and the Masculine "New World" of Darwinized Science

> Darwin opened a new world to me. . . . The Unity of Nature was the core of the creed to which I owe my subsequent mental progress—the Doctrine of Evolution that by which I have come to peace.
>
> (Linton [1885] 1976, 3:79)

The young Eliza Lynn was a poorly educated, strong-willed, but sensitive girl, who in the 1840s went to London from the obscurity of a country vicarage and in defiance of a patriarchal father to become the first salaried professional woman journalist in Britain. An intense, bookish woman, Eliza moved in the circles of the radical intelligentsia. In 1858, when she was thirty-six years old, she married the radical artisan William Linton. Linton was an engraver of considerable artistry who subordinated his talents to the active promotion of his political views. He was prominent in the national Chartist movement, but in some respects, particularly in his views on education, the liberating powers of science, and female emancipation, Linton shows the influence of the Owenite socialists (Smith 1973; Taylor 1983).

When Eliza first met him, he was living with the consumptive Emily Wade and their seven children, all of whom, girls and boys alike, were dressed in long blue flannel blouses, with shoulder-length hair and identical broad-brimmed hats. Initially, Eliza was charmed by Linton, his republicanism, his "moral purity," and his strange "Bohemian" household. She took over the family, helping the impractical Linton and Emily financially and imposing a certain middle-class order on the chaotic household. The children's diet, table manners, hair length, and accents were reformed to Eliza's exacting standards. After Emily's death, she decided to "legalize" her position with the children by marrying Linton (Anderson 1987, 72–81).

All her life, Eliza Lynn Linton was riven by a contradiction she never managed to resolve and which is reflected in all her writings on the woman question. She earned her own living, associated with leading radicals and intellectuals, and lived the life of an independent, strong woman, yet she clearly yearned for middle-class respectability and the more conventional Victorian role and rewards of wife and mother. Anderson's recent psychological portrait of Lynn Linton makes a strong case for her lifelong inability to transcend the difficulties of her formative years, her subsequent conflicted self-hatred, and the intense male identification that fueled her criticisms of female character and women's rights (Anderson 1987). Anderson, however, fails to recognize the extent to which Eliza's Darwinism sustained these tensions in her personal life and her relation to Victorian feminism.

Eliza's attempt to acquire a ready-made family and live out her idealized Victorian role of wife and mother was an abysmal failure. Her choice of husbands is an instance of the contradiction that dominated her life. She married a committed radical activist and tried to turn him into a conventional Victorian husband, a "capable and successful *doer*" (Anderson 1987, 80). Her failure to achieve this precipitated one of those Victorian crises of faith so characteristic of the period.

By 1865 she and Linton had separated, and the intelligent, hard-working writer could not accept the public humiliation she attached to her own violation of her socially assigned womanly role. After a period of despair, she found spiritual and social redemption in the certainties of science and the scientific meetings she eagerly attended at every opportunity:

> Those Friday Evening Lectures at the Royal Institution, when Tyndall experimented or Huxley demonstrated . . . what evenings in the Court of Paradise those were! How I pitied the poor wretches who did not come to them! . . . I do not think there was one in the whole audience who drank in the wine of scientific thought with more avidity than I. . . . It strengthened, warmed, exhilarated and almost intoxicated me. (Linton 1976, 3:83-84)

Eliza's new creed was the doctrine of scientific naturalism: "In science were FACTS, and these were of the kind to make a new mental era—a new departure of thought for the whole world, as well as for myself individually." In the "substitution of the scientific method for the theological," she saw the emancipation of the human intellect from superstition, and she pinned her faith in human progress and her own moral redemption on the new Darwinism (Linton 1976, 3:79-81).

Eliza's association with Linton and the radicals probably prepared the ground for her ready conversion to Darwinism. Transmutationism was widely popular among radical artisans who believed that it served their republican and materialist platform (Desmond 1989). Around the time of her marriage to Linton, Eliza favorably reviewed the pre-Darwinian arguments of Robert Chambers and Herbert Spencer for the "progressive improvement" of life and society (Linton 1858). Spencer's later role as foremost Social Darwinist was to defuse the revolutionary appeal of transmutationism with the more socially acceptable mechanism of continuous social progress through elimination of the "unfit." The regeneration of society was now guaranteed by the "fixed laws" of capitalist competition, and here Eliza found a substitute for William Linton's republican idealism that was more consistent with her growing political and social conservatism. Further, she palliated her unconventional agnosticism and materialism and her self-perceived anomalous social situation with an obsessive outward conformity with Victorian prudery and propriety. All this found expression in the famous series of articles she published in the *Saturday Review* in early 1868, which became known collectively by the title of one of them: "The Girl of the Period."

With attention-catching titles and vivid prose, she vituperatively attacked and caricatured just about everything nineteenth-century feminism represented in articles such as "The Girl of the Period" ("a creature who dyes her hair and paints her face . . . who lives to please herself . . . bold in bearing . . . masculine in mind"), "Modern Mothers" ("this wild revolt against nature, and specially this abhorrence of maternity"), "What Is Woman's Work?" ("professions are undertaken and careers invaded which were formerly held sacred to men; while things are left undone which, for all the generations that the world has lasted, have been naturally and instinctively assigned to women to do"), "Wild Women," "Modern Man Haters," and so on (Linton 1883).

Eliza's essays caused a sensation and ensured her professional success. They inspired cartoons, fashions in clothing, a satirical journal, *The Girl of the Period Miscellany*, and several other publications (Anderson 1987, 120-25). The "girl of the period," or "GOP," became fixed in the Victorian vocabulary as a catchphrase or acronym for a "modern" or "fast" girl. Against this unnatural hussy of her creation, Eliza held up the ideal of the inherently modest, domestically oriented girl who "when she married, would be her husband's friend and companion, but never his rival; one who would consider his interests as identical with her own . . . who would make his house his true home and place of rest" (Linton 1883, 1). The fact that her own practice of these feminine virtues had driven Linton from the marital home was beside the point, and the contradiction between this ideal and her own circumstances was generally ignored.

Obviously, Eliza's reassertion of traditional Victorian values was highly marketable in a context of middle-class antipathy toward the threatening economic and political independence of women. But it would be simplistic to dismiss her as a gifted writer, onetime radical and emancipated woman, who sold out to a reactionary antifeminism through personal disappointment and for professional gain. Despite her exaggerated concern for the proprieties and her diatribes against the "shrieking sisterhood," Eliza consistently held to three issues she regarded as the "core of this question of woman's rights." They were women's right to an education "as good as . . . but not identical with, that of men" (this, she thought, should include some science education), their right to property, and their right to divorce and custody of their children. These "rights" were "just and reasonable" and, above all, did not conflict with Eliza's insistence on the "natural limitation of sphere . . . included in the fact of sex" (Linton 1870, 224-38; Linton 1976, 3:2-4).

Like the Darwinians, Eliza assumed naturalistic limits to women's aspirations and based these firmly within a traditional rendering of Victorian femininity. Her insistence on the essential domesticity and modesty of women was grounded in her unshakable materialism and her Darwinism. Women, she held, could no more emancipate themselves from the laws of biology than the earth could free itself from the law of gravitation, and it was this rigid scientific certainty that underpinned and undermined her stance on the woman question. All the contradictions of that stance—personal, professional, and scientific—are manifest in her eight-page petition to Huxley over the issue of the exclusion of women from the meetings of the Ethnological Society in 1868 (Huxley Papers 21.223-26).

Three years earlier, Huxley had publicly declared himself on the "irrepressible" woman question by asserting his Darwinian certainty that women would remain the natural inferiors of men. Nevertheless, he had argued, it was the liberal man's duty to see that "not a grain is piled upon that load beyond what Nature imposes; that injustice is not added to inequality" (Huxley 1968, 74).

Huxley's "Emancipation—Black and White" of 1865 served a number of purposes, but it was aimed primarily at the rival and rabidly racist Anthropological Society, which had broken away from the Ethnological Society in 1863 over the ostensible issue of the admission of women. The Anthropologicals quickly built up a large, enthusiastic, and exclusively masculine membership devoted to the dissemination of antifeminist and racist propaganda in the guise of physical anthropology. Their anatomical method of describing, measuring, and classifying racial and gender differences allegedly proved the natural inferiority of women and blacks. John Stuart Mill's claim for black and female suffrage was therefore a scientific absurdity, contradicted by the "facts of human nature" as revealed by the researches of the anthropologist. The Anthropologicals endorsed slavery and the more racist manifestations of British imperialism. Primarily medical men who felt themselves particularly threatened by the professional aspirations of middle-class feminists, they also vented their spleen against those unnatural women who sought to deny their natural mission of motherhood and make themselves ridiculous by "meddling in and muddling men's work." Professedly anti-Darwinian, the Anthropologicals even on occasion adapted evolutionary rhetoric to their antifeminist stance: Women possessed less than men of that "combativeness which is necessary not only in political life, but even in the ordinary struggles for existence." Woman's subordination to man was "natural and eternal," and any attempt to "revolutionize the education and *status* of woman on the assumption of an imaginary sexual equality" would induce a "perturbation in the evolution of the races" (Richards 1989, 261–70).

There was little difference between such arguments and Huxley's denial to women of any natural equality, existing or potential. His extension to women of their right to legal and political emancipation was offered on the understanding that they would not be able to overcome their biological limitations and compete with men on equal terms. Huxley was as certain as any Anthropological of the crucial cerebral differences between men and women and ranked women's intelligence with that of the "lower races." He was also forcefully opposed to the admission of women to scientific societies. For all his liberal rhetoric, Huxley's personal views of women were remarkably consistent with the publicly expressed opinions of the GOP author. With few exceptions, he viewed women as mostly frail, religious creatures, stuck at the "doll stage of evolution." To the careerist Huxley, women were ipso facto amateurs, fit for the classroom but utterly out of place in the cut and thrust of professional scientific forums, where their amateur presence threatened that Darwinian expertise and status to which he was so committed (Huxley 1900, 1:211-12; Richards 1989, 225-61; Desmond 1994, 310-63).

In 1865 the Anthropologicals were in the ascendant over the Darwinians who had colonized the moribund Ethnological Society and were vying with the Anthropologicals for control of the strategically significant science of "man." "Emancipation—Black and White," with its prohibition on denying their liberal rights to blacks and women on anatomical or any other grounds, was Huxley's attempt to refuse scientific authority to the Anthropologicals while asserting it for the Darwinians through the subordination of black and female equality to the inescapable struggle for existence. When the unruly Anthropologicals proved recalcitrant to Darwinian control, Huxley's strategy became one of amalgamation of the two societies under the "proper direction" of the Darwinians. One of the tactics he deployed as its newly elected president was to initiate the exclusion of women from Ethnological Society meetings (Richards 1989, 267-70).

This, then, was the context in which Eliza Lynn Linton, who, through her interest in human evolution and need for communion with "clever men," had become an assiduous attender of Ethnological Society meetings, was forced to step outside her paid professional role of deriding and attacking the "girl of the period" to plead passionately on behalf of her right to a better education and opportunities. "You know how few opportunities we women have for getting any serious or valuable talk with men," she told Huxley in 1868 (Huxley Papers 21.223-26).

> We meet you in "Society" with crowds of friends about & in an atmosphere of finery & artificiality. Suppose I, or any woman—let her be as fascinating as possible—were to bombard you with scientific talk—would you not rather go off to the stupidest little girl who had not a thought above her pretty frock, than begin a discussion on the Origin of Species? (Huxley Papers 21.223-26)

If women were not to talk of science in society, and were excluded from scientific societies, how were they to learn about science? There were very few scientific meetings open to women, and it was not easy to obtain the favor of an invitation to these more popular and fashionable events—"I have been [to the Royal Institution] only thrice in my life."

In paraphrase of Huxley's own "Emancipation—Black and White," Eliza sought to remind him of his liberal Darwinian obligations:

> What are the facts of woman's personal condition? We are thrown into an active hand to hand struggle for existence all the same as men—we of the middle classes have to earn our own bread—with very badly trained hands & brains it must be sorrowfully confessed. . . . The battle of life is a very serious matter to some of us, and we are frequently hindered and heavily weighted. . . . It is not fair to exclude us from the means of knowledge & of active thought,

> of extended views—such as we get from attending learned discussions—on the simple plea of our womanhood. (Huxley Papers 21.223-26)

Although Eliza made her case according to her own precepts, with proper regard for the proprieties and without shrieking, in the form of a personal letter to Huxley, even in (unconscious?) parody of feminine intellectual incompetence, "muddling up" her reasons ("like a woman!"), her powerful but womanly plea did not meet with the anticipated fair treatment from the ruthless Huxley. He and the Darwinian-dominated council came up with the ingenious compromise of demarcating "Ordinary Meetings," which would be for "scientific" discussions to which "ladies will not be admitted," from larger, popular "Special Meetings," to which "ladies" might be admitted "by special invitation" (Richards 1989, 275).

With one timely stroke, this admirable (and typically Huxleyan) solution reconstituted the Ethnological Society as a "gentlemen's society" and paid lip service to the liberal principle of female admission. All must have been well satisfied except Eliza (and those she represented), who was now inexorably relegated to the more frivolous "popular element" she deplored and exiled from the serious scientific discussions she craved. But, as a leading public advocate of the "separate spheres" ideology, she was hardly in a position to complain.

The exclusion of women served Huxley's purposes: it upgraded the professional status of the Ethnologicals and it removed one of the major impediments to their amalgamation with the Anthropologicals, which he achieved in 1871, the same year in which Darwin's *Descent of Man* consolidated the Darwinian endorsement of many aspects of the Anthropologicals' platform. Darwin was as insistent as any Anthropological (and Huxley) on the biological basis of the continuing intellectual inferiority of women and blacks, and as much opposed to Mill's environmentalist explanations. Above all, he followed Huxley's lead, by arguing on evolutionary grounds that the higher education of women could have no long-term impact on their social evolution and was, strictly speaking, a waste of resources (Darwin 1871, 2:326-29; Richards 1983; 1989, 276).

II. Frances Power Cobbe, the "Duties of Women," and the Dissent from Darwinism

> To those amongst us who have not bowed to the new moral system of Darwin and Spencer, there is something almost pathetic in the ignorance both of the passions and of the spiritual part of human nature which these philosophers unconsciously betray.
>
> (COBBE 1881a, 70 n)

Frances Power Cobbe provides an illuminating contrast to Eliza Lynn Linton. She believed absolutely that women were more chaste, generous, and moral than men and was adamant that they must nurture their offspring, look after the home, and not succumb to selfishness and loose, "Bohemian" manners and morals (Cobbe 1881*a,* iv). But unlike Lynn Linton, Cobbe did not dissociate herself from the professional and other liberal goals of the women's movement. Asserting that the "cause of the emancipation of women is identical with that of the purification of society," Cobbe actively sought the extension of woman's existing social role and a proper recognition of its importance (Cobbe 1881*a,* 11; Caine 1992, 103-49).

Cobbe also experienced the prevalent crisis of faith and was an early convert to evolutionary naturalism. But after a short period of agnosticism, the intolerable sense of severance caused by the death of her much-loved mother persuaded her to opt for a form of theism based on the idea of a just and rational God whose moral law was evident to all people through their own intuition, not through revelation. It also provided a basis for her rebellion against the moral authority of her domineering father, although she continued to serve him as a dutiful daughter. It was her theism, her absolute belief in the moral autonomy of women, and her strong sense of their mental and moral difference from men that constituted the core of Cobbe's feminism (Caine 1992, 115-19, 131-32).

While she paid lip service to the Victorian imperative of marriage for women, Cobbe was highly critical of the institution. She inveighed vigorously against the married woman's loss of legal identity, of property or earnings, and the domestic tyranny and misery to which so many were subjected (Cobbe 1862, 1863). She never married and, after her father's death, traveled widely, pursued an independent, hardworking writing vocation, and lived in domestic harmony and comfort for thirty-four years with Mary Lloyd, a painter. It seems to have been one of those Victorian "female marriages" that provided affection and commitment for its participants without any necessary sexual involvement (Caine 1992, 120-25).

Cobbe's attitude toward Darwinism was shaped by her feminism and her theism. She knew Darwin personally and, initially at any rate, greatly admired him. In her autobiography she recounts how, on encountering Darwin while out walking, they held a shouted discussion across a bramble patch about the significance of Mill's views on women for Darwin's forthcoming *Descent of Man.* Mill, Darwin asserted, could learn some things from science. Women's nature, like men's, was rooted in their biology, and it was through the "struggle for existence and (especially) for the possession of women that men acquire their vigor and courage" (Cobbe 1894, 2:124-25). Cobbe got the opportunity to set him right when she reviewed Darwin's *Descent.* She had no theological difficulties with tracing "Man to the Ape." But, while she did not dissent from Darwin's views on the evolution of the physical differences between men and women, Cobbe forcefully opposed his "Simious Theory of Morals," his "most dangerous" utilitarian interpretation of the evolution of the human mind and morality from animal instincts (Cobbe 1872, 14; 1894, 2:127). To the "Atheistic Morals" and materialism of the evolutionists, she opposed her doctrine of "Theistic Ethics," which asserted the existence of free will and the immortality of the soul, championed love over knowledge, and stressed the special duties of women to "those who have no free-will—the lower animals." This neatly removed the mental and moral differences between men and women from the biological to the spiritual domain and so, in Cobbe's view, guaranteed the moral autonomy and authority of women (Cobbe 1881a, 17, 67).

The realpolitik as far as Cobbe was concerned was less the campaigns for the suffrage and higher education than her antivivisectionist crusade. It was into this highly controversial movement that she channeled most of her abundant energy, intellect, and consummate political skills, and it was her passionate involvement in this cause that brought her into direct conflict with the Darwinians, Huxley and Darwin in particular.

Cobbe publicly became involved in the antivivisection movement in 1875 when she circulated a memorandum urging the Royal Society for Prevention of Cruelty to Animals to mobilize to restrict the practice of experimentation on live animals. Among the many eminent persons she approached was Darwin. Darwin refused to sign Miss Cobbe's "foolish paper" but reacted with alarm to the "many powerful names" who had indicated their support for her proposal. He alerted Huxley to the need for action lest the House of Commons, "being thoroughly non-scientific," should pass some "stringent law, enough to check or quite stop the revival of physiology in this country" (Darwin Archive 97:37.8). Huxley backed Darwin's suggestion of a counterproposal from "eminent physiologists and biologists" for "reasonable" legislation "as the best method of taking the wind out of the enemy's sails." He added for good measure, "My reliance as against that 'foolish fat scullion' & her fanatical following is not in the wisdom and justice of the House of Commons, but in the large number of foxhunters therein" (Darwin Archive 166:338).

For Darwin and Huxley, antivivisectionists were the "enemy" who threatened the progress and prestige of British science. They were characterized by their foolishness (i.e., irrationality) and their fanaticism (i.e., emotionalism), and they were also, as Darwin and Huxley well understood, female. Not only were they led by the redoubtable Cobbe, who came to personify the antivivisection movement, but, as Darwin put it, it was women "who from the tenderness of their hearts and from their profound ignorance" were the "most vehement opponents" of vivisection (Darwin 1994, letter 10546). On the home front, Darwin found himself called to account for his provivisectionist stance by his daughter Henrietta (Darwin 1888, 3:202–3). When the physiologist Romanes came to visit, Darwin warned him not to talk about experiments on animals "when in presence of my ladies" (Darwin 1994, letter 9916). Women, then, were the enemy of the progress of rational science, and they were so because of those very feminine traits that otherwise made them such desirable and conforming wives and daughters. As antivivisectionists they were doubly subversive, trading on their femininity to unman the scientist on his own ground.

The commitment of so many women to the antivivisection movement is another illustration, like that of spiritualism, of how concepts of femininity and moral superiority could be used to legitimate a range of public and quasi-public activity not usually associated with the traditional female role (French 1975, 240-41; Elston 1987; Owen 1990). Women like Cobbe could thus adopt a leadership role and lay claim to special skills and knowledge without doing damage to those qualities that constituted the Victorian ideology of femininity. At the same time, paradoxically, their involvement in the antivivisection movement provided an opportunity for women to use their very femininity to achieve and wield power, especially over male scientists and doctors, to subvert female subservience with the conventional feminine tools of sentimentality and womanly concern for suffering. But this interpretation does not account for the extraordinary identification of Cobbe and many other Victorian women with animals and their sufferings, particularly at the hands of medical scientists.

Lansbury's intriguing thesis is that for these women the vivisected animal was the surrogate of woman, humiliated, exposed, and threatened by the gynecologist's knife and by the pornographer's whip, stirrups, and other paraphernalia taken from the stables and kennels. She argues that the full impact of the antivivisection movement on Victorian culture cannot be understood without recognition of the fusion of imagery from three major areas: gynecological, pornographic, and literary. The pornographic literature of the period dwelt repetitively on the ritual bestialization of women. Women were "broken to the bit," "mounted," made to "show their paces," collared, chained, bound, flogged, and seduced into grateful submission to their "masters." Lansbury points to the "uneasy similarity" between the devices made to hold women for sexual pleasure in such male fantasies and the gynecological table and "stirrups" that came into general use around 1860. Women doctors like Elizabeth Blackwell and Anna Kingsford, both committed antivivisectionists, deplored the "degrading cruelty" with which poor women were treated in the major hospitals of the day. Antivivisection literature routinely conflated the plight of such women, who allegedly were made the victims of cruel and unnecessary experimental surgery, with that of the dogs, cats, and monkeys who were the piteous, defenseless objects of the merciless vivisector. The symbols evoked by women antivivisectionists were "all the more potent because they were drawn from a muffled context of reticence and ambiguity" (Lansbury 1985; see also Moscucci 1990, 112-27; Elston 1987).

Lansbury's thesis is given greater plausibility by Ritvo's explorations of the political and social resonances of Victorian relations with animals. Their relation to dogs, for example, exemplified many of the tensions in Victorian class and sexual ideologies. Dogs, like lower-class women and prostitutes, possessed dangerous sexualities that necessitated control and were a potential source of contagion to middle-class humans. The identification of woman's sexuality and nature with dogs and other domestic animals was made most explicit in the discourse of breeders. When, for instance, they discussed the difficulty of getting a prize bitch to mate exclusively with a selected male, their discussion was transparent to assumptions about the sexuality of human females (Ritvo 1987, 3-4, 180-86; 1988). The dog, above all, signified the loved but subservient being that thoroughly understood and accepted its inferior position. Even its body proclaimed its profound submission to humanity. It was the most malleable of all man's domestic productions, its shape and size responding most readily to the caprice of the breeders (Ritvo 1987, 20-23).

There is more than a degree of coincidence between these traits associated with the domesticated dog and those accorded to women within the terms of Victorian domestic ideology. Women were to serve, to obey, to be pliant to masculine whim and will, to stay where men commanded or follow where they led. The young Darwin made the direct connection: marriage was analogous to pet keeping; a wife was an "object to be beloved & played with.—better than a dog anyhow" (Darwin [1838] 1986, 2:444). Darwin's semifacetious musings were not innocent. The stereotyping of women as domestic animals was deeply entrenched in Victorian culture.

Darwin's immersion in the literature of the breeders guaranteed his full exposure to such metaphorical discourse. In any case, his argument for the evolution of human mind and morality by the same natural agencies as the struggle for existence and for mates was dependent on breaking down the traditional theological distinctions between animal and human mentality. The pages of *The Descent of Man* bristle with anthropomorphic descriptions of animals, with loyal dogs and brave monkeys, proud peacocks, coy bitches, aggressive, promiscuous stags and cocks—tropes that when analogically reapplied to human behavior and social institutions provided naturalistic corroboration of Victorian values. His concept of sexual selection, largely dependent on his studies of the observations and activities of contemporary animal breeders, was inescapably anthropomorphic, transferring Victorian social values and stereotypes back onto his conception of human biological and social evolution (Richards 1983).

When the aesthetic choice he attributed to female animals could not be made to fit this proper Victorian's conception of the submissive sexuality and inferior intelligence of human females, Darwin simply overturned it and put into men's hands the modifying and shaping power of human sexual selection. Man, he claimed, being "more powerful in body and mind," had seized the power of selection from woman. The differing standards of beauty of the various races offered the explanation, via male aesthetic preferences, of racial and sexual differentiation. "Monstrous" as it might seem, Darwin was convinced that the "jet-blackness of the negro" had been gained through the process of male selection, just as had the supposedly more pleasing secondary sexual characteristics of European women—sweeter voices, long tresses, and greater beauty (Darwin 1871, 2:368-84). For Darwin, the human male was the analogue of the animal breeder who exercised his caprice in varying the appearance of the breed, and woman's body, like the dog's, was pliant to male manipulation (Richards 1983, 76-79).

Talking about animals, therefore, offered those like Darwin who "would have been reluctant or unable to avow a project of domination directly a way to enact it obliquely" (Ritvo 1987, 6). The other side of the coin is Lansbury's claim that for women, the subjugated, to protest against vivisection "was to challenge a world of male sexual authority and obscenity which they sensed unconsciously, even if they had no direct experience of it" (Lansbury 1985, 422).

For Cobbe, the parallels were obvious. Women, like animals, were subject to the power of doctors and scientists and to the brutality of many men. Her powerful article "Wife Torture in England" (1878) was written on the crest of her involvement in the antivivisection campaign. It was an indictment of the endemic domestic violence that Cobbe recognized as not being confined to the working class; of the culture that expected female service in the home, that condoned and even drew entertainment from wife beating; and of the system in which women lacked legal and political rights and were regarded as the property of their husbands (Cobbe 1878*a;* Caine 1992, 135-38). At the same time Cobbe attacked the arrogance and cruelty of the male medical profession, which she saw as exerting an increasingly oppressive control over women's lives, turning healthy women into a "whole sex of Patients" constantly liable to illness and dependent on the incompetent and callous attentions of doctors (Cobbe 1878*b*). Doctors were "doubly treacherous" to women, a reference to the connection Cobbe made between the way doctors treated women as patients and their fight to exclude women from the profession of medicine (Cobbe 1881*b,* 325).

Doctors and medical scientists were of course her targets in the antivivisection campaign, and she believed that those who engaged in vivisection were brutalized by the suffering they inflicted so that, like those husbands who beat and assaulted their wives, they came to find excitement and even pleasure in the infliction of pain. Medicine, like marriage, exacerbated women's oppression. The unspeakable ambivalences of sexuality and cruelty that Cobbe evoked in her denunciations of vivisectors, medical men, and wife torturers found expression in the outrage and distress that many women, inhibited from articulating them on their own behalf, evinced on behalf of vivisected animals. The extreme behavior, the "emotionalism, floods of tears, and fainting," that characterized antivivisection meetings is legendary (French 1975, 248; Lansbury 1985).

How the sadosexual associations of vivisection meshed with the reactions of Darwin and Huxley to Cobbe and her "fanatical following" is not easy to judge. It is unlikely that even these highly respectable family men were unaware of the pervasive masculine world of Victorian pornography. However, even if they perceived it, it is inconceivable that either Huxley or Darwin would have flouted Victorian convention to contest openly the darker sexual imagery conjured up by Cobbe's allusions. Their overt reaction to the emotionalism of the antivivisectionists was to deride and dismiss it as so much female hysteria and irrationality. Darwin thought the antivivisectionists to be "half mad" (Darwin 1888, 3:210), while Huxley ridiculed their "fanaticism of philozoic sentiment" and contrasted it with the "rational" basis of experimental physiology (Huxley, 1900, 1:434).

The ostensible issue for both men was the progress of British science. The reiterated danger was that the attacks of the antivivisectionists would undermine the already precarious status of the new science of experimental physiology. This was the issue over which Darwin was prepared to go public and, for the first and only time in his long career, engage in direct political action. It was an issue that concerned Huxley, the professional scientist, even more. While he was busily promoting the social standing and moral responsibility of medical scientists, one of Cobbe's tactics was to downgrade their status and morality. Medical men, the upper-class Cobbe alleged, were not gentlemen; they tended to come from the lesser ranks of society and were a "parvenu profession, with the merits and the defects of the class." Their social origins explained their defective morality, their trade unionism, and the sordid materialism that pervaded their ranks and made them unwontedly ambitious, motivated by monetary gain, and careless of suffering (Cobbe 1881*b*). Huxley strongly contested such charges, while Darwin was greatly offended by Cobbe's "monstrous" attribution of sadistic pleasure in animal suffering to leading scientists of the day (Darwin 1888, 3:200–203; Huxley 1900, 1:427–34). A cruel scientist was an anomaly, a contradiction of the gentlemanly image and high moral standards they claimed for British science and its practitioners. Cobbe's accusations of scientific cruelty and assumption of the higher moral ground on behalf of women antivivisectionists were, therefore, in direct conflict with Huxley's professionalization strategy and his promotion of the scientist as the appropriate moral arbiter of important social questions.

For almost twenty years, from her headquarters of the Victoria Street Society, Cobbe fought a sustained but inevitably losing campaign against the growing power and authority of a science and medicine that, in her view, oppressed both women and animals. She came to abhor the "priest-like arrogance of some representatives of the modern scientific spirit." Her disillusion was complete when the "great naturalist who has revolutionized modern science" became the center of an "adoring *clique* of vivisectors." She clashed with Darwin in the *Times* and pursued him through the pages of her journal, the *Zoophilist* (Cobbe 1894, 2:123-29, 269-70, 408-9; 1881c, 17-19; Darwin 1888, 3:205-8). Darwin, for his part, thought it only "fair to bear [his] share of the abuse poured in so atrocious a manner on all physiologists" by Miss Cobbe. He helped initiate the Science Defence Association in 1881, gave it financial support, and even briefly considered accepting its presidency (Darwin 1888, 3:206-10; Darwin 1903, 2:437-41). This became the highly influential Association for the Advancement of Medicine by Research, a powerful coalition of leading British biologists and medical men who successfully lobbied behind the scenes on behalf of experimental medicine against the interventions of the antivivisectionists (French 1975, 200-219).

In the end, Cobbe was brought down by that very emphasis on feminine moral superiority that underpinned her feminism and antivivisectionism. In 1892, she was made personally responsible for the willful distortions of a major publication compiled by her Victoria Street Society that failed to acknowledge the routine use of anesthetics in the "brutal" animal experiments it otherwise quoted directly from the research literature. The medical and lay press gloated over this evidence of moral fallibility from one who had "assumed a superior morality, a higher scientific knowledge, and a pontifical right to anathematise medicine and all her most honoured followers throughout the world." Was Cobbe's perversion of truth now to be discounted as the "privileges of womanhood"? Was this where the vaunted superior moral sense of women led (Hart 1892, 710-11)?

The media assault seriously damaged Cobbe's credibility and that of her movement, and had more general repercussions for the participation of women in science and public affairs (French 1975, 249-50; Cobbe 1894, 2:306-11). *Punch,* that arbiter of establishment opinion, summed up the sexual and scientific politics of Cobbe's public humiliation with the adjoining cartoon, which says it all (figure 6.1).

III. Conclusion

Lynn Linton, Cobbe, and the Darwinians all drew the qualities they attributed to women from the same model of femininity, an illustration of the extent to which evolutionary science and feminism were both bound by the ideology of their time and place. This inevitably set the parameters of their debate on women's condition and future prospects. But what is striking is the variety of ways in which the individuals concerned negotiated or reworked the assumption of sexual difference for different ends: the Darwinians to assert their social and professional hegemony, Lynn Linton in

Figure 6.1 *Punch's* comment on the sexual and scientific politics of Cobbe's antivivisection crusade. (*Punch* 103 [5 November 1892]: 205.)

biological determinist opposition to feminist aims but also to argue the necessity of female participation in scientific societies, and Cobbe to promote female agency through moral superiority. Cobbe rejected evolutionary justification of women's subordination, while the Darwinian Lynn Linton endorsed it in contradiction to her own situation as a woman intellectual in Victorian England. Cobbe's theistic ideology was more compatible with the leadership role she assumed for women in the antivivisection campaign, but it could not be sustained in a context of religious decline and the growing authority and prestige of a science geared to the needs of a capitalist economy and the gendered nature of the public sphere. In the late Victorian period it was the Darwinians who articulated the dominant constructions of femininity, sexuality, and science and naturalized the barriers against feminine intellectual and social equality in order to protect Darwinian institutional and social interests against the threat posed by the burgeoning women's movement.

The Darwinian redrawing of traditional boundaries was made all the more devastating by the problem that many feminists themselves were deeply committed to naturalistic scientific explanations and to the new Darwinism. In the face of the evolutionary onslaught, a number of them retreated from the egalitarian ideal to claim for woman a biologically based "complementary genius" to man's—a "genius" that was rooted in her innate maternal and womanly qualities (Alaya 1977). The American feminist Antoinette Brown Blackwell, for instance, did not dispute Darwin's view that the mental differences between men and women were biologically based and the product of evolution; rather, she disputed whether woman's innate mental differences could properly be called inferior to man's. She balanced man's greater strength, reasoning powers, and sexual love against women's greater endurance, insightfulness, and parental love and argued that social progress was dependent upon the evolution and perpetuation of these sexually divergent traits (Tedesco 1984). Such argumentation had a dangerous tendency to reinforce traditional stereotypes and cater to the drawing of biological limits to feminine potentiality.[1]

Eliza Lynn Linton, the contradictions notwithstanding, is best located among such advanced women, whose confidence in the liberating powers of science, and whose opposition of naturalistic interpretations of human nature and society to conventional theological wisdom and authority, ultimately betrayed them when science, especially Darwinism, gave a naturalistic basis to the class and sexual divisions of Victorian society (Richards 1989, 279–80).

Thus, Lynn Linton, in old age (still campaigning indefatigably against the "shrieking sisterhood"), came to endorse her own Huxley-engineered expulsion from organized science on Darwinian grounds. In 1885 she published a bizarre novel, *The Autobiography of Christopher Kirkland,* which was a dramatization of her own life in a male persona. Here, she travestied William Linton as a radical feminist whom Kirkland, against his better judgement, marries and tries to reform to the Victorian ideal of womanhood. His attempt to wean his wife from the "platform" to the "fireside"

---

1. It should be noted that there was potential within Darwinism for female agency, for females as sexual selectors and the main agents of social progress, and this form of Darwinism was promoted by some socialist feminists, notably the American visionary Eliza Burt Gamble, the English secularists and socialists Annie Besant and Edward Aveling, and Alfred Russel Wallace, cofounder with Darwin of the theory of natural selection. But such attempts to radicalize sexual selection and give women an active and central role in evolutionary theorizing received little attention or support from mainstream feminists and Darwinians (Richards 1995; Gamble 1916; Wallace 1890; 1913, 125–49).

fails, and Kirkland finds solace in Darwinism. He associates with leading Darwinian scientists and frequents the meetings of the learned societies. But here we find no hint of the arguments for women's right to knowledge and membership in scientific societies with which Eliza petitioned Huxley all those years ago. Kirkland endorses Eliza's claims of women's rights to property, divorce, and an education different from but equal to men's, but on the whole he subscribes to the doggerel that "Women's Rights are Men's Lefts." He is a rigid biological determinist who defends his antifeminism on the Darwinian grounds that "unless we accept the creed . . . that the moral sense is as much a matter of evolution as is the intellectual—we are lost in a sea of contradictions." For Kirkland, the stereotypically Victorian intellectual and moral differences between men and women are the products of evolution and are grounded in "the material fact of sex." They are the foundations of society and morality, "the division of labour and function, against which women revolt [in vain], and men must fare forth while they bide within" (Linton 1976, 3:5, 12, 167-72).

Here, in her masculine alter ego, Eliza finally won the entry into science that was denied her in real life. But her fictional victory is achieved only at the cost of the acceptance into a science from which her alter ego Kirkland, logically, would also have excluded her as a woman. Kirkland is the ultimate Darwinian fellow traveller, a Spencerian, from the likes of whom the more subtle Huxley was soon carefully to dissociate himself. But Eliza/Kirkland has simply taken Huxley's position on the woman question to its logical extreme. It was Huxley, after all, who both excluded women from science in the name of science and redefined that science to ratify their exclusion.

Cobbe, by contrast, offered her followers a radical critique of Victorian science, based on her opposition of spiritual and feminine values to the materialism and masculine tyranny that constituted that "exquisite kind of vice" that found expression in vivisection and the maltreatment of women (Caine 1992, 145-49). However, while her rejection of the Darwinian naturalization of mind may have provided ideological support for those opposing the imposition of naturalistic limits on women's aspirations, Cobbe's doctrine of theistic ethics could not provide a real political alternative for those confronting the antifeminist applications of Darwinism. The only solution she offered was the unrealizable goal of the individual moral and religious reform of those scientists like Huxley who had constituted themselves the new secular priesthood of Victorian society. Furthermore, in certain significant respects, Cobbe's theism and antivivisectionist stance conduced to the Victorian feminization of feeling and the masculinization of reason, to the legitimation of Huxley's exclusion of women from science and the Darwinian definition of feminine nature as essentially incommensurate with the masculine pursuit of science.

Bibliographical Note

The significant role played by leading Darwinians, including Darwin himself, in imposing naturalistic, scientific limits on the claims by nineteenth-century feminists for political and social equality has been well documented by feminist scholars. The more notable of these, Conway (1970), Fee (1974), Rosenberg (1975), Alaya (1977), Mosedale (1978), Rosser and Hogsett (1984), Russett (1989), and Jann (1994) have all contributed to our understanding of the concerted Darwinian refutation of the natural equality of men and women. As well, Love (1983), Tedesco (1984), Egan (1989), and Erskine (1995), among others, have undertaken studies of the response to Darwinism by some prominent nineteenth-century feminists, notably Charlotte Perkins Gilman, Olive Schreiner, and Antoinette Brown Blackwell.

However, many of these studies, especially the earlier ones, useful as they are, fall into the category of "feminist empiricism," i.e., they maintain the integrity of the standard view of science as objective and value-free, and argue that androcentrism in science is socially caused and is, therefore, the result of "bad" science. The solution to such sexist science is to be found in a closer adherence to the proper methodologies of scientific enquiry, and nineteenth century evolutionists and anthropologists are censured for their failure to conform to these standards. There is a disjuncture between such studies and the contextualist or constructivist evolutionary historiography pioneered by Robert Young in the 1960s, which was, in its turn, regrettably gender-blind (Young 1985), and more recent contextual feminist analyses (e.g., Outram 1989; Haraway 1989; Hall 1992). My own studies of Darwin, Huxley and Victorian feminism (Richards 1983, 1989, 1995) are attempts to integrate feminist insights into Victorian science and society with contextual Darwin historiography. Readers should also consult the recent biographies and studies by Desmond, Moore and others (Desmond 1989, 1994; Desmond and Moore 1991; Moore 1989).

There is a sympathetic portrait of Eliza Lynn Linton by Nancy Fix Anderson (1986) and an excellent study of Frances Power Cobbe by Barbara Caine (1992, 103–49), both of which have enriched our understanding of the individual struggles of these important but neglected Victorians to give meaning to their lives and circumstances within the confines of the Victorian sphere of femininity.

References

Alaya, Flavia. 1977. "Victorian Science and the 'Genius' of Woman." *Journal of the History of Ideas* 38:261–80.

Anderson, Nancy Fix. 1987. *Woman against Women in Victorian England: A Life of Eliza Lynn Linton*. Bloomington: Indiana University Press.

Caine, Barbara. 1992. *Victorian Feminists*. Oxford: Oxford University Press.
Cobbe, Frances Power. 1862. "What Shall we do with our Old Maids?" *Fraser's Magazine* 66:594-610.
———. 1863. "Celibacy v. Marriage." In *Essays on the Pursuits of Women*. London: Emily Faithfull, 38-57.
———. 1872. "Darwinism in Morals." In *Darwinism in Morals, and Other Essays*. London: Williams and Norgate, 1-33.
———. 1878a. "Wife Torture in England." *Contemporary Review* 32:56-87.
———. 1878b. "The Little Health of Ladies." *Contemporary Review* 31:276-96.
———. 1881a. *The Duties of Women: A Course of Lectures*. London: Williams and Norgate.
———. 1881b. "The Medical Profession and its Morality." *Modern Review* 2:296-326.
———. 1881c. "Mr. Darwin on Vivisection." *The Zoophilist*, special supplement, 1:17-19.
———. 1894. *Life of Frances Power Cobbe By Herself*. 2 vols. London: Richard Bentley and Son.
Conway, Jill. 1970. "Stereotypes of Femininity in a Theory of Sexual Evolution." *Victorian Studies* 14:47-62.
Darwin, Charles. Archive. Cambridge University Library.
———. [1838] 1986. "Darwin's Notes on Marriage." In *The Correspondence of Charles Darwin*, edited by Frederick Burckhardt and Sydney Smith. Cambridge: Cambridge University Press, 2:443-45.
———. 1871. *The Descent of Man, and Selection in Relation to Sex*. 2 vols. London: John Murray.
———. 1888. *Life and Letters of Charles Darwin*. Edited by Francis Darwin. 3 vols. London: John Murray.
———. 1903. *More Letters of Charles Darwin*. Edited by Francis Darwin. 2 vols. New York: D. Appleton and Company.
———. 1994. *A Calendar of the Correspondence of Charles Darwin, 1821-1882*. Edited by Frederick Burckhardt and Sydney Smith. Cambridge: Cambridge University Press.
Desmond, Adrian. 1989. *The Politics of Evolution*. Chicago: University of Chicago Press.
———. 1994. *Huxley: The Devil's Disciple*. London: Michael Joseph.
Desmond, Adrian, and James Moore. 1991. *Darwin*. London: Michael Joseph.
Egan, Maureen L. 1989. "Evolutionary Theory in the Social Philosophy of Charlotte Perkins Gilman." *Hypatia* 4:102-19.
Elston, Mary Ann. 1987. "Women and Anti-vivisection in Victorian England, 1870-1900." In *Vivisection in Historical Perspective*, edited by Nicolaas A. Rupke. London: Croom Helm, 259-94.
Erskine, Fiona. 1995. "*The Origin of Species* and the Science of Female Inferiority." In *Charles Darwin's "The Origin of Species": New Interdisciplinary Essays*, edited by David Amigoni and Jeff Wallace. Manchester and New York: Manchester University Press, 95-121.
Fee, Elizabeth. 1974. "The Sexual Politics of Victorian Social Anthropology." In *Clio's Consciousness Raised: New Perspectives on the History of Women*, edited by Mary Hartman and Lois Banner. New York: Harper, 86-102.

French, Richard D. 1975. *Antivivisection and Medical Science in Victorian Society.* Princeton, N.J.: Princeton University Press.

Gamble, Eliza Burt. 1916. *The Sexes in Science and History: An Inquiry into the Dogma of Woman's Inferiority to Man.* New York: G. P. Putnam's. (Revised edition of *The Evolution of Woman,* 1894.)

Hall, Catherine. 1992. "Feminism and Feminist History." In *White, Male and Middle Class: Explorations in Feminism and History.* Cambridge: Polity Press, 1-40.

Haraway, Donna. 1989. *Primate Visions.* New York and London: Routledge.

Hart, Ernest. 1892. "Women, Clergymen, and Doctors." *New Review* 7:708-18.

Huxley, Thomas Henry. Papers. Huxley Archives. Imperial College.

———. 1900. *Life and Letters of Thomas Henry Huxley,* edited by Leonard Huxley. 2 vols. London: Macmillan and Company.

———. [1865] 1968. "Emancipation—Black and White." In *Collected Essays.* New York: Greenwood Press, 3:66-75.

Jann, Rosemary. 1994. "Darwin and the Anthropologists: Sexual Selection and Its Discontents." *Victorian Studies* 37:287-306.

Lansbury, Coral. 1985. "Gynaecology, Pornography, and the Antivivisection Movement." *Victorian Studies* 28:413-37.

Linton, Eliza Lynn. 1858. "The Unities of Nature." *National Magazine* 5:52-7.

———. 1870. *Ourselves: A Series of Essays on Women.* London: G. Routledge and Sons.

———. 1883. *The Girl of the Period and Other Social Essays.* London: Richard Bentley and Son.

———. [1885] 1976. *The Autobiography of Christopher Kirkland.* 3 vols. New York: Garland.

Love, Rosaleen. 1983. "Darwinism and Feminism: The 'Woman Question' in the Life and Work of Olive Schreiner and Charlotte Perkins Gilman." In *The Wider Domain of Evolutionary Thought,* edited by David Oldroyd and Ian Langham. Dordrecht: D. Reidel, 113-31.

Moore, James R., ed. 1989. *History, Humanity and Evolution.* Cambridge: Cambridge University Press.

Moscucci, Ornella. 1990. *The Science of Woman: Gynaecology and Gender in England, 1800-1929.* Cambridge: Cambridge University Press.

Mosedale, Susan Sleeth. 1978. "Science Corrupted: Victorian Biologists Consider 'The Woman Question.'" *Journal of the History of Biology* 11:1-55.

Outram, Dorinda. 1987. "The Most Difficult Career: Women's History in Science." *International Journal of Science Education* 9:409-16.

Owen, Alex. 1990. *The Darkened Room: Women, Power, and Spiritualism in Late Victorian England.* Philadelphia: University of Pennsylvania Press.

Richards, Evelleen. 1983. "Darwin and the Descent of Woman." In *The Wider Domain of Evolutionary Thought,* edited by David Oldroyd and Ian Langham. Dordrecht: D. Reidel, 57-111.

———. 1989. "Huxley and Woman's Place in Science: The 'Woman Question' and the Control of Victorian Anthropology." In *History, Humanity and Evolution,* edited by James R. Moore. Cambridge: Cambridge University Press, 253-84.

———. 1995. "'The new woman worship': T. H. Huxley and Victorian Feminism." Paper presented at the international conference "T. H. Huxley: Victorian Science and Culture," Imperial College of Science, Technology and Medicine, London.

Ritvo, Harriet. 1987. *The Animal Estate*. Cambridge, Mass.: Harvard University Press.

———. 1988. "Sex and the Single Animal." *Grand Street* 1 (spring): 124–39.

Rosenberg, Rosalind. 1975. "In Search of Woman's Nature: 1850–1920." *Feminist Studies* 3:141–54.

Rosser, Sue V., and A. Charlotte Hogsett. 1984. "Darwin and Sexism: Victorian Causes, Contemporary Effects." In *Feminist Visions: Toward a Transformation of the Liberal Arts Curriculum,* edited by Diane L. Fowlkes and Charlotte S. McClure. University, Ala.: Alabama University Press, 42–52.

Russett, Cynthia Eagle. 1989. *Sexual Science: The Victorian Construction of Womanhood*. Cambridge, Mass.: Harvard University Press.

Smith, F. B. 1973. *Radical Artisan: William James Linton*. Manchester: Manchester University Press.

Taylor, Barbara. 1983. *Eve and the New Jerusalem: Socialism and Feminism in the Nineteenth Century*. New York: Pantheon Books.

Tedesco, Marie. 1984. "A Feminist Challenge to Darwinism: Antoinette L. B. Blackwell on the Relations of the Sexes in Society and Nature." In *Feminist Visions: Toward a Transformation of the Liberal Arts Curriculum,* edited by Diane L. Fowlkes and Charlotte S. McClure. University, Ala.: Alabama University Press, 53–65.

Wallace, Alfred Russel. 1890. "Human Selection." *Fortnightly Review* 48:325–37.

———. 1913. *Social Environment and Moral Progress*. London: Cassell and Company.

Young, Robert M. 1985. *Darwin's Metaphor*. Cambridge: Cambridge University Press.

7

Satire and Science in Victorian Culture

JAMES G. PARADIS

I. Satire and the Boundaries of Knowledge

In a lively satire titled "Vestiges of Creation" appearing in *Punch* in September 1859—two months before the publication of Charles Darwin's *Origin of Species*—development theory is burlesqued in cartoons of duck heads superimposed on fashionably dressed female forms promenading along the Serpentine in Hyde Park (figure 7.1). Departing from an observation by Sir Samuel Morton Peto in the *Times* that "The Serpentine, and the whole of Belgravia, were formerly a Lagoon of the Thames," quatrains of rhymed doggerel yoke images of a remote prehistoric world of geological and biological forms with the fashionable contemporary world associated with Belgravia society dwelling on the fringes of the Palace grounds:

> The slimy reptile here, no doubt,
> Wriggled and crawled in greed or malice:
> Now see the Courtier creep about—
> Near as he dares to yonder Palace.
> (*Punch* 1859, 37:100)

Laced with references to celebrities of science and engineering like Richard Owen, William Buckland, and Sir William Cubitt (civil engineer of the Cardiff docks and the Southeastern Railway), the doggerel neatly combines verbal and visual irony to produce a witty farce on metamorphosis. The laughter is directed at Robert Chambers's anonymous *Vestiges of the Natural History of Creation,* at geological and evolutionary speculation in general, and at the effete denizens or "vestiges" found in polite London society.

This cartoon, with its elaborate bric-a-brac of ideas, reveals the unique potential of irony and satire to support a kind of swashbuckling intellectual comedy based on absurd yet somehow telling associations. Relaxing artistic and social restraint to the verge of foolery and personal libel, irony—

144   *Satire and Science*

Figure 7.1 An example of *Punch*'s comedy of ideas. ("Vestiges of the Natural History of Creation," *Punch* 37 [1859]: 100.)

considered one of the four foundational literary "myths" by Northrop Frye—assumes a unique capacity to explore paradox and dualism. What we see in a bald piece of hackwork like the "Vestiges" doggerel is the impulse of irony and satire to enframe within a common literary field the incoherence of human experience that other forms of representation cannot capture.[1] Irony and its militant form, satire, Frye observes, attempt "to give

---

1. Literary studies of irony approach it as one of the primary literary architectures, uniquely able to capture conflict and discontinuity. Two nineteenth-century analyses, Thirlwall [1833] 1878 and Kierkegaard [1841] 1989, examined the classical origins of irony in the Sophoclean stage and Socratic dialogue. Kierkegaard associated irony with the Hegelian dia-

form to the shifting ambiguities and complexities of unidealized existence" (Frye 1957, 223). The ironist uses figurative language to bind the views and terms of one habit of mind with those of another. The physical world erected by natural knowledge in Chambers's *Vestiges,* a world of alien landscapes and biological forms, scarcely fits into the reality of Hyde Park promenading by polite society. Duck heads on forms in crinoline gowns symbolize the stark boundaries of incongruence. From the "slime" of ancient forms, we slip by ironic double meanings and punning to the finery of the "courtier." As the irony builds into full-fledged burlesque, the dualistic yoking of incongruous materials inspires amusement, as well as the reader's humorous recognition that both "worlds" represented in the cartoon may have something in common after all. The reader, an essential part of the formula of satire, is situated in yet another, outside world of reasoned perspective. From this external, more comprehending viewpoint, we laugh at the absurd associations Mr. Punch has joined together in his comic theater of ideas, ideas that are out of context and out of control.

The cartoons in the pages of *Punch* are by no means the earliest examples of the satirical construction of a scientific worldview. Among the many targets in the long history of English satire—political life, social manners, religious orthodoxy—science and its institutions occupied a prominent place. As Shaftesbury wrote, "[When] the minute examiner of nature's works proceeds with Zeal in the Contemplation of the Insect-Life, the Conveniencys, Habitation, and Oeconomy of a Race of Shellfish, when he has directed a Cabinet in due form and made it the real Pattern of his Mind . . . he then indeed becomes the Subject of a sufficient Raillery . . . the Jest of common Conversations" (Shaftesbury [1711] 1732, 3:156–60). This idea of the zealous myopic philosopher lavishing his intellectual powers on the world of trivia was personified for the broad theatergoing public in such farcical characters as Sir Nicholas Gimcrack of Thomas Shadwell's *Virtuoso* (1676). The zealous Gimcrack, scornful of utility, is introduced to the audience in the absurd pose of studying the idea of swimming by writhing on top of a laboratory table in imitation of a frog in a tub (Shadwell [1676] 1966). With deadpan literalism, Shadwell incorporates into his farcical world the descriptions of actual experiments performed by Robert Boyle, Robert Hooke, and others in the chambers of the Royal Society. Fatal transfusions of sheep blood to humans, the weighing of air, and brutal respira-

---

lectic. Northrop Frye treats irony with satire as one of four literary categories (with the romantic, tragic, and comic) that are "broader than, or logically prior to, the ordinary literary genres" (Frye 1957, 162, 223–39). Frye's treatment of satire as a special case of "militant irony" makes assumptions I am also making in this essay. Excellent general discussions of irony can be found in Muecke 1969 and Wellek 1955. But see also Booth (1974), who restricts irony more narrowly to its verbal form. Rorty (1989) identifies pragmatism with an ironic approach to the problem of reality, in which numerous worldviews, some more valid than others, exist as "incommensurate" vocabularies. Glicksberg (1969) also examines philosophical implications of irony.

tion experiments on living dogs are all part of Gimcrack's lunatic, self-indulgent pursuit of useless, fanciful knowledge (Shadwell 1966).

By the early Augustan period, experimentation and associated antiquarian interests in the zealous collection of facts had inspired a full-blown knowledge controversy (Hunter 1981; J. Levine 1991). Tory wits like Jonathan Swift, John Arbuthnot, and Alexander Pope, seeking to discredit the new knowledge associated with the Royal Society, used the reductive power of satire in the *Memoirs of the Extraordinary Life, Works, and Discoveries of Martinus Scriblerus* (1714) to caricature the accumulation of factual detail and historical artifact as vulgar pursuits by philistines of mere trivia and novelty (J. Levine 1977; Kerby-Miller 1988). Using a dazzling language of burlesque, caricature, and mock-heroism, Swift followed Shadwell in making fun of the experimentalist and the specialized language he used within his community. The Royal Society became the Academy of Lagado in *Gulliver's Travels* (1726). Pope devoted the fourth book of his magnum opus, *The Dunciad* (1743), to the Virtuosos or "minute philosophers" who specialized in the study of antiquities and natural history trifles such as "Butterflies, Shells, Bird's Nests, Moss, etc." (Pope 1993, 514). Satire thus became an indispensable part of a cultural strategy by which the finest literary talents of the day undertook to erect a comic theater of ideas to weigh, publicize, and censor what they perceived to be a series of intellectual abuses that had gained an institutional footing.

Although Victorian science has been the subject of several recent literary studies, historians have said little about the extensive use of irony and satire in the Victorian treatment of science.[2] Why did some Victorians use caricature to help situate science within the broader cultural context? The satirist's representations, to be sure, were full of the distortion that attends the use of caricature, parody, and reductio ad absurdum. Still, reduction has its uses, and literary like scientific reduction can place its object in an entirely new and revealing light. Victorian satire on science was a literature of contested ideas, disciplines, and epistemologies. The contrasts between the broad syncretic traditions of the humanist and the analytical approaches of the natural philosopher had greatly sharpened with the growth of organized science. Societies like the British Association for the Advancement of Science took shape in the context of extensive defining activity by intellectuals like Charles Babbage, John Herschel, and William Whewell, who struggled with the nature of scientific knowledge, the boundaries of scientific legitimacy, and the relations of the sciences to each other and to other forms of human knowledge (Morrell and Thackeray 1981, 273–95; Schweber 1983). This intellectual and institutional growth generated friction and an extraordinary range of argument as Victorian society sought ac-

---

2. Students in literature and science in the Victorian period should see the bibliographical note.

commodation with the rapid consolidation of the sciences. As the natural sciences grew into a constellation of progressive fields, the cultural resources for reconciling an expanding naturalistic picture of the world with a range of traditional views included the British periodical press and an extraordinary variety of literary works produced by an emergent class of intellectuals (Ellegard 1990; Heyck 1982). British irony and satire operated at the cultural divides by calling attention to the conflicts in a striking imagery that was at once colorful and controversial. In this remarkable imagery, we find contending interests vying for the defining cultural representations of Victorian science.

Irony, including its militant form, satire, was an important Victorian choice for expressing the difficulty of assimilating science and its trends. Science-inspired irony and satire appear in literary texts, the comic periodical press, and extensive ephemera of occasional pamphlets, personal notes, diaries, small journals, and correspondence. Victorian science-inspired irony is found in Thomas Carlyle's *Sartor Resartus* (1833–34), Charles Kingsley's *Water-Babies* (1863), Matthew Arnold's *Culture and Anarchy* (1869), and Thomas Henry Huxley's *Lay Sermons, Addresses, and Reviews* (1870). We find much science-inspired irony and satire in the prolific comic press, an example of which we have seen in the "Vestiges" cartoon in *Punch* (1841–1992). Similar periodicals like *Figaro in London* (1831–38) and the *Comic Almanack* (1835–53) provide a satirical literature of ideas that reveals popular Victorian interest in contemporary science (Vann and VanArsdel 1994). This literature became an important conduit for conveying scientific ideas of the day to the broad public (Desmond 1989; Desmond and Moore 1991; Rushing 1990). There is also extensive satire on scientific controversy, student angst, and popular taste (Browne 1992). Satirical caricatures of personalities and intellectual castes are found in the correspondence of Darwin, Huxley, and others. Broadside satires of men of science were issued in pamphlets like *Protoplasm, Powheads, and Porwiggles* (1875), an attack on Huxley, occasioned by his Edinburgh lecture "On the Physical Basis of Life." In cartoons and doodles, circulated in letters, scientific elites used caricature and humor as instruments of scientific infighting to contrast reform platforms with orthodox resistance (Rudwick 1975, 1985; Desmond 1982).

Victorian ironists and visual caricaturists have left us a provocative commentary on their extensive efforts to situate science in its contemporary culture. This commentary runs both on the middle road of periodicals like *Punch* and on the high road of works by Victorian intellectuals like Carlyle, Arnold, Kingsley, and Huxley. For these authors, the reductions of the scientific specialist operated in interesting ways with the caricature of the satirist. Using irony to establish distinct boundaries between the materials of received and progressive culture, these authors typically reduced one side of the contrast and represented the other side as broader and more flexible.

Irony and satire provided a meeting ground that an author with sufficient wit could tilt in order to draw out the limitations of the antagonist. In the satirist's self-conscious use of distortion to achieve an end, we encounter a potent imagery of incongruent forms that juxtaposes the symbols of one worldview with those of another while making no commitment to reconciling these diverse materials. Indeed, the energies of irony and satire as a literary form are derived from their failure to resolve. This freedom—indeed, irresponsibility—which empowers the ironist to use reduction and to record conflict without resolving it, also made possible the widespread participation of Victorians at different levels in the science-generated intellectual traffic of the day. The bits and pieces of irony and satire remain as vestiges of cultural formation that help us understand science-related cultural conflicts as Victorians saw them.

II. Science in Bohemia: *Punch* and Early Victorian Comic Periodicals

The most influential of the many fleeting circles of artists and journalists who created the Victorian comic press was the circle of Londoners that included George Cruikshank, Henry Mayhew, Gilbert à Beckett, Douglas Jerrold, Mark Lemon, and William Thackeray. These writers all wrote and illustrated on demand for the theater and Grub Street and, with the exception of the teetotaller Cruikshank, were at home in the London Bohemian life of taverns, social nonconformity, radical politics, and debt (Cross 1985, 102-6). They were all associated with one or more of the closely related serials *Figaro in London* (1831-38), the *Comic Almanack* (1835-53), and *Punch* (1841-). They were on close terms with the same London street life that Charles Dickens, who lived at the edge of their circle, was gathering into his own magazines and voluminous novels (Price 1957, 83). Most of them shared a variety of loosely defined middle-class commitments to social reform, some like Jerrold more radically, others like Lemon and Thackeray more cynically. Neither institution-bound professionals nor academics, these comic serialists assumed a spectator's view of establishment politics and institutions.

Mayhew, whose fertile and grandly whimsical intellect is generally credited with the inspiration for *Punch* in 1841, saw the comic periodical as an instrument of witty commentary with a message of social justice and reform (Spielmann 1895, 12-13, 17; Price 1957, 27).[3] Although he edited

---

3. Along with Lemon, a successful Grub Street playwright and an intimate of Dickens, Mayhew conceived and wrote a prospectus for *Punch* in June 1841 (Price 1957, 353-54; Spielmann 1895, 240; Prager 1979, 36-37). They were attended by the printer Joseph Last, the engraver Ebenezer Landells, and the writer Sterling Coyne. Mayhew also recruited his prolific collaborator à Beckett, as well as his father-in-law Jerrold and illustrator John Leech. For var-

Punch for only six months, from June through December, being replaced by the more reliable and productive Lemon, Mayhew's vision and staffing did most to create the new form (Prager 1979, 41–42). Collectively, Mayhew and his fellow writers—à Beckett, Jerrold, Lemon, and Joseph Coyne—were already churning out scores of farces, comedies, melodramas, and histories for the contemporary theater. With their capacity for invention on demand, a necessity of Grub Street survival, they continually sought more stable sources of employment for their talents. Thus, they united serial publication, the political and social cartoon, and contemporary Grub Street comedy in a new hybrid—a dispersed theater of farce that used dramatic conflict and dialogue to burlesque the forms and ideas of contemporary society. The same ironic puns, reversals, and double entendre that created the substance of the comic theater now furnished *Punch* and its many imitators with a flood of whimsical news events and witty asides. Victorians reading the comic press were thus presented with the arresting prospect of everyday life as farce.[4]

The comic writers and artists of *Punch* were familiar with scientific developments of the day. Thackeray, the only member of the early *Punch* staff who had been to university, had, as Browne has noted, developed a low opinion of the Cambridge academic natural philosopher, which he turned to account in a thinly veiled caricature in *Punch* of William Whewell as a pugilist (Browne 1992, 179; *Punch* 1848, 15:201). Mayhew, on the other hand, with no exposure to science during his days at Winchester School, had become an enthusiastic amateur experimenter and an admirer of Davy. Mayhew maintained some kind of a laboratory, spending long hours in the 1830s and 1840s experimenting with electrical apparatus (Price 1957, 27; Humpherys 1984). He had educated himself in the natural sciences, which he turned to account in the elementary experiments in physics and chemistry in his 1855 book of self-help for youths, *The Wonders of Science; or Young Humphry Davy*. It was Mayhew who assembled the "medical trio" of *Punch* in its first months, a group whose expertise on a variety technical subjects was available during the weekly editorial sessions at the Mahogany table. This trio included Percival Leigh, Albert Smith, and John Leech, who had all known each other at St. Bartholomew's Hospital, the institution where Richard Owen, who was destined to make many an appearance in *Punch*, had studied medicine and raised himself to a lectureship in compar-

---

ious historical accounts of Punch, see Spielmann 1895; A. Mayhew 1895; Adrian 1966; Price 1957; and Prager 1979.

4. This growth of the humorous potential of contemporary culture, as Athol Mayhew noted in his history of *Punch*, came as something of a surprise to his father and the founders of *Punch*. "By the time the many oppositions to *Punch* began to appear, there was only stifled talk to be heard of the impossibility of sustained humorous effort—of the madness of attempting to be funny for fifty-two weeks, all year around" (A. Mayhew 1895, 132).

ative anatomy. Smith had studied medicine at Middlesex Hospital and in 1838 had become a licentiate of the Society of Apothecaries and a member of the College of Surgeons. Leigh was educated at St. Bartholomew's, where he met Smith and Leech; he became a licentiate of the Society of Apothecaries in 1834 and a member of the Royal College of Surgeons in 1835, while writing humor under the pen name of Paul Prendergast. Leech, one of the greatest of the early *Punch* cartoonists, had also studied medicine at St. Bartholomew's, where he earned distinction as an anatomical sketcher, but he was forced to drop out for lack of funds (Prager 1979, 83, 87).

Science satire and whimsy were part of the *Punch* formula from the start in 1841, when it assumed a variety of overt and subtle forms. In the earliest volumes of *Punch,* medical subjects were a major feature, with many farces by Albert Smith examining various abuses in medical education and medical practice. The medical student as Bohemian was a favorite theme; two of John Leech's earliest cartoons showed smoking, inebriated students mumbling about medicine and metaphysics (*Punch* 1841, 1:71, 149). As science institutions like the Royal Institution, the Geological Society of London, the Royal College of Surgeons, and the British Association for the Advancement of Science gained the visibility they so actively sought, their personalities, institutional structures, and reductive methods furnished *Punch*'s verbal caricaturists, themselves skilled in the arts of reduction, with a rich source of satire and whimsy. Science provided Londoners a source of novelty that was turned into impressive theater by Humphry Davy, Michael Faraday, John Tyndall, and others at institutions like the Royal Institution (Altick 1978, 363-74). This theatrical combination of personality and physical show appealed to the dramatic sensibilities of the *Punch* staff, theater being the great obsession of London's Bohemia. Unexpected, often sensuously spectacular, science fascinated cartoonists like Cruikshank, Leech, and George du Maurier, who juxtaposed its forms and institutions with a bizarre variety of Victorian social and political forms. As the drama of science shifted into the contentions of evolutionary thinking, acted out by a star cast of confrontational clerics and men of science—not to mention other simians—*Punch* and the comic press were delivered a mesmerizing source of amusement, made to order for the Bohemian fascination with social caste, convention, and political power. Caricaturists like du Maurier and Linley Sambourne, absorbed by the sheer strangeness of Darwinian metamorphosis, furnished a steady stream of morphological eccentricities that revealed a profound comic impulse in the materials of evolution (Culler 1968). Their interpolations made for stunning intellectual satire and visual punning (see figure 7.2). Science thus took shape in periodicals like the *Comic Almanack* and *Punch* in a great number of ways that caricatured its many complex Victorian personal, institutional, and methodological manifestations. These caricatures were sometimes mildly whimsical, invoking

PUT ON A TAIL COAT, STAND ON THE ROOF, DRAW YOUR BREATH AND WAVE YOUR HANDS GENTLY UP AND DOWN FOR A FEW GENERATIONS. BY AN EXTENSION OF MR. DARWIN'S THEORY YOU WILL GRADUALLY FIT YOURSELF FOR INDEPENDENT VOLITATION. (THIS PLAN REQUIRES MUCH PATIENCE AND SELF-DENIAL.)

Figure 7.2 Detail from George du Maurier's satire on evolution and aerial navigation. ("Suggestions for Aerial Navigation," in *Punch's Almanack for 1871, Punch* 60 [1871]: xii.)

farcical amusement. They were at other times brutally satirical, directing scorn against a variety of political and social targets, including science itself.

Caricaturization, especially when associated with evolutionary themes, often reinforced crude prejudices against Jews, women, Africans, and the Irish. Mr. Punch was a notorious wife beater and murderer, as Mayhew's informant noted casually in the descriptions of London street entertainers in *London Labour and the London Poor* (H. Mayhew 1865, 51–55). Punch and Judy shows amused Victorian street audiences everywhere with Punch's beating his argumentative wife and child to death with a stick and then outwitting the hangman and the demon Shalla-Ba-La. These themes were also an integral part of the *Punch* philosophy, as Mark Lemon outlined it in the initial issue's "The Moral of Punch," which intertwined the themes of social justice and freedom from imprisonment with the amusements of social stereotype (*Punch* 1841, 1:1). Mr. Punch's close ally, as Lemon noted, was that other popular London entertainer, Jim Crow, the street Negro, whose dance and song routines Mr. Punch often appropriated for himself (H. Mayhew 1865, 129). Thus, the social content of *Punch*'s scientific satire often reinforced—and was reinforced by—caricatures derived from a variety of powerful Victorian racial and sexual stereotypes (Prager 1979, 75).

Well before the formation of *Punch*, George Cruikshank had hit upon the idea of a humorous periodical that, making extensive use of visual material, would use scientific themes as an important part of its content. Cruikshank's annual *Comic Almanack* (1835–53) was organized by the months of the year. Reminiscent of a great number of other natural history series bursting upon the publishing scene (Allen 1994, 85–87), including Gilbert White's newly revived and popular *Natural History and Antiquities of Selborne,* Cruickshank's journal was a parody of natural history reporting. Full of Cruikshank's eccentric illustrations, the *Almanack* played

Monster discovered by the Ourang Outangs.

Figure 7.3 George Cruikshank's "Monster discovered by the Orang Outangs" as a burlesque of the natural historian's tunnel vision (Cruikshank 1852, 353.)

with the public craze for natural history and the growing use of statistics and taxonomy as an approach to the understanding of nature. In an 1852 cartoon, titled "Monster discovered by the Ourang Outangs," Cruikshank presented a naturalist in the midst of being discovered by a tribe of tittering Orangutans, with an accompanying report from the *Ouran-outan Town Journal and Monkey World Gazette* titled "The 'What is It?' " (see figure 7.3). The report, a study in Orangutan natural history reporting, concluded that the monster is a "debased and degenerate breed of some savage Ouran-outan race, who, cut off from civilization and refinement, offer now a humiliating example of what a monkey may come to" (Cruikshank 1852, 352). These representations were reminders that science and its pursuits are matters of perspective, still very much the reflections of the idols of the tribe. In this manner, the *Almanack* used the language conventions and perspectives of natural history as the framework of human behavior, comically reducing it to caricature. The *Almanack* thus discovered a rich source of humor by applying the forms and quantifications of natural history to human social circumstance.

Typological satire flooded the pages of *Punch* and the *Comic Almanack*, often directed at some social or political injustice, sometimes directed at scientific reduction itself. *Punch* routinely used the false categorization or taxonomy as a source of satirical reduction. The humor is typically twofold. It is partly a matter of organizing that which cannot be organized, of squeezing the chaos of things into the neat ledger of scientific reduction. The reduction, however, often reveals a surprising truth. One of

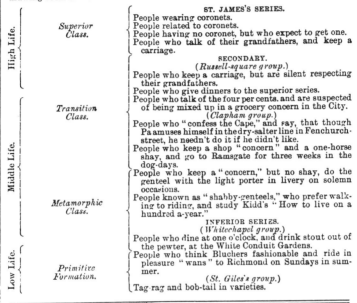

Figure 7.4 John Oxenford's discovery of a new use for stratigraphy. ("The Geology of Society," *Punch* 1 [1841]: 157.)

the great examples of this kind of satire is found in John Oxenford's early geological table, titled "The Geology of Society," modeled on the stratigraphy formats of contemporary geological texts like Charles Lyell's *Elements of Geology* (figure 7.4). Oxenford's social geology organized society into a hierarchy of three great strata, which included "High Life," "Middle Life," and "Low Life" (*Punch* 1841, 1:157). These categories were further divided into a series of classes topped by the "Superior Class" and moving down through "Transition Class" and "Metamorphic Class," to "Primitive Formation." The classes neatly integrated the social sectioning and geographical mapping of London on the basis of trite differentiae. "People wearing coronets" lived at the surface in the "St. James's series" and "Tag-rag and bob-tail in varieties" lived in the cellar in the "St. Giles's group." Oxenford's social strata thus juxtaposed the spatial forms of science with those of society. Not only did this irony reflect the stiffness of the scientific form itself, but the reduction also revealed the frivolity and stiffness of the geosocial self-arrangements of Londoners. In true Bohemian spirit, social form becomes the subject of amusement. Most interesting of all, however, was the way in which a common methodology furnished the basis of scientific categorization and humorous caricature. The semantics of science can be made to furnish the schemata of satire.

The distortions of satire released significant truths as the reductions of caricature gave fresh insight into the subject matter. One of the most successful caricatures in the early years of *Punch* was the anonymous series of Albert Smith, "Physiology of the London Medical Student," which drew directly upon Smith's inside experience of medical education and licensing and the profession he had abandoned. This series ran in twelve lengthy pieces on the fortunes of the medical student Joseph Muff. Number 8 of the series was titled "Of the Examination at Apothecaries' Hall," and it described the farce of licensing would-be medical practitioners in the Society of Apothecaries (*Punch* 1841, 1:225). This licensing process stemmed from the Apothecaries Act of 1815, which required students wishing to practice medicine to attend lectures and to pass an examination in the use of medical herbs and the like (Desmond 1989, 154). Smith's "Physiology" number 8 gave advice to the nervous student on obtaining one's "testimonials of attendance to lectures and good moral conduct in his apprenticeship," otherwise known as the "morality ticket." Smith's observation was that only "clever manoeuvring" could secure this necessary signature. The lecturer "should always be caught flying—either immediately before or after his lecture—in order that the whole business may be too hurried to admit of investigation." This advice, along with some hints on judicious forgery to fill out the testimonial, was followed with a "code of instructions" on how to take the apothecaries' examination itself. Smith's code provided acting instructions for proper exam-taking costume and for "feigning nervousness" in order to invite the examiner's sympathy (*Punch* 1841, 1:225). This account underscored the widespread corruptions in medical licensing associated with the Apothecaries Act (Allen 1994, 94-95). "Physiology" number 12 concluded with Joseph Muff's sitting for the examination at the Royal College of Surgeons at Lincoln's Inn Fields (*Punch* 1841, 1:265). The series was very like the Grub Street farces Smith wrote.

Punch and the *Comic Almanack*—and presumably their loyal British audiences—rarely tired of the farcical scientific proceeding, which was used from the late 1830s through the 1870s to burlesque the organizations, personalities, language, and subject matter of the sciences. Although it rarely achieved the brilliance of Swiftian satire—Bohemian being no match for Scriblerian standards of language—this humor was very much the stuff of the old Shadwellian stage farce. It gravely meted out gibberish and nonsense, according to Scriblerian themes of so much earnest zeal being misapplied to the minutiae of existence. In one of its numerous British Association pieces, *Punch* devoted more than a thousand words to the Bath meeting of September 1864 in an article titled "Under Hydrothermal Influence" that dwelled on the hot baths of Bath, on an observation in Lyell's presidential address that "the inhabitants of sea and land before and after the grand development of ice and snow were nearly the same," and on a proposal by a Dr. Grusselback of the University of Uppsala, who wanted to experiment on freezing and thawing convicts (*Punch* 1864, 47:139). The article speculated on the merits of freezing Gladstone and Disraeli for later historical appearances. Another thousand-word report titled "Punch's Scientific Register" followed in December 1864, summarizing papers, all nonsensical, delivered at the Geological, Zoological, Geographical, Photographic, and Astronomical Societies (*Punch* 1864, 47:232–33). The article played off the farcical contrast in the notion of a social register for the elite and the reality of a group of dusty, squinting, hairsplitting virtuosos. Much of this material did not rise to very sophisticated levels of satire. The nervous humor, seeking to domesticate the grave subject matters of scientific inquiry, often strained for effect: "*Zoological Society:* Prof. Porpus in the chair. Mr. Stratelace read a paper on the exceedingly vulgar language used by some of the cockatoos in the Society's Gardens, and upon the probable origins of it." In these often silly episodes, mere farce rarely cut to the bone. The humor often seems gratuitous, quite wide of any mark. What we see here, as in much of the science humor of *Punch,* is less the engagement of scientific materials than loss of capacity to engage them, the language and subject matter of science having begun to pull away from the comprehension of the larger culture.

Punch often played with the progressive ideology of science by contrasting the grand ambition with a meager reductive result. It sometimes lampooned the idea that science is a selfless pursuit of knowledge by associating it with vanity and self-aggrandizement. This old Scriblerian theme was brilliantly illustrated in John Tenniel's genially sarcastic visual celebration of James Grant and John Speke's discovery of the source of the Nile. Tenniel dramatized the encounter as mock-heroic farce, with an astonished, embowered Nilus glancing into the beaming, self-approving face of Britannia (figure 7.5). The cartoon was accompanied by four stanzas of doggerel, "The Nile Song," said to have been "sung at the Meeting of the Royal Geographical Society, May 25, 1863, when it was announced that 'the Nile was Settled.'"

> Hail to the chiefs who in triumph advancing
> Bring us as trophy the Head of the Nile!
> Light from the African Mystery glancing
> Brightens the name of our Tight Little Isle.
> Honour to Speke and Grant, Each bold hierophant
> Tells what the Ages have thirsted to know:
> Loud at the R. G. S.
> Sets out their great success,
> Roderick vich Murchison, ho, ieroe!
> (*Punch* 1863, 44:232)

Figure 7.5 John Tenniel's genial lampoon of the spirit of discovery, on the occasion of James Grant and John Speke's discovery of the source of the Nile. ("Aha, Mr. Nilus," *Punch* 44 [1863]: 233.)

As Britannia blurts "Aha, Mr. Nilus! So I've found you at last," the irony rebounds back against the progressive worldview associated with discovery, whether geographical or scientific. The Nile has been reduced to a discovered trickle of water. "Discovery" becomes self-promotion associated here with Murchison and the Royal Geographical Society. Although the cartoon is amusing, its wit is rather stinging and unpleasant.

Liberties were often taken with the image of the man of science and the personalities who increasingly personified science. Whewell, William Benjamin Carpenter, Murchison, Lyell, Owen, Louis Agassiz, Hermann von Helmholtz, Huxley, Tyndall, and Darwin all earned ironic mention and sometimes satirical citations of their work in the annals of *Punch*. Lyell, Huxley, Tyndall, and Darwin were among the most frequently featured, although each was dealt with differently. Tyndall, an Irishman, was flatly lampooned, the *Punch* staff, notwithstanding John Leech's Irish ancestry, having a penchant for anti-Irish humor. *Punch* writers dwelled on Tyndall's intense earnestness, which sought continually to coin deeply mystical significance out of thin air. They had an avid ear for Tyndall's facile, Panglossian sense of wonder and adopted Tyndall as a favorite. In a piece titled "Frankenstein's Chemistry," *Punch* amused itself with Tyndall's speculations in *Fragments of Science* about the possibility of a human baby resulting from a certain fortuitous combination of the chemicals (*Punch* 1871, 61:41). Perhaps under certain circumstances, the combination of "sugar, and spice, and everything nice" might produce "not a plum-pudding, but little girls." Tyndall's zeal for the commonplace supernatural became a rich source, as well, for the scientific grandiose. Tyndall was "Prof. Petgoose," a "Prof. Tindall" filled with air, and "Democritus at Belfast" (*Punch* 1863, 44:213; 1863, 45:164; 1874, 67:85). The latter reference appeared just after Tyndall's presidential address at the British Association at Belfast in 1874. The series of quatrains began

> Tyndall, high-perched on Speculation's summit,
> May drop his sounding-line in Nature's ocean,
> But that great deep has depths beyond *his* plummet,
> The springs of law and life, mind, matter, motion."
> (*Punch* 1874, 67:85)

The stanzas went on to compare Tyndall with Democritus, Plato, Epicurus, and Milton, and concluded with the suggestion that matter was the wise man's (i.e., Democritus's) God and the crowd's (i.e., Tyndall's) "chatter."

Two numbers later, *Punch* parodied Huxley's equally sensationalistic Belfast paper titled "Hypothesis that Animals are Automata." An article titled "British Automata; or, the hopelessly Unconscious," gave an account of a Mr. Robinson, who for two months every year lapsed into brainless activity as he was dragged on holiday by his family (*Punch* 1874, 67:105). Huxley was treated in *Punch* more like the loose cannon that he was. In 1861, *Punch* had celebrated Huxley for his pugnaciousness and his brawling wit, which manufactured its own comedy of ideas. This was the thrust of the much-celebrated "Monkeyana" cartoon, in which Sir Phillip Edgerton undertook to review—anonymously—Huxley's indecorous fight with Richard Owen over the relationship of humans to other primates. A gorilla with a sign hung on its chest, asking the question "Am I a man and a brother?" sang its lament from the Zoological Gardens of London, in a series of thirteen stanzas that began "Am I satyr or man?/ Please tell me who can/ And settle my place in the scale" (*Punch* 1861, 40:206). The voice went on to review the various issues of the place it occupied in the hierarchy of nature, the last six stanzas being devoted to the controversy between Huxley and Owen over human brain anatomy. That this formidable controversy turned on, among other items, minute differences in the barely noticed *hippocampus minor* was perfect for *Punch* satire, which always delighted in the old Scriblerian tactic of associating obscure trivia with grave scientific discussion. *Punch* editors, ever fascinated with hierarchy, saw the addition of the gorilla as a sensational new motif in the long-standing British obsession with social station. Gorilla imagery dramatically shifted the social center, for, next to the hierarchy of biological ancestry, other hierarchies now became trivial by comparison.

Simian satire, playing on striking visual confusions of the human form, could generate a nearly endless stream of ironic humor as one attempted to place oneself on one side or other of the irony. As Rushing has noted in his exhaustive study of the so-called gorilla wars, *Punch* exploded with gorilla satire, running more than twenty items on gorilla themata in 1861 alone (Rushing 1990; Rupke 1994, 298–309). Not only did the rising ambiguities over human origins introduce new ironies into the discussion of human ancestry, but the rational process of scientific inquiry was destabilizing the human form itself, forcing one to think of oneself metamorphically.

This science-generated flood of irony supported—and was supported by—the three-way, interlinking dualisms of ape, European, and African, as backed by a morass of controversies associated with Paul Du Chaillu, Huxley, Owen, Samuel Wilberforce, and many others concerning science, slavery, racism, cultural relativity, and human identity itself. In a sensationalistic cartoon, probably by John Leech, titled "The Lion of the Season," an alarmed Irish-looking servant at the door of an evening party announces a surprising new guest, a "Mr. G-G-G-O-O-O-rilla," dressed in evening coat and tails (figure 7.6). The immense gorilla hesitates at the door, as the unsuspecting gentry within conduct their soiree (*Punch* 1861, 40:213). We the audience see what they do not—the gorilla at the door. This is quintessential *Punch* theater, with a plexus of ironic linkages too complicated to be decisively untangled. Startling contrasts of racial, cultural, and indeed biological stereotype generate a multileveled, omnidirectional irony that suggests middle-class complacency and wild scientific theorizing, as well as human biological lineage, scientific infighting, anthropological speculation, sociopolitical reform, and casual racism (see also Desmond and Moore 1991, 511; Brantlinger 1988, 184–88; Rupke 1994, 314–22). Like viewers of the stage, we see a perpetual moment of ironic encounter between appearance and reality, a larger truth that the unsuspecting subjects in the illustration do not grasp. This dramatic irony is voyeuristic and appeals to the popular mentality in search of spectacle and novelty.

In essence, Mr. Gorilla has become Mr. Punch, a symbol of irreducible irony whose placement in any locale was sufficient to invoke the satirical reflex. In recognition of this historical and irreversible merger, the editors of *Punch* dedicated the preface of volume 40 (1861) to the gorilla and showed the indomitable Mr. Punch playing leapfrog with his equally indomitable alter identity, Mr. Gorilla (figure 7.7).

The ironic perspectivism of Victorian periodical satire was widely symbolized by the rebirth of the old crotchet and misogynist, Mr. Punch, as Victorian humorist and wit. For their perspectives on British society, the journalists of *Punch* made theater a worldview, and dramatic irony became a framework for contemporary Victorian urban life. Mr. Punch made the

Figure 7.6 An example of Victorian intellectual and social conflict, as reflected in Bohemian stage irony. ("The Lion of the Season," *Punch* 40 [1861]: 213.)

dualistic spirit of irony a daily drawing room presence for his immense weekly audience of 40,000 readers among the middle class and intellectuals of England (Ellegard 1990). He was followed avidly, as well, by many politicians and members of the gentry (Cruse 1935). By merely occupying the same spaces with his subjects, Mr. Punch signaled a wider, more complicated, reflexive perspective on them. *Punch* thus used the serial format brilliantly to develop an evolving ironical commentary on contemporary science. The irony producing the farce and satire of *Punch* was best demonstrated, in many respects, by Cruikshank's natural history observer being himself observed by a society of orangutans and the entire situation being observed by a Victorian viewer. These reductive reversals, when applied to such themes as natural history observation, brought important new insights on the scientific habit of mind. In some respects, *Punch* did not, in its initial three decades, escape the old Bohemian suspicion that science was

𝔉𝔬𝔯𝔱𝔦𝔢𝔱𝔥 𝔙𝔬𝔩𝔲𝔪𝔢.

Figure 7.7 Mr. Punch and the gorilla as alter egos in the age of simian satire. (Detail from *Punch's Almanack for 1861, Punch* 40 [1861]: iv.)

Brahmin, rigid, and dangerously abstract and theoretical. Yet *Punch* also retained Mayhew's old whimsical fascination for the physical show of science. In the most satirical representations of science, there was always a fraternal undercurrent of admiration for the sheer wit and ingenuity of scientific material. It was appreciated that the ironic view of *Punch* often emerged from the same reductions that produced the specialized views of physical science.

III. Scientific Naturalism as a Source of Victorian Irony

The humor of *Punch* was in many respects the product of the dramatic vision of its architects. Yet the sources of the ironic conflicts existed outside the writers in the materials of culture itself. Farce and satire worked well with scientific materials, because scientific reduction offered subject matter that was rigid, narrow, and mechanical—the kind of material on which the satirist could build. "Irony," Connop Thirlwall wrote in *The Philological Museum* (1833-34), a journal he edited with fellow Trinity philologist Julius Hare, is like the "calm, grave, respectful judge" in a case with two vigorous contending parties, full of deep feeling and excitement:

> What makes the [ironic] contrast interesting is, that the right and the truth lie on neither side exclusively: that there is no fraudulent purpose, no gross imbecility of intellect, on either side; but both have plausible claims and specious reasons to allege, though each is too much blinded by prejudice or passion to do justice to the views of his adversary. (Thirlwall 1878, 8)

Thirlwall, the translator of Barthold Niebuhr and Friedrich Schleiermacher and author of the eight-volume *History of Greece* (1839–44), saw irony as the historical perspective itself. Irony does not originate in the judge's way of looking at things. Rather, irony is structural. Irony is a disconnection, a differential between two views of reality. Such large ironic views of the world, Thirlwall held, could be found in the Socratic irony of the Platonic dialogues, in Pascal's association of mystical ambiguity with the precision of his mathematics, and in the dramatic irony of the Sophoclean theater in which an outside viewer, knowing the larger reality, watches the actors struggle with the invisible forces of their destinies. "Like a transparent vesture closely fitted to every limb of the body," Thirlwall argued, irony is a garment cut from a larger reality to fit the contours of appearance (Thirlwall 1878, 2).

The contrast between appearance and reality identified by Thirlwall was associated by many Victorian intellectuals with the emergence of scientific naturalism as an experiential standard or worldview (Turner 1974; Lightman 1987; Paradis 1978). Many writers who felt the force and value of material progress groped for ways to situate science and industrial civilization within the broader culture. Like *Punch* but with more discipline—and sobriety—writers like Thomas Carlyle and Matthew Arnold used irony as a way of incorporating the more rigid position into the larger, more open and fluid view. Carlyle followed Pascal in clothing the empirical realm of physical experience in a transcendental vesture of higher truth, a natural supernaturalism. The so-called clothes philosophy of Carlyle's *Sartor Resartus* is ironic, as Carlyle himself acknowledged, the irony being his bridge beyond the world of fact (Carlyle [1833–34] 1937, 128–29, 258–60). In *Culture and Anarchy,* Arnold, too, used irony to contrast machinery with culture. The empirical world of operational fact can no more suffice as the basis for virtuous action, Arnold holds, than faith in machinery can deliver a higher conception of the self (Arnold [1869] 1963, 44–50). Arnold's term "machinery," while not directly mentioning science, stood for the rigidity, literal-mindedness, and passion for law that Arnold felt inspired those who consider the methods and results of science a sufficient model for the cultural ideal. Irony also appealed as a literary mode to Charles Kingsley and Thomas Henry Huxley, who were occupied by the same problem of science as an epistemological touchstone for human experience. Similarly struck by the power and intellectual richness of Carlyle's vision, both Kingsley and Huxley made irony the key tool of their explorations of scientific naturalism in the 1860s, invoking the same contrasts between appearance and reality. But, where Carlyle's irony tended toward the Romantic cosmic laughter, the irony of Kingsley and Huxley was more domestic and satirical.

Kingsley, although clearly supportive of scientific progress associated with the work of Lyell, Darwin, Huxley, and others, was occupied in finding a new post-Darwinian equivalent to natural theology. He had written to his friend Frederick Denison Maurice in April 1863, "Now that Huxley, Darwin, and Lyell have got rid of an interfering God—a master Magician as I call it—they have to chose between an absolute empire of accident and a living, immanent, ever working God" (F. Kingsley 1887, 337). In exploring the meaning of science as a worldview, he wrote one of his generation's most imaginative critiques of scientific naturalism. This issued in one of the least probable of places—a serially published fairy tale in *Macmillan's Magazine*. In *The Water-Babies* (1863), a richly ironic parable and childrens' story, Kingsley developed his plot around the humorous portrait of the scientific naturalist as the rigid, doctrinaire authority, bent on reducing human experience to the terms of his naturalistic vocabulary. He thought of his series as a "parable" critical of the emergent scientific worldview that was incomplete and unnecessarily rigid.

Working with the materials of scientific naturalism, Kingsley caricatured the scientific worldview to show its limits as a philosophy of life. In *The Water-Babies*, we find a catalog of abuses in the extraordinary language, institutions, personalities, and worldviews of science. These abuses all develop out of the driving efforts of a Professor Ptthmllnsprts ("put-them-all-in-spirits") to reduce the experiential world to the sensory terms of his narrow scientism (Rushing 1990, 456–60; Blinderman 1961). Kingsley compares a world that allows for spirit and spontaneity with a world that allows only for matter and force. The former is more fluid and open to the possibilities of human experience than the latter. The main plot of Kingsley's story, to be sure, is insipid and moralistic. The dirty little chimney sweep, Tom, is obsessed with becoming clean and runs away from Grimes, his abusing master. Tom is transformed into a newtlike water-baby, who then struggles to earn a conscience and to become clean. Thus, the first Tom is the biologically derived self, struggling, driven, selfish, waiflike; the second Tom is the spiritually evolving self, emerging from pupa to a fully developed Carlylean soldier of the middle-class economy. In his parable, Kingsley develops a humorous allegory of the "Doasyoulikes," a culture that stops striving and, so, begins to decline in a reverse evolution to become apelike creatures (figure 7.8). This imagery of reversal and devolution, with its powerful message of cultural loss, was recaptured by Matthew Arnold in his second chapter of *Culture and Anarchy*, "Doing as One Likes."

What rescues Kingsley's parable is the wonderful countercurrents associated with his humorous manipulation of science. Unlike the harsh Juvenalian cut of Swiftian satire, Kingsley's volume belongs to the moderate Menippean satirical tradition of the critique of the grand style (Frye 1957, 309–11). In Kingsley's volume, the grand style is Victorian science, with its vigorous personalities, hegemonic language, and formidable institutions. Kingsley's man of science is the brittle egoist, organizing the world around himself. Scientific naturalism is identified with Professor Ptthmllnsprts, the great-man naturalist, professor of "necrobiopalaeonthydrochthonanthro-

Figure 7.8 Devolution in the Doasyoulikes: Linley Sambourne's illustration of Charles Kingsley's *Water-Babies* (Kingsley 1885).

popithekology" at a newly established university in the Cannibal Islands (C. Kingsley 1863, 149–50). Kingsley plays with the problem of the official biological existence of water-babies, which the professor rejects on a seaside amble with the little milady, Ellie, even as he snares poor Tom in his net. The professor resolutely rejects as "contrary to nature" all forms that have not received the official sanction of science. Tom's water-baby status is resolutely denied as Ptthmllnsprts cannot decide how to induct him into canonical biology—whether by pronouncing him a "large pink Holothurian" or by giving him the new binomial name "Hydrotechnon Ptthmllnsprtsianum" (C. Kingsley 1863, 156–57). Abstract technical terminology, hair-splitting controversies over dubious anatomical structures like the "hippopotamus major," professional rivalry, excessive earnestness, and driving reductionism add up to a worldview that seriously limits the imaginative world of the child and, by implication, the human imagination itself. Like the gray, sharp-edged world of Gradgrind that Dickens had painted in *Hard Times,* the world of Ptthmllnsprts is beset with an intimidating rigidity. Kingsley's comic man of science emerges as the disciple of a closed worldview in which the professor's imagination is limited to the forms of existence that have binomial names. His belief system, despite its progressive ideology, asserts a power of censorship over the realm of imagination.

In Huxley's hands, these materials delivered a very different result. In Huxley, the Swiftian projector rose up to take his revenge on the witty tyranny of the scoffer. Huxley's satire became a potent literary tool for overturning the conventions and orthodoxies that had all too often been used to burlesque the modern. Huxley's rapier wit was widely acknowledged by his contemporaries to be a formidable weapon that needed constant monitoring. His reputation for rapid ironic reversals under pressing circumstances was well established early in his career by the series of brilliant popular essays he collected in *Lay Sermons, Addresses and Reviews.* This iconoclasm was further demonstrated in the celebrated encounter with Bishop Samuel Wilberforce at the Oxford British Association meeting of 1860. This much-discussed event achieved the larger-than-life status of a myth in which the David of science toppled the Goliath of orthodoxy by a transcendent ironic reversal (Lightman 1987; Jensen 1991). By his own account, Huxley sees as ironic the essential character of the event, in which the bishop, a man of great gifts and possessing many privileges, stoops to hearsay trivialization in the presence of a body of grave inquirers.

> If, then, said I, [Huxley wrote to Dyster] the question is put to me would I rather have a miserable ape for a grandfather or a man highly endowed by nature and possessed of great means and influence and yet who employs those faculties and that influence for the mere purpose of introducing ridicule into a grave scientific discussion—I unhesitatingly affirm my preference for the ape. (Bibby 1960, 69)

The irony of a bishop's being less worthy intellectually and morally than an ape is not lost on Huxley's contemporaries. Huxley thus clothes orthodoxy in the ironic cloth of his earnest scientific naturalism by affirming his preference for the ape. The triumph of Huxley over Wilberforce thus transcends the actual details of the event to become a resilient, colorful cultural formation consistent with the ironic liberal view that science, when compared to orthodoxy, is more serious about and open to the human experience.

Although Huxley was a gifted plain speaker with an impressive command of metaphor (Houghton 1949) and narrative (Block 1986), the power of *Lay Sermons* lies in the richness of his ironic vision. Irony and satire, Huxley discovered, could be used to privilege the emergent institutions of science. In the fluid, omnidirectional domain of satire, Huxley showed in convincing ways that progressive knowledge could be more flexible, less rigid, than orthodoxy. Orthodoxy, in turn, could be mined as a rich source of ironic contrast with a scientific worldview. These contrasts between worldviews identified with science and orthodoxy become the unifying theme of Huxley's *Lay Sermons*. Made up of Huxley's occasional essays over some fifteen years, this volume placed an older, more established, static belief system into the larger, more open perspective of a progressive intellectual culture identified with the insights of scientific naturalism. In using wit and humor to clothe the body of traditional culture, to invoke his mentor Carlyle's sartorial terms, Huxley used parody, caricature, and ironic literalization with stunning virtuosity. Much of this wit reductively identified orthodoxy with what Huxley held to be historically primitive forms of culture, for example, "Hebrew cosmology" (Huxley 1870*b*, 281). Just weeks before his encounter with Wilberforce at the British Association meeting of 1860, Huxley had written in his explosive review of Darwin's *Origin:*

> Extinguished theologians lie about the cradle of every science as the strangled snakes beside that of Hercules; and history records that whenever science and orthodoxy have been fairly opposed, the latter has been forced to retire from the lists, bleeding and crushed, if not annihilated; scotched, if not slain. (Huxley 1870*b*, 278)

Orthodoxy is the serpent's tongue. Theologian and clergy are symbols of evil casuistry in an imagery that is rescued from blunt insult only by its brilliant satirical reversal. Huxley's visceral language of battle to the death belongs to the stark Juvenalian tradition of two of his favorite authors, Voltaire and Swift. The personification of science as Hercules strangling the snakes captured all the themes of youth, strength, and courage while rendering the work of Darwin and, by extension, that of the sciences as forces of destiny.

In *Lay Sermons,* Huxley's vision of everyday life is deeply ironic in the Sophoclean dramatic sense of human agency struggling amid only partly understood forces. This vision is memorably conveyed in the chess metaphor of "A Liberal Education," which appeared in *Macmillan's Magazine* in 1868 — roughly at the time Arnold's articles on culture and anarchy were appearing in *Cornhill Magazine*. The two ironists were on distinctly different sides of the divide. Speaking to an audience at the South London Working Men's College, Huxley told his working men that they were all operating on the chessboard of the world as participants in a contest. Their antagonist was only partly known, and the rules of the game were not totally clear. Despite their sense of freedom, much was determined for them in ways they could only partly understand.

> The chess-board is the world, the pieces are the phenomena of the universe, the rules of the game are what we call the laws of Nature. The player on the other side is hidden from us. We know that his play is always fair, just, and patient. But also we know, to our cost, that he

> never overlooks a mistake, or makes the smallest allowance for ignorance. To the man who plays well, the highest stakes are paid, with that sort of overflowing generosity with which the strong shows delight in strength. And one who plays ill is checkmated—without haste, but without remorse. (Huxley 1870b, 31-32)

This has much in common with Thirlwall's "calm, grave respectful judge," watching the irony of the world unfold. Huxley's "calm, strong angel," however, is destiny itself, the naturalistic force of a physical struggle. Huxley has removed his audience from its privileged view and put them on the Sophoclean stage as participants in a struggle to the death with forces they only partly comprehend. The scientific naturalist sees the ongoing struggle by which Darwin has forever ironized his contemporaries.

Huxley's satire draws on the extensive terminological reversals of verbal irony. In verbal irony, a statement is made in order to emphasize some contrast between its literal meaning and an alternate, often opposite meaning (Muecke 1969, 20-21). The contrast or double meaning is itself surprising and may be a source of humor. A classic example of Huxley's verbal irony can be found in his popular essay "On the Physical Basis of Life," which appeared in the *Fortnightly Review* (1868). This essay makes thorough use of ironic literalization to reduce life agency to mechanical process.[5] A spirit of paradox emerges from Huxley's terminological manipulations of physiological and technological materials. Life forms are not ends in themselves but are rather "disguises" of protoplasm: "Under whatever disguise it takes refuge, whether fungus or oak, worm or man, the living protoplasm is always dying and, strange as the paradox may sound, could not live unless it died" (Huxley 1870b, 32). Huxley speaks of the "catholicity" of protoplasmic assimilation as a form of "transubstantiation" in which mutton and lobster may be converted to man and man to lobster. He invokes a series of contrastive meanings, where "catholic" is an adjective both for "universal" and for "Roman Christian" and physiological assimilation is metaphorically linked with the sacrament of transubstantiation in which bread is changed into the body of Christ (Huxley 1870b, 133-34). This unexpected convergence of literal crustacean terminology with the abstract categories of religious vocabulary is both absurd and humorous, at the expense, to be sure, of the Roman Catholic Church, which is deftly ridiculed. Huxley's verbal play thus pits the satirist against a great institutional authority.

The verbal irony in Huxley's lobster example reflects the larger structural irony, in which the two worldviews of materialism and spiritualism are brought into stinging contrast. Like Thirlwall's metaphor of the ironic

---

5. In Swiftian satire, one set of terms is commonly reduced to its literal meanings and then contrasted with another, more flexible set of terms in a process of "ironic literalization" (Lund 1983).

garment, Huxley's protoplasmic language clothes the world in the language of physiology, the material terms ultimately displacing the spiritual, sacramental world of transubstantiation. These contrasts are epistemological in that they invite comparison between the empirical regime, with its self-contained terminological system, and a belief system based on established dogma—between two incommensurate world vocabularies.[6] In these early formulations of dualistic philosophical standoff lay the seeds of Huxley's agnostic formulations some few years later at the Metaphysical Society, when he would permanently bracket the category of final cause. Although Huxley's agnosticism had deep roots in philosophical skepticism, it was also a doctrine of philosophical irony in which two worldviews existed in a permanent standoff (Lightman 1987). Such contrasts belong to a tradition of ironic satire that includes Cervantes's *Don Quixote* and Voltaire's *Candide,* which juxtapose the material and ideal orders as a source of humorous contrast that is directed against the world of idealistic self-absorption. In Huxley's instance, the material regime that drives the satire has become the professional world of contemporary progressive science. This latter world is presented as open to new experience, whereas the dogmatic world of theology is by definition nonprogressive.

In *Lay Sermons,* Huxley, like Kingsley, spoke of contemporary Victorian culture in the deeply ironic terms of Victorian anthropological imagery. Adopting the imagery of primitive cultures to characterize contemporary orthodoxy, Huxley drew on the widely available anthropological terminology of the day to present the proponents of orthodoxy as ironically tribal in a modern, progressive age, given to the gross superstitions of closed ideologies that develop out of fear and lack of comprehension. One of his main caricatures is that of the modern savage or barbarian as a persistent primitive type, still given to gross superstition and irrational fear. In his "Physical Basis" essay, he spoke of the experience of many of his brightest contemporaries, who watched the advancing tide of materialism "in such fear and powerless anger as a savage feels" when seeing an eclipse of the sun (Huxley 1870*b,* 142). Positivism was "a gigantic fetish" and "sheer popery" (Huxley 1870*b,* 148–49). In "The Origin of Species," modern Christian cosmography was that of the "semi-barbarous Hebrew" (Huxley 1870*b,* 278). In "On the Study of Zoology," he noted that a "Christian Roman boy"

---

6. In his "Private Irony and Liberal Hope" (Rorty 1989, 73–95) Richard Rorty identifies the "ironist" with an individual who is conscious that some vocabulary systems are more comprehensive than others and who manipulates vocabulary systems in order to reveal his or her ironic consciousness of a distance between his or her own vocabulary (appearance) and a vocabulary that presents a different, perhaps better version of reality. "Ironists . . . see the choice between vocabularies as made neither within a neutral and universal metavocabulary nor by an attempt to fight one's way past appearances to the real, but simply by playing the new off against the old" (73). See also Levine's wide-ranging volume on realism, science, and literature (G. Levine 1993).

of a wealthy Roman citizen of fifteen hundred years ago "could be transplanted into one of our public schools, and pass through its course of instruction without meeting a single unfamiliar line of thought" (Huxley 1870*b*, 117). In these and many additional references, Huxley deftly bound the scientific worldview with the mind of the informed, progressive liberal.

In the notorious episode of his lecture on the frog's soul at the Metaphysical Society on 8 November 1870, Huxley decisively fused the scientific and ironic worldviews. This fusion was the essence of the agnostic position, in which the ironic formation consisted of a permanent dualistic standoff between material reality and the spiritual question of the soul. The lecture, titled "Has A Frog a Soul, and of What Nature is That Soul, Supposing it to Exist?" was without doubt one of the most bizarre given by a Victorian. Huxley undertook a jarring juxtaposition of the imagery of clinical inquiry and the language of spiritual transcendence. Drawing on a history of ideas in the work of Descartes and Robert Whytt, an eighteenth-century Edinburgh physician, Huxley searched for the locus of the soul in graphic physiological language, tracing for his audience the sectionings of frog nerve, exposure of limbs to acids, removal of portions of brain. "Soul-inquiry" involved the destruction of the organism, which Huxley described in the controlled and ironic language of the dissecting laboratory for his no-doubt incredulous audience of intellectuals, artists, and critics. "If the leg of a living frog be cut off, the skin of the foot may be pinched, cut, or touched with a red-hot wire, or with a strong acid, and it will remain motionless" (Huxley 1870*a*, 1). In familiar physiological language of his own teaching, Huxley droned on, probing each part of the slowly disintegrating organism for signs of a soul. But the soul remained undetected and undefined, and the question of consciousness was abandoned as unknowable.

To reduce the search for the soul to probing in a frog was itself absurdity in the highest degree. We seem to be back with Shadwell's Sir Nicholas Gimcrack, studying the idea of swimming by studying a frog, with infinite trust in the laboratory as the setting of truth. That the inquiry proceeded without record of laughter or outburst suggests not that Huxley's audience had lost its sense of irony, but that the clinical language of the dissecting room was, in essence, no different from the clinical language of the philosophical ironist. Thus, for a brief hour, the Metaphysical Society was converted to a dissecting theater of the absurd, as the audience, which included the duke of Argyll, Father Dalgairns, Richard Holt Hutton, William George Ward, Dr. Manning, John Ruskin, and Mark Pattison, groped with Huxley for the locus of the soul (Brown 1947). Such an experiment, worthy without a doubt of Swift's Academy of Lagado, was now performed in the sober light of a Victorian meeting room. It was a performance worthy of a full-page cartoon in *Punch,* with Huxley as surgeon, assisted at the scalpel by Argyll and watched by a crowd of admiring medical students—Ruskin, Hutton, Ward, Pattison—learning the craft of soul surgery. Only the merest wall of decorum stood between earnest inquiry and the most caustic satire for an audience that could not have been unaware that they were in the presence of a master ironist.

IV. Conclusion: Science and the Ironic Worldview

In his master's thesis, *The Concept of Irony, with Continual Reference to Socrates* (1841), Søren Kierkegaard identified irony with dialectic. Irony, he argued, is the displacement of one actuality with another actuality that contains it, and the result is the incorporation of the old into the new (Kierkegaard 1989, 262). This incorporation could be seen in a historical context, as the absorption of one general view of the world into another. For Kierkegaard, the great instance of this historical process was the Socratic irony, which recast the older, less reflexive belief system of classical Greece into a new philosophical formation.

We can see Huxley and Kingsley as part of a much larger Victorian cultural transformation in which the older concerns of natural theology are seeking to keep pace with the rapidly consolidating institutions of science. As the scientific components of natural theology burst beyond the confines of Revelation and the old partnership—once imagined as goals of the Bridgewater Treatises—became untenable, the spirit and method of irony presided over the emergent dualistic conflicts that resulted. In the instances of Kingsley and Huxley, we see the irony of the spiritual and material vocabularies unable to find mutual accommodation. We also see this irony moving in and out of satire and privileging one half of the dualism, a privileging that is presented as dialectical, as something that is emergent in the sense of a process of change in which a concept passes over into and is fulfilled by its opposite.

Thus, the ironical worldview of *Lay Sermons* is emergent, as in Huxley's own words, "matter and law devour spirit and spontaneity" (Huxley 1870*b*, 142). This is an ongoing cultural process, Huxley argues, that is the "honest," "truthful" result of the innocent progress of knowledge, a classical Socratic position (Huxley 1870*b*, 271; Kierkegaard 1989, 264-65). In Kingsley's world of parables, irony is put to a different use. The spiritual, mythical experience is presented as a consciousness parallel but superior to that of the scientific naturalist by Kingsley's drawing out the fundamental humor of the reductive man of science pretending to be the authority of reality. The contrast is comical, because the man of science, woefully tunnel visioned and nearsighted, misses the more vibrant world that swarms around him.

These structural ironies, seen in the examples of Huxley and Kingsley, have been characterized as contrasts between "closed" and "open" ideologies (Muecke 1969, 125-28). The ironist presents the open worldview as continually unfolding, receptive to change, and therefore less rigid and mechanical than the closed worldview. This is the essential contrast in Arnold's *Culture and Anarchy* between "machinery" and "culture," the contrast, as well, of Augustan satire between the more enlightened ancients and their minimalist modern cousins. For Victorians, the ironic contrast was increasingly ambidextrous, as we see in *Punch*. It was the irony of John Tenniel's satire on discovery, in which the investigator was presented in a reductive light, but it was also the irony of the gorilla at the door, in which the evening party goer is blandly unaware of a surprising new guest. The literary "art" of satire concerns a cultural struggle to construct convincing —which is to say humorous—contrasts between closed and open ideologies.

Scientific naturalism, as Huxley used it, however, was not simply a philosophical inquiry; it had social objectives. Frank Turner has argued, for example, that scientific naturalism was furnishing Huxley, Tyndall, Joseph Hooker, and their colleagues a polemical basis to "displace alternative intellectual groups" that were identified with existing institutions (Turner 1981, 174). It should be added that establishment intellectual groups were trying just as hard to maintain their positions. We can see the social polemic as a heavily ironized discourse, often shading into the old Scriblerian technique of applying satire to displace the belief system of an opponent. This is almost certainly what was motivating the satire of Huxley's *Lay Sermons*, which was very much in the tradition of the eighteenth-century Augustans.

Bergson, in his 1900 essay *Laughter*, observed that the deepest intention of laughter is social (Bergson [1900] 1956). One who laughs not only directs criticism at the object of his laughter, but also invites his companions to share his sentiments. Irony and satire from the 1840s to the 1860s had increasingly become tools in the scientific community for shaping a minority cultural vision. We see this private use, for example, in correspondence among in-groups, in which irony, caricature, and humor serve to establish common representations of various cultural, intellectual, and social divides. This in-group irony is found, for example, in the private correspondence of members of the so-called X-Club (Barton 1990). As Barton and Desmond and Moore have demonstrated, running jokes, caricature, and lampoon, aimed at figures like Owen and Wilberforce as well as at ideas associated with orthodoxy, are common in the private interactions among members of this rising group of new scientists (Barton 1990; Desmond and Moore 1991; see also Rudwick 1975). Even the sober Darwin turned to irony and satire, writing such things as "What a book a Devil's Chaplain might write on the clumsy, wasteful, blundering low & horribly cruel works of nature" (Desmond and Moore 1991, xv).

Huxley and his scientific colleagues clearly saw the way in which satire often attached to the public imagery of science; indeed, most of them had made their debuts in *Punch* by the early 1860s, if not sooner. This experience of being projected socially in the context of humor to audiences of thousands of readers must have been electrifying. One response was that of their extensive private in-group irony and satire. This irony, although it had roots in deep philosophical divides, could also be exploited socially. It could be turned to humor in order to build consensus and to gain the comparative advantage of associating science with the open, politically liberal worldview. What made Thomas Huxley unique was his brilliant literary skill as an ironist and his success in exporting this association of science and the open worldview to the public arena. In his satirical reductions of orthodoxy, his ridicule of religious rigidity, Huxley used his gifts as an ironist-aphorist to turn the direction of the irony against received tradition and to seize the moral high ground for a progressive intellectual culture associated with the sciences.

Bibliographical Note

"Literature and Science" is an immense, rather loosely defined subject, covering many historical periods. Although the lengthy record of commentary on it begins more than a century ago, there is no good history of the subject. The recent annotated bibliography of Schatzberg, Waite, and Johnson (1987) provides an excellent sampling of work between 1880 and 1980. This bibliography is annually updated in the triannual journal *Configurations* (1993-), which also carries articles on the subject and is published by Johns Hopkins University Press for the Society for Literature and Science. See also Rousseau's discussion of literature and science as a field (1978).

Studies in Victorian literature and science are too numerous to list here. They are based on a variety of approaches, including cultural studies, literary history and theory, rhetorical analysis, and the history of ideas. Cultural background to the many topics of Victorian literature and science may be found in Houghton (1957), Cannon (1978), and Heyck (1982), as well as in *Victorian Studies*. Undergraduate students seeking useful introductions to the subject may consult Cosslett (1982) and Chapple (1986). Several essay collections explore topics in the relations of Victorian science and literature, including Paradis and Postlewait (1983), Jordanova (1986), G. Levine (1987), and Christie and Shuttleworth (1989). Of the many full-length studies, six of the best are Beer's groundbreaking study (1983) of narrative structure in works of Victorian science and literature, Morton's study (1984) of biological metaphor in late Victorian fiction, Shuttleworth's study (1984) of George Eliot and science, G. Levine's study (1988) of Darwin and Victorian fiction, Dale's study (1989) of the Victorian idea of scientific culture, and Merrill's survey (1989) of literary natural history. Levere (1981) and Peterfreund (1990) provide good starting points for early-nineteenth-century issues in literature and science. Also useful is the work of Gross (1990) and G. Levine (1993), who examine rhetorical and philosophical issues related to Victorian scientific and literary representation.

References

Adrian, Arthur. 1966. *Mark Lemon: The First Editor of "Punch."* New York: Oxford University Press.

Allen, David Elliston. 1994. *The Naturalist in Britain: A Social History.* Princeton, N.J.: Princeton University Press.

Altick, Richard. 1957. *The English Common Reader: A Social History of the Mass Reading Public, 1800-1900.* Chicago: University of Chicago Press.

———. 1978. *The Shows of London.* Cambridge, Mass.: Harvard University Press.

Arnold, Matthew. [1869] 1963. *Culture and Anarchy.* Edited by J. Dover Wilson. Cambridge: Cambridge University Press.

Barton, Ruth. 1990. "'An Influential Set of Chaps': The X-Club and Royal Society Politics, 1864-85." *British Journal for the History of Science* 23:53-81.

Beer, Gillian. 1983. *Darwin's Plots: Evolutionary Narrative in Darwin, George Eliot, and Nineteenth-Century Fiction.* London: Routledge.

Bergson, Henri. [1900] 1956. *Laughter.* In *Comedy,* edited by Wylie Sypher. New York: Doubleday.

Bibby, Cyril. 1960. *T. H. Huxley: Scientist Humanist, Educator.* New York: Horizon.

Blinderman, Charles. 1961. "Huxley and Kingsley." *Victorian Newsletter* 20:25-28.

Block, Ed. 1986. "T. H. Huxley's Rhetoric and the Popularization of Victorian Scientific Ideas, 1854-1874." *Victorian Studies* 29:363-86.

Booth, Wayne C. 1974. *A Rhetoric of Irony.* Chicago: University of Chicago Press.

Brantlinger, Patrick. 1988. *Rule of Darkness: British Literature and Imperialism, 1830-1914.* Ithaca, N.Y.: Cornell University Press.

Brown, Allan Willard. 1947. *The Metaphysical Society: Victorian Minds in Crisis, 1869-1880.* New York: Columbia University Press.

Browne, Janet. 1992. "Squibs and Snobs: Science in Humorous British Undergraduate Magazines around 1830." *History of Science* 30:165-97.

Burke, Kenneth. 1957. *The Philosophy of Literary Form: Studies in Symbolic Action.* New York: Vintage.

———. 1969. *A Rhetoric of Motives.* Berkeley: University of California Press.

Cannon, Susan. 1978. *Science in Culture: the Early Victorian Period.* New York: Dawson and Science History Publications.

Carlyle, Thomas. [1833-34] 1937. *Sartor Resartus: The Life and Opinions of Herr Teufelsdrockh.* Edited by Charles Harrold. New York: Odyssey Press.

Chapple, J. A. V. 1986. *Science and Literature in the Nineteenth Century.* London: Macmillan.

Christie, John, and Sally Shuttleworth, eds. 1989. *Nature Transfigured: Science and Literature, 1700-1900.* New York: Manchester University Press.

Cosslett, Tess. 1982. *The "Scientific Movement" and Victorian Literature.* New York: St. Martin's Press.

Cross, Nigel. 1985. *The Common Writer: Life in Nineteenth-Century Grub Street.* Cambridge: Cambridge University Press.

Cruikshank, George. 1835-53. *The Comic Almanack: An Ephemeris in Jest and Earnest, Containing Merry Tales, Humorous Poetry, Quips, and Oddities.* With [William] Thackeray, Albert Smith, Gilbert à Beckett, the Brothers Mayhew, with many Hundred Illustrations by George Cruickshank and other Artists. London: John Hotten.

Cruse, Amy. 1935. *The Victorians and Their Reading.* Boston: Houghton Mifflin.
Culler, Dwight. 1968. "The Darwinian Revolution and Literary Form." In *The Art of Victorian Prose,* edited by George Levine and William Madden. New York: Oxford University Press, 224-46.
Dale, Peter Allan. 1989. *In Pursuit of a Scientific Culture: Science, Art, and Society in the Victorian Age.* Madison: University of Wisconsin Press.
Desmond, Adrian. 1982. *Archetypes and Ancestors: Palaeontology in Victorian London, 1850-1875.* Chicago: University of Chicago Press.
———. 1989. *The Politics of Evolution: Morphology, Medicine and Reform in Radical London.* Chicago: University of Chicago Press.
———. 1994. *Huxley: The Devil's Disciple.* London: Michael Joseph.
Desmond, Adrian, and James Moore. 1991. *Darwin.* New York: Norton.
Ellegard, Alvar. 1990. *Darwin and the General Reader: the Reception of Darwin's Theory of Evolution in the British Periodical Press, 1859-1872.* Chicago: University of Chicago Press.
Frye, Northrop. 1957. *Anatomy of Criticism: Four Essays.* Princeton, N.J.: Princeton University Press.
Gilbert, Nigel, and Michael Mulkay. 1984. *Opening Pandora's Box: A Sociological Analysis of Scientists' Discourse.* Cambridge: Cambridge University Press.
Glicksberg, Charles. 1969. *The Ironic Vision in Modern Literature.* The Hague: M. Nijhoff.
Gross, Alan. 1990. *The Rhetoric of Science.* Cambridge, Mass.: Harvard University Press.
Heyck, T. W. 1982. *The Transformation of Intellectual Life in Victorian England.* Chicago: Lyceum.
Houghton, Walter. 1949. "The Rhetoric of T. H. Huxley." *University of Toronto Quarterly* 19:159-75.
———. 1957. *The Victorian Frame of Mind: 1830-70.* New Haven, Conn.: Yale University Press.
Humpherys, Anne. 1984. *Henry Mayhew.* Boston: Twayne.
Hunter, Michael. 1981. *Science and Society in Restoration England.* Cambridge: Cambridge University Press.
Huxley, T. H. 1870a. "Has a Frog a Soul; and of What Nature is that Soul, Supposing it to Exist?" Metaphysical Society Papers (unpublished collection). 2 vols. Bodleian Library, Oxford, 2657e.1.
———. 1870b. *Lay Sermons, Addresses, and Reviews.* New York: D. Appleton.
Jensen, Vernon. 1991. *Thomas Henry Huxley: Communicating for Science.* Newark: University of Delaware Press.
Jordanova, Ludmilla, ed. 1986. *Languages of Nature: Critical Essays on Science and Literature.* London: Free Association Books.
Kerby-Miller, Charles, ed. 1988. *Memoirs of the Extraordinary Life, Works, and Discoveries of Martinus Scriblerus.* Written in collaboration by the members of the Scriblerus Club: John Arbuthnot, Alexander Pope [and others]. New York: Oxford University Press.
Kierkegaard, Søren. [1841] 1989. *The Concept of Irony, with Continual Reference to Socrates.* Edited and translated by Howard V. Hong and Edna H. Hong. Princeton, N.J.: Princeton University Press.

Kingsley, Charles. 1863. *The Water-Babies: A Fairy Tale for a Land Baby.* London: Macmillan.

———. 1885. *The Water-Babies: A Fairy Tale for a Land-baby.* New edition, with one hundred illustrations by Linley Sambourne. New York: Macmillan and Company.

Kingsley, Francis. 1887. *Charles Kingsley: His Letters and Memories of His Life.* Abridged ed. New York: Scribner's.

Levere, Trevor. 1981. *Poetry Realized in Nature: Samuel Taylor Coleridge and Early Nineteenth-Century Science.* New York: Cambridge University Press.

Levine, George. 1987. *One Culture: Essays in Science and Literature.* Madison: University of Wisconsin Press.

———. 1988. *Darwin and the Novelists: Patterns of Science in Victorian Fiction.* Cambridge, Mass.: Harvard University Press.

———, ed. 1993. *Realism and Representation: Essays on the Problem of Realism in Relation to Science, Literature, and Culture.* Madison: University of Wisconsin Press.

Levine, Joseph M. 1977. *Dr. Woodward's Shield: History, Science, and Satire in Augustan England.* Berkeley: University of California Press.

———. 1991. *The Battle of the Books: History and Literature in the Augustan Age.* Ithaca, N.Y.: Cornell University Press.

Lightman, Bernard. 1987. *The Origins of Agnosticism: Victorian Unbelief and the Limits of Knowledge.* Baltimore, Md.: Johns Hopkins University Press.

Lund, Roger. 1983. "*Res et Verba:* Scriblerian Satire and the Fate of Language." In *Science and Literature,* edited by Harry Garvin and James M. Heath. Toronto: Associated University Presses, 69–72.

Mallock, William Hurrell. [1877] 1950. *The New Republic: or Culture, Faith, and Philosophy in an English Country House.* Edited by J. Max Patrick. Gainesville: University of Florida Press.

Mayhew, Athol. 1895. *A Jorum of Punch.* London: Downey.

Mayhew, Henry. 1865. *London Labour and the London Poor.* 2d ed. London: Charles Griffin.

Merrill, Lynn. 1989. *The Romance of Victorian Natural History.* Oxford University Press.

Morrell, Jack, and Arnold Thackeray. 1981. *Gentlemen of Science: Early Years of the British Association for the Advancement of Science.* Oxford: Oxford University Press.

Morton, Peter. 1984. *The Vital Science: Biology and the Literary Imagination, 1860–1900.* London: Allen and Unwin.

Muecke, D. C. 1969. *The Compass of Irony.* London: Methuen.

Paradis, James. 1978. *T. H. Huxley: Man's Place in Nature.* Lincoln: University of Nebraska Press.

Paradis, James, and Thomas Postlewait, eds. 1983. *Victorian Science and Victorian Values: Literary Perspectives.* New Brunswick, N.J.: Rutgers University Press.

Peterfreund, Stuart, ed. 1990. *Literature and Science: Theory and Practice.* Boston: Northeastern University Press.

Pope, Alexander. 1993. *Alexander Pope* [selections]. Edited by Pat Rogers. New York: Oxford University Press.

Prager, Arthur. 1979. *The Mahogany Tree: An Informal History of Punch.* New York: Hawthorn.

Price, Richard. 1957. *A History of Punch*. London: Collins.
Protoplasm, Powheads, Porwiggles; and the Evolution of the Horse from the Rhinoceros; Illustrating Professor Huxley's Scientific Mode of Getting up the Creation and Upsetting Moses. 1875. Aberdeen: A Brown and Company.
Punch, Or the London Charivari. 1841-1992. Vols. 1-301. London: Punch Publications.
Rorty, Richard. 1989. *Contingency, Irony, and Solidarity*. New York: Cambridge University Press.
Rousseau, George S. 1978. "Literature and Science: The State of the Field." *Isis* 69:581-91.
Rudwick, Martin. 1975. "Caricature as a Source for the History of Science: De La Beche's Anti-Lyellian Sketches of 1831." *Isis* 66:534-60.
———. 1985. *The Great Devonian Controversy: The Shaping of Scientific Knowledge among Gentlemanly Specialists*. Chicago: University of Chicago Press.
Rupke, Nicolaas A. 1994. *Richard Owen: Victorian Naturalist*. New Haven, Conn.: Yale University Press.
Rushing, Homer. 1990. "The Gorilla Comes to Darwin's England: A History of the Impact of the Largest Anthropoid Ape on British Thinking from Its Rediscovery to the End of the Gorilla War, 1846-1863." M.A. thesis, University of Texas at Austin.
Schatzberg, Walter, Ronald A. Waite, and Jonathan K. Johnson, eds. 1987. *The Relations of Literature and Science: An Annotated Bibliography of Scholarship, 1880-1980*. New York: Modern Language Association of America.
Schweber, S. S. 1983. "Scientists as Intellectuals: The Early Victorians." In *Victorian Science and Victorian Values: Literary Perspectives*, edited by James Paradis and Thomas Postlewait. New Brunswick, N.J.: Rutgers University Press, 1-37.
Shadwell, Thomas. [1676] 1966. *The Virtuoso*. Edited by Marjorie Hope Nicolson and David Stuart Rodes. Lincoln: University of Nebraska Press.
Shaftesbury, Anthony Ashley Cooper, Earl of. [1711] 1732. *Characteristicks of Men, Manners, Opinions, Times*. 5th ed. 3 vols. London: n.p.
Shuttleworth, Sally. 1984. *George Eliot and Nineteenth-Century Science: The Make-Believe of a Beginning*. New York: Cambridge University Press.
Spielmann, Marion Harry. 1895. *The History of "Punch."* London: Cassell.
Thirlwall, Connop. [1833] 1878. "On the Irony of Sophocles." In *Remains, Literary and Theological of Connop Thirlwall*. London: Daldy, Isbister, and Company, 3:1-57.
Turner, Frank M. 1974. *Between Science and Religion: the Reaction to Scientific Naturalism in Late Victorian England*. New Haven, Conn.: Yale University Press.
———. 1981. "John Tyndall and Victorian Scientific Naturalism." In *John Tyndall: Essays on a Natural Philosopher*, edited by W. H. Brock et al. Dublin: Royal Dublin Society, 169-80.
Vann, J. Don, and Rosemary VanArsdel. 1994. *Victorian Periodicals and Victorian Society*. Toronto: University of Toronto Press.
Wellek, Rene. 1955. *The Romantic Age: A History of Modern Criticism, 1750-1950*. Vol. 2. New Haven, Conn.: Yale University Press.
Williams, Raymond. 1983. *Culture and Society: 1780-1950*. New York: Columbia University Press.

PART TWO

Ordering Nature

8

Ordering Nature: Revisioning Victorian Science Culture

BARBARA T. GATES

Victorians located science in many places, not just in the laboratory, or in the rooms where scientific theory was debated by members of learned societies, or in the texts written by the scientists themselves. Large and small public lectures and scientific demonstrations, textbooks, atlases, dozens of popular magazines and pamphlets, and even the literature of science fiction provided hosts of learners with insights into the discoveries of science. There was, of course, much to rethink and then much to reorder in nineteenth-century Britain. Victorian scientific culture, like Victorian culture in general, was marked by change. Not just the Darwinian revolution, with its complex implications for Victorian religion and Victorian values, but discoveries in medicine, mathematics, and physical science altered the way people might understand life or locate themselves in the universe. As Thomas Carlyle pointed out in 1831 when characterizing his age, for Victorians change had become "the very essence of our lot and life in this world" (Carlyle 1899, 39).

Changes in perceptions of the natural order shook Victorian culture to its core. Nature, once seen as a hallmark of God's hand (as in Deism and natural theology) or as a sister category or replacement for God (as in Romanticism) now seemed mutable in ways unforeseen. The insights of science forced constant reassessments of self, society, and nature, both by scientists and by members of a science-hungry and science-fearful public. Thus, as scientific discovery reordered the ways people saw nature, new ideas of how nature might be ordered in turn suggested ways in which society also needed to be revisioned. Social Darwinism, with its application of ideas of species survival to human social and economic survival, is again only the most obvious case in point. The scientific enterprise had an impact on Victorian society, but social change also shaped the ways in which scientists continually reordered nature. Here the notion of "ordering" nature takes on ideological connotations. To contextualist historians of science it implies the way in which scientists came to their study of nature, often unknowingly, with a range of prior assumptions; it also points to ways in which nature became a resource for those Victorians who consciously wished to put forward biased ideas or particular visions of society. In this sense, there was nothing natural about the conception of the natural order presented by Victorian scientists and intellectuals; the social and natural order existed together in dynamic tension.

But if scientific revolution reordered the way scientists understood nature, an even greater variety of individuals, institutions, and texts in turn revisioned scientific discovery by reinterpreting its insights. The essays in this section offer inroads into such reinterpretations of nature and society. By departing from the usual emphasis on well-known scientists and their professional scientific audience and looking instead at the relationship between mainstream science and those who stood outside its well-defined professional borders, they provide new contexts for science study. By reopening less well-known Victorian texts, they help recuperate a Victorian popular culture of science.

Written texts that helped disseminate Victorian science have certainly not become invisible through scarcity. Evidence of them is still easy to garner. We can, for example, still find dozens of Victorian natural history books and illustrations in the bookstalls along the Thames and in what were once British colonial outposts worldwide, still come across all manner of Victorian textbooks in a multiplicity of fields for a few pence or cents on either side of the Atlantic and beyond, and locate the little-known medical novels in hundreds of libraries. Because we are now beginning to realize the multifaceted nature of Victorian science, we are rethinking such documents, not as peculiar ephemera—sport for the odd collector here and there—but as items worthy of detailed historical and literary study. They offer more than just a contextualization for "high" science, for they too are aspects of science study. They are not illegitimate sources of scientific knowledge but legitimate aspects of cultural knowledge. Looking at Victorian science through the lenses of history and literature has often rendered a valorization of eminent scientists and their writing. This ignores what occurs every time any text or theory is reinterpreted: it is recreated or reinvented, and a new and different text appears, with new applications for a new set of readers. And this applies not just to written texts but to texts in the larger sense of the word: to museums, laboratories and their equipment, and other objects of material culture. As a result of our tunnel vision, we have until recently been painting a limited picture of Victorian scientific culture, both in terms of what science was and in terms of its audience (Cooter and Pumfrey 1994).

Take, for example, the case of Victorian natural history. For years we have told ourselves a bifurcated tale of what might have happened in this area. Off somewhere in the land of science sat the "real" interpreters of nature, theorizing and arguing the merits of species competition and sexual selection, assessing collections and producing books and position papers. Meanwhile, out in the field were numberless amateurs, women and men avidly collecting butterflies, marine animals, ferns, and rocks and filing their discoveries away in Wardian cases, or aquaria, or collecting drawers, or notebooks. We have seen their representations in Victorian magazines like *Punch,* in mad pursuit by sunlight and lanternlight. But with a few exceptions, like David Allen's work on Victorian naturalists (Allen 1969, 1976, 1995), earlier historians of culture decided not to examine just how these Victorian men and women were learning, and why, but instead to side unequivocally with those of their contemporaries who spoofed their avocations and interests. And so until recently we have by and large opted to discount those ordinary Victorians who were smitten with the new worlds of natural science. If, in jest, they were constructed by some of their contemporaries as in need of psychiatric help, twentieth-century scholarship has continued that construction, or looked at Victorians as silly for bringing nature into the "boudoir" (Barber 1980). Such domestic language reminds us that even if we have had an interest in the cultural phenomenon of the natural history craze, it has probably been a gendered interest. Charles Kingsley's *Glaucus,* for example, a book written in 1854 to inform a hypothetical London merchant about the wonders of the seashore, assumes the merchant will have belittled his daughters' "pteridomania" (Kingsley 1859, 4-5). It is this kind of ridicule that Ann Shteir's pioneering work about women and botany attempts to remedy (Shteir 1984, 1987, 1996).

For what were pteridomaniacs but people learning to reorder nature, retraining their eyes to look as never before to witness what was around them in their everyday worlds, or in the wider worlds of the British Isles, or the British Empire? Their kind of seeing became a hallmark of Victorian culture, a culture obsessed with sight. Retraining the eyes led to the excitement of a personal rediscovery of the everyday world, but it also aided the scientific enterprise. Boundaries dividing amateur from professional scientist were fluid. Charles Darwin was indebted to the women in his own family and his circle of acquaintances for much of his information about individuals' and species' behaviors. And astronomical observers like the Earl of Rosse and William Huggins were at the same time amateurs and at the forefront of scientific discovery. Amateur observation also led to the production of countless nature journals carefully compiled by their observers and was of course not confined to the animal, vegetable, and mineral in the conventional sense of those categories. George Eliot, who did keep a nature journal detailing her time at the seaside at Ilfracombe, also wrote her essay "The Natural History of German Life" (1856) believing that the new science of positivism demanded the kind of careful scrutiny of people and peoples that the natural and physical sciences were demanding of the seashores. The rage to see and then help classify in order to understand and reorder came to dominate social science as it did "hard" science.

This central focus on basic observation, rather than on theory, set Victorian natural history apart from the growing scientific professionalism of the middle to later nineteenth century. In his introduction to the *Origin of Species,* Darwin took great care to distinguish his own theoretical work from that of the natural historians when he suggested that they look beyond the visible and external into the unseen mechanisms of species building (Darwin 1859). Often ignorant of scientific authority, and preferring to revision through particularizing and scrutinizing, many students of Victorian natural history did not choose Darwin's course. Instead they elected to push the borders of natural history—as they saw it—extending the location and scope of Victorian natural science. As they did so they often became enamored of popular inventions: the telescope, which allowed many ordinary Victorians to see and speculate about Mars, as Paul Fayter points out in his chapter, "Strange New Worlds of Space and Time"; and the microscope, which permitted John George Wood, whom Bernard Lightman discusses as one of the best-known practitioners of science popularization, to teach his students to see more effectively. And, even more significantly, they refocused on the human instrument behind those mechanisms—the eye, which James Krasner has so effectively discussed in his insightful book *The Entangled Eye* (1992), and whose powers of observation educators of the working classes were attempting to school through formal training.

What all the chapters in this section, "Ordering Nature," have set out to do is to begin to correct for our own cultural shortsightedness, to look at the institutions and pedagogy that reordered Victorian science and society, and to examine cultural discourses that were different from the professional scientific discourse that we have quite myopically focused upon all too exclusively in the past. They recover and explore texts and institutions that spoke to a general audience. Many of these mediators of knowledge may be unfamiliar because popular purveyors of scientific culture have often been, like the people for whom they wrote, the butt of derision. As Bernard Lightman reminds us in "The Voices of Nature," until recently they have derogatorily been considered the "hacks" of scientific writing. Yet, as Lightman also shows, then as today the popularizers of science often had a greater influence on their culture than did scientific professionals. We cannot ignore the fact that Victorian popularizers stood positioned between the secular implications of scientific naturalism and the theological underpinnings of the culture. In a culture hostile to materialism, they helped initiate the acceptance of science by reconfiguring its message. Lightman's chapter, along with those of Shteir and Fayter, reveals how the public was not simply educated but romanced and controlled through the efforts of popularizers who knew how to read public concerns. Shteir reminds us just how science was pedagogically reconstructed and domesticated for Victorian women so as to both include and exclude them from understanding.

Science fiction too provided avenues for exploring and even extending the insights of science. If some science fiction writers like Herbert George Wells subverted or resisted scientific orthodoxies as they reinterpreted them, others, including scientists like Francis Galton, utilized the new genre to restate or reinvent their own ideas. For them, popularization was another discourse that offered another audience but was not something entirely separate from the business of disseminating science. Science fiction, like scientific popularizations, domesticated the unfamiliar; but it also went a step further and defamiliarized the domestic, unsettling while informing its Victorian audience.

In disseminating science all of the mediators discussed in this section performed other cultural work, often reinforcing biases inherent in Victorian culture. Eliza Brightwen's sentimental and proprietary attitudes toward animals, for example, reflected the urge to domesticate and thus control species other than humans (Turner 1980). Brightwen's unexamined assumption that the educated classes had a right to control inferior species related to class and racial biases that also surface in some aspects of science popularization. Thomas Henry Huxley, for all his devotion to the interests of the working class, remained committed to a bourgeois program of education that offered just so much knowledge as was necessary to help people do their current jobs better and no more. And racism was codified in popularizations of anthropology like those of John George Wood and Robert Brown, in whose work Douglas Lorimer finds notions of cultural hegemony mixed with stereotypes of savages. Despite—or rather because of—their attempts at scientific classification, such texts, with their rage to order, actually contributed to the stereotyping of human subjects.

Stereotyping takes us into one corner of scientific reinterpretation. But in Victorian Britain occasions arose when the disorder of nature challenged stereotypes. Take, for example, an anomaly, a man written about and also pictured in the *Lancet* in January of 1866, Jean Battista dos Santos, who defied all stereotypes. Dos Santos possessed an extra leg and two penises.[1] (see figures 8.1 and 8.2). His human body became a cultural site, subject of interest for the professional medical person and the curious public alike. After being featured in *Lancet,* dos Santos's story was continued in the pages of the *British Medical Journal,* which labeled *Lancet*'s sensational representation of dos Santos "pornographic." At the same time the *British Medical Journal* itself betrayed a deep sense of uneasiness about who con-

---

1. For my information about dos Santos, I am indebted to Lisa Kochanek, who rediscovered his story and will have her own version of it printed in *Victorian Periodicals Review* (forthcoming).

Figure 8.1 "A Remarkable Case of Double Monstrosity in an Adult." (*Lancet*, January 1866, 71.)

trolled the Victorian medical gaze; cases like dos Santos's were a sight for medical eyes, not for the general public's. His body was the property of science, not sensationalism. But that body nevertheless severely challenged cultural as well as sexual norms. It contributed to fears about where nature might be leading human beings via evolution. Once exposed, it was hard to hide.

Medical anomalies like dos Santos were often featured not just in medical publications but also at Victorian freak shows. These were common around the seedy, bustling, Haymarket-Leicester Square area at midcentury, which is where material reinterpretations of anthropological orthodoxy like those discussed by Douglas Lorimer were also on public display. For this area was the site of Reimer's Anatomical and Ethnological Museum, "consisting of upwards of 300 superb and nature-like Anatomical figures, in wax. For gentlemen only. Admission, One shilling" (Altick 1978, 341). Also located in the area was one of the incarnations of Dr. Kahn's museum, which among other curiosities displayed a model of the body of "Duplex Boy," who had a double torso and two sets of legs and arms and was written up in a detailed pamphlet on sale in the museum for sixpence.

Such museums, like the written texts discussed in the essays in this section, were mediators of knowledge. They offer yet another place to look for the implications of the reordering of nature and its impact on nonprofessionals, an impact that must have matched the effect of the medical journals on medical professionals. In Dr. Kahn's museum, people from a variety of Victorian subcultures could cross paths for a moment and gaze on the wax

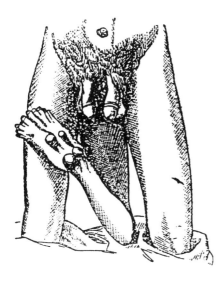

Figure 8.2 A closeup of Jean Battista dos Santos's unusual anatomical arrangement. (*Lancet*, January 1866, 72.)

dummies of "monstrosities" that defied easy categorization and cried out for further scientific explanation. Thus commodification opened the doors of science to nonprofessionals in a way quite different from the ways in which schools, or science fiction, or botany manuals, or John Wood's popularizations might have.

Stories located in a specific moment of the history of science and in multiple kinds of texts, like the story of dos Santos, force us to interrogate even more aspects of Victorian culture to find out just how and for whom science was being reordered. Their unfolding reminds us that we too can afford to be seized with a mania—a mania for the history of science, an enthusiasm for scrutinizing even more closely not only museums but stuffed bird collections, the apparatuses of the laboratory, the sketches of John Wood's lectures, Marianne North's interesting assembly of her gallery of botanical paintings at Kew, and hundreds of other places where science met people. They remind us too that we still do not really know the Victorian audience for science. We need to look further at letters of the people who went to the museums on Leicester Square and attended Huxley's school and John Wood's lectures and Mary Kingsley's public talks, to reread journals like Emily Shore's (Shore 1991) to find out what educated people were reading and visiting and how and why. And we might, when we are prompted to review Alfred Russel Wallace's illustrious career as a scientist, also review Wallace's observations on the "Hall of Science" near Tottenham Court Road, where he spent nights when he was a builder's apprentice in London. As we begin to fill the lacunae that remain in our story of Victorian science, as the chapters in this section have done, we may begin better to understand the effects of the dissemination of that science—not just on the audience for whom it was intended but on its other audiences, including ourselves.

References

Allen, David. 1969. *The Victorian Fern Craze: A History of Pteridomania.* London: Hutchinson.

———. 1976. *The Naturalist in Britain: A Social History.* London: A. Lane.

———. 1995. *The Naturalist in Britain.* 2d ed. Princeton, N.J.: Princeton University Press.

Altick, Richard. 1978. *The Shows of London.* Cambridge, Mass.: Harvard University Press.

Barber, Lynn. 1980. *The Heyday of Natural History: 1820-1870.* Garden City, N.Y.: Doubleday and Company.

Carlyle, Thomas. 1899. "Characteristics." In *Critical and Miscellaneous Essays,* vol. 3. London: Chapman and Hall, 1-43.

Cooter, Roger, and Stephen Pumfrey. 1994. "Separate Spheres and Public Places: Reflections on the History of Science Popularization and Science in Popular Culture." *History of Science* 32:237-67.

Darwin, Charles. 1859. *On the Origin of Species by Means of Natural Selection, or the Preservation of Favoured Races in the Struggle for Life.* London: J. Murray.

Kingsley, Charles. 1859. *Glaucus; or the Wonders of the Shore.* Cambridge: Macmillan and Company.

Krasner, James. 1992. *The Entangled Eye: Visual Perception and the Representation of Nature in Post-Darwinian Narrative.* Oxford: Oxford University Press.

Shore, Emily. 1991. *Journal of Emily Shore.* Edited by Barbara Timm Gates. Charlottesville: University Press of Virginia.

Shteir, Ann B. 1984. "Linnaeus's Daughters: Women and British Botany." In *Women and the Structure of Society: Selected Research from the Fifth Berkshire Conference on the History of Women,* edited by Barbara J. Harris and JoAnn McNamara. Durham: Duke University Press, 67-73.

———. 1987. "Botany in the Breakfast Room: Women and Early Nineteenth-Century British Plant Study." In *Uneasy Careers and Intimate Lives: Women in Science, 1789-1979,* edited by Pnina G. Abir-Am and Dorinda Outram. New Brunswick, N.J.: Rutgers University Press, 31-43.

———. 1996. *Cultivating Women, Cultivating Science: Flora's Daughters and Botany in England, 1760-1860.* Baltimore: Johns Hopkins University Press.

Turner, James. 1980. *Reckoning with the Beast: Animals, Pain, and Humanity in the Victorian Mind.* Baltimore: Johns Hopkins University Press.

9

"The Voices of Nature": Popularizing Victorian Science

BERNARD LIGHTMAN

In the past twenty years the Western public has developed a voracious appetite for information on the discoveries of modern science. The circulation of established magazines like *Science Digest, Scientific American,* and *The New Scientist* has increased significantly, while new publications, such as *Discover, Omni,* and *Physics Today,* have begun to line the magazine racks. Many book-length popularizations of science have appeared at the same time, written by scientists of stature, including Stephen Jay Gould, Lewis Thomas, Edward O. Wilson, Stephen Hawking, and Ilya Prigogine. The success of Carl Sagan's television series *Cosmos* has spawned a host of science documentaries, many featuring lavish, high-tech special effects, catering to the public fascination with the fantastic wonders of cutting-edge scientific discovery (Fahnestock 1993, 18). It is not possible to overestimate the importance of current popularizations of science, in all their varied forms, for our understanding of the relationship between contemporary science and culture. Can the same be said for the Victorian period, or is the popularization of science a phenomenon of significance only in the twentieth century? Who wrote the best-selling books on science for a popular audience—who were the Goulds and Sagans of the latter half of the nineteenth century?

Professional scientists such as Thomas Henry Huxley and John Tyndall account only for a small portion of the works of Victorian popularizers of science. As science became professionalized during the Victorian period and professional scientists began to pursue highly specialized research, the need arose for nonprofessionals, who could convey the broader significance of many new discoveries to a rapidly growing Victorian reading pub-

---

The author would like to thank Alisa Klinger, Suzanne Le-May Sheffield, Jim Secord, Anne Secord, and Adrian Desmond, whose comments on various drafts of this essay made the piece stronger. The work for the essay was done while the author held a Social Sciences and Humanities Research Council of Canada research grant.

lic. Some periodical editors even preferred to recruit journalists, rather than professional scientists, to write on scientific subjects. William Thomas Stead, editor of *Cassell's Magazine,* warned fellow editors never to employ an expert, scientific or otherwise, to write a popular article on his own area of research, for "he will always forget that he is not writing for experts but for the public and will assume that they need not be told things which, although familiar to him as ABC, are nevertheless totally unknown to the general reader" (Stead 1906, 297). Stead believed it was far better to use an ignorant journalist, who could tap the expert's brains to write the piece, and then send the proof to the expert to correct.

But there were knowledgeable amateurs and journalists in the latter part of the nineteenth century, many prolific and wildly successful, who produced books aimed at the mass market. Seldom mentioned by scholars until very recently, these popularizers of science may have been more important than the Huxleys and Tyndalls in shaping the understanding of science in the minds of a reading public composed of children, teenagers, women, and nonscientific males. Their success as popularizers was partially due to their ability to present the huge mass of scientific fact in the form of compelling stories, parables, and lessons, fraught with cosmic significance. Popularizers not only found the cosmic in the awe-inspiring infinite space of the heavens, they also detected it within the structure of the tiniest living organism. Though the common context provided by natural theology for the middle and upper classes was fragmented in part by the appearance of Charles Darwin's *Origin of Species* in 1859, many middle-class popularizers of science perpetuated a revised form of natural theology in their works. While professional scientists moved toward scientific naturalism during the Victorian period, middle-class popularizers of science and their audiences remained enthralled by the traditional moral, aesthetic, teleological, and divine qualities of the natural world. There were radical popularizers who produced a subversive science repudiating all of these qualities during the early Victorian period (Desmond 1987), but the focus of this chapter will be on a specific middle-class context.

If these popularizers of Victorian science were so important in their own day, why do we know so little about them? The relative neglect of popularizers by scholars is indicative of the success of the campaign waged by Victorian scientific naturalists to convince future generations that scientists were the authoritative guides to deciphering the meaning of natural things—that they alone gave voice to mute nature. Until recently, the concept of popularization has been dependent on a two-stage historiographical model (Hilgartner 1990, 519). Relying on the epistemological purity guaranteed by the scientific method, a scientific elite produces genuine, privileged knowledge. Popularizers then disseminate simplified accounts to a passive readership. Referred to by two historians as "the positivist diffusion model," this approach to popularization excludes both popularizers and the reading public from the production of knowledge (Cooter and Pumfrey 1994, 251). Popularization can be relegated to a low status, to be left to "non-scientists, failed scientists or ex-scientists as part of the general public relations effort of the research enterprise" (Whitley 1985, 3). While any differences between genuine and popularized science are attributed to a process of distortion for which the popularizer is held responsible, the scientist is given the final authority to determine which simplifications are distortions (Hilgartner 1990, 520).

Since the 1980s, scholars have offered telling criticisms of the positivist diffusion model of popularization. Hilgartner, Whitely, and Cooter and Pumfrey point out that we should be suspicious of any model that, in granting to scientists the sole possession of genuine scientific knowledge, serves to support their epistemic authority. The idea that popularization is merely a simplification of pure knowledge is itself a simplification. Distinguishing appropriate simplification from distortion in popularizations of science is not straightforward. Similarly, the boundary between genuine knowledge and popularized knowledge is often difficult to find (Hilgartner 1990, 524-29). As Cooter and Pumfrey so acutely observe, we cannot adopt the positivist diffusion model as a heuristic guide to research because it uncritically assumes the existence of two independent, homogeneous cultures, elite and popular, and forces the latter into a purely passive role. Popular culture can actively produce its own indigenous science, or can transform the products of elite culture in the process of appropriating them, or can substantially affect the nature of elite science as the price of consuming the knowledge it is offered (Cooter and Pumfrey 1994, 249-51).

In addition to recent criticism of the traditional historiographical model for approaching the popularization of science in general, scholars have noted the paucity of studies of Victorian popularizers in particular. In his important article "Natural Theology, Victorian Periodicals, and the Fragmentation of a Common Context," first published in 1980 but written much earlier, Robert Young argues that the breakup of the common intellectual context informed by natural theology led to the development of specialization and increasing professionalization. Though Young confines his attention in this piece to elite intellectual circles, he asks, "Who was left to interpret science to the layman and to discuss the large issues raised by science" once scientists had withdrawn from the common intellectual culture? With the exception of professional scientists like Huxley, Wallace, and Tyndall, who were self-consciously involved in popularization, "the field was left to pretentious hacks and to more or less competent amateurs." Young issues a call for "detailed study of this new sort of interpreter" but does not himself undertake the project (Young 1985, 156).

Young has been, of course, one of the early proponents of contextualist history of science, and we would expect to find a keen interest in science, popular culture, and the popularization of science among historians influenced by his work. However, as Cooter and Pumfrey have noticed, the shift toward an interest in the social and cultural context of science ironically "tended further to close off the space for considering the dissemination and cultivation of science in popular culture." Young's call for a study of "this new sort of interpreter" went unheeded, largely because scholars believed that if all science was culturally situated, then it was not necessary to examine popularization in particular to uncover how science was shaped by its social and cultural context (Cooter and Pumfrey 1994, 241–42). To many contextualists, it seemed far more important to focus on Darwin, Huxley, Kelvin, and other major scientific figures, since internalist accounts of the history of science depended so heavily on the alleged purity of elite science.

It is only in the 1990s that scholars have begun to make a concerted effort to formulate a new historiographical model that treats popularizations of science as "sophisticated production of knowledge in its own right," to borrow a phrase from McRae's introduction to a collection of essays on twentieth-century popular scientific writing (McRae 1993, 10). In his study of science in mass-circulation family magazines in Britain in the late nineteenth and early twentieth centuries, Broks has drawn from the field of media studies to deal with themes such as the struggle over meaning and the production of consent (Broks 1993). Topham looks to the history of books for clues on how to recover the agency of readers in his fine essay on the communication circuit running from the authors of the *Bridgewater Treatises* through the publishers, printers, binders, distributors, and booksellers, to the audience (Topham 1994). Drawing upon the history of popular culture, Cooter and Pumfrey recommend that we pay more attention to "a greater plurality of the sites for the making and reproduction of scientific knowledge" (Cooter and Pumfrey 1994, 254). This means going beyond a narrow focus on the laboratory or the scientific society toward an investigation of science in such sites as the pub, as Anne Secord does in her superb article on artisan botanists (A. Secord 1994). Cooter and Pumfrey also urge us to move away from the idealist and textual products of authorized science and to be more open to "a greater plurality of signifiers of scientific activity," such as museums, world fairs, photography, and natural history (Cooter and Pumfrey 1994, 255).

There are three primary reasons why a study of Victorian popularizers of science is vitally important for our understanding of the social and cultural contexts of Victorian science. First, the topic of popularization offers scholars numerous opportunities to examine the rich interaction between Victorian science and culture. Perhaps the cultural dimension of science is nowhere more evident. During the latter half of the nineteenth century a series of overlapping cultural and social developments shaped the trajectory of science popularization. The growth of an educated middle class, and therefore a large reading audience, and the invention of new printing technologies made possible the birth of a mass market. But why did the reading audience choose to read about science? Commercial science journals, for example, flourished, increasing from five in 1815 to over eighty by 1895 (Brock 1980, 95). Is it merely coincidental that the births of mass media and professional science both took place during the second half of the nineteenth century (Broks 1993, 123)? Since science was now considered to provide important insight into the truth of things, the reading public wanted to know the implications of new scientific discoveries for the crucial issues of the day. What did science have to say about the controversies over the role of women in society? Could science provide a solution to economic and social upheaval, particularly in large urban centers prone to labor unrest? Did science throw any light on the question of the existence of God? The relationship between science, gender, society, and religion in Victorian culture are central issues in the works of the popularizers.

But to whom did the reading public go in order to learn about the ultimate meaning of modern science, the professionals or the popularizers? This brings us to the second important reason for investigating the Victorian popularizers of science: during that period they may very well have been more important than the professionals in shaping the public image of science. The success of scientific naturalists like Huxley and Tyndall in secularizing science dismantled the bridge between elite science and public discourse. Scientific naturalists worked to cleanse scientific thought of those elements that previously had connected public and scientific culture, including anthropomorphic, anthropocentric, teleological, and ethical views of nature. The resulting fragmentation of a common cultural context linking scientists, clerics, and laypersons in the 1870s and 1880s left the public in a precarious position. The professionals claimed to be the only experts with "a legitimate interest in, and with legitimate rights to pronounce upon, the domain of secularised nature" (Shapin 1990, 997–1000). The public was given the role of supporting the programs of work undertaken by the professionals from which they were to expect substantial utilitarian benefits. But did the public accept the role provided for it by the professionals? In the past, the public had been interested in what religious and moral lessons could be drawn from nature, not just the technical and economic utility of natural knowledge (Shapin 1900, 1005). The popularizers catered to this interest and continued to give the public a sense that they participated in the production of knowledge. The publishing success of popularizers indicates that there was resistance to the claims of professional scientists to provide the only legitimate voice of nature and to their attempt to secularize science.

The popularizers of Victorian science not only provided an alternate voice to be heard by the reading public, but also offered different ways of speaking about nature. Herein lies the third reason for pursuing an analysis of the popularization of science during the Victorian period: in examining the attempts of popularizers to experiment with narrative form, the storytelling quality of all science is illuminated. Used by Galileo in his *Discourse Concerning the Two New Sciences* (1638) and by Robert Boyle in his *Sceptical Chymist* (1661), the dialogue was a conventional form for reporting scientific theories previous to the nineteenth century. Since the dialogue introduced a fiction to teach about facts, it explicitly embodied science in a narrative form. However, by the mid-nineteenth century the dialogue form rarely appeared in books dealing with scientific matters, even among popularizers, and the use of the dialogue by literary authors such as Charles Kingsley in *Madam How and Lady Why* (1869) and John Ruskin in his *Ethics of the Dust* (1865) to call to mind earlier views of nature represents the end of a tradition (Myers 1989). But the gradual disappearance of the dialogue did not bring to an end the narrative dimensions of modern science. Both popularizers and professionals have continued to tell stories about the ultimate meaning of things as revealed by science, though this characteristic of science has been more concealed in the scientific reports and papers of professional scientists (Locke 1992). The Victorian popularizers present us with a continuous spectrum of narrative form, from the most "fictional" parables to the least "fictional" imitations of the narrative of professional scientists, all of which tell the story of how science reveals the cosmic in the commonplace.

First appearing 1855, *The Parables of Nature* was an immense publishing success. In its eighteenth edition by 1882, the book was reissued many times by different publishers right up until 1950 and translated into German, French, Italian, Russian, Danish, Swedish, and Esperanto (*Dictionary of National Biography,* s.v. "Gatty, Margaret"). According to Rauch, *The Parables* was familiar to almost every middle-class child in the latter half of the nineteenth century (Rauch 1997). The author was Margaret Gatty (1809–73), the daughter of a clergyman, the Reverend Alexander John Scott, Lord Nelson's chaplain, and the wife of a Low Church clergyman, the Reverend Alfred Gatty, vicar of Ecclesfield, Yorkshire. Though the majority of her many works fall into the category of children's literature, she had more than a passing interest in science. Her passion for marine biology led to the publication of *British Seaweeds* (1863), a well-regarded introductory textbook. Gatty's scientific activity and her domestic life were virtually inseparable. She first collected seaweeds as an antidote to the boredom she experienced during a winter at Hastings recovering from the birth of her seventh child (Drain 1994, 6). On subsequent occasions, the entire family joined her at the seashore to help in the search for rare specimens, and her third daughter became a minor authority on seaweeds at the age of eight (Maxwell 1949, 97). For Gatty, the home was an important site for the production of scientific knowledge.

Gatty's *Parables from Nature* consists of a series of fictional short stories for children about the world of nature. She did not necessarily lose an adult audience by choosing to write for children, since parents, teachers, or governesses would read her stories to their children. Gatty's natural world was not that of the scientific naturalist, stripped of moral and divine significance. Rather, it was the nature with which the public was so familiar, where moral dramas were enacted and from which moral lessons could be learned, whether the characters in the story were human or animals with human characteristics (Shapin 1990, 1005). In "Law of the Wood," for example, selfish spruce-firs, whose ethical "rule is to go our own way, and let everybody else do the same," don't realize that this rule would work only if everyone lived in a separate field. Their death as a result of growing too closely together is confirmation that "mutual accommodation is the law of the wood" (Gatty [1855] 1861, 86). Similarly, in the story "The Circle of Blessing," the generous vapors of the sea, who give of themselves to thirsty flowers, tumbling waterfalls, and the earth, illustrate through their "labours of love" how ethical goodness in the global circuit of the winds benefits the entire creation (Gatty 1861, 80).

For Gatty, the natural world was also charged with religious significance in the tradition of natural theology. In the story "Waiting," the only unhappy creatures on the prehuman earth are the crickets, who cannot understand their place in the scheme of things. A wise mole counsels patience. Wait and "everything will fit in and be perfect at last," the mole declares (Gatty 1861, 56). Sure enough, a future generation of crickets discovers that their purpose is to sing by the side of hearthstones in human houses. The teleological character of nature is also emphasized in "A Lesson of Hope," when a human impressed with the fury of a violent storm begins to think of disorder as the law of nature. A wise owl sets him straight by expounding on the lessons of natural theology. Disorder, death, and destruction are transitory, have no law or being in themselves, and exist only as disturbances within a purposeful scheme. "Life, order, harmony, and peace; means duly fitting ends; the object, universal joy. This is the law," the owl teaches (Gatty 1861, 64).

Though the teleological nature of things is often only dimly perceived by humans, Gatty believed that science offered the means for ascertaining the true meaning of God's works. Nature, she declared, held out to us "wonderful adumbrations of divine truths" in the many "similitudes and analogies between physical and spiritual things" (Gatty 1861, 192). Miraculous transformations in nature—the metamorphosis from caterpillar to butterfly or grub to dragonfly—gave rational individuals license to conceive of the existence of a higher spiritual reality. The resurrection of vegetable life out of decayed seed was analogous to the resurrection of the body; both St. Paul and Sir Thomas Browne had argued in such a fashion (Gatty 1861, 156). But to really understand the spiritual, and the analogy between the physical and the spiritual, it was absolutely essential to have a scientific grasp of the physical. Gatty therefore made her children's tales as scientifically accurate

Figure 9.1 "Inferior Animals," from *Red Snow and Other Parables from Nature* (Gatty 1864). The illustrations are by Gatty and her daughters.

as possible and even added in later editions of *The Parables* a lengthy section of notes that included detailed information on the scientific theories informing each of the stories. Even though Gatty's stories contained talking animals and plants, they were based on the observable and the empirical (Rauch 1997). The science is not merely incidental to the story. The analogy that underpins the point of the story can hold only if the scientific understanding of the physical is accurate. The happy song of the crickets, after discovering the purpose of their existence, becomes for Gatty a metaphor for the way analogies in nature can teach us about the human condition. Though we can recognize neither speech nor language in the crickets' song of hope fulfilled, "there is yet a voice to be heard among them by all who love to listen, with reverent delight, to the sweet harmonies and deep analogies of nature" (Gatty 1861, 60).

Gatty's perpetuation of the natural theology tradition brought her into opposition with professional scientists who espoused evolutionary naturalism (Katz 1993, 47–48). Her satirical story "Inferior Animals" (see figure 9.1) added to a later edition of *The Parables,* lodged a protest against the arrogance of evolutionists who claimed that Darwin's theory was ultimate truth (Rauch 1997). In this way she was able to participate in the controversy even though most women were excluded from the debate. Similarly, Gatty managed to cross the lines beginning to be drawn in the mid-nineteenth century between amateur and professional by cultivating the acquaintance of experts like William Henry Harvey, who became chair of botany at Dublin in 1857, and George Johnstone, an authority on marine biology (Maxwell 1949, 93). Gatty became Harvey's unofficial assistant, and each benefited from their informal arrangement. In return for answering the questions of ignorant amateurs who wrote Harvey, helping him in the identification of seaweeds, and sharing with him anything unusual, Gatty received answers to her scientific queries, books and materials unavailable to her, and Harvey's help in correcting the proofs of her publications (Drain 1994, 7).

Like Gatty, the naturalist Eliza Brightwen (1830-1906) drew upon the natural theology tradition and conveyed scientific information to a popular audience by telling stories about the natural world. Brightwen was recognized in her time as one of the most popular naturalists; her *Wild Nature Won by Kindness,* first published in 1890, was in its fifth edition by 1893 (*Dictionary of National Biography Supplement 1901-1911*, s.v. "Brightwen, Mrs. Eliza"). Her other works included *More About Wild Nature* (1892), *Glimpses into Plant-Life* (1897), *Rambles with Nature Students* (1899), *Quiet Hours With Nature* (1904), and *Last Hours With Nature* (1908). Brightwen was raised by her uncle, Alexander Elder, one of the founders of the publishing house Smith, Elder and Company, after her mother's death in 1837. Plagued by a fear of abandonment, a feeling of loneliness, and an exaggerated sense of her own sinfulness, Brightwen could find comfort only in the study of nature (Brightwen 1909, 105-6). She married banker and businessman George Brightwen in 1855, and they settled in Stanmore on a beautiful, secluded estate, where Brightwen resided for the rest of her life surrounded by a menagerie of pet animals. In 1872 a physical illness led to complete debilitation, and only the death of her husband in 1883 roused her from her inactivity. Seven years later she began to write and publish her books.

Brightwen's purpose in *Wild Nature Won By Kindness* is to foster "the love of animated nature" in her audience, especially "in the minds of the young" (Brightwen 1890, 13). Reaching children is not a difficult task, according to Brightwen, for they "have a natural love of living creatures, and if they are told interesting facts about them they soon become ardent naturalists" (Brightwen 1890, 15). But Brightwen also simplifies her task by establishing a warm rapport with her readers through the use of a conversational mode of communication. She describes the chapters in her book as "quiet talks with my readers" in which she will "tell them in a simple way about the many pleasant friendships I have had with animals, birds, and insects" (Brightwen 1890, 12).

In contrast to Gatty's fictional parables based on scientific fact, Brightwen offered anecdotal stories, told from the first person point of view. These stories focused on her real experiences taming animals, conveying in the process scientific information on their habits, diet, and physiology. She referred to these stories as "life histories of my pets"—or in the case of her pet robin Robert the Second, a "biography"—which began at the point when they were found as babies, recounted their memorable escapades, and then ended with their unfortunate deaths (Brightwen 1890, 16, 182). Each animal emerges as an individual, with its own personality. In "Richard the Second," Brightwen describes her relationship with her pet starling, Richard, who was part of her "home-life" for more than five years (Brightwen 1890, 42). With obvious relish, she recalls his mischievous pranks, his close brush with death when he went off to hobnob with wild birds, and his ability to speak some words. Even a snail, often thought of as slimy and ugly, "is a wonderfully curious creature" to Brightwen, with its own special characteristics (Brightwen 1890, 143). Seldom leaving the bounds of her estate, Brightwen came to view the abundant wildlife there as her dearest friends. Birdie, a nightingale, was her daily companion for fourteen years. "Never," Brightwen declares, "was there a closer friendship" (Brightwen 1890, 85). For his part, Birdie became so attached to Brightwen that he adopted her as a "kind of mate," constructing a nest for her and trying to put flies into her mouth (Brightwen 1890, 83).

Brightwen's anthropomorphizing of animals, her treatment of them as individuals rather than members of a species, and the fact that her work was done in her secluded country estate and not a laboratory (though this never damaged Darwin's reputation) flew in the face of the scientific naturalists' conception of proper science. But even worse, from their point of view, Brightwen was advocating an alternative, nonexperimental approach to gathering knowledge of nature in her instructions on how to tame wild animals. Brightwen advised that the "little wild heart" could be won only "by quiet and unvarying kindness," that "there are no secrets that I am aware of in taming anything, but love and gentleness" (Brightwen 1890, 12, 74). Brightwen is suggesting how to draw closer to living things—how to enter into a relationship with nature. While scientific naturalists could be seen to adopt the experimental model for knowing nature, with its emphasis on questioning nature so as to force it to reveal its secrets, Brightwen's experiential knowledge comes from a personal encounter with nature based on love. It is quite striking that Brightwen's books contain no references to authoritative scientific experts and borrow nothing from established scientific writers, even though she enjoyed the friendship during her life of several of the leading men of science, in particular, Philip Henry Gosse (whose second wife was her sister-in-law), Sir William Flower, and Sir James Paget. Her closer relationship to nature establishes her as an independent authority, and her books provide her readers with the method for obtaining the same status for themselves.

However, Brightwen's loving relationship to nature not only leads to scientific knowledge, it also leads to knowledge of God's existence and wisdom. Brightwen's strong evangelical leanings manifest themselves throughout *Wild Nature Won By Kindness*. In the introduction, she hopes that her work will "tend to lead the young to see how this beautiful world is full of wonders of every kind, full of evidences of the Great Creator's wisdom and skill in adapting each created thing to its special purpose" (Brightwen 1890, 17). In the conclusion, titled "How to Observe Nature," she discusses the two great books given to us by God for our instruction. While the Scriptures are widely read, "how many fail to give any time or thought to reading the book of nature" (Brightwen 1890, 205). Brightwen shares with Gatty the firm belief that nature is designed by God to teach us moral lessons. "The whole realm of nature is meant, I believe," Brightwen announces, "to *speak to us,* to teach us lessons in parables—to lead our hearts upward to God who made us and fitted us also for our special place in creation" (Brightwen 1890, 204). Gatty's *Parables in Nature* are no more didactic than Brightwen's "lessons in parables" in *Wild Nature Won By Kindness:* both claim to attune their readers to the divine voice of nature.

Like Gatty and Brightwen, Arabella Buckley (1840–1929) popularized science in such a way as to draw attention to its storytelling nature. But whereas Gatty wrote fictional parables based on scientific fact and Brightwen related anecdotal stories about real experiences with nature, Buckley conveyed scientific information in the form of children's fairy tales. Daughter of the Reverend J. W. Buckley, vicar of St. Mary's, Paddington, she was in touch with the leading scientists of the day through her position as Sir Charles Lyell's secretary from 1864 until his death in 1875 (Kirk 1965, 592). Buckley's popular *Fairyland of Science* (1879) was published by no fewer than seven publishers in both England and the United States, the last edition appearing in 1905. Her other publications include *A Short History of Natural Science* (1876), *Botanical Tables for the Use of Junior Students* (1877), *Life and Her Children* (1880), *Winners in Life's Race; or, the Great Backboned Family* (1882), *Through Magic Glasses* (1890), *Moral Teachings of Science* (1891), and *Insect Life* (1901).

Buckley's avowed aim in *The Fairyland of Science* is to awaken "a love of nature and of the study of science" in "young people" who more than likely "look upon science as a bundle of dry facts" (Buckley 1879, v, 1). In order to undermine this uninspiring misconception of science, Buckley draws upon her audience's love of the magic and imagination of fairy tales. Science, Buckley promises, tells us about an enchanted natural world that, like fairyland, "is full of beautiful pictures, of real poetry, and of wonder-working fairies" (Buckley 1879, 2). To illustrate her point, Buckley draws attention to the storytelling nature of science in the opening chapter of the book. "Let us first see for a moment what kind of tales science has to tell," Buckley suggests, "and how far they are equal to the old fairy tales we all know so well" (Buckley 1879, 2). In "Sleeping Beauty" the spellbound inhabitants of the castle are frozen until the valiant prince kisses the princess and everything comes to life again. Is there less magic in the scientific tale of frozen water, spellbound by "the enchantments of the frost-giant who holds it fast in his grip," until a sunbeam kisses the ice and sets the water free (Buckley 1879, 3)? Or compare the magical powers of the man in the fairy tale "Wonderful Travellers," whose sight is so keen he can hit the eye of a fly sitting on a tree two miles away, to the "wonderful instrument" the spectroscope, which enables you to tell one gas from another in the far-distant stars (Buckley 1879, 4). "We might find hundreds of such fairy tales in the domain of science," Buckley asserts (Buckley 1879, 5).

The stories of science have an affinity to fairy tales because in nature, as in fairyland, things happen "so suddenly, so mysteriously, without humans having anything to do with it" due to the magical actions of invisible fairies ceaselessly at work. "There are *forces* around us, and among us," Buckley writes, "which I shall ask you to allow me to call *fairies,* and these are ten thousand times more wonderful, more magical, and more beautiful in their work, than those of the old fairy tales" (Buckley 1879, 5-6). The first chapter of *The Fairyland of Science* deals briefly with the fairies heat, cohesion, gravitation, crystallization, and chemical attraction. The remainder of the book is devoted to explaining how the science fairies do their work in nature, particularly in sunbeams, gases, water, sound, plants, coal, and beehives. Buckley insists that any common object, "the fire in the grate, the lamp by the bedside, the water in the tumbler, . . . anything, everything, has its history and can reveal to us nature's invisible fairies" if "touched with the fairy wand of imagination" (Buckley 1879, 13). Entrance to the fairyland of science, then, is especially easy for children, who have the "glorious gift" of imagination that must be cultivated in adults (Buckley 1879, 7).

Despite Buckley's emphasis on the narrative quality of science, her book is less "fictional" than Gatty's *Parables* or even Brightwen's anecdotal *Wild Nature*. Buckley's sustained exploration of the analogies between fairies and natural forces functions more as a hook to capture the interest of her audience, and less as an element that disturbs the content of the story science tells. "With the exception of the first of the series," Buckley declares in her preface, "none of them have any pretensions to originality, their object being merely to explain well-known natural facts in simple and pleasant language." She acknowledges that she has availed herself freely of "the leading popular works on science" and that all of the material she presents has "long been the common property of scientific teachers" (Buckley 1879, v). Furthermore, Buckley refers several times with approval to the works of scientific naturalists like Tyndall and Huxley, praising the latter in particular for his ability to get beyond the dry facts of a scientific subject (Buckley 1879, 87, 128, 21, 23).

However, Buckley's scientific fairy tales present a challenge to the tales of scientific naturalists, not only in the moral lessons that we are to draw from them, but also in the teleological message they convey. In a number of her works, Buckley reinterprets the story of evolution in a way that emphasizes the moral dimensions of the process. The purpose of evolution was not, as Darwin had argued, merely the preservation of life, it encompassed the development of mutuality as well (Gates 1997). When Buckley deals with this theme in *The Fairyland of Science,* she connects it closely with the will of God. The mutual adaptation of bees and flowers "teaches the truth that those succeed best in life who, whether consciously or unconsciously, do their best for others." This leads her to conclude that from "our wanderings in the Fairy-land of Science" we "shall learn how to guide our lives" and we will see "that the forces of nature, whether they are apparently mechanical, as in gravitation or heat; or intelligent, as in living beings, are one and all the voice of the Great Creator, and speak to us of His Nature and His Will" (Buckley 1879, 237). Though Buckley is post-Darwinian in her emphasis on natural law—the invisible fairies are, after all, secondary natural causes—the world is no less a pre-Darwinian arena of divine design. Buckley's fascination with the "wonderful contrivances" in the relationship between bees and flowers, her perception that everything has a purpose (even those ancient plants that later became coal in order to make England great), and her belief that a child who gazes at nature with open eyes "must rise in some sense or other through nature to nature's God," all mark her out as a part of the natural theology tradition (Buckley 1879, 233, 192, 25). The wonder of fairyland is the same wonder perceived by the natural theologian.

Though more conventional in his selection of a narrative form, Richard Anthony Proctor (1837–88) had no reservations about indulging in daring speculations on the existence of extraterrestrial life in his many popular astronomical works. Thousands of members of the public were introduced to astronomy by Proctor's writings (Crowe 1986, 377). The youngest son of a wealthy solicitor, Proctor entered St. John's College, Cambridge, in 1856, where he studied theology and mathematics. To pay off a huge debt, incurred when an investment failed, Proctor turned to a career in journalism. Though his literary career was never a resounding financial success, he was able to develop a writing style that eventually won him recognition from both professionals and the public. In 1866 he was elected to the Royal Astronomical Society, later filling the office of honorary secretary, while his first major success, *Other Worlds Than Ours* (1870), was followed by triumphant lecture tours of America and Australasia. His other major works, all of which were published in three or more editions, include *Lessons in Elementary Astronomy* (1871), *Light Science for Leisure Hours* (1871), *The Sun* (1871), *The Orbs Around Us* (1872), *The Moon* (1873), and *Transits of Venus* (1874).

Proctor catered to the reading public rather than the expert astronomer. A number of his books were easy-to-follow guides for budding young astronomers, such as *A New Star Atlas for the Library, School and the Observatory* (1870), which by 1895 had sold nineteen editions. Proctor stressed hands-on astronomy, for he who takes his astronomy at second hand from books "may lightly disregard the grand lesson which the heavens are always teaching, and find only the grotesque and the incongruous, where in reality there is the perfect handiwork of the Creator." But the astronomer, Proctor declared, "imbued with the sense of beauty and perfection which each fresh hour of world-study instills more deeply into his soul, reads a nobler lesson in the skies" (Proctor 1870, 158). Proctor therefore saw himself as leading his readers to God through the lessons of astronomy.

Proctor's most popular book, *Other Worlds Than Ours,* which by 1909 was in its fourth edition, cast his science into a teleological framework. When considering the glowing mass of Jupiter, which can sustain no life, readers are invited to find a "raison d'être," for Proctor cannot accept the idea that God would create something for no purpose at all. The "wealth of design" in Saturn is so striking in Proctor's eyes that we cannot question but "that the great planet *is* designed for purposes of the noblest sort," though we may be unable to fathom those divine purposes. And Proctor enthuses as if he were a Bridgewater Treatise author over the recent discoveries of science, which "are well calculated to excite our admiration for the wonderful works of God in His universe" (Proctor 1970, 154, 159-60, 21). Proctor even structured *Other Worlds Than Ours* along the lines of a cosmic, post-Darwinian natural theology. The beginning chapters, "What Our Earth Teaches Us" and "What We Learn from the Sun," set the didactic tone for the entire book. Here nature's lessons concerning God's intentions and will are revealed by the telescope, spectroscope, and the other tools of the astronomer's trade. These first two chapters are a part of the nine-chapter section on the solar system, which leads into a series of three chapters on the stars and nebulae, extending the discussion of how God instructs us through nature to the rest of the universe. The concluding chapter, titled "Supervision and Control," is designed to teach the public how to read the lessons to be found by examining astronomy and the province of God. Proctor's story is a familiar one—it is the same cosmic story of purpose and design told by natural theologians, though it is validated by the findings of the most up-to-date astronomical science (Lightman 1996).

The Reverend John George Wood (1827-89) found that the cosmic story of natural theology was as appropriate for speaking to a popular audience of the minuscule wonders of the microscopic world as it was for conveying the majesty of the heavens. His *Common Objects of the Microscope,* published in 1861, was so popular that it eventually required a third edition. Wood was a prolific writer whose publications included *Bees* (1853), *Common Objects of the Sea Shore* (1857), *The Boy's Own Book of Natural History* (1860), *Animal Traits and Characteristics* (1860), *The Natural History of Man* (1868-70), *Common Moths of England* (1870), *Insects at Home* (1872), *Insects Abroad* (1874), *Half Hours with a Naturalist* (1875), *Half Hours in Field and Forest* (1875), *Common British Beetles*

(1875), *Common British Insects* (1882), *Illustrated Natural History for Young People* (1887), and *The Romance of Animal Life* (1887).

Celebrated as a great popularizer of his day for his long list of publications and his lecturing, Wood was an Oxford man, receiving his B.A. in 1848 and an M.A. in 1851 (Gates 1993, 304). He was appointed to a series of ecclesiastical and academic posts, including curate of the parish of St. Thomas the Martyr, Oxford, in 1852, chaplain to St. Bartholomew's Hospital in 1856, and reader at Christ Church, Newgate Street, but ill health in 1858 forced him to resign from all three. The success of his voluntary work with a parish choir led to his appointment as precentor of the Canterbury Diocesan Choral Union, whose annual festivals he conducted from 1869 to 1875. Wood later took up lecturing as a second profession, delivering a series of lectures from 1879 to 1888 throughout England and America. His lectures, particularly the Lowell Lectures at Boston in 1883–84, were renowned for their inclusion of color chalk illustrations (*Dictionary of National Biography,* s.v. "Wood, John George").

Many of Wood's books were designed as introductory works to a particular field of scientific study. Wood's *Common Objects of the Microscope,* like Gatty's *British Seaweeds,* is meant to be a catalogue of the basic facts for the "young and inexperienced observer" (Wood 1861, 37). In the preface Wood explains that his book has been produced to satisfy "a general demand" for "an elementary handbook upon the Microscope and its practical appliance to the study of nature" (Wood 1861, iii). Wood leads his readers through a series of microscopic observations of vegetable cells in plant hairs, starch grains, pollen, seeds, and algae, and of animal structures such as fish scales, insect antennae, feathers, and human skin, nails, bone, teeth, and muscle. After introducing readers to the different types of microscopes available, Wood instructs them to compare objects they view under the microscope to the illustrations provided and to check the accuracy of their observations (see figure 9.2). He then informs the audience of the conclusions to be drawn from such an exercise.

Though *Common Objects of the Microscope* appears to amount to little more than a list of different images viewed under the microscope, Wood nevertheless has a tale to tell his audience. It is the story of the divine wonders of the microscopic world that exist all around us but, until recently, remained unknown. Drawings, Wood declares, cannot do justice to the "lovely structures revealed by the microscope." Form and color can be indicated, "but no pen, pencil, or brush, however skillfully wielded, can reproduce the soft, glowing radiance, the delicate pearly translucency, or the flashing effulgence of living and ever-changing light with which God wills to imbue even the smallest of his creatures, whose very existence has been hidden for countless ages from the inquisitive research of man, and whose wondrous beauty astonishes and delights the eye, and fills the heart with awe and adoration" (Wood 1861, iv). In Wood's eyes, the microscope is a

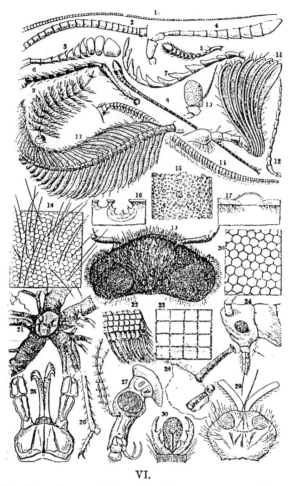

Figure 9.2 Plate 6 from *Common Objects of the Microscope* (Wood 1866?). The illustrations are by Tuffen West.

tool that allows access to a new world of wonders testifying to the existence and wisdom of God, a new revelation of his immense power.

Such an accessible tool was too important as an aid to faith to be left in the hands of the professional scientists for use in their private laboratories. Furthermore, the microscope was not one of those expensive scientific instruments that only the wealthy could afford to buy. Wood intended to restrict his observations "to that class of instrument which can be readily obtained and easily handled, and to those supplementary pieces of microscopic apparatus which can be supplied by the makers at a cost of a few shillings, or extemporized by the expenditure of a few pence and a little ingenuity on the part of the observer" (Wood 1861, 1). In the second chapter, on different types of microscopes, Wood goes into great detail on the least expensive, but most adequate, microscopes and gives tips on how to construct important apparatus on the cheap. It was not "the wealthiest, but the acutest and most patient observer who makes the most discoveries," Wood affirmed, "for a workman is not made, nor even known by his tools, and a good observer will discover with a common pocket-magnifier many a secret of nature which has escaped the notice of a whole array of *dilettanti* microscopists in spite of all their expensive and accurate instruments" (Wood 1861, 7).

In fact, once the amateur was armed with a decent microscope, there was no telling what important discoveries would result. As long as the amateur had an "observant mind" and the discipline to study "the commonest weed or the most familiar insect, he would, in the course of some years' patient labour, produce a work that would be most valuable to science and enrol the name of the investigator among the most honoured sons of knowledge" (Wood 1861, 5). As encouragement to his readers, Wood recounted the story of an old lady who, through her study of her own tiny backyard in the suburbs of London, contributed many "valuable original observations" to his notebook (Wood 1861, 4). There was no need to have access to a laboratory or to travel to the ends of the earth for exotic specimens to study. "So richly does nature teem with beauty and living marvels," Wood insisted, " . . . there is not one who may not find an endless series of Common Objects for his microscope within the limits of the tiniest city chamber" (Wood 1861, 3). Since the cosmic could be found within all common objects, anyone could use the microscope to conduct useful research in any place. Wood's entire series of books on commonplace objects in nature, whether they be moths, beetles, insects, or marine life, represents an open invitation to amateurs to become producers, not just consumers, of knowledge.

Similarly, Agnes Mary Clerke (1842–1907), a late Victorian popularizer of astronomy, summoned amateurs to contribute to the collection of astronomical data. Astronomy is "the science of amateurs," Clerke announces, and "there is no one 'with a true eye and a faithful hand' but can do good work in watching the heavens" (Clerke 1885, 7). Like Wood, Clerke was convinced that, with her help, the reader's encounter with nature would lead "towards a fuller understanding of the manifold works which have in all ages irresistibly spoken to man of the glory of God" (Clerke 1885, vi). The daughter of a bank manager with a keen interest in science, Clerke was educated entirely at home as a child. At the age of thirty-five, she embarked on a writing career and produced a series of important works, including *A Popular History of Astronomy During the Nineteenth Century* (1885), *The System of the Stars* (1890), *The Herschels and Modern Astronomy*

(1895), *Problems in Astrophysics* (1903), and *Modern Cosmogonies* (1905), that gained her partial admission into the male-dominated astronomical world.

In her *Popular History of Astronomy*, which reached a fourth edition in 1902 in addition to being translated into German, Clerke explained to the reading public how the new astronomical information generated by the spectroscope and camera had revealed a divinely designed universe full of complexity. A devout Catholic, Clerke perceived the hand of God in the most spectacular astronomical phenomena. Whether it be the evolution of the planets, whose growth is guided "from the beginning by Omnipotent Wisdom"; or the "sequence of Divinely decreed changes" by which nebulae are transformed into star clusters; or even gigantic galactic rifts of starless space, wherein "Supreme Power is at work in dispersing or refashioning" star clouds, Clerke saw the hand of God (Clerke 1885, 348; 1905, 297; 1903, 541). Though the picture of the cosmos emerging from the "new astronomy" of the late nineteenth century emphasized complexity and inexhaustible variety, Clerke nevertheless asserts that no matter where the telescope is pointed, it reveals the same pattern of design in the limitless regions of space that was so evident on the earth (Clerke 1885, 24). Even at the end of the nineteenth century, the natural theology tradition within popular scientific works was perpetuated by Clerke.

Clerke had no interest in experimenting with narrative form. Her scholarly works, written from the impersonal, objective point of view, imitate the form adopted by professional scientists. Clerke's high standing within the astronomical community, relative to other popularizers, also can be attributed to her attempt to interpret the larger meaning of recent astronomical discoveries to the professional astronomers themselves. Though contributing no original research, Clerke took the discoveries of isolated specialists and synthesized them. In her later works, Clerke often ended her review of the most recent research in a particular area with suggestions on the future work to be done by astronomers to answer the remaining questions. For some astronomers, like Richard A. Gregory, Norman Lockyer's protégé and assistant editor of *Nature*, Clerke represented a major problem. When Clerke began to work on projects that were less accessible to a popular audience and more technical in nature, Gregory wrote a series of vicious attacks on her scientific credentials in *Nature* pointing to her gender as grounds for refusing to take her work seriously. In presuming to instruct the experts as to the direction of their research, Clerke had, in Gregory's mind, crossed the line separating female popularizers of science from male professionals (Lightman 1997).

Two important observations have emerged from our study of Victorian popularizers of science. First, the question of who should participate in the making of science was still unresolved during the Victorian period. As Anne Secord has demonstrated, "the contest over science in the early nineteenth century was a contest about who could participate and on what terms" (A. Secord 1994, 299). By the mid-nineteenth century, popular science was becoming increasingly marginalized, and clergymen, women, artisans, and "nonprofessionals" in general were excluded by professionals. But science continued to be contested territory in the latter half of the nineteenth century. Wood's invitation to his readers to engage in the production of scientific knowledge is a theme lying latent in the works of popularizers, though participation in science likely meant something different to each of them. Do-it-yourself guides like Gatty's *British Seaweeds,* Proctor's *New Star Atlas* and Wood's entire series on common objects and animals encouraged the reader to actively observe nature and become familiar with basic scientific facts. But only Wood strongly encouraged his readers to seek out new knowledge. Popular science periodicals, in particular, mechanics' magazines and natural history periodicals, also encouraged amateur scientific activity (Sheets-Pyenson 1985, 553-54). The immensely successful *English Mechanic,* for example, a cheap mass-circulation science journal founded in 1865, was run cooperatively with its largely working-class readers, who used the pages of the publication to exchange views and information on a wide range of topics (Brock 1980, 111-13). The number of women engaged in popularizing science in the latter half of the nineteenth century is also indicative of the continuing efforts of marginalized groups to be a part of the scientific world. Buckley and Clerke accepted the traditional responsibility of women to educate and teach morality to the uneducated and the young, but both also represent a new confidence among women popularizers of science in their ability to speak with authority and to make contacts with leading scientists (Gates 1993, 298). But by the end of the century women began to lose their status as popularizers, not only because male popularizers perceived them as competitors, but also because of the introduction of natural history education into the schools, which reduced the need for science books in the home (Gates 1993, 305).

The unresolved question of who participates in the making of science was raised by popularizers in tandem with a second concern about what kinds of stories should be told about nature. For professional scientists, the answer was clear. The story should describe the operation of nature according to secondary law, particularly the law of evolution, avoiding all reference to supernatural causes. Professionals, like Huxley, Tyndall, and Herbert Spencer all tried their hand at writing popular works. Perhaps the most famous attempt at codifying and popularizing scientific knowledge in a systematic fashion to a wide reading public, *The International Scientific Series,* appeared in the United States and five European countries in over 120 titles between 1871 and 1910. Written for the most part by professional scientists, directed in its early years by an advisory committee composed of Huxley, Tyndall, and Spencer, and devoted, particularly in the eighties, to exploring the wider implications of evolutionary theory, the series stands as a monument to the efforts of professionals to control the public's understanding of modern science (MacLeod 1980). Popularizers like Gatty, Brightwen, Buckley, Proctor, Wood, and Clerke rarely sought to engage professionals in controversy, but they were not passive conveyors of the story of scientific naturalism. Their emphasis on the teleological, aesthetic, moral, and divine quality of nature connects them to the earlier natural theology tradition. Their alteration of the story told by scientific naturalists was not the result of ignorance or simplification—it was an intentional refashioning of recent scientific discovery into a form full of meaning for their audience. Competing interpretations of the cosmic significance of science were offered by popularizers committed to natural theology and popularizers and professionals grounded in scientific naturalism.

Perhaps it is more accurate to characterize the competition as existing between two groups of professionals, professional scientists and professional writers. Cross has analyzed the formation of writers into an occupational group during the nineteenth century (Cross 1985). As the mass reading public grew in numbers, it was possible for more and more "common writers" to make a living in the publishing industry. Certainly Gatty, Proctor, and Clerke devoted much of their time to their craft and depended heavily on their writing as an important source of their total income. They and the countless writers who supplied newspapers and journals with endless copy on scientific topics saw themselves as professional writers and therefore could draw strength from their link to the profession as a whole. The professionalization of science took place during the same period. The clash between two groups of recently established professionals may therefore be an important factor in the relationship between scientists and popularizers of science.

Scholars have barely scratched the surface in their attempts to understand the popularization of Victorian science. We still know very little about the major popularizers. Books, of course, were only one medium for the popularization of science. We need to know far more about how science was popularized during the Victorian period in magazines, journals, textbooks, children's literature, encyclopedias, and newspapers, and we need to go beyond the written word to popular lectures, museums, fairs, and exhibitions. But even so, concentrating on the thoughts and methods of the popularizers does not bring us into direct contact with the audience for whom these popularizations of science were intended. How did they read the message directed at them, and how was the message read and appropriated in different ways in different local settings by different social groups, whether they be aristocratic, middle class, or working class? This would lead us to examine the relationship between the popularization of science and elite and popular science.

In the sixth lecture of her *Fairyland of Science,* Buckley instructed her readers on "The Voices of Nature and How We Hear Them." Nature speaks to us, Buckley asserts, through sound waves, in a voice that "is sharp or tender, loud or gentle, awful or loving" (Buckley 1879, 159). Listen to these voices, Buckley advises the reader, "and ponder how it is that we hear them" (Buckley 1879, 166). Though Buckley has been dealing here with the physics of sound and the physiology of the human ear, her book, and the books of the other popularizers, are intended to be "voices of nature." The popularizers claimed, as did the professionals, to speak for a mute nature, or at least to interpret the true meaning of what seems to be a cacophony of noise for the reader whose ears are not properly attuned to the voices of nature. But behind these voices, Buckley and the others heard the "voice of the Great Creator" (Buckley 1879, 258). The voices of nature spoke to them of God's purpose, of his moral and natural laws, and of the place of humanity in the grand scheme of things. Their books were therefore designed to be reflections of the second revelation of God's will in nature, of the wonder to be found in the limitless heavens as well as the tiniest microbe, as Brightwen put it, "lessons in parables," or as Gatty says, "lessons of analogy." The cosmic stories of these popularizers testify to the continuing importance of religion to the reading public in the latter half of the nineteenth century and the belief that science was still an aid to faith, no matter what the Huxleys, Tyndalls, or Darwins said to the contrary.

Bibliographical Note

The best theoretical and historiographical studies on the concept of popularization of science in general are Whitley (1985), Hilgartner (1990), and Cooter and Pumfrey (1994). No published account of the popularization of science in nineteenth-century England exists. However, two dissertations focus on specific periods, Kitteringham (1981) from 1800 to 1830, and Hinton (1979) from 1830 to 1870. For information on Huxley, a professional scientist who was also an important popularizer, see Jensen (1991), Paradis (1978), and Block (1986). MacLeod (1980) explores the role of *The International Scientific Series* in popularizing the scientific naturalism of professional scientists. Myers (1985) looks at the career of a particular scientific metaphor within popular writing and culture in his essay on nineteenth-century popularizations of thermodynamics, which includes sections on such professional scientists as Tyndall, Kelvin, James Clerk Maxwell, and Balfour Stewart.

Moving on to the "nonprofessional" popularizers, Robert Chambers and his vastly popular *Vestiges of the Natural History of Creation* (1844) are examined in Hodge (1972), Millhauser (1959), J. Secord (1989, 1994), and Yeo (1984). In her book on Mary Somerville, Patterson (1983) focuses on another significant popularizer from the first half of the century. Useful secondary sources on the popularizers discussed in this chapter include Drain (1994), Katz (1993), Maxwell (1949), and Rauch (1997) on Gatty, Lightman

(1997) and Brück (1991, 1993, 1994) on Clerke, Lightman (1996) on Procter, and Gates (1997) on Buckley. There are a few short studies that deal with several popularizers by looking at a particular aspect of the popularization of Victorian science. Gates (1993) examines the way female popularizers retold the story of science, touching on Margaret Bryan, Jane Marcet, Buckley, Alice Bodington, Wood, and Brightwen. Myers's (1989) essay on scientific dialogues for children and women investigates Maria Edgeworth, Kingsley, and Ruskin.

As Myers (1994) points out, science existed in many forums and forms during the nineteenth century, not just in books. However, scholars are only beginning to explore these various forums. The popularization of science in periodicals has received attention from Sheets-Pyenson (1985) and Broks (1988, 1990, 1993), while Brock (1980) has drawn attention to the development of commercial science journals. Yeo's essay on encyclopedias (1991) does not address the popularization theme directly. The theme of science and its publics, which is closely connected to the issue of popularization, has also generated some interest. Shapin (1990) delivers a useful overview of the relationship between science and the public in the West from the seventeenth century to the present. Turner's (1993) chapter on public science in Britain from 1880 to 1919 dwells on the body of rhetoric, argument, and polemic produced by professional scientists to persuade the public or influential sectors thereof that science was worthy of support. Finally, Topham (1994) is one of the few who attempts to move from the authors of popular scientific works to their readers.

References

Block, Ed. 1986. "T. H. Huxley's Rhetoric and the Popularization of Victorian Scientific Ideas, 1854-1874." *Victorian Studies* 29:363-86.
Brightwen, Mrs. 1890. *Wild Nature Won By Kindness.* London: T. Fisher Unwin.
———. 1909. *Eliza Brightwen: The Life and Thought of a Naturalist.* Edited by W. H. Chesson. London: T. Fisher Unwin.
Brock, W. H. 1980. "The Development of Commercial Science Journals in Victorian Britain." In *Development of Science Publishing in Europe,* edited by A. J. Meadows. Amsterdam: Elsevier Science Publishers, 95-122.
Broks, Peter. 1988. "Science and the Popular Press: A Cultural Analysis of British Family Magazines 1890-1914." Ph.D. thesis, University of Lancaster.
———. 1990. "Science, the Press and Empire: 'Pearson's' Publications, 1890-1914." In *Imperialism and the Natural World,* edited by John M. MacKenzie. Manchester: Manchester University Press, 141-63.
———. 1993. "Science, Media and Culture: British Magazines, 1890-1914." *Public Understanding of Science* 2:123-39.
Brück, Mary T. 1991. "Companions in Astronomy: Margaret Lindsay Huggins and Agnes Mary Clerke." *Irish Astronomical Journal* 20:70-77.
———. 1993. "Ellen and Agnes Clerke of Skibbereen, Scholars and Writers." *Seanchas Chairbre* 3:23-42.

———. 1994. "Agnes Mary Clerke, Chronicler of Astronomy." *Quarterly Journal of the Royal Astronomical Society* 35:59-79.
Buckley, Arabella B. 1879. *The Fairyland of Science*. London: Edward Stanford.
Clerke, Agnes M. 1885. *A Popular History of Astronomy During the Nineteenth Century*. Edinburgh: Adam and Charles Black.
———. 1903. *Problems in Astrophysics*. London: Adam and Charles Black.
———. 1905. *The System of the Stars*. 2d ed. London: Adam and Charles Black.
Cooter, Roger, and Stephen Pumfrey. 1994. "Separate Spheres and Public Places: Reflections on the History of Science Popularization and Science in Popular Culture." *History of Science* 32:237-67.
Cross, Nigel. 1985. *The Common Writer: Life in Nineteenth-Century Grub Street*. Cambridge: Cambridge University Press.
Crowe, Michael J. 1986. *The Extraterrestrial Life Debate 1750-1900: The Idea of a Plurality of Worlds from Kant to Lowell*. Cambridge: Cambridge University Press.
Desmond, Adrian. 1987. "Artisan Resistance and Evolution in Britain, 1819-1848." *Osiris* 3:77-110.
Drain, Susan. 1994. "Marine Botany in the Nineteenth Century: Margaret Gatty, the Lady Amateurs and the Professionals." *Victorian Studies Association Newsletter* 53:6-11.
Fahnestock, Jeanne. 1993. "Accommodating Science: The Rhetorical Life of Scientific Facts." In *The Literature of Science: Perspectives on Popular Scientific Writing*, edited by Murdo William McRae. Athens: University of Georgia Press, 17-36.
Gates, Barbara T. 1993. "Retelling the Story of Science." *Victorian Literature and Culture* 21:289-306.
———. 1997. "Revisioning Darwin with Arabella Buckley." In *Natural Eloquence: Women Reinscribe Science,* edited by Barbara T. Gates and Ann B. Shteir. Madison: University of Wisconsin Press.
Gatty, Mrs. Alfred. [1855] 1861. *Parables from Nature*. London: Bell and Daldy.
———. 1864. *Red Snow and Other Parables from Nature*. 3d ser. London: Bell and Daldy.
Hilgartner, Stephen. 1990. "The Dominant View of Popularization: Conceptual Problems, Political Uses." *Social Studies of Science* 20:519-39.
Hinton, D. A. 1979. "Popular Science in England, 1830-1870." Ph.D. thesis, University of Bath.
Hodge, M. J. S. 1972. "The Universal Gestation of Nature: Chamber's 'Vestiges' and 'Explanations,' " *Journal of the History of Biology* 5:127-51.
Jensen, J. Vernon. 1991. *Thomas Henry Huxley: Communicating for Science*. Newark: University of Delaware Press; London and Toronto: Associated University Presses.
Katz, Wendy R. 1993. *The Emblems of Margaret Gatty: A Study of Allegory in Nineteenth-Century Children's Literature*. New York: AMS Press.
Kirk, John. 1965. *A Supplement to Allibone's Critical Dictionary of English Literary and British American Authors*. Vol. 1. Detroit: Gale Research Company.
Kitteringham, Guy Stuart. 1981. "Studies in the Popularisation of Science in England, 1800-30." Ph.D. thesis, University of Kent at Canterbury.
Lightman, Bernard. 1996. "Astronomy for the People: R. A. Proctor and the Popularization of the Victorian Universe." In *Facets of Faith and Science,* edited by

Jitse van der Meer. Lanham, Md.: Pascal Center for Advanced Studies in Faith and Science; University Press of America, 3:31-45.

———. 1997. "Constructing Victorian Heavens: Agnes Clerke and Gendered Astronomy." In *Natural Eloquence: Women Reinscribe Science,* edited by Barbara T. Gates and Ann B. Shteir. Madison: University of Wisconsin Press.

Locke, David. 1992. *Science as Writing.* New Haven, Conn.: Yale University Press.

MacLeod, Roy M. 1980. "Evolutionism, Internationalism and Commercial Enterprise in Science: The International Scientific Series 1871-1910." In *Development of Science Publishing in Europe,* edited by A. J. Meadows. Amsterdam: Elsevier Science Publishers, 63-93.

Maxwell, Christabel. 1949. *Mrs Gatty and Mrs Ewing.* London: Constable Publishers.

McRae, Murdo William. 1993. "Introduction: Science in Culture." In *The Literature of Science: Perspectives on Popular Scientific Writing,* edited by Murdo William McRae. Athens: University of Georgia Press, 1-13.

Millhauser, Milton. 1959. *Just before Darwin: Robert Chambers and Vestiges.* Middletown, Conn.: Wesleyan University Press.

Myers, Greg. 1985. "Nineteenth-Century Popularizations of Thermodynamics and the Rhetoric of Social Prophecy." *Victorian Studies* 29:35-66.

———. 1989. "Science for Women and Children: The Dialogue of Popular Science in the Nineteenth Century." In *Nature Transfigured: Science and Literature, 1700-1900,* edited by John Christie and Sally Shuttleworth. Manchester: Manchester University Press, 171-200.

———. 1994. "Forms and Forums for Popular Science in the 1830s." In *Science and British Culture in the 1830s: Papers to Be Presented at a Conference to Mark the Bicentenary of the Birth of William Whewell, 1794-1866.* Stanford in the Vale: British Society for the History of Science, 47-56.

Paradis, James G. 1978. *T. H. Huxley: Man's Place in Nature.* Lincoln: University of Nebraska Press.

Patterson, Elizabeth Chambers. 1983. *Mary Somerville and the Cultivation of Science, 1815-1840.* Boston: Martinus Nijhoff Publishers.

Proctor, Richard A. 1870. *Other Worlds than Ours.* New York: A. L. Fowle.

Rauch, Alan. 1997. "Parables and Parodies: Margaret Gatty's 'Parables from Nature.'" In *Natural Eloquence: Women Reinscribe Science,* edited by Barbara T. Gates and Ann B. Shteir. Madison: University of Wisconsin Press.

Secord, Anne. 1994. "Science in the Pub: Artisan Botanists in Early Nineteenth-Century Lancashire." *History of Science* 32:269-315.

Secord, James A. 1989. "Behind the Veil: Robert Chambers and *Vestiges.*" In *History, Humanity and Evolution: Essays for John C. Greene,* edited by James R. Moore. Cambridge: Cambridge University Press, 165-94.

———. 1994. "Introduction." In *Vestiges of the Natural History of Creation and Other Evolutionary Writings,* by Robert Chambers. Edited by James A. Secord. Chicago: University of Chicago Press, ix-xlv.

Shapin, Steven. 1990. "Science and the Public." In *Companion to the History of Modern Science,* edited by R. C. Olby, G. N. Cantor, J. R. R. Christie, and M. J. S. Hodge. London: Routledge, 990-1007.

Sheets-Pyenson, Susan. 1985. "Popular Science Periodicals in Paris and London: The Emergence of a Low Scientific Culture, 1820-1875." *Annals of Science* 42:549-72.

Stead, W. T. 1906. "My System." *Cassell's Magazine,* August, 293-97.

Topham, Jonathan R. 1994. "Beyond the 'Common Context': The Readership of the 'Bridgewater Treatises.'" In *Science and British Culture in the 1830s: Papers to Be Presented at a Conference to Mark the Bicentenary of the Birth of William Whewell, 1794-1866.* Stanford in the Vale: British Society for the History of Science, 57-66.

Turner, Frank M. 1993. "Public Science in Britain: 1880-1919." In *Contesting Cultural Authority: Essays in Victorian Intellectual Life.* Cambridge: Cambridge University Press, 201-8.

Whitley, Richard. 1985. "Knowledge Producers and Knowledge Acquirers: Popularisation as a Relation between Scientific Fields and Their Publics." In *Expository Science: Forms and Functions of Popularisation,* edited by Terry Shinn and Richard Whitley. Dordrecht: D. Reidel Publishing Company, 3-28.

Wood, Rev. J. G. 1861. *Common Objects of the Microscope.* London: George Routledge and Sons.

———. 1866?. *Common Objects of the Microscope.* London: George Routledge and Sons.

Yeo, Richard. 1984. "Science and Intellectual Authority in Mid-Nineteenth-Century Britain: Robert Chambers and *Vestiges of the Natural History of Creation.*" *Victorian Studies* 28:5-31.

———. 1991 "Reading Encyclopedias: Science and the Organization of Knowledge in British Dictionaries of Arts and Sciences, 1730-1850." *Isis* 82:24-49.

Young, Robert M. 1985. *Darwin's Metaphor: Nature's Place in Victorian Culture.* Cambridge: Cambridge University Press.

10

Science and the Secularization of Victorian Images of Race

DOUGLAS A. LORIMER

Victorian images of race were diverse in their origin and complex in their meaning, yet histories of scientific racism often underestimate that complexity. Studies of scientific racism seek to account for biological determinist explanations of racial inequality within Victorian science, whereas works on nineteenth-century racism attempt to account for the role of science within an ideology shaped by the broader cultural and political context. This latter approach seeks to establish the historical conditions that were conducive to the dissemination of racism and looks to a rather different range of sources to trace the form and chronology of the influence of science on popular images of race.

Within the history of science, the origins and institutional foundation of a pervasive scientific racism is most commonly identified with the 1850s and 1860s. The intriguing contest of personalities and ideas associated with the birth pangs of anthropology and the new Darwinian synthesis also witnessed the emergence of theories of the inequality and separate origins of human races that challenged the received Christian orthodoxy of the common origin and common nature of human beings. In identifying the origins of modern racism with these ideas derived from the natural sciences, historians, either explicitly or implicitly, claim a significant influence for an elite, but nonetheless limited, circle of intellectuals. Questions of intellectual influence are notoriously difficult to answer, and yet if we dodge this issue, we run the risk of a selective Victorianism that presents a distorted picture of nineteenth-century racism. The logic of biological determinism may be less important than the transformation of what the Victorians viewed as sentiment and what they viewed as knowledge. To come to a fuller understanding of the complexity and power of scientific racism, it needs to be

> The author is indebted to the following archives for permission to cite sources: the Anti-Slavery Papers, Rhodes House, Oxford; College Archives, the Library, University College London; the Huxley Papers, Imperial College, London; Archives of the Royal Anthropological Institute, London. Funding for research and travel was provided by Wilfrid Laurier University.

considered as part of the broader cultural and social process of "secularization" (Chadwick 1975; Heyck 1982).

I. Racial Stereotypes in Scientific Discourse

The flourishing field of cultural and literary analysis of colonial discourse has added greatly to our understanding of the binary connection between the stereotyped attributes assigned to the colonial other and the identity of the Victorian self. Much of this literature, born of the postmodern linguistic turn and seeking the source of the fetishism of the stereotypes in the psychology of the observer, stresses the continuity of racism over time (Brantlinger 1988; Said 1985, 1993; J. Richards 1989a; Gilman 1985; Malchow 1993; R. Young 1990; Kovel 1988; Gay 1993, 68-95). Historical studies are premised on the study of change over time and in the case of nineteenth-century scientific racism explore how developments in science effected changes in the Victorian discourse on race.

From the 1830s through to the 1870s, Victorian racial discourse took place within a common context in which scientific papers presented at learned societies were indistinguishable from the books and articles seeking to address an educated public. The Royal Geographical Society sponsored travels of exploration, most notably the quest for the source of the Nile, and upon their return, travelers first presented papers to learned societies, and those presentations later appeared as chapters in their best-selling travel accounts (R. M. Young 1985; Helly 1969; Livingstone 1984, 1992; Stafford 1989). Pride of place as men of science went to the medical practitioners with an interest in comparative anatomy. The purpose of their studies was to establish a correlation between anatomical features and mental traits and social behavior. In this task, the comparative anatomists were dependent upon the context of the common culture. They presumed that the psychological traits and social behavior of various races, as encapsulated in commonplace stereotypes, were known. The "new knowledge" they hoped to establish was the correlation of these traits with particular anatomical features.

Racial stereotypes derived their power and utility from their ambiguous and even contradictory character, as in the common depiction of colonized peoples as having the attributes of children and of savages. This ambivalence was particularly evident in early Victorian abolitionist and missionary literature inspired by evangelical Christianity. Exploiting the symbolism of black and white, this literature delighted in the polarities of good and evil. It depicted the sinful and degraded condition of peoples defined as savages and yet held forth the possibility of a spiritual conversion that allowed these same savages to express an innate Christianity. In a more secular form, this duality persisted in juvenile adventure stories appearing in popular serials from the 1860s onward. These narratives featured sinister and brutal villains of a variety of non-European origins, but also individuals cast as noble savages or as faithful subordinates in loyal support of the valiant English hero defending virtue and civilization (Bhabha 1986; Hannabuss 1989; J. Richards 1989*b;* Dunae 1977, 1980; Lorimer 1978, 75-81).

The scientific discourse on race involved a process of selection of attributes from existing racial stereotypes. This selection served to secularize images of race by defining what stereotypical attributes were matters of objective knowledge and what features were mere expressions of sentiment. In this process of selection, as will be seen in the discussion of popular authors, the racial discourse of the scientists retained the negative attributes of peoples designated as sinners, or savages, and redefined the more positive affirmations of abolitionists and missionaries as pious sentimentality. Similarly, those who claimed the authority of scientific knowledge rejected the possibility of conversion, or a transformation in the lives of colonized peoples, and advanced the deterministic case that the existing characteristics and inequalities of race were a fixed product of nature. Before Alfred Russel Wallace and Charles Darwin presented the case for natural selection, historical experience of the conflict between European settlers and aboriginal peoples seemed to demonstrate that change occurred principally by destruction. As the case for natural and cultural evolution was more fully articulated, from the 1860s onward, the possibility of change was admitted, but change was not effected by the deliberate actions of human agents. Under the premises of scientific naturalism, change occurred by long-term, uniform, and incremental forces comparable to those that operated in the natural world (Turner 1974, 17-30; Lightman 1987, 28-29).

II. The Politics of Race and Science

The scientists' quest for secularized "objective" knowledge about race also involved a transformation in their stance toward troubling ethical and political issues associated with colonial and other forms of racial oppression and conflict. Although the historiography of evolution has long abandoned a crude conflict between science and religion, accounts of the origins of anthropology are still premised on a conflict between science and humanitarianism. In 1844, Dr. Thomas Hodgkin and some friends in the Aborigines Protection Society (APS) founded the Ethnological Society of London in order to pursue scientific studies free of the political purposes of the parent body (Rose 1981; Stocking 1987, 240-47). Even though the APS was an important agency of cultural imperialism, we need to be cautious in simply dismissing it as an arm of the civilizing mission. From the APS's formation in 1837 until it amalgamated with the antislavery society in 1909, it defended the treaty entitlement of aboriginal peoples to land and their standing as equals before the law, and it acted as the principal political agency through which colonized peoples made their voices heard in Westminster (APS 1837; Bourne 1899).

Undoubtedly, institutional and intellectual tensions existed between those agencies that claimed to be concerned for the welfare of colonized peoples and the learned societies that claimed to study those same peoples, as physical objects defined by their anatomy or as cultural objects defined as survivals from the remote past of human evolution. Nonetheless, to describe those tensions as a conflict between humanitarianism and science is to remain trapped within the ideological pretensions of science. The political disengagement of science from humanitarian causes was part of a larger process that transformed ideas about the place of human beings in nature and created new uncertainties about ethical and social relationships between peoples differing in race and culture. This process of secularization may well represent a liberation of reason from the religious and cultural authority of the past. The disturbing question is why this liberation weakened existing forces of resistance to racism and, at the same time, strengthened the forces of colonial oppression.

Much of the literature on scientific racism focuses on the mid-Victorian discussion of theories of polygenesis and monogenesis dramatized by the controversial polemics of the Anthropological Society of London (ASL) led by Dr. James Hunt, its publicity-seeking founder. The controversy marked the birth pangs of a newly named science of anthropology claiming to liberate itself from the humanitarian associations of the Ethnological Society. In the 1860s, Hunt and his associates, inspired by the racial theories of Robert Knox, launched a vigorous assault on religious and humanitarian agencies such as the missionary and antislavery movements. Eventually, in 1871, the Darwinian establishment led by Thomas Henry Huxley managed to reunite the two rival societies in the newly created and more respectable Anthropological Institute (Hunt 1863*a;* Knox [1850] 1862; ASL 1869; P. M. Duncan to T. H. Huxley, 8 September 1868, 25 September 1868, and 2 June 1869 [Huxley Papers]; copy Hyde Clarke to James Hunt, 21 August 1868 [Huxley Papers]; Report of James Hunt to President and Council of ASL, 7 June 1869 [Huxley Papers]; Huxley [1870] 1901; Biddiss 1976; Burrow 1970, 118–36; Rainger 1978; Stocking 1971*b;* 1987, 246–57; Lorimer 1978, 137–61; E. Richards 1989).

Knox and Hunt may share a fate common among many Victorian authors. The two racial determinists have had far more readers between 1963 and 1995 than between 1863 and 1895.[1] The historiographical revival of

---

1. In a study of the *Journal of the Anthropological Institute,* from its foundation in 1871 to 1900, I was able to identify only one reference to Knox and one to Hunt, and both came from former members of the ASL (Lorimer 1988, 412); see also the discussion of popular authors below.

Knox, Hunt, and the ASL began in the 1960s, when scholars started to explore the origins of modern racism. Nonetheless, a closer look at the post-1870 period indicates that monogenesis remained the orthodox view among British professional scientists. Rather than define scientific racism by the racial typologies derived from comparative anatomy, a more fruitful approach may be to probe the broader implications of scientific naturalism for the direction of scientific discourse on race in the later nineteenth century.[2]

Part of the interest in anthropology grew out of the quest for "natural man" living in a state untouched by Western colonialism. Such specimen peoples were increasingly difficult to locate, and often Victorian travelers underestimated how extensively such peoples' lives had already been altered by the colonial encounter. Evolutionary anthropologists treated their chosen objects of study as living fossils providing through their systems of kinship, social organization, and religion, as well as the material culture of their art and artifacts, evidence of the state of nature out of which civilized Victorians had evolved. As president of the Anthropological Institute and an advocate of the collection of material artifacts on evolutionary principles, Colonel Augustus Lane-Fox feared that the impact of the West would soon make it impossible to observe non-European cultures in "their pristine condition." As the nation that had, according to Lane-Fox, "done more than any other to destroy all those races and to obliterate their culture," the English had a special duty to keep a "scientific record of that which we destroy." Accepting that the destruction of aboriginal peoples and cultures was part of a natural rather than a historical process subject to human will and agency, ethnographic description and collection became a matter of archival conservation (Lane-Fox 1877, 178; see also Lane-Fox 1874–75; van Keuren 1984).

Just as ideas of biological evolution were part of the social fabric of the age, so too ideas of social and cultural evolution were part of the fabric of the ideology of empire (Burrow 1970; Stocking 1987; R. M. Young 1982). The comparative assessment of cultures strengthened the sense of British preeminence. The search for natural causes of change eliminated the historical agency and the moral responsibility of the colonizers and similarly denied the colonized a historical role in determining their own fate. In this naturalistic vision, primary resistance movements were the irrational outbreaks of savages, and the political demands of Western-educated nationalists were the imitative voices of mission converts (Fabian 1983; Robinson 1972; Atmore 1984). The most evident change in perception, though, was

2. Both Stepan (1982, 83–110) and Stocking (1971, 42–68) rely on American and French scientists for evidence of late-nineteenth-century polygenesis; cf. Lorimer 1988, 412–21. Evelleen Richards (1989) usefully points to the influence of scientific naturalism, though in the post-1870 discussions of race, in my opinion, it was articulated more fully through evolutionary theories than through comparative anatomy.

in the differing scale of time. Early Victorian humanitarians expected a conversion or an emancipation to occur within their own lifetime. Scientific naturalism imposed not a human, historical time frame, but a natural or geological time frame. Consequently, the British Empire might well be a force for evolutionary progress, but change would be at the pace of glaciers, and the ice age of the British Empire would endure long past the life of Victorian scientists, or even the lives of their children's children.

The marriage of scientific naturalism and imperial ideology was clearly evident in the political orientation of the Anthropological Institute. In his treatment of the institutional development of Victorian anthropology, George Stocking argues that the formation of the institute in 1871 represented a political compromise. On the one hand, the leaders of the new institute rejected the older Ethnological Society's links with humanitarian causes, and on the other, they equally repudiated the political advocacy of racial determinism by James Hunt and the ASL (Stocking 1987, 269-73).

In the larger perspective of Victorian racism, this political compromise has some disturbing implications. After the death of Thomas Hodgkin in 1866, no leading member of the Aborigines Protection Society played a significant role in anthropological studies or in the Anthropological Institute. Similarly, no leading figure in the institute belonged to the APS, or participated in its public meetings, or signed APS petitions to protest against colonial abuses. In their anniversary addresses, presidents of the institute—for example, Francis Galton, who served in that office from 1885 to 1889—extolled the practical lessons of anthropological science for Britain's imperial enterprise. The institute's most ambitious if unsuccessful effort was to create a state-funded Imperial Bureau of Ethnology designed to assist colonial administrators by providing anthropological knowledge of subject races. While this exercise failed, the Anthropological Institute did succeed in gaining the title "Royal" in 1907. Ironically, the newly amalgamated Aborigines Protection and Anti-Slavery Society lost its royal patronage in 1910. The society's protest against racial discrimination in the newly constituted Union of South Africa made it partisan and an unsuitable agency for the favors of a constitutional monarch (RAI 1907, 1909, [1911], 1912; "Mr. Asquith and Anthropology," *Times,* 12 March 1909, 10d; Anti-Slavery Society 1910, entries 1798 [5 August 1910], 1811 [7 October 1910], and 1830 [4 November 1910]).

The political compromise of the Anthropological Institute satisfied the test of scientific detachment and objectivity. It avoided the contrary extremes of humanitarian agencies, which at times defended aboriginal rights and criticized the excess of colonialism. It also avoided the crude racist determinism of Knox and Hunt, which championed racial subordination and oppression. Nonetheless, the compromise was not apolitical, as the institute's leading members actively promoted the practical benefits of anthropology for the colonial state.

III. The Popularization of Scientific Racism

The historiographical focus on Knox, Hunt, and the ASL has led scholars less aware of developments after 1870 to view these racial determinists as popular and influential authors. A brief review of the works of four popular writers—Reverend John George Wood, Robert Brown, Edward Clodd, and Professor Augustus Henry Keane—will suggest that the popularization of scientific racism in the later nineteenth century was not simply an extension of mid-Victorian racial determinism, but a consequence of broad transformations within the metropolitan culture. Racial stereotypes derived from the common cultural context persisted in a secularized form, while the professionalization of intellectual endeavor gave science a new authority. Through the application of the ideas of scientific naturalism, popular writers also claimed a detached objectivity reflective of the political compromise of Victorian anthropology. In effect, the popular science of race detached itself from the historical and ethical issues of human agency and conflict and defended the imperial enterprise and racial subordination as products of nature best understood by the lessons of professional science.

If we are going to deal with the tricky question of the popular influence of scientific ideas, then the reception of these ideas and their publishing history need to be taken into account. For example, James Hunt faced hisses and catcalls at the British Association in 1862. His principal critic, William Craft, a fugitive African American slave, received enthusiastic cheers from the same audience (Hunt 1863b; Ripley 1985; Blackett 1983, 191-93). According to the accounts of his publisher, five hundred copies of Hunt's pamphlet *The Negro's Place in Nature* (1863) were printed. At the most, 250 copies were sold, and after 1865 the remainder gathered dust in the warehouse (Kegan, Paul, Trench, Trubner, Henry King [1858-1912] 1973).

IV. Reverend John G. Wood (1827–89)

For a popular presentation of scientific images of race in the 1860s, we should begin with Reverend John George Wood, an Anglican clergyman and much-beloved lecturer and widely read author on natural history, who produced an illustrated serial, *The Natural History of Man*, in thirty-two parts between 1868 and 1870 (J. Wood 1868-70; T. Wood 1890; Mumby 1934, 75-79; *Dictionary of National Biography* [1885-1900], s.v. "Wood, John George"; *Times*, 5 March 1889, 9). George Routledge put up five thousand pounds for its production and in the initial contract based the author's royalties on a sale of fifteen thousand copies per issue. This estimate proved somewhat optimistic, and after six months, six thousand copies became its normal run (Routledge and Sons [1867] 1973a; [1869] 1973b; [1853-1902] 1973c, vol. 1, Jan. 1867-Jan. 1870; T. Wood 1890, 76-79; Mumby 1934, 78).

While not as great a success as some of Wood's other popular science works, *The Natural History of Man* reached a larger readership than works more frequently referred to in the history of scientific racism. At the price of a shilling an issue, however, the serial aimed at a middle-class readership. It was favorably assessed in an illustrated review in *Nature,* and Edward Burnett Tylor, despite his reservations about the distortions of Wood's illustrators, recommended Wood as the best and most readily available compendium on the arts and culture of primitive man (Power 1870; Tylor 1874*a*, 1874*b*, 1881, bibliography).

Coming out of a natural history tradition that kept human beings apart from nature, Wood defined his subject as dealing with those peoples "who have not as yet lost their individuality by modern civilization" (J. Wood 1868-70, 1:v). He included no discussion of comparative anatomy but relied on travel accounts, working feverishly in the British Museum's reading room to extract forty-eight pages of text for each monthly issue. Nor was Wood an evolutionist, as he refused to accept the application of Darwin's ideas to human beings (T. Wood 1890, 113-14).

The absence of comparative anatomy and of evolutionary theory did not mean that Wood's text, and its accompanying engravings, were free of racial stereotyping. Wood generalized from particular examples to present a universal, conventional stereotype of the "savage." For instance, his description of the Makoba peoples in the northern reaches of the Kalahari Desert incorporated these universal yet contradictory attributes:

> Their character seems much on a par with most savages, namely, impulsive, irreflective, kindly when not crossed, revengeful when angered, and honest when there is nothing to steal. To judge from the behaviour of some of the Makoba men, they are crafty, dishonest, and churlish; while, if others are taken as a sample, they are simple, good-natured, and hospitable. Savages, indeed, cannot be judged by the same tests as would be applied to civilized races, having the strength and craft of man with the moral weakness of children. (J. Wood 1868-70, 1:376)

Like many of his contemporaries, Wood specifically rejected any notion of "noble savages" and claimed their common vices—"drunkenness, cruelty, immorality, dishonesty, lying, slavery"—were all firmly established prior to contact with Europeans (J. Wood 1868-70, 1:338, 2:47, 195, 336, 343). Contact with white settlers only exacerbated these "savage" tendencies. Wood presented the decline of aboriginal peoples as a natural process, but he made no reference to Darwin, Herbert Spencer, or natural selection. In the case of Australian Aborigines, their failure was more biblical than Darwinian. According to Wood, they failed "to exercise dominion over the beasts and the birds," and "although they inherited the earth, they did not subdue it, nor replenish it" (J. Wood 1868-70, 2:105).

V. Robert Brown (1842-95)

In the 1870s, Reverend Wood's main rival as a popularizer was Robert Brown, an Edinburgh-trained botanist who made his reputation as head of the Vancouver Island Exploring Expedition in 1864. After returning from the Pacific Northwest, he failed in his quest for a university post as a biologist. He then turned to journalism, eventually, in 1878, joining the editorial board of the *Standard,* a Conservative London daily. In addition to his journalism, Brown became an author of popular science works for Cassells. Between the 1870s and 1890s, he wrote a number of weekly and monthly serials that were illustrated compilations of travel accounts describing peoples and places around the world (*Dictionary of National Biography* [1885-1900], s.v. "Brown, Robert"; *Times,* 29 October 1895, 8c; Nowell-Smith 1958, 104; Hayman 1989, 3-28, 198-201).

Like Wood, Brown made limited use of comparative anatomy. In his six-volume *Peoples of the World,* the thirteen-page introduction borrowed its typology from Robert Gordon Latham (1812-88), a medical doctor, influential philologist, and a member of the Ethnological Society in the 1850s. While Brown, in common with many Victorian commentators, ranked races hierarchically, he had little use for the technicalities of comparative anatomy or for its practitioners, whom he called "closet naturalists" and "untravelled anthropologists":

> They find in their museums a shelf of skulls labelled more or less accurately; they compile a few vocabularies from the travels of voyagers not much better informed, and even less scientific, than themselves; . . . and call the result an "ethnological scheme," any objections to which are overwhelmed with a cloud of fragments of speech, mixed with the names of bits of bone. (Brown 1881-86, 1:7)

Affirming monogenesis as the orthodox scientific view, not from religious doctrine, but from evidence of the fertility of racial crosses, Brown saw that the real value of his volumes lay in the description of the social and psychological attributes of the world's peoples.

His limited use of comparative anatomy in no way inhibited Brown's recourse to racial stereotypes. Despite the fact that Brown was well traveled, his xenophobia informed the binary contrasts of virtue and vice in his description of both European and non-European peoples. His description of what he called the conservative virtues of the English, including "truth, fidelity and sincerity," contrasted with his list of the failings and vices of the Irish, Italians, French, and Germans (Brown 1881-86, 6:309-10, 319-20, 5:178-79, 269, 6:15, 264). He followed the conventional distinction between "semi-civilised" and "savage" peoples, but dishonesty, treachery, and cruelty were as much a characteristic of the peoples of India and China as of the indigenous inhabitants of the Americas, Australasia, and Africa. (Brown 1873-76, 4:110-15, 173; 1884-89, 5:44-45).

With some insight he observed that conflict, suspicion, and treachery were endemic in colonial situations. In a chapter "On the Decline of the Wild Races," he treated the advance of white colonists as inevitable, but with some sympathy toward Native Americans, he enumerated the specific causes of their decline rather than relying on an appeal to a Darwinian law or to Spencer's "survival of the fittest" (Brown 1873-76, 3:198-221). During his career as a popular writer, Brown became a more ardent advocate of British imperialism, especially in Africa, as he praised Sir George Goldie's Niger Company and Cecil Rhodes's British South Africa Company (Brown 1892-95, 4:191, 246-58; see also 1873-76, 4:117; 1884-89, 5:155-56). While he was aware of the suffering experienced by peoples facing the onslaught of colonial expansion, Brown dodged the issue of moral responsibility. Appealing to scientific objectivity, he observed that "we, as men of science, are only concerned with the fact that these things are brought about, and their cause; but with the ethical side of the question we are, fortunately, not called upon to deal" (Brown 1873-76, 3:221).

While he enumerated the natural vices of various peoples, Brown paid particular attention to their capacity as productive and compliant laborers. He praised the industry and quiescence of migrant Chinese laborers in California and British Columbia but reiterated the conventional charges of idleness against the ex-slave populations of the southern United States and the West Indies (Brown 1873-76, 4:208-15, 3:195-96; 1884-89, 2:167-73, 183-90). Relying on Sir Samuel Baker, an African traveler from a West Indian slave-holding background, Brown offered what became the conventional wisdom in the late nineteenth century. While Britain had pursued the morally worthwhile goal of ending the slave trade and slavery, emancipation itself had been a failure (Brown, 1873-76, 3:177-98, 298-307). The lesson of this failure was that colonial development in Africa would have to be under white direction. By necessity it would depend upon African labor, but that labor would require compulsion through the law or taxation. The failure of the abolitionists and missionaries was that they tried to assimilate non-Europeans to English conditions, whereas scientists, free of humanitarian sentiments, knew that colonial conditions had to be differently constituted to take into account the racial inheritance of colonized peoples (Brown 1873-76, 3:198; see also Brown 1870, 532-41; 1892-95, 4:158).

Neither Robert Brown nor Reverend John Wood made any claim to originality; their significance lay in the form in which they conveyed conventional ideas to a more popular readership. Both authors claimed a status as detached communicators reproducing observed facts or knowledge from their own travels or from recognized authorities. For Wood, his selections from travel literature aimed to present human beings as natural objects and, in the tradition of natural history, to draw moral lessons from what he presented as accurate observations. Brown claimed a scientific objectivity free from the humanitarian commitments of the past, and yet his surveys defended colonialism and racial subordination in a fashion indistinguishable from that of editorials in the *Standard* ([Brown] 1889). Neither author made much use of comparative anatomy, yet their racial stereotypes incorporated the binary opposition of the virtues of the Victorian self and the vices of the colonial other. In the secularization of Victorian images of race, Wood and Brown facilitated the process of detaching observations about race from divisive ethical and political issues and confirmed that existing prejudices about the psychological and social traits of races were matters of "scientific fact."

Wood and Brown attempted to describe the world's peoples rather than to offer an explanation of human physical and cultural varieties, whereas two other popular writers, Edward Clodd and Professor Augustus Henry Keane, defined their task in a different fashion. They aimed to popularize the ideas of professional science, and consequently they devoted greater attention to evolutionary ideas and racial typologies. The enormous expansion of the size of the reading public, as a consequence of the impact of state education, and the reduction of costs of books for a growing middle-class, lower-middle-class, and respectable working-class public, afforded Edward Clodd, Augustus Keane, and others opportunities that simply did not exist for mid-Victorian racial determinists such as Knox and Hunt.

VI. Edward Clodd (1840–1930)

Edward Clodd, a banker from a Liberal Nonconformist background, was a popular science writer and an activist in promoting secular education and a number of other progressive causes including votes for women. Clodd assisted Richard Proctor in editing *Knowledge,* a popular science magazine in competition with Norman Lockyer's *Nature* but seeking a less well-educated readership among the growing ranks of schoolteachers and office clerks of the lower middle class. Clodd, a popular lecturer on astronomy and on primitive man, became best known as an author of science books for juveniles that promoted a secular reconciliation of science with Christian morality. In the 1860s, partly as a result of reading Edward Burnett Tylor's *Primitive Culture,* Clodd experienced a crisis of religious faith. Owing to this crisis and his varied intellectual and educational interests, he became active in the National Sunday School Association and in the Folk-lore Society, serving as its president in 1895. By the late 1880s, Clodd had become an agnostic and actively promoted his secular views, serving as chair of the Rationalist Press Association in 1906 (McCabe 1932, 13-32; *Dictionary of National Biography Supplement* [1922-30], s.v. "Clodd, Edward").

Clodd's rational and secular viewpoint gave his writings a cultural relativist slant. In *The Childhood of Religions,* first published in 1875, and according to his publisher reaching eleven thousand copies in its edition of 1889, he instructed his juvenile readers that the account of creation in Genesis was comparable to similar myths in other religions. Furthermore, if it was read not as a literal account but as a metaphor, Clodd maintained, much to the consternation of his religious critics, that Genesis was quite compatible with evolutionary science (Clodd [1875] 1889, 10-44, 230-40; Catholicus 1880).

Clodd associated the major religions of south Asia with the eastern branch of the Aryan race, and identified Islam as Semitic in origin. He defended Mohammed as a great religious leader and yet still offered a conventional and contradictory stereotype of the prophet's followers, advising his readers that "the Arabs are to-day what they were hundreds of years ago; lovers of freedom, temperate, good-hearted; but withal crafty, revengeful, dishonest" (Clodd 1889, 210, 222).

In *The Childhood of Religions,* Clodd also discussed "ethnology," or the study of human races. He offered no detailed racial typology, affirmed the common origins of human beings, and stressed the contrast between races, which had evolved over long periods of time:

> Our present knowledge strengthens the early belief that man first arose in one part of the earth, but the result of many causes, such as changes in climate, removal to new lands, different food, working through long ages, has been to create wide varieties in his descendants, such as we see between an Englishman and a Negro, and between a Hindu and a Chinaman. (Clodd 1889, 78-79)

In *The Childhood of the World: A Simple Account of Man's Early Years,* which sold twenty thousand copies within four years of its publication in 1873 and included among its bedtime readers the Prince of Wales and Queen Victoria's grandchildren, Clodd evoked human life in the earliest Stone Age. He called upon his young readers to imagine life "far across the wind-tossed seas, far away in such places as Australia, Borneo, and Ceylon" where "there live at this day creatures so wild that if you saw them you could scarcely believe that they were human beings and not wild animals" (Clodd 1873, 5; McCabe 1932, 29-32).

In 1887, in *The Story of Creation,* which sold two thousand copies in its first week of publication and five thousand copies within three months, Clodd reiterated the claim that human races had a common origin. In this popular work, and in its abridged juvenile version, *A Primer of Evolution* (1895), he offered a Darwinian explanation for racial varieties, which occurred "through the potent agency of natural and sexual selection acting upon variations induced by diverse conditions—conditions which have surrounded man in virtue of his migrations from pole to pole, and which have called his industry and resource into full play" (Clodd [1887] 1888, 203; see also Clodd 1895, 144-45; McCabe 1932). Incorporating the Malthusian logic of human fertility exceeding the means of subsistence, Clodd stressed the evolutionary role of "the wholesale destruction of communities by wars, pestilences, famines, and catastrophes." According to Clodd, "Man's normal state is therefore one of conflict," and yet out of this fierce struggle of "savage instincts" humans evolved "curiosity, the mother of knowledge," and "in the conflicts between tribes, patriotism, morals and the hardy virtues" (Clodd 1888, 211-13).

This struggle facilitated the evolution of a modern social and political order, as conflict dictated the rise of the ablest leaders and the "specialization of peoples into classes" (Clodd 1888, 214). Globally, this evolutionary process determined that higher civilizations, a product of the challenging environments of temperate climates, dominated over the less highly developed (Clodd 1887, 221). Under these modern conditions, conflict continued not so much on the battlefield as in the marketplace. "Among advanced nations," Clodd observed, "the military method may be more or less superseded by the industrial, and men may be mercilessly starved instead of being mercifully slain, but, be it war of camps or markets, the ultimate appeal is to force, and the hardiest and craftiest win" (Clodd 1887, 221; 1888, 212).

As a liberal, secular, progressive educationalist, Edward Clodd fulfilled his chosen role as a popularizer of the ideas of professional science. Without recourse to racial typologies, the technicalities of comparative anatomy, or the unorthodox thesis of polygenesis, his accounts of human evolution incorporated the conventional binary opposition of savage and civilized life. In this view, the inequalities of races in power, wealth, and status were natural conditions, and change occurred by a form of natural selection in which conflict sanctioned the existing order within the metropolitan society, and globally in the colonial world.

VII. Professor Augustus Henry Keane (1833–1912)

Whereas Clodd's primary aim was to popularize evolution rather than theories of race, Professor Augustus Henry Keane addressed the question of race directly. Keane made his reputation as a popularizer of anthropology, for he wrote an astonishing range of articles for magazines, encyclopedias, and other reference works, collaborated with other scientists in producing illustrated serials, and authored textbooks for secondary school and university students (*Who Was Who, 1897-1915*, s.v. "Keane, Augustus Henry"; Brabrook 1912, 53; Johnston 1908, 143; Parkyn 1908).[3]

---

3. Keane was a regular reviewer for *Nature,* claimed to have written 110 articles for the *Academy,* wrote numerous entries on ethnology for the *Encyclopaedia Britannica* and *Chambers's Encyclopaedia,* and contributed eight hundred entries for *Cassell's Storehouse of General Information.*

As a linguist, classics master, and author of school grammars, Keane became interested in anthropology when, in 1878, he assisted Alfred Russel Wallace in preparing vocabularies for the volume on Australasia for *Stanford's Compendium of Geography and Travel.* In a paper first presented before the Anthropological Institute in 1879, and then published as a four-part article in *Nature,* Keane offered a speculative account of the independent origins of his three primary black, yellow, and white races (Keane 1879-80, 1880-81; Lorimer 1988, 413-16; 1990, 379-83). Keane later identified his thesis as a form of "unorthodox monogenesis" in which human beings had a common place of origin rather than a common ancestor. Even though the polygenetic implications of Keane's theory were clear, in his text *Ethnology* (1896), he affirmed the specific unity of human beings. In his review of racial classifications since Georges Cuvier, he made no mention of James Hunt, associated Robert Knox with the discredited American polygenist school, and attributed greater significance to the ethnology of James Cowles Prichard and Robert Gordon Latham (Keane, 1880-81, 199, 274; 1896, 142-43, 150-56, 165-66).

Keane's extensive knowledge of Asiatic languages, his published vocabularies and articles, and his acquaintance with leading members of the Anthropological Institute gained him the appointment of professor of Hindustani at University College, London, in 1883 (University College 1883). As a philologist, he believed that races could be classified both by physical attributes and by mental and psychological characteristics. In his textbook *Man: Past and Present,* Keane headed each chapter with a brief racial taxonomy. Much like labels attached to exhibits at zoos or museums, the taxonomy included a brief description under the following headings: Primeval Home; Present Range; Physical Characteristics—Hair, Colour, Skull, Nose, Eyes, Stature, Lips, Legs, Arms, Feet; Temperament; Speech; Religion; Culture. Under "temperament," Keane provided a brief summary of mental attributes and abilities. For example, his synopsis described the "Southern Mongols" as "somewhat sluggish, with little initiative, but great endurance, cunning rather than intelligent; generally thrifty and industrious, but most indolent in Siam and Burma; moral standards low, with slight sense of right and wrong" (Keane 1899, 170). These profiles of psychological traits—for example, of the "Mongols" or of the "Negro"—reappeared in Keane's contributions to encyclopedias and magazines as well as in his own and other authors' textbooks. He defended these descriptions against the charge that individuals within races might vary in physique and temperament by the claim that he described only "ideal types" (Keane 1890-94, 367; 1894, 729; 1899, 36; 1908, 16).

The form in which he presented his racial typologies was as important as their content. In the brief but definitive prose of encyclopedia entries or in the taxonomies of his textbooks, he intended his typologies, presented in a more systematic fashion than in Wood or Brown, to be learned by rote by captive student readers. Alfred Cort Haddon, who held the first lectureship in anthropology at Cambridge, claimed that one of the skills to be learned from the subject was that of "race discrimination." This skill at identifying racial types would be an important test of anthropological knowledge for prospective colonial administrators. In 1906, the examinations for the diploma of anthropology at Oxford required students to be proficient in anthropometry and in the identification of the characteristics of racial types from representative photographs (Haddon 1910, 13, 47-48; Read 1906, 56-59).

Keane took a lead in using photographs to illustrate his racial types. In 1905, he helped edit *The Living Races of Mankind*, a fortnightly serial, with eight hundred photographs and twenty-five colored plates and maps, and costing sevenpence an issue. He included 270 photographs in his text *The World's Peoples*, published by Hutchinson in 1908. The photograph had a number of advantages in objectifying its subject as an ideal racial type. To readers unaware of the art of the photographer, and of the author's role in selecting the picture to illustrate the text, the photograph represented reality in a more authoritative manner than either the written word or an artist's engraving. The photograph also froze its subject in time. Ethnographers invariably sought to capture the "native" subject in states of dress—or to late Victorian and Edwardian readers, states of undress—representative of a natural or savage state free of Western influences. In addition, the photograph offered the possibility of presenting racial types without the dull, incomprehensible technicalities of comparative anatomy. The editors of *The Living Races of Mankind* reassured readers that even though anthropology was known as "a dry and difficult—not to say repellant—science," it was now accessible through the photograph (Lydekker 1905, i). Of course, ethnographic photos may have a scientific purpose when taken in the field, but that purpose was transformed by the publisher's marketplace. Ethnography had long had an exotic, not to say erotic, appeal. When the covers of *The Living Races of Mankind* displayed bare-breasted African and Polynesian women on Edwardian newsstands, ethnography became pornographic (Green 1984; Edwards 1992; see also Stocking 1987, 197-208, 216-19; Kuklick 1991, 13).

As an acknowledged, popular authority on anthropology, Keane was sought even by those former antagonists of the science of man—the missionary movement. In 1894, he contributed to the Church Missionary Society's *Missionary Atlas*, giving his usual description of racial types. He emphasized his belief in monogenesis and the harmony between science and Scripture while omitting his unorthodox claim that modern races had separate places of origin. Nonetheless, a hint of polygenesis remained, for Keane suggested that religious differences corresponded to racial differences—blacks were "non-theistic," yellows were "polytheistic," and whites were "monotheistic" (Keane 1894, 721-30). Racial typologies were also thought fit knowledge for children at Sunday school, as can be seen from the seven-part series on racial types included in 1896 in *News from Afar,* a juvenile magazine of the London Missionary Society (Horne 1896).

In 1897, in recognition of his role as a popularizer of ethnology, Keane received a civil list pension (Brabrook 1912, 53). Keane's text *Ethnology* (1896), according to Grant Allen, was "the first systematic treatise on Ethnology as a whole that has appeared since the general acceptance of the evolutionary theory" (Allen 1896, 159-60). Nonetheless, some reviewers worried about the distorting process by which the knowledge of experts was conveyed to the general public. In 1908, the reviewer of *The World's Peoples* in *Man,* the Anthropological Institute's new popular magazine, warned of the danger of simplifying questions still in need of research but hoped that Keane would continue "to apply his unrivalled knowledge to the popularisation of a science of such great practical import to our worldwide Empire" (Parkyn 1908, 190-91). Sir Harry Johnston defended popular reference works against the condescension of the experts. He poked fun at the *Times* and other publishers who attempted "to invigorate knowledge by hypnotising the British public into the purchase of encyclopedias, histories, and self-educators." Nonetheless, from his dismay at the widespread ignorance about the peoples of the empire, Johnston claimed that such publications had "increased the general education of the upper and middle classes by at least one-fifth." He specifically noted that "Popular anthropology—I mean anthropology popularised—owes much to the labours and researches of Professor Augustus Henry Keane" (Johnston 1908, 143-45).

Keane and others benefited from the educational reforms of the late nineteenth century. A captive market now existed among the growing ranks of teachers and students in both secondary and postsecondary institutions. The popular texts and serials pioneered by Keane and others were reissued after 1918 and shaped the presentation of images of race in the emerging social sciences of psychology, anthropology, and human geography until World War II.[4] For Keane and other publicists, the study of race was an applied science with important practical political lessons. Keane applied his racial science to political topics of the day, sanctioned imperial adventures, and derided "those sentimental philanthropists who go about preaching the doctrine of the inherent equality of all mankind" (Keane 1884, 99). In *The Boer States: Land and People* (1900), he attributed the longer-term origins of the South African War to the misguided efforts of

4. There are no reliable statistics for the number of students in secondary education prior to 1900, but the number of children in attendance in inspected day schools grew from 2,751,000 in 1880 to 4,666,000 in 1900 (Cook and Keith 1975, 195). In 1900/1901, there were twenty thousand students in universities and five thousand in teacher training colleges (Butler and Sloman 1979, 313). With A. H. Quiggin, A. C. Haddon reissued Keane's *Man, Past and Present* in 1920, and Haddon's own *Races of Man and their Distribution* (1912) appeared in revised form in 1929.

abolitionists such as Lord Glenelg, the colonial secretary (1835-39), who, according to Keane, was "unfortunately a member of the Aborigines Protection Society" (Keane 1900, 194-202).

VIII. Conclusion

By 1900, the relationship between science and humanitarianism had moved full circle. In 1844, humanitarian efforts to protect aboriginal peoples from the onslaught of colonialism awakened a scientific curiosity in human physical and cultural varieties that led to the foundation of the Ethnological Society. By the beginning of the twentieth century, members of its parent organization, the Aborigines Protection Society, realized that they had lost both the hearts and minds of the broader public (Dilke 1901, 3-5). As an opponent of the South African War and of imperialism in general, Leonard Trelawney Hobhouse expressed his dismay at how "humanitarianism is now dismissed as sentimentality." He particularly noted how the application of biological theory undermined the belief in human rights, and doubted whether "the governing race" could distinguish between "a just application of evolutionary principles" and their own material self-interest (Hobhouse [1904] 1972, 59-60, 84-96).

In assessing the influence of science in effecting this regressive change in Victorian attitudes to race, it needs to be recognized both that the community of scientists inherited the racism of the common context and that, as science gained greater intellectual authority and institutional and professional status, it came to reshape that inheritance. The quest for the origins of scientific racism in mid-Victorian comparative anatomy presents an oversimplified view and underestimates the potency and ubiquitous character of racism within the culture. Insofar as science came to shape popular views of race, this influence occurred not as a direct consequence of the controversies of the 1860s, but as a result of changes in the external and domestic context between the 1880s and the outbreak of World War I.

By that time, external changes, most notably the new imperialism of the late nineteenth century, together with new features of the metropolitan society, created conditions for the spread of ideas about race more specifically rooted in the discourse of the natural sciences. Of particular importance were changes associated with the broader process of secularization, especially the professionalization of science, the expansion of education at the secondary and postsecondary levels, and a transformation in the ways in which images of race were communicated to a larger public. These conditions facilitated the application of the ideology of scientific naturalism to the unprecedented patterns of race relations, most evident in the United States and South Africa, emerging at the end of the nineteenth century. Premised on an objectivity that detached the science of race from an older association with humanitarianism, this ideology treated human beings as natural objects subject to forces analogous to those operating in the natural world. Consequently, it excluded the possibility of radical change and denied the role of human will and agency in both the oppression and the liberation of colonized peoples. In this fashion, the ideology of scientific racism came to sanction the world order of the 1920s and 1930s and faced its own crisis of legitimacy from World War II through the 1960s. In looking at the history of racist ideology not simply as the creation of Victorian scientists, but as a product of the interplay between developments within science and the broader cultural and social context, we are in a better position to assess the enduring Victorian legacy not only for the racism of the 1920s and 1930s, but for the racism of today.

Bibliographical Note

The new scientific orthodoxy of the 1950s, which rejected biological theories of racial inequality and, in the case of population genetics, abandoned the concept of race altogether, awakened historical interest in the origins of the "pseudoscience" of race in the nineteenth century. In his influential *Image of Africa* (1965), Philip Curtin found in Dr. Robert Knox, the discredited Edinburgh anatomist and advocate of polygenesis, a fitting conclusion to his study of the regressive change in British views of West Africa between the 1780s and the 1850s. Following from his work on Joseph Gobineau, often identified as the father of European racism, Michael Biddiss (1976), found in Knox a comparable British figure. Michael Banton (1977), as well as Bolt (1971) and Stepan (1982) in two their surveys, and more recently Evelleen Richards (1989), have also attributed a significant influence to Knox and his followers, whereas I have questioned the importance of polygenesis after 1870 (Lorimer 1988).

Other scholars have endeavored to place scientific theories of race within the broader social and cultural context of Victorian England and its global empire. Kiernan (1972) provided a broad survey of European attitudes shaped by colonial encounters and the task of imperial administration. From the experience of racial minorities within the United Kingdom, principally peoples of African descent, one of my own studies (Lorimer 1978) treated the scientific debate as a facet of the politics of race and social class within the metropolitan culture. The intellectual horizons of Victorian racism have also been enlarged by literary studies of colonial discourse, best exemplified by Brantlinger (1988) and Edward Said, whose *Orientalism* (1985) and *Culture and Imperialism* (1993) have influenced the assessment of the role of Western science, scholarship, and literature in constructing an imperialist vision of the world. Studies that more specifically address scientific as distinct from literary discourses include, most notably, George Stocking's *Victorian Anthropology* (1987), which examined the discipline's institutional history and how images of the "savage" reflected the ideology of the metropolitan society. David Livingstone (1984, 1992) has similarly explored the moral economy of space, climate, and race as constructed by geographers in the nineteenth and early twentieth centuries.

References

Allen, Grant. 1896. "Review of A. H. Keane's *Ethnology.*" *Academy* 49:159-60.

Altick, Richard D. 1957. *The English Common Reader: A Social History of the Reading Public, 1800-1900.* Chicago: University of Chicago Press.

Anti-Slavery Society. 1910. Minute book of the British and Foreign Anti-Slavery Society. Anti-Slavery Papers, Rhodes House, Brit. Emp. S. E2/12.

APS (British and Foreign Aborigines Protection Society). 1837. *Regulations of the Society and Address.* London.

ASL (Anthropological Society of London). 1869. "Report of the Committee of Investigation." *Journal of the Anthropological Society of London* 7:i-xxi.

Atmore, A. E. 1984. "The Extra-European Foundations of British Imperialism: Towards a Reassessment." In *British Imperialism in the Nineteenth Century*, edited by C. C. Eldridge. London: Macmillan, 106-25.

Banton, Michael. 1977. *The Idea of Race.* London: Tavistock.

Bhabha, Homi K. 1986. "The Other Question: Difference, Discrimination and the Discourse of Colonialism." In *Literature, Politics and Theory: Papers from the Essex Conference, 1976-84,* edited by Francis Barker, Peter Hulme, Margaret Iversen, and Diana Loxley. London: Methuen, 148-72.

Biddiss, Michael D. 1976. "The Politics of Anatomy: Dr. Robert Knox and Victorian Racism." *Proceedings of the Royal Society of Medicine* 69:245-50.

Blackett, R. J. M. 1983. *Building the Anti-Slavery Wall: Black Americans in the Atlantic Abolitionist Movement, 1830-1860.* Baton Rouge: Louisiana State University Press.

Bolt, Christine. 1971. *Victorian Attitudes to Race.* London: Routledge and Kegan Paul.

Bourne, H. R. Fox. 1899. *The Aborigines Protection Society: Chapters in its History.* London: P. S. King.

Brabrook, Sir Edward. 1912. "A. H. Keane." *Man* 12:53.

Brantlinger, Patrick. 1988. *Rule of Darkness: British Literature and Imperialism, 1830-1914.* Ithaca, N.Y.: Cornell University Press.

British Publishers Archives. 1973. Bishop Storford: Chadwick-Healey. Microfilm.

Brown, Robert. 1870. "Mission Work in British Columbia." *Mission Life*, 1 September, 532-41.

―――. 1873-76. *The Races of Mankind: Being a Popular Description of the Characteristics, Manners and Customs of the Principal Varieties of the Human Family.* 4 vols. London: Cassell, Petter, and Galpin.

―――. 1881-86. *The Peoples of the World.* 6 vols. London: Cassell, Petter and Galpin.

―――. 1884-89. *The Countries of the World: Being a Popular Description of the Various Continents, Islands, Rivers, Seas, and Peoples of the Globe.* 6 vols. London: Cassell.

[———]. 1889. "The Relapse of the Negro." *Standard,* 20 September, 3a.
———. 1892-95. *The Story of Africa and its Explorers.* 4 vols. London: Cassell.
Burrow, J. W. 1970. *Evolution and Society: A Study in Victorian Social Theory.* Cambridge: Cambridge University Press.
Butler, David, and Anne Sloman. 1979. *British Political Facts, 1900-1979.* 5th ed. London: Macmillan.
Catholicus. 1880. *A Caution Against the Educational Writings of Edward Clodd.* London: John F. Shaw.
Chadwick, Owen. 1975. *The Secularization of the European Mind in the Nineteenth Century.* Cambridge: Cambridge University Press.
Clodd, Edward. 1873. *The Childhood of the World: A Simple Account of Man's Early Years.* London: Macmillan.
———. [1875] 1889. *The Childhood of Religions: Embracing a Simple Account of the Birth and Growth of Myths and Legends.* London: Kegan Paul Trench.
———. 1887. "Social Evolution." *Knowledge,* 1 August, 220-21.
———. [1887] 1888. *The Story of Creation: a Plain Account of Evolution.* London: Longmans.
———. 1895. *A Primer of Evolution.* London: Longmans Green.
Cook, Chris, and Brendan Keith. 1975. *British Historical Facts, 1830-1900.* London: Macmillan.
Curtin, Philip D. 1965. *The Image of Africa: British Ideas and Action, 1780-1850.* London: Macmillan.
Dilke, Sir Charles. 1901. "Remarks of Sir Charles Dilke to the Annual Meeting of the Aborigines Protection Society." *Aborigines' Friend,* April, 3-5.
Dunae, Patrick. 1977. "Boys' Literature and the Idea of Race." *Wascana Review* 12:84-107.
———. 1980. "Boys' Literature and the Idea of Empire." *Victorian Studies* 24:105-21.
———. 1989. "New Grub Street for Boys." In *Imperialism and Juvenile Literature,* edited by Jeffery Richards. Manchester: Manchester University Press, 12-33.
Edwards, Elizabeth, ed. 1992. *Anthropology and Photography, 1860-1920.* New Haven, Conn.: Yale University Press.
Fabian, Johannes. 1983. *Time and the Other: How Anthropology Makes Its Object.* New York: Columbia University Press.
Gay, Peter. 1993. *The Cultivation of Hatred.* Vol. 3 of *The Bourgeois Experience: Victoria to Freud.* New York: W. W. Norton.
Gilman, Sander. 1985. "Black Bodies: White Bodies: Toward an Iconography of Female Sexuality in Late Nineteenth Century Art, Medicine, and Literature." *Critical Inquiry* 12:204-42.
Green, David. 1984. "Photography and Anthropology: The Technology of Power." *Ten.8* 14:30-37.
Haddon, A. C. 1910. *History of Anthropology.* London: Watts.
Hannabuss, Stuart. 1989. "Ballantyne's Message for Empire." In *Imperialism and Juvenile Literature,* edited by Jeffery Richards. Manchester: Manchester University Press, 53-71.
Hayman, John. 1989. *Robert Brown and the Vancouver Island Exploring Expedition.* Vancouver: University of British Columbia Press.

Helly, Dorothy O. 1969. " 'Informed' Opinion on Tropical Africa in Great Britain, 1860-1900." *African Affairs* 68:195-200.

Heyck, T. W. 1982. *The Transformation of Intellectual Life in Victorian England*. London: Croom Helm.

Hobhouse, L. T. [1904] 1972. *Democracy and Reaction*. Edited by P. F. Clarke. Brighton: Harvester.

Horne, Leonard T. 1896. "Race Types." *News from Afar,* n.s., 2:41-42, 56-57, 86-87, 121-22, 138-39, 152-53, 184-85.

Hunt, James. 1863a. *Introductory Address on the Study of Anthropology*. London: Trubner.

———. 1863b. *The Negro's Place in Nature*. London: Trubner.

Huxley, Thomas Henry. Papers. Imperial College, London.

———. [1870] 1901. "Anniversary Address of the President of the ESL, May 24, 1870." In *Scientific Memoirs of T. H. Huxley,* edited by M. Foster and E. R. Lankester. London: Macmillan, 3:554-63.

Johnston, H. H. 1908. "The Empire and Anthropology." *Nineteenth Century* 64:133-46.

Keane, A. H. 1879-80. "On the Relations of the Indo-Chinese and Inter-Oceanic Races and Languages." *Journal of the Anthropological Institute* 9:254-89.

———. 1880-81. "The Indo-Chinese and Oceanic Races—Types and Affinities." *Nature* 23 (December 30): 199-203, (January 6): 220-23, (January 13): 247-51, (January 20): 273-74.

———. 1884. "The Haitian Negroes." *Nature* 31 (December 4): 99.

———. 1890-94. "Mongols." In *Cassell's General Storehouse of Information*. London: Cassell, 367.

———. 1894. "The World: Population, Races, Languages, and Religions." *Church Missionary Intelligencer* 45, n.s. 19: 721-30.

———. 1896. *Ethnology*. Cambridge: Cambridge University Press.

———. 1897. *Anthropological, Philological, Geographical, Historical and Other Writings Original and Translated by A. H. Keane*. London: privately printed.

———. 1899. *Man: Past and Present*. Cambridge: Cambridge University Press.

———. 1900. *The Boer States: Land and People*. London: Methuen.

———. 1908. *The World's Peoples*. London: Hutchinson.

Keane, A. H., Richard Lydekker, et al. 1905. *The Living Races of Mankind*. London: Hutchinson.

Kegan, Paul, Trench, Trubner, Henry King. [1858-1912] 1973. Book Accounts. Vol. 2, 30 June 1865, 29 March 1884, reel 25. In *British Publishers Archives*. Bishop Storford: Chadwick-Healey. Microfilm.

Kiernan, V. G. 1972. *The Lords of Human Kind: European Attitudes to the Outside World in the Imperial Age*. Harmondsworth, England: Penguin.

Knox, Robert. [1850] 1862. *The Races of Men: A Philosophical Enquiry into the Influence of Race over the Destinies of Nations*. London: Henry Renshaw.

Kovel, Joel. 1988. *White Racism: A Psychohistory*. London: Free Association.

Kuklick, Henrika. 1991. *The Savage Within: The Social History of British Anthropology, 1885-1945*. Cambridge: Cambridge University Press.

Lane-Fox, Col. Augustus. 1874-75. "On the Principles of Classification adopted in the arrangement of his Anthropological Collection, now exhibited in the Bethnal Green Museum." *Journal of the Anthropological Institute* 4:293-308.

———. 1877. *Journal of the Anthropological Institute* 6:178.
Lightman, Bernard. 1987. *The Origins of Agnosticism: Victorian Unbelief and the Limits of Knowledge.* Baltimore: Johns Hopkins University Press.
Livingstone, David N. 1984. "Natural Theology and Neo-Lamarckianism: The Changing Context of Nineteenth-Century Geography in the United States and Britain." *Annals of the Association of American Geographers* 74:9-28.
———. 1992. "A 'Sternly Practical Pursuit': Geography, Race and Empire." In *The Geographical Tradition,* edited by D. Livingstone. Oxford: Blackwell, 216-59.
Lorimer, Douglas A. 1978. *Colour, Class and the Victorians: English Attitudes to the Negro in the Mid-Nineteenth Century.* Leicester: Leicester University Press.
———. 1988. "Theoretical Racism in late Victorian Anthropology, 1870-1900." *Victorian Studies* 31:405-30.
———. 1990. "*Nature,* Racism and Late Victorian Science." *Canadian Journal of History* 25:369-85.
Lydekker, Richard. 1905. "Introduction." In *The Living Races of Mankind,* by A. H. Keane, Richard Lydekker, et al. London: Hutchinson.
MacLeod, Roy M. 1969. "Is It Safe to Look Back?" *Nature* 224:417-76.
Malchow, H. L. 1993. "Frankenstein's Monster and Images of Race in Nineteenth-Century Britain." *Past and Present* 139:90-130.
McCabe, Joseph. 1932. *Edward Clodd.* London: Bodley Head.
Mumby, F. A. 1934. *The House of Routledge, 1834-1934.* London: G. Routledge and Sons.
Nowell-Smith, Simon. 1958. *The House of Cassell, 1848-1958.* London: Cassell.
Parkyn, E. A. 1908. "Review of Keane, *The World's Peoples.*" *Man* 8:190-91.
Power, H. 1870. "The Natural History of Man." *Nature* 3 (November 3): 9-13.
RAI [Royal Anthropological Institute]. 1907. "Anthropological Institute: Augmentation of Title." *Man* 7, no. 70:112.
———. 1909. "Anthropology and the Empire: Deputation to Mr. Asquith." *Man* 9:85-87.
———. [1911]. "Memorial on Imperial Bureau of Anthropology." Printed copy, no date. Royal Anthropological Institute Archive A56.
———. 1912. "Report of the Council for 1911." *Journal of the Anthropological Institute* 42:5.
Rainger, Ronald. 1978. "Race, Politics and Science: The Anthropological Society of London in the 1860s." *Victorian Studies* 22:51-70.
Read, C. H. 1906. "Anthropology at the Universities." *Man* 6, no. 38:56-59.
Richards, Evelleen. 1989. "The 'Moral Anatomy' of Robert Knox: The Inter-Play between Biological and Social Thought in Victorian Scientific Naturalism."*Journal of the History of Biology* 22:373-436.
Richards, Jeffery, ed. 1989*a*. *Imperialism and Juvenile Literature.* Manchester: Manchester University Press.
———. 1989*b*. "With Henty to Africa." In *Imperialism and Juvenile Literature,* edited by Jeffery Richards. Manchester: Manchester University Press, 72-106.
Ripley, C. P., ed. 1985. "Exchange by William Craft and Dr. James Hunt, British Association, 27 August 1863." In *The Black Abolitionist Papers.* Vol. 1, *The British Isles, 1830-1865.* Chapel Hill: University of North Carolina Press, 537-43.
Robinson, Ronald. 1972. "Non-European Foundations of European Imperialism: a

Sketch for a Theory of Collaboration." In *Studies in the Theory of Imperialism*, edited by R. Owen and B. Sutcliffe. London: Longman, 117-42.

Rose, Michael. 1981. *The Curator of the Dead: Thomas Hodgkin (1778-1866)*. London: Peter Owen.

Routledge, George, and Sons. [1867] 1973*a*. Contract between George Routledge and Sons and Rev. John G. Wood for *The Illustrated Natural History of Man*. George Routledge and Company, 1853-1902, reel 2: Contracts, vol. 3: R-Z, fol. 355, [March 1867]. *British Publishers Archives*. Bishop Storford: Chadwick-Healey. Microfilm.

———. [1869] 1973*b*. Addendum to the Wood Contract. George Routledge and Company, 1853-1902, reel 2: Contracts, vol 3: R-Z, 1 October 1869, no folio number. *British Publishers Archives*. Bishop Storford: Chadwick-Healey. Microfilm.

———. [1853-1902] 1973*c*. *Publication Books*. George Routledge and Company, 1853-1902, reel 4. *British Publishers Archives*. Bishop Storford: Chadwick-Healey. Microfilm.

Said, Edward W. 1985. *Orientalism*. Harmondsworth, England: Penguin.

———. 1993. *Culture and Imperialism*. New York: Alfred A. Knopf.

Stafford, Robert A. 1989. *Scientist of Empire: Sir Roderick Murchison, Scientific Exploration, and Victorian Imperialism*. Cambridge: Cambridge University Press.

Stepan, Nancy. 1982. *The Idea of Race in Science: Great Britain, 1800-1960*. Hamden, Conn.: Archon Books.

Stocking, Jr., George W. 1971*a*. "The Persistence of Polygenist Thought in Post-Darwinian Anthropology." In *Race, Culture and Evolution*, edited by G. W. Stocking. New York: Free Press, 42-68.

———. 1971*b*. "What's in a Name? The Origins of the Royal Anthropological Institute (1837-71)." *Man*, n.s., 6:369-90.

———. 1987. *Victorian Anthropology*. New York: Free Press.

Turner, Frank M. 1974. *Between Science and Religion: The Reaction to Scientific Naturalism in late Victorian England*. New Haven, Conn.: Yale University Press.

Tylor, E. B. 1874*a*. "The Races of Mankind." *Nature* 9 (February 12): 279-80.

———. 1874*b*. "Fritch's South African Races." *Nature* 9 (April 23): 479-82.

———. 1881. *Anthropology*. London: Macmillan.

University College, London. 1883. Testimonials for A. H. Keane and the report of the Committee on the lectureship in the vernacular languages of India, Senate, April 12, and Council, April 14. University College Archives.

van Keuren, David K. 1984. "Museums and Ideology: Augustus Pitt Rivers, Anthropological Museums, and Social Change in Later Victorian Britain." *Victorian Studies* 28:171-89.

Wood, John G. 1861. *The Boy's Own Book of Natural History*. London: Routledge, Warne, and Routledge.

———. 1868-70. *The Natural History of Man: Being an Account of the Manners and Customs of the Uncivilised Races of Men*. Vol. 1, *Africa* (1868). Vol. 2, *Australia, New Zealand, Polynesia, America, Asia and Ancient Europe* (1870). London: George Routledge.

Wood, Theodore. 1890. *The Rev. J. G. Wood: His Life and Work*. London: Cassell.

Young, Robert. 1990. *White Mythologies: Writing History and the West.* London: Routledge.

Young, Robert M. 1982. "Darwinism *Is* Social." In *The Darwinian Heritage*, edited by David Kohn. Princeton, N.J.: Princeton University Press, 609-38.

———. 1985. "Natural Theology, Victorian Periodicals, and the Fragmentation of a Common Context." In *Darwin's Metaphor: Nature's Place in Victorian Culture.* Cambridge: Cambridge University Press, 126-63.

11

Elegant Recreations? Configuring Science Writing for Women

ANN B. SHTEIR

When Mary Somerville dedicated her book *On the Connexion of the Physical Sciences* to Queen Adelaide in 1834, she expressed a wish "to make the laws by which the material world is governed more familiar to my countrywomen." Over the next decades, while many science books and essays for general readers included women in their intended audience, another considerable body of writing explicitly designated a female audience. Scientific work was adapted with girls, women, and "ladies" in mind, and periodicals, textbooks for informal and formal learning, and coffee-table books articulated science for women of different ages, classes, and levels of expertise. Recently, scholars examining relationships between women and Victorian science culture have studied complex links among actual women and gendered ideas at that time (Russett 1989; Richards 1983, 1989; Phillips 1990). Historical work has shown that women were an important part of the cultural map of Victorian science, as audience, readers, writers, cultivators of science, investigators, and helpmates. During the same period, however, institutional and social changes led to more exclusionary relations between women and science culture. Over the course of the Victorian decades, professionalizing directions led to bifurcations in science practices, and women were pushed to the periphery, relegated to arenas of "amateur" and "popular" science. As this essay will show, textual changes within science culture took place in tandem with these institutional changes. Academic and professionalizing writers produced books and essays to represent their emergent disciplines and shape the next generation of scientists, and Victorian science writing for women helped in important ways to clear a space for science writing for men.

Science writing for women, a series of literary, historical, social, and ideological strands woven into the complex fabric of Victorian science culture, is part of the larger gendered history of both popularizing science and professionalizing science. But what does it mean to earmark science writing "for women"? Was this a matter of politics and publishing within Victorian science? Was it a psychosocial matter based in notions of "natural" difference? Is science writing for women discernible by its subject matter? Are there marked differences in the form and tone of science writing for women? This discussion of science writing for women features three configurations of the topic: women as a designated audience for books and periodicals about science; science writing as an area of women's cultural work—that is, women as themselves science writers; and women, gender, and narrative form in science writing. To illustrate issues of audience, authorship, and narrative form in Victorian science writing for women, examples will be drawn principally from botany, the science with which women were particularly associated during much of the nineteenth century.

I. Women as Audience

The Young Lady's Book: A Manual of Elegant Recreations, Arts, Sciences, and Accomplishments (1859) illustrates nicely how assumptions about women, gender, and science shaped women's engagement in science culture (figure 11.1). This popular miscellany, intending to "present a capital epitome of a young lady's best pursuits, exercises and recreations," is framed by discussion among aristocratic women. Chapters on topics such as embroidery, female deportment, photography, and archery are interspersed with chapters on science. There are expositions about geology, ornithology, and conchology written by "distinguished professors" that encourage young women to study sciences such as these. Yet substantive chapters are positioned literally alongside discussions about "natural" female delicacy. A long section entitled "The Botanist," for example, teaches about plant classification according to the Linnaean and the natural systems and provides an extended and substantial discussion of the class Cryptogamia (flowerless plants). Another chapter on plants, entitled "The Florist," encourages female knowledge of plants because "there is something peculiarly adapted to feminine tenderness in the care of flowers"; at the same time that we read, however, that a garden "offers many light and graceful occupations to a young lady," we also find the author declaring it inappropriate for her to "dig up the earth, study the modes of manuring it, or prepare compost" (33-34). *The Young Lady's Book* does not exclude women from knowledge of nature, but gives women's relationship to nature a location in terms of naturalized notions about gender and class that accorded with "sexual science" and Victorian constructions of womanhood (Russett 1989).

In the previous generations, going back into the mid-eighteenth century, science writing for women had been part of the Enlightenment culture of improvement across the middle and upper ranks of society. Periodicals such as Eliza Haywood's *Female Spectator* from the 1740s and Charlotte

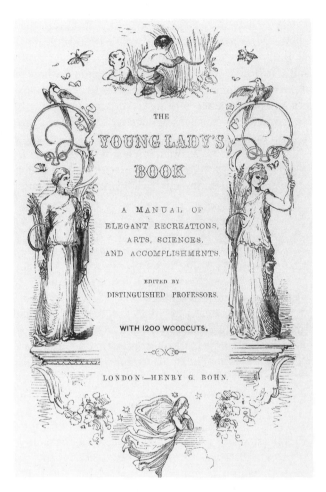

Figure 11.1 Title page of *The Young Lady's Book: A Manual of Elegant Recreations, Arts, Sciences and Accomplishments* (1859). (By permission of the British Library [BL1608/5949].)

Lennox's *Lady's Museum* from the 1760s, as well as conduct books such as Ann Murry's *Sequel to Mentoria* (1799), introduced girls and women to the sciences of the day. Textbooks such as Margaret Bryan's *Lectures on Natural Philosophy* (1806) and *Comprehensive Astronomical and Geographical Class Book* (1815) provided material for informal learning as well as for formal schoolroom use at home or in boarding schools or day schools.

Science writing for women continued into the early Victorian decades and beyond. Topics ranged over entomology, botany, and conchology during the earlier decades, and moved into geology, microscopy, and natural science during the later decades. In some cases, science writing for women was part of the general education movement; in other cases, it belonged to "polite science" and "elegant recreations," or, conversely, to substantial school science. An essay by a male popular science lecturer in *The Young Lady's Magazine of Theology, History, Philosophy and General Knowledge* in 1838 called for more science education for girls. Lamenting "that the elements of popular science are not taught [to girls] in their schools," the author argued that "every respectable ladies' school ought in fact to be a minor college, or university, where the sciences of Astronomy, Geography, Botany, Chemistry and Natural Philosophy, Phrenology etc. should be taught by competent Professors" (Dewhurst 1838, 35).

Science writing that was directed to women in England during the fraught decades from the revolutionary and counterrevolutionary 1790s, across the Romantic decades, and into the tumult of the 1830s served a variety of social, political, moral, and religious purposes (Benjamin 1991*b*). It was congruent with some religious education, relating to evangelical teachings as well as to natural theology. It satisfied a call for rational recreations, as part of a roster of improving activities. It also accorded with ideas about women's maternal responsibilities as teachers of actual school subjects and also as shapers of moral and religious values. Exhortations to women to study science were inflected in many ways, depending on class and religious orientations. In their recent essay on the problematics of popularization, Cooter and Pumfrey pointed out that historical study of the popularization of science needs a less static model than those conventionally used until now. Instead of "false coherence" and a diffusionist model of popularization, they call for "the disaggregation of . . . different groups, interests and forms of knowledge" (Cooter and Pumfrey 1994, 251). Such a decentering approach to science writing for women opens possibilities for telling stories from the vantages of various communities of practitioners and their likely divergent interests. Science writing for women thus can be interpreted as serving differential purposes for some women and for some men, according to circumstances linked to class, religion, gender ideologies, and individual stories.

The great variety and abundance in Victorian print culture offered readers many formats and points of entry into scientific information, and authors and publishers articulated their target audiences. George William Francis, for example, dedicated his *Little English Flora* (1839) to "the young ladies of England, whose occupations, tastes and sensibilities, render the science of botany so peculiarly a proper object for their study." Charles Alexander Johns positioned his *Flowers of the Field* (1853), a widely reprinted illustrated handbook about plants in the Linnaean and natural systems, as aiming "to teach the unscientific how to find out the names of the flowers they may happen to fall in with in the course of their country rambles." Popular science and natural history periodicals addressed readerships of varying ages and social stations and sometimes shaped their own alternative "low scientific culture" (Sheets-Pyenson 1985; Brock 1994). Thus, *Hardwicke's Science Gossip* (1865–1900) sought to encourage "popular Natural History" and microscopy as "scientific hobbies." In a note to readers at the conclusion of volume 4 (1868), the editor reflected on the magazine's mandate this way: "It is our aim to gossip freely, in as untechnical a manner as possible, on scientific subjects, so that even those of our readers who have not had a scientific training, who have no systematic knowledge of Zoology or Botany, or the cognate sciences, and who have no adequate leisure for such pursuits, may read with pleasure and understanding."

Women figured among the audiences who were encouraged to read about science and were cultivated as readers of books and periodicals. Victorian advice manuals suggested titles, and "Scientific Reading" was among the genres recommended for "The Girl's Library" (Flint 1993, 85). In a period hungry for information, with burgeoning technologies for supplying that information, it is not surprising that women who were already interested in science, or were being directed toward reading that included science topics, would find ready access to materials in the form of periodicals and books. The science writer Mary Kirby, compiler of *A Flora of Leicestershire* (1850), recalled being well supplied with books and periodicals as part of her middle-class upbringing in the Midlands during the second quarter of the nineteenth century: "Every Saturday night, a number of the Penny and Saturday Magazine used to come out of our father's coat pocket. . . . and by and by, 'Chamber's Miscellany,' and [Charles] Knight's publications" (Kirby 1887, 13).

Readers of the periodical *The Ladies' Companion at Home and Abroad*, edited by the popular writer Jane Loudon, found a wide range of topics in the arts and sciences. Jane Loudon articulated highly gendered beliefs about sex differentiation in roles for women and men yet equally claimed space for women to study science. Her periodical was meant to enforce "the necessity of mental cultivation . . . not to make women usurp the place of men, but to render them rational and intelligent beings" (Loudon 1849–50, 1:8). She situated science writing within a broader educational philosophy of "introducing women to information and to subjects now generally confined to men" and wrote that "most of [the] contributors on serious subjects are men, and they will communicate their views in no other way than they would take in addressing men, though their choice of subjects will be determined by the exclusively feminine character of certain duties and employments" (1:76). In line with this educational mandate, male scientific authorities contributed essays on geology, entomology ("The British Insects of Spring"), and chemistry ("The Chemistry of Everyday Life"), and Jane Loudon herself supplied essays on botany.

Within the print culture of Victorian natural history, women were a ready audience for books about plants. Since the later eighteenth century, botany had been a science to which women had relatively easy access. They came into botanical activities through polite science, and horticulture and artistic work in flower painting also eased entry for women into plant-related work. Many books taught botany to women, and authors and publishers acknowledged a female readership (Shteir 1996). George William Francis's *Grammar of Botany* (1840), for example, explains plant structure and classificatory systems in a volume adapted to "young females." George Bentham meant his *Handbook of the British Flora* (1858) to be "for the Use of Beginners and Amateurs" and, mindful of women readers, deliberately simplified terms "for the ladies" (Allen 1995, 139). Shirley Hibberd's *Field Flowers* (1870) encouraged female botanizing and "outdoor study of British plants" as good for the intellect, and constructed an active and intelligent ideal female reader who was a robust plant collector. In *British Sea-Weeds* (1862) Margaret Gatty acknowledged the keen interest among women in collecting marine flora and addressed the material reality of how women should dress in order to work along the shore in search of specimens. Many other books were written for juvenile readers, at times specifically to be read by mothers or other older women. Emily Ayton's *Words by the Way-Side; or the Children and the Flowers* (1855) is one of many examples of an influential juvenile format in which narratives were directed to a dual audience of adults and children.

At the same time that Victorian science writing brought information and scientific practices to women, children, and general readers across the middle and working classes, other communities of scientific interests were establishing themselves. For example, within botanical culture there were constituencies of artisanal botanists and physiological botanists (Secord 1994; Stevens 1995). Most influential were professionalizers and specialists who set out to delineate a disciplinary culture for the science of botany. Some early-nineteenth-century botanists worked to defeminize botany, and their efforts show the degree to which gender is part of the history of professionalizing science. The history of discipline formation in botany reveals various maneuvers to edge women out of science by demarcating a realm for male botanists that is at a remove from activities tagged as "feminine" (Shteir 1996, chap. 6).

The gendered professionalizing project within Victorian botanical culture was made manifest institutionally and textually. A sociologist studying patriarchy and the professions has identified strategies of occupational closure through which boundaries in professions have been historically created and controlled. Her findings apply to the history of Victorian botanical culture. Demarcative work in professionalizing occupations, she explains, can turn "not upon the exclusion, but upon the encirclement of women within a related but distinct sphere of competence in an occupational division of labour" (Witz 1992, 47). In the history of botanical culture, scientific or professional botany was increasingly demarcated from the profile of mental improvement and politeness that had guided plant study in earlier generations. Institution builders within botany aimed to channel botanical training and work toward a world of utility, specialization, and expertise. Increasingly, literary discourses in science writing were separated out from technical and academic discourses.

The botanical teacher and institution builder John Lindley played an important role in demarcating audiences for science, and women and gender were part of this story. As the first professor of botany at the newly formed London University, he set out to put botany on a new footing as a science rather than as an area of polite culture. In his inaugural lecture at the university, he distinguished between botany as "accomplishment" and as science, and between botany as "an amusement for ladies" and "an occupation for the serious thoughts of man" (Lindley 1829, 17). Lindley endeavoured by his teaching and publications to introduce a new type of systematics in botany, based on the Continental taxonomic theories of Augustin-Pyramus de Candolle, that separated itself from Linnaean ideas of classification based upon the reproductive parts of flowers and took its organizational point from a variety of morphological features. Lindley worked to move botany away from its representation as a Linnaean, aristocratic, and polite science and to shape it into a more rigorous and utilitarian pursuit. He worked to defeminize botany. Lindley's project formed part of larger agendas at the secular London University. Like other secular, middle-class scientific naturalists who contested aristocratic, Tory, and clerical interests, he wanted to define a new "man of science" and bring in a more naturalistic approach to studying the order of nature (Turner 1993). To extirpate Linnaean botany and all that it represented to him, Lindley published books on the natural method of classification for botanical specialists, schoolboys, university students, the general public, and women.

Lindley's *Ladies' Botany* (1834–37) introduces the new post-Linnaean botany in a lavishly illustrated two-volume work written in the form of letters (figure 11.2). Lindley's book aimed to create an audience for his new configuration of science, one that differed markedly from Linnaean, fashionable, and "feminine" botany. It is addressed to a mother who would like to be able to teach her children about the botanical systematics of Candolle. Lindley's choice of narrative form for his popular botany book imitates that in Jean-Jacques Rousseau's *Lettres élémentaires sur la botanique*, the influential text translated into English by Thomas Martyn and published with additional essays under the title *Letters on the Elements of Botany* (1785). Lindley's book displays a clear pedagogical mandate to teach women the natural system of plant classification. Within the swirling currents of scientific culture at the time, his book has a larger purpose too, in that it delineates a realm for female botanizing that is different from male botanizing. Lindley does not exclude women from new-style botany, but accords them

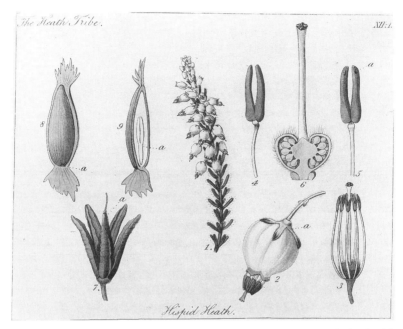

Figure 11.2 Plate 12.1 of *Ladies' Botany, or A Familiar Introduction to the Study of the Natural System of Botany* (Lindley 1834-37).

a clear niche in botanical practice as mothers and teachers. He gives women science, but science in a separate sphere, where they can disseminate knowledge to the next generation as part of their maternal role.

In a discussion of popular books that brought the Newtonian philosophy to female learners, John Mullan has argued that science writing for women during the eighteenth century served actual women and also carried symbolic purposes. In Benjamin Martin's *The Young Gentleman and Lady's Philosophy* (1759), for example, the female pupil in scientific dialogues represents encouragement for women readers to emulate her interest in Enlightenment science. In addition, the female pupil represents the untutored mind, the ingenuous witness, the potential convert to new knowledge. By coming to understand the scientific arguments, she demonstrates the transparency of the theories being explained. The Woman Reader matters, in this regard, more than actual women readers. Thus, female readers or pupils in scientific dialogues or other popularizations of science serve as "symbolic as well as actual agents" (Mullan 1993, 56). In a similar way, Lindley can be seen as having directed his *Ladies' Botany* to women as both actual audience and symbolic agents. In the defeminizing campaign that Lindley waged against polite science, writing a certain kind of book for a female audience was one way to demarcate kinds of scientific readers and practitioners. Lindley's *Ladies' Botany* thereby brings into sharp relief a gender dimension in the history of science writing and the history of the professionalization of Victorian science.

II. Women as Science Writers

During the decades when women were being edged out of a professionalizing science culture, they continued to be active as science writers. Science writing was, in fact, an area of professional practice for Victorian women writers, part of their generic repertory. The growth of popular science culture during the eighteenth century had shaped an audience of women and children who participated in home-based science education and scientific activities. Women's authorship of science books accorded with ideas about maternal responsibility and with home-based education, and books such as Priscilla Wakefield's *Introduction to Botany* (1796) and Jane Marcet's *Conversations on Chemistry* (1806) and *Conversations on Natural Philosophy* (1819) continued in print deep into the nineteenth century.

Within the gender configurations of Victorian authorship, women took a path into public writing by assuming the mantle of mothers, teachers, or guides through the elementary stages of science learning (Benjamin 1991b; Gates 1993). During the early and mid-Victorian years, science writing directed explicitly to women and children found a place on the lists of many publishers. Among well-known authors, Anne Pratt wrote *Wild Flowers* (1852), and many other natural history books "for popular use and general interest," Mary Ward, author of *A World of Wonders Revealed by the Microscope* (1858), wrote about entomology and microscopy, and Isabella Gifford compiled an introductory guide to seaweeds in *The Marine Botanist* (1848).

Science writing provided an area of productive labor for women. Financial need led some women to struggle to establish careers for themselves as writers on scientific topics for children and general readers. Sarah Bowdich Lee looked to her pen to support herself and her children in a career that began during the 1820s and continued for thirty years. *Elements of Natural History* (1844), for example, presents vertebrate zoology "for the use of schools and young persons," as a "stepping-stone" to monographs and more detailed productions of "deep science." The publication list of Mary Roberts, who also had a writing career that dated back to the 1820s, includes *A Popular History of the Mollusca* (1851) and *Voices from the Woodlands; or History of Forest Trees, Ferns, Mosses, and Lichens* (1850) (Gould 1993). Books by Mary Roberts formed part of the midcentury Popular Natural History Series published by the naturalist Lovell Reeve. Other authors in that series were Agnes Catlow (*The Conchologist's Nomenclator* [1845]) and her sister Maria Catlow, who published *Popular Geography of*

Plants (1855) and *Popular British Entomology* (1848), "for the use of young beginners."

Jane Loudon, as wife and then as widow of the horticultural writer John Claudius Loudon, assiduously produced a large body of books about botany and gardening. She directed some writing to women and some to children. Her botany books for children included *The First Book of Botany* (1841), designated "for Schools and Young Persons." Her aim was to explain botanical terms, "the alphabet of the science," so as to facilitate study of systems of classification. At a time when the Linnaean system was increasingly being marginalized and replaced in serious professional practice by the natural system of Bernard de Jussieu, as altered by Candolle, Loudon shaped an inclusive book that suggested the continuing utility of the Linnaean system for botanical novices.

During the Victorian decades women writers were part of the culture of introducing and teaching science to various and distinctive kinds of audiences. They cast scientific material into vocabulary suitable for audiences that had little or no technical expertise. In addition to producing books at an elementary level for those not yet initiated into scientific knowledge, some women writers produced accounts that explained high-level specialist work in science to those who were scientifically literate but uninformed about specialist findings. Thus, as Bernard Lightman (1997) shows, the encyclopedist of astronomy Agnes Clerke was an important mediator of knowledge about the Victorian heavens.

Work by the botanical writer Phebe Lankester represents popular science writing of another kind, neither on the diffusionist model of popularization nor as an example of mediating among the scientifically literate. In the 1860s a new, third edition of James Edward Smith's *English Botany* began to appear. Smith's foundational manual of identification and classification (first published 1790–1814) was revised according to the natural system of classification rather than the Linnaean system, and the technical material was augmented by a "Popular Portion." The "purely technical matter" was distinguished from "the popular part," also referred to as the "literary" part. This important reworking enshrined the bifurcated path that botanical practice and languages of nature took during the early Victorian period. For example, after a highly technical discussion of a forget-me-not (*Myosotis palustris:* "Rootstock elongate, creeping, oblique, stoloniferous. . . . Stem decumbent. . . . Pedicels rather slender, in fruit horizontal or reflexed-divaricate"), Lankester reflects upon this "pretty plant . . . peculiarly the favourite of poets and sentimentalists. . . . it is a household favourite, and reminds us that there is in the human mind a deep and close association between the external beauty of nature and the strongest feelings of our heart" (Syme [1863–86] 1867, 7:98–101). Phebe Lankester contributed nearly four hundred entries to eleven volumes of the third edition of *English Botany*, with information about the uses and medical properties of plants as well as poetic and historical remarks. She was a successful botanical author of books on ferns and wildflowers; with her husband, the doctor and reforming public health officer Edwin Lankester, she also wrote articles for *The Popular Science Review* (English 1990, 132-34).

Other publications illustrate the versatility of women writers who were attentive to audiences outside the middle class. Elizabeth Twining, a keen botanist, artist and lecturer on botanical subjects, and author of several books on plants, is best known now for her *Illustrations of the Natural Orders of Plants* (1849-55), a two-volume collection of drawings and botanical descriptions. But Twining's books from the 1850s, 1860s, and 1870s show her to have been an important popularizer of scientific knowledge, especially for the working class. *Short Lectures on Plants, for Schools and Adult Classes* (1858), for example, is a small publication about the natural method of plant classification. The book developed from lectures that Elizabeth Twining gave to young women at the Working Men's College in Great Ormond Street, London. Founded in 1854, this pioneering adult education institution declared its interest in bringing liberal education to the wives and the children of the working men who were its students. The Working Men's College held separate classes for women and for girls between 1855 and 1860; starting in 1858, a girls' school was conducted daily, from 10 A.M. to 1 P.M. (Harrison 1954, 106). Elizabeth Twining was affiliated with the college at that time and commented in the preface to her book that "it was . . . very satisfactory to observe the attention of the class, and to be assured that the history of plants was capable of affording pleasing instruction to scholars who have hitherto had neither time nor opportunity for such a study" (x-xi). *Short Lectures on Plants,* addressed partly to students and partly to teachers, assumes no previous knowledge of botany and contains no technical botanical language. "The science of Botany was, until lately, so peculiarly enveloped in technical scientific language, that young persons were almost entirely deterred from learning it. I have, therefore, endeavoured in a humble manner to clear a path for young students, and to show that plants may be admitted amongst the subjects of the highest and yet simplest interest to all classes of learners" (Twining 1858, x). Later, Twining folded the same botanical material into a larger publication, *The Plant World* (1866), adding chapters on the parts of plants as well as on types of plants.

Elizabeth Twining's work as a science writer also was yoked to a powerful sense of religious and philanthropic duty. Sister to the social reformer Louisa Twining, she superintended a Ragged School, developed Mothers' Meetings for poor women in her London parish, and issued publications based on talks and lectures delivered in mission rooms. *A Lecture on Plants as Water-Drinkers* (1872), for example, explains about the physiology of plant tissue. Issued as a one-penny pamphlet by the National Temperance

Society, it makes the argument that water is more important for plants than anything else and develops an analogy between plants and human beings — "all those who belong to the great class of *water-drinkers*" (Twining 1872, 15). Elizabeth Twining belongs in the company of numerous Victorian women writers who made science writing a tool for themselves in educational, philanthropic, political, and religious campaigns. Like them, she melded scientific information and religious conviction and wrote with an eye to shaping the values of her chosen audiences.

III. Women, Gender, and Narrative Form in Science Writing

In a discussion about twentieth-century science, Greg Myers delineates types of textual practices that often distinguish articles about popular science from those about professional science. He gives the label "narratives of nature" to popular, nontheoretical accounts that focus on plants or animals themselves and avoid technical terminology; "narratives of science," by contrast, embody generically the work of modern discipline-based science and are committed to methodology and concepts, at a remove from the organisms themselves (Myers 1990). These categories have resonance for understanding developments in Victorian science writing for women, particularly when augmented by the category "narratives of natural theology," proposed by Barbara Gates to designate the large body of Victorian popular natural history books that were shaped by belief in the natural theology teachings of William Paley (Gates 1993, 290–91). From the 1790s through the 1830s and 1840s, women were prominent as authors of narratives of nature and narratives of natural theology. Their popular narratives of nature and natural theology often located scientific activities within cultural and religious mandates for middle-class families. They incorporated discussions of moral and spiritual topics and also gave authority to mothers, or maternal surrogates, as teachers and to girls as engaged learners. Their writings conventionally took the form of educational conversations, dialogues, or letters featuring girls of the "rising generation" and an older female teacher. Thus, Priscilla Wakefield constructed after-dinner family conversations about insects, plants, and animals in *Mental Improvement, or the Beauties and Wonders of Nature and Art* (1794–97) and featured girls developing scientific expertise. Likewise, Jane Marcet embodied a pedagogy as well as a cultural location for learning and practicing science when she created home-based conversations that featured Mrs. B. as the narrative teacher and scientific authority in *Conversations on Chemistry* (1806).

By the mid-nineteenth century, however, science books in conversational formats came to be generally characterized as "feminine," and narratives of nature and natural theology were increasingly seen as problematic.

As the place of women and polite science was contested within influential professional scientific communities, narratives of science became the touchstone of textual practice, and science books in conversational and epistolary formats were considered old-fashioned. When Jane Loudon, professional horticultural and botanical writer, shaped her introductory *Botany for Ladies* (1842), she moved away from the conversational form that had been commonplace for didactic writing. (She herself had used a family-based narrative in one of her earliest pieces of popular science writing, *Conversations on Chronology and General History* [1830], which featured a mother and two daughters.) *Botany for Ladies* was Jane Loudon's alternate version of Lindley's *Ladies' Botany*, an introduction to the natural system of plant classification written in language that she considered suited to her women readers. The book gives a full synopsis of Candolle's ideas in language that, while technical, succeeds in being accessible. Loudon positioned the book as being written for women readers like herself who once were botanical beginners: "It is so difficult for men whose knowledge has grown with their growth, and strengthened with their strength, to imagine the state of profound ignorance in which a beginner is, that even their elementary books are like the old Eton Grammar when it was written in Latin—they require a master to explain them" (Loudon 1842, vi). *Botany for Ladies,* deliberately female-specific and gender-conscious in its title, nevertheless is part of a transitional moment in the history of women and science writing. By the 1850s, book titles designated specifically for women began to disappear from the lists of publishers. When *Botany for Ladies* was issued in a new edition in 1851, the title was changed to *Modern Botany*. Within the climate of the Great Exhibition, this change reflected new realities for publishers and authors alike. Certainly, a sex-neutral title that promised new-style botany made eminent sense for a widowed professional science writer who needed to capture the widest possible audience.

Yet conversational formats continued to exist as a way to teach science. Sarah Tomlinson taught science to children through conversations in *The Vegetable Kingdom* (1856), part of a multivolume compilation on topics in science and natural history issued by the Society for Promoting Christian Knowledge. The narrative features a daughter and two sons accompanying their father on his morning walk during the early spring and receiving instruction about stages of vegetable life and the uses of plants. Tomlinson teaches moral and religious lessons as well; thus, when the children learn about cruciferous plants such as the passion-flower, they are told that "these flowers are like us. . . . every Christian, even a little child, has a cross to bear" (S. Tomlinson 1856, 102). Sarah Tomlinson worked with her husband, a lecturer on experimental science at King's College, in producing popular science treatises during the 1850s–1860s as well as essays for the weekly *Saturday Magazine* in the format of "Easy Lessons" (M. Tomlinson

1900). John Ruskin, a better-known writer, also adopted the conversational mode for a science book. In *Ethics of the Dust: Ten Lectures to Little Housewives on the Elements of Crystallisation* (1866), Ruskin deliberately chose an anachronistic form for writing about mineralogy. He flouted professionalizing directions in mid-Victorian science culture by appealing to a style of scientific narrative that by then had lost cachet as a way to address serious general readers (Myers 1989).

By midcentury, however, gender-tagged titles and science writing in the form of conversations and letters had largely disappeared. Narratives set at home and featuring mothers and children were construed as not in line with the modern spirit in science. More broadly, cultural mechanisms were at work, regulating women's power at home, in their families, and in society at large. It is notable, for example, that, in marked distinction to the intellectual authority assigned to mothers in many earlier juvenile expositions of science, scientific authority in Sarah Tomlinson's midcentury *Vegetable Kingdom* is invested solely in the paternal teacher; the mother does not participate in scientific instruction in her narrative, but figures in the background in a more narrowly domestic and maternal role.

After the 1850s, opposition to family-based science writing designated specifically for women came not only from professionalizers within science culture, but also from some women reformers, feminists, and educators. Among women working to improve the level of female education, gendered, sex-specific, and domestic-based practices were considered problematic in the classroom and also on the page in science writing. Increasingly, the model of boy's education was adopted as the model for girls as well. For example, Lydia Becker, an activist in suffrage campaigns as well as a keen botanist and botanical writer, objected to female-specific books or pedagogies and linked science writing for women to larger debates about equity and women's education. In an essay in the *English Woman's Review,* she argued that "the mind has no sex," and that gendered features ("the conventional masculine type of mind") appear in both male and female bodies; thus, "there is no necessary, or even presumptive connexion between the sex of a human being, and the type of intellect and character he possesses" (Becker 1868, 491).

This analysis was the basis for enunciating a principle of equality about women and education in Lydia Becker's important essay "On the Study of Science by Women" (1869). There Becker posed a question of continuing resonance: "Why are there fewer scientific women than scientific men?" Her essay, a fervent account of the advantages that scientific study affords men, and could bring to women, itemizes the mental benefits of intellectual pursuits. "Many women might be saved from the evil of the life of intellectual vacuity, to which their present position renders them so peculiarly liable, if they had a thorough training in some branch of science, and the opportunity of carrying it on as a serious pursuit, in concert with others having similar tastes" (Becker 1869, 388). Becker deplores the lack of common ground for women to pursue scientific work, and contrasts the exclusion of women from scientific societies in England with an account of the Irish Royal College of Science, which, from the mid-1850s, had invited women students into lectures and laboratory work as full participants. The high level of success recorded by women there in examinations and other competitions buttresses Becker's argument about giving women access to existing, coeducational institutions. "It proves, by practical experiment, that men and women can associate with as much mutual advantage in the classroom and examination-room as in the home and the drawing-room . . . ; and the honours and rewards attaching to intellectual attainments, such as scholarships, fellowships, university degrees, and membership in learned societies, ought to be within the reach of either man or woman who has the taste to desire and the ability to earn them" (Becker 1869, 399). Becker goes on to oppose educational initiatives in her day to establish separate classes for women. She labels these efforts "well-meant but ill-advised" and insists, instead, that the success of coeducational local examinations offered through the University of Cambridge demonstrates the principle of equality much better than any system of "separation and exclusion." Becker warns that men will downgrade the achievements of those who pass "the women's examination," such as that offered at the University of London. She therefore argues at the level of both theory and pragmatics against single-sex education for girls, especially in connection with studying the sciences.

Lydia Becker herself wrote a science book that embodied her approach to female science study. *Botany for Novices* (1864) carries no gender tags in the title or the preface. An account of the natural system of classification for those with no prior acquaintance with "botanical science," the book is an agreeable but narratively spare exposition of Candolle's division of flowering plants into monocotyledons, dicotyledons, and acotyledons, with explanations of the terminology and illustrations of the parts of plants. Becker's book probably developed from lectures she gave on botany in girls' schools. In the context of a female tradition of science writing for women, there is nothing that would identify the work as a lineal descendent of earlier women science writers such as Priscilla Wakefield or Jane Marcet.

Changes like these in the form of science writing for women echo broader shifts in mid-Victorian science culture. With the spread of formal schooling, and particularly following upon criticisms of the inadequacy of girls' education by the Schools Inquiry Commission in 1868, the scientific education of girls became a high priority. It would be intriguing to know what textbooks were used to teach science to girls at the Ladies' College in Bedford Square (later Bedford College) in the natural science classes taught by William Benjamin Carpenter during 1849–50, or in classes taught there within the departments of botany, chemistry, or physics from the 1870s on (Tuke 1939). While information on textbooks in the new girls' schools is meager, school archives could facilitate research into the selection of science textbooks at pathbreaking institutions such as the North London Collegiate School for Ladies (where the founder, Frances Mary Buss, claimed to be offering "complete courses of natural science" by 1864) or the Girls' Public Day School Trust during the 1870s (Scrimgeour 1950; Bryant 1979, 94; 1986). Certainly, textbooks became a topic in themselves, as scientists and educators debated about pedagogy and textbook styles. The book review section of the *Journal of Botany,* for example, shows ongoing discussion during the 1870s and 1880s about topics such as object lessons and levels of language for elementary instruction.

It is likely that science textbooks chosen for girls to use in schoolrooms during the 1860s and 1870s more closely resembled twentieth-century textbooks than the styles of science writing that had brought their sisterly counterparts into science during the Enlightenment and Romantic eras. Later books embody a narrative of science, in contrast to earlier narratives of nature and narratives of natural theology. Yet, even though Lydia Becker asseverated the "principle of equality" in female education and shaped a text style in *Botany for Novices* that is ostensibly gender-neutral, it remains the case that her narrative of science was fundamentally as gendered as the feminized conversational format of an earlier generation. Like John Lindley during the 1830s, Lydia Becker during the 1860s wanted to defeminize science and science writing. Lindley wanted to defeminize botany in order to reposition it as a discipline for secular, middle-class, male experts rather than as a pursuit for parson-naturalists or an elegant recreation for women and aristocrats. Lydia Becker wanted to defeminize botany because she believed that women's access to science was best secured by arguments about equality rather than complementarity. Her science writing thereby represents the contestational feminist practice of the early suffrage movement.

Taken together, Lindley and Becker show that Victorian science writing for women is a fruitful field for both gender analysis and a broader cultural history of science. Women and gender issues were part of the history of Victorian science writing, for women read and wrote about science, and women also served a symbolic function for some shapers of science culture who used women and the "feminine" to distinguish in a binary way between what was professional and what was not. Male scientists did not completely erase women from textual science; rather, they positioned women (and Woman). Gender therefore surely factors into analyses of Victorian science writers and readers who were male as much as into those who were female. Among male science writers, Charles Kingsley encouraged men to study natural history as a way to enlarge themselves, by means of what we now would call enhancing their "feminine" side, but clearly addresses a fear that natural history study will lead men into "effeminacy." In *Glaucus; or, The Wonders of the Shore* (1855) the ideal naturalist is characterized as a male figure of chivalry, a knight-errant on a moral quest, "able to haul a dredge, climb a rock, turn a boulder, walk all day" (Kingsley 1855, 39). In like manner, when the editor of *Hardwicke's Science Gossip* suggests in the opening number in 1865 that his ideal scientific hobbyist would find "solace, . . . mental enjoyment, [and a] feeling of manhood elevated in the pursuit of ornithology or entomology," his language shows the degree to which science writing is implicated in larger gender economies.

Bibliographical Note

Analysis of women, gender, and Victorian science writing is still sparse. For discussion of historiographical issues, see Jordanova (1993); see also Cooter and Pumfrey (1994). For contextualizing analyses of women's relationships to Victorian science culture, see Russett (1989), Phillips (1990), and Richards (1983, 1989). For attention to the social history within which women were encouraged to participate in natural history and other sciences, see Allen (1995), Phillips (1990), and Scourse (1983). Studies in the popularization of science are beginning to integrate a gender dimension; see Gates and Shteir (1997). Benjamin has edited two collections that explore issues in the history of women, writing, and science (1991*a* and 1993); see also essays in Shuttleworth and Christie (1989). As well, see the special issue "Women and Science" in the journal *Women's Writing: The Elizabethan to Victorian Period* (Shteir 1995).

References

Allen, David Elliston. 1995. *The Naturalist in Britain*. 2d ed. Princeton, N.J.: Princeton University Press.

Ayton, Emily. 1855. *Words by the Way-Side; or the Children and the Flowers*. London: Grant and Griffith.

Becker, Lydia. 1864. *Botany for Novices: A Short Outline of the Natural System of Classification of Plants*. London: Whittacker.

———. 1868. "Is There Any Specific Distinction between Male and Female Intellect?" *English Woman's Review* 8:483-91.

———. 1869. "On the Study of Science by Women." *Contemporary Review* 10:386-404.

Benjamin, Marina, ed. 1991*a*. *Science and Sensibility: Gender and Scientific Enquiry, 1780-1945*. Oxford: Basil Blackwell.

———. 1991*b*. "Elbow Room: Women Writers on Science, 1790-1840." In *Science and Sensibility: Gender and Scientific Enquiry, 1780-1945*, edited by Marina Benjamin. Oxford: Basil Blackwell, 27-29.

———, ed. 1993. *A Question of Identity: Women, Science, and Literature*. New Brunswick, N.J.: Rutgers University Press.

Bentham, George. 1858. *Handbook of the British Flora.* London: Lovell Reeve.
Brock, William H. 1994. "Science." In *Victorian Periodicals and Victorian Society,* edited by J. Don Vann and Rosemary T. VanArsdel. Toronto: University of Toronto Press, 81-96.
Bryant, Margaret. 1979. *The Unexpected Revolution: A Study in the History of the Education of Women and Girls in the Nineteenth Century.* London: University of London Institute of Education.
———. 1986. *The London Experience of Secondary Education.* London: Athlone.
Catlow, Agnes. 1845. *The Conchologist's Nomenclator.* London: Lovell Reeve.
Catlow, Maria. 1848. *Popular British Entomology.* London: Reeve, Benham and Reeve.
———. 1855. *Popular Geography of Plants.* London: Reeve, Benham and Reeve.
Cooter, Roger, and Stephen Pumfrey. 1994. "Separate Spheres and Public Places: Reflections on the History of Science Popularization and Science in Popular Culture." *History of Science* 32:237-67.
Dewhurst, H. W. 1838. "An Essay on Female Education." *Young Lady's Magazine of Theology, History, Philosophy and General Knowledge* 1:34-36.
English, Mary P. 1990. *Victorian Values: The Life and Times of Dr. Edwin Lankester M.D., F.R.S.* Bristol: Biopress.
Flint, Kate. 1993. *The Woman Reader 1837-1914.* Oxford: Clarendon Press.
Francis, G. W. 1839. *The Little English Flora.* London: D. Francis.
———. 1840. *The Grammar of Botany.* London: D. Francis.
Gates, Barbara T. 1993. "Re-telling the Story of Science." *Victorian Culture and Literature* 21:289-306.
Gates, Barbara T., and Ann B. Shteir. 1997. *Natural Elegance: Women Reinscribe Science.* Madison: University of Wisconsin Press.
Gatty, Margaret. 1862. *British Sea-Weeds.* London: Bell and Daldy.
Gifford, Isabella. 1848. *The Marine Botanist.* London: Darton.
Gould, Stephen J. 1993. "The Invisible Woman." *Natural History* 102, no. 6:14-23. Reprinted in *Dinosaur in a Haystack: Reflections on Natural History,* by Stephen J. Gould (New York: Harmony Books, 1996), 187-201, and in *Natural Eloquence: Women Reinscribe Science,* edited by Barbara T. Gates and Ann B. Shteir (Madison: University of Wisconsin Press, 1997).
Harrison, J. F. C. 1954. *A History of the Working Men's College, 1854-1954.* London: Routledge and Kegan Paul.
Hibberd, J. Shirley. 1870. *Field Flowers.* London: Groombridge.
Johns, C. A. 1853. *Flowers of the Field.* London: Society for Promoting Christian Knowledge.
Jordanova, Ludmilla. 1993. "Gender and the Historiography of Science." *British Journal for the History of Science* 26:469-83.
Kingsley, Charles. 1855. *Glaucus; or, the Wonders of the Shore.* Cambridge: Macmillan.
Kirby, Mary. 1850. *A Flora of Leicestershire.* London: Hamilton, Adams.
——— 1887. *"Leaflets from my Life": A Narrative Autobiography.* London: Simpkin and Marshall.
Kirby, Mary, and Elizabeth Kirby. 1873. *Chapters on Trees: A Popular Account of their Nature and Uses.* London: Cassell.
Lee, Sarah Bowdich. 1844. *Elements of Natural History.* London: Longman.

Lightman, Bernard. 1997. "Constructing Victorian Heavens: Agnes Clerke and the 'New Astronomy.'" In *Natural Eloquence: Women Reinscribe Science,* edited by Barbara T. Gates and Ann B. Shteir. Madison: University of Wisconsin Press.

Lindley, John. 1829. *Introductory Lecture Delivered in the University of London on Thursday, April 30, 1829.* London: John Taylor.

———. 1834-37. *Ladies' Botany, or A Familiar Introduction to the Study of the Natural System of Botany.* London: Ridgway.

Loudon, Jane [Webb]. 1830. *Conversations upon Chronology and General History.* London: Longman.

———. 1841. *The First Book of Botany.* London: Bell and Daldy.

———. 1842. *Botany for Ladies; or A Popular Introduction to the Natural System of Plants.* London: John Murray.

———, ed. 1849-50. *The Ladies' Companion at Home and Abroad.* Vol. 1.

Marcet, Jane. 1806. *Conversations on Chemistry.* London: Longman.

———. 1819. *Conversations on Natural Philosophy.* London: Longman.

Mullan, John. 1993. "Gendered Knowledge, Gendered Minds: Women and Newtonianism, 1690-1760." In *Science and Sensibility: Gender and Scientific Enquiry, 1780-1945,* edited by Marina Benjamin. Oxford: Basil Blackwell, 41-56.

Myers, Greg. 1989. "Science for Women and Children: The Dialogue of Popular Science in the Nineteenth Century." In *Nature Transfigured: Science and Literature, 1700-1900,* edited by Sally Shuttleworth and J. R. R. Christie. Manchester: Manchester University Press, 171-200.

———. 1990. *Writing Biology.* Madison: University of Wisconsin Press.

Phillips, Patricia. 1990. *The Scientific Lady: A Social History of Women's Scientific Interests, 1520-1918.* London: Weidenfeld and Nicolson.

Pratt, Anne. 1852. *Wild Flowers.* London: Society for Promoting Christian Knowledge.

Richards, Evelleen. 1983. "Darwin and the Descent of Woman." In *The Wider Domain of Evolutionary Thought,* edited by D. Oldroyd and I. Langham. Dordrecht: D. Reidel, 57-111.

———. 1989. "Huxley and Woman's Place in Science: The 'Woman Question' and the Control of Victorian Anthropology." In *History, Humanity and Evolution: Essays for John C. Greene,* edited by James R. Moore. Cambridge: Cambridge University Press, 253-84.

Roberts, Mary. 1850. *Voices from the Woodlands; or History of Forest Trees, Ferns, Mosses, and Lichens.* London: Reeve, Benham, and Reeve.

———. 1851. *A Popular History of the Mollusca.* London: Reeve and Benham.

Russett, Cynthia Eagle. 1989. *Sexual Science: The Victorian Construction of Womanhood.* Cambridge, Mass.: Harvard University Press.

Scourse, Nicolette. 1983. *The Victorians and Their Flowers.* London: Croom Helm.

Scrimgeour, Michael, ed. 1950. *The North London Collegiate School, 1850-1950: A Hundred Years of Girls' Education.* London: Oxford University Press.

Secord, Anne. 1994. "Science in the Pub: Artisan Botany in Early Nineteenth-Century Lancashire." *History of Science* 32:269-315.

Sheets-Pyenson, Susan. 1985. "Popular Science Periodicals in Paris and London: The Emergence of a Low Scientific Culture, 1820-1875." *Annals of Science* 42:549-72.

Shteir, Ann B., ed. 1995. Special issue: "Women and Science." *Women's Writing* 2, no. 2.
———. 1996. *Cultivating Women, Cultivating Science: Flora's Daughters and Botany in England, 1760 to 1860.* Baltimore: Johns Hopkins University Press.
Shuttleworth, Sally, and J. R. R. Christie, eds. 1989. *Nature Transfigured: Science and Literature, 1700–1900.* Manchester: Manchester University Press.
Somerville, Mary. 1834. *On the Connexion of the Physical Sciences.* London: John Murray.
Stevens, P. F. 1995. "Natural History and Botany in the Nineteenth Century: a Multiplicity of Constituencies." Unpublished paper, Harvard University Herbarium.
Syme, J. T. Boswell. [1863–86] 1867. *English Botany, or Coloured Figures of British Plants.* 3d ed. London: Hardwicke.
Tomlinson, Mary. 1900. *The Life of Charles Tomlinson.* London: Elliot Stock.
Tomlinson, Sarah. 1856. *First Steps in General Knowledge.* Pt. 4, *The Vegetable Kingdom.* London: Society for Promoting Christian Knowledge.
Tuke, Margaret J. 1939. *A History of Bedford College for Women, 1849–1937.* London: Oxford University Press.
Turner, Frank M. 1993. "The Victorian Conflict between Science and Religion: A Professional Dimension." In *Contesting Cultural Authority: Essays in Victorian Intellectual Life.* Cambridge: Cambridge University Press, 171–200.
Twining, Elizabeth. 1849–55. *Illustrations of the Natural Order of Plants, with Groups and Descriptions.* London: Joseph Cundall.
———. 1858. *Short Lectures on Plants, for Schools and Adult Classes.* London: David Nutt.
———. 1866. *The Plant World.* London: T. Nelson and Sons.
———. 1872. *A Lecture on Plants as Water-Drinkers.* London: W. Tweedie.
Wakefield, Priscilla. [1794–97] 1995. *Mental Improvement, or the Beauties and Wonders of Nature and Art.* Edited by Ann B. Shteir. East Lansing, Mich.: Colleagues Press.
———. 1796. *An Introduction to Botany in a Series of Familiar Letters.* London: Newberry.
Ward, Mary. 1858. *A World of Wonders Revealed by the Microscope.* London: Groombridge.
Witz, Anne. 1992. *Professions and Patriarchy.* London: Routledge.
The Young Lady's Book: A Manual of Elegant Recreations, Arts, Sciences, and Accomplishments. 1859. London: Bohn.

12

Strange New Worlds of Space and Time: Late Victorian Science and Science Fiction

PAUL FAYTER

The stories had titles like *The Time Machine* and *Journey to Mars.* They offered readers disturbing portraits of scientists like *Doctor Moreau* and *The Invisible Man.* They described future histories, astounding inventions, subterranean and prehistoric worlds, apocalyptic nightmares, feminist and socialist utopias, and extraterrestrial cultures. They presented reconstructions and reconciliations of religion and science, as well as sly commentaries on contemporary class, racial, and sexual politics. They were typically serialized in the new periodicals, including the *Strand,* and *Pearson's Weekly.* They were known as "scientific romances," a term coined by the mathematician Charles Howard Hinton in 1884 and made legendary by Herbert George Wells in the last decades of what Alfred Russel Wallace rightly called *The Wonderful Century* (Hinton 1884-85).

However, Victorian science fiction has been largely excluded from the social history of nineteenth-century science. This essay aims to contest that exclusion and to illustrate with a few examples the two-way traffic between Victorian science and science fiction. One might think that science fiction constitutes a legitimate subject for scholarly attention because it must have served to *popularize* the sciences that respectable historians study. Note the faintly pejorative and patronizing whiff of that word, "popularize." It is almost as if Thomas Huxley were anticipating this new literature when, in a letter to his friend Joseph Dalton Hooker, he related his suppository theory of popular science education: "The English nation will not take science from above," he wrote, "so it must get it from below. We, the doctors, who know what is good for it, if we cannot get it to take pills,

---

The author gratefully acknowledges the generous and helpful editorial assistance of Bernard Lightman. The citation from the Hooker Papers is used by permission of the Library and Archives, Royal Botanic Gardens, Kew.

must administer our remedies par derriere" (Huxley to Hooker, 6 October 1864 [Hooker Papers]).[1]

But science fiction's importance is not limited to popularization. Recalling the problematic connotations of the French equivalent—*vulgarisation*—we should be aware of the implicit privileging of scientific discourse as uniquely authorized, autonomous, and transcendent, apparent in most accounts of popularized science (Sheets-Pyenson 1985; Lancashire 1988; Broks 1988; Cooter and Pumfrey 1994). That is, we need to take seriously the allegedly low or marginal literature of science fiction as an integral part of the historical study of the contingent, social context and content of natural knowledge.

Why explore the noncanonical literature that is Victorian science fiction? Because it offers a new angle on the institutionalization, mobilization, and legitimation of scientific knowledge and practice. It reveals the ambivalence of attitudes toward science, invention, women, scientists, and social change. It helps us see how science was integral to Victorian culture and suggests the need to reconceive the history of scientific popularization. The mutual traffic between science fiction and science is most obvious within popular culture, among commercial novelists for working- and middle-class readers (including children and women), and in the work of scientific popularizers, journalists, and amateur naturalists. But it was not limited to these groups. Professional scientists not only helped shape science fiction, in many cases their work was shaped by it. In a time of rapid industrialization, professionalization, and specialization, this late Victorian literature played a role in maintaining a common popular scientific culture. If we pay attention to the content and diverse cultural locations of both science and science fiction, to who was writing it and who was reading it, we will notice a fluid exchange of ideas—not only across national and disciplinary boundaries, but across lines traditionally separating amateur and professional, highbrow and lowbrow, established knowledge and speculation, science and fiction.

Contrary to repeated claims that modern science fiction was created in the American pulps of the 1920s, the genre was an offspring of the nineteenth-century Age of Science. According to Wells, the century underwent two major transformations, one technological, the other intellectual and spiritual. First was the appearance of new forms of transportation, especially railroads, which were symbolic of a host of engineering marvels with material benefits and other social transformations. Second was the crisis signaled by Thomas Malthus's *Essay on the Principles of Population* (1798), and climaxing in Charles Darwin's *Origin of Species* (1859). For Wells, Malthus shattered any dream of utopia through social reconstruction

---

1. This passage was trimmed from the sanitized version in Huxley's *Life and Letters*, volume 1. Wells (1894a) objected that the "popularised" science produced by professionals was too jargon laden and condescending by half.

without ruthless control over reproduction. And in inspiring Darwin's secular evolutionism, he helped set "such forces in motion as have destroyed the very root-idea of orthodox righteousness." Wells believed that the discovery of deep geological time and astronomical space set humanity in a cosmic context, reducing the biblical story of Creation and Fall from high historical drama to a quaint and local folktale. Indeed, the nineteenth century "lost the very habit of thought from which the belief in a Fall arose. It is as if a hand had been put upon the head of the thoughtful man and had turned his eyes about from the past to the future" (Wells 1901b, 288-90).

While it cannot be limited to secular or future fiction, science fiction can be most simply described as the literature of social change as initiated or mediated by technology and science. This is not to imply that the only relationship science fiction has to "real science" is one-way and parasitic, that is, that science is always prior and primary, with fiction merely adopting whatever passes for scientific knowledge at the time. Science fiction not only reflected contemporary trends, but in suggesting new scientific and technical possibilities and applications, it helped create the expectation of change (Clarke 1979). One classic survey of the field in the 1870s-1890s publicized the predictive character of this literature by noting dozens of ideas and machines that appeared in advance of real-world discovery, invention, or application (Bailey 1947, chaps. 4 and 9). In this work, *Pilgrims through Space and Time*—the first scholarly discussion of the subject—Bailey defined science fiction as "a narrative of an imaginary invention or discovery in the natural sciences and consequent adventures and experiences." By imaginary invention, he meant such things as spaceships or atomic bombs: imaginable, but beyond technological realization at the time of writing. "The discovery," he continued, "may take place in the interior of the earth, on the moon, on Mars, within the atom, in the future, in the prehistoric past, or in a dimension beyond the third; it may be a surgical, mathematical, or chemical discovery." But, whatever it is, it must be at least rationalized as scientifically possible. Such writing does not end with mere anticipation, though. The author must also consider the impact of an extraordinary discovery on society, and try "to foresee how mankind may adjust to the new condition" (Bailey 1947, 10-11).

Among science fiction critics the most influential definition is Darko Suvin's: it is, he says, the literature of "cognitive estrangement" (Suvin 1979, chap. 1; cf. Alkon 1994, chap. 1). That is, it invites the kind of fresh, take-nothing-for-granted social analysis that the proverbial Martian anthropologist would do upon visiting Earth for the first time. Science fiction both defamiliarizes (or makes strange) the world of quotidian life and encourages a critical awareness (or cognition) of the world's underlying values, beliefs, and assumptions.

To think about the known and mundane as unfamiliar is to recognize its contingency. The way things are is not the way things have to be. Such a perspective, along with faith in the scientific intelligibility of nature's order, implies that if the laws of this world were uncovered, then, instead of accepting without question some thing as natural and therefore inevitable, it could be controlled or even changed. The legendary ability of a Galileo or a Newton to be "naive" enough to regard a swinging chandelier or falling apple as strange enough to require explanation are good examples. Studying science fiction gives us access to a part of British culture that, while often written by romanticizing dreamers and deceivers deserving skeptical interrogation, maintained a critical relationship to the emerging secular late Victorian society. One of the primary locations for exploring this relationship is the mass fiction magazines of the period.

In the 1890s, and roughly coinciding with the birth of commercial science fiction and the rise of the "new imperialism" and the "new journalism," many new, inexpensive magazines began appearing on both sides of the Atlantic. In the United Kingdom many of these magazines published science fiction short stories and serials, which joined the over two hundred American dime novels of fantastic scientific invention and adventure in the Frank Reade, Jack Wright, Tom Edison, Jr., and Electric Bob series.

The creation of a commercial audience for science fiction encouraged an explosion of subgenres that popularized, exploited, and even forecast the latest scientific theories and technological marvels. Following the earlier gothic scientific romances of Mary Shelley and Edgar Allan Poe, and alongside the poorly translated *voyages extraordinaires* of Jules Verne, and future utopias from Mary Griffith to Edward Bellamy, there came stories of future wars and lost worlds, interplanetary travels and end-of-the-world catastrophes. The world of Victorian ideas and inventions, from the domestic (e.g., vacuum cleaners) to the astonishing (e.g., spaceships) was celebrated in mass magazine fiction and illustration, creating a visual and imaginative vocabulary of the future (Evans 1976; Frewin 1988; see figure 12.1).

Why did mass science fiction not "take off" until the 1890s? Part of the answer lies in the story of Victorian publishing. After such developments as the invention of the steam press, cheap wood-pulp paper, and the stereotyping process, it became both possible and profitable to publish a variety of early and mid-Victorian newspapers and periodicals such as the *Illustrated London News* (from 1842). There were penny weeklies, shilling magazines, half-crown almanacs, six-shilling quarterlies, and guinea annuals. But science fiction in print or picture was rare in such outlets, as were such novels in book form. This was the age of the "three-deckers," three-volume novels, few of which were science fiction.

The reading public in England grew to depend on the relatively expensive three-decker format for fiction thanks to an alliance between book

Figure 12.1 A futuristic flying machine over St. Paul's. The late Victorian skies were filled with such machines. (From Smale 1900. Illustration by R. W. Wallace.)

publishers and circulating libraries. It was cheaper, of course, for a person to acquire unlimited borrowing privileges for a small yearly fee than to purchase three-deckers on her own. So people naturally supported their local libraries. In turn, the popularity of libraries made them an attractive market for publishers and their lines of long, three-decker novels. This cycle tended to discourage English publishers from producing the less expensive single volumes more common in other countries.[2]

From the early 1870s through early 1890s, science fiction stories existed mainly in pamphlet and single-volume formats, that is, in an alternative reading network outside the three-decker circulating library system, and

---

2. See Alkon (1994, 40–41), who follows Stableford's (1985) argument about the flourishing of science fiction whenever there is a commercial market for shorter fiction. The massive social and physical detail in the novels of a Charles Dickens would be difficult to reproduce out of whole cloth in stories of unknown worlds and imaginary cultures. And yet there were three-decker science fiction novels, beginning with the 1818 prototype, Mary Shelley's *Frankenstein*. The argument further breaks down if we recall the stunningly detailed accounts of Martian culture in the 1890s. (In focusing on periodicals here, I am leaving out the effects of such innovations in the book trade as "net retail pricing," pioneered by Macmillan in the 1890s.)

the penny-serial circuit. If this suggests that science fiction was not originally recreational or self-improving literature aimed at working families of the lower and middle classes, the as yet unanswered questions are: What was it for and who was it aimed at (Suvin 1983)? In any case, Wells, modern science fiction's godfather and midwife, began his professional writing career in the 1890s just as the three-decker was being replaced by thinner novels and mass-market magazines with hundreds of thousands of readers that, along with more respectable periodicals, began carrying short stories, including science fiction.

So late Victorian science fiction participated in a wave of change in publishing. Small-circulation magazines for middle- and upper-middle-class readers were joined by publications that—owing to recent changes in typesetting, printing, engraving, photography, distribution, and advertising—could lower prices and appeal to growing numbers of readers looking for information and escape in the years before film and radio were transformed from scientific curiosities into mass media.

The *Strand*, an illustrated monthly that George Newnes began publishing in 1891, was in the forefront of this commercial expansion (Moskowitz 1968). To sell profitably at sixpence required a large readership; articles on popular, topical subjects from ladies' fashions to London opium dens soon pushed circulation to the half-million mark. *Pearson's Weekly*, a tabloid that had started up in 1890, soon began printing science fiction stories (as did the *Idler*, a monthly from Chatto and Windus that began in 1892); its science fiction stories were often reprinted in the United States by *McClure's Magazine*. When rivals to the *Strand* began appearing, the competition prompted changes in the magazine's content, especially its fiction. This opened the door to Arthur Conan Doyle's scientific detective Sherlock Holmes (Smith 1994, chap. 7) and even stranger characters. In May 1893, priced at a shilling, Routledge's *Pall Mall Magazine* began aiming for higher, more refined literary tastes. But it was getting harder to ignore popular science fiction. *Pearson's Magazine*, a monthly, began publishing both fact and fiction in January 1896, following its companion, the *Weekly*, and in competition with the the *Strand*. At sixpence, it needed, in Arthur Pearson's words, a "colossal circulation" to stay in business. To that end, it soon serialized Wells's shocking tale of Martian invasion, *The War of the Worlds*. Once *Pearson's Magazine* began regularly featuring science fiction, the *Strand* followed suit. To such stock characters as the lady, the colonel, the detective, and the explorer was added the scientist (see Haynes 1994, chaps. 8–11).

The strange new worlds of space and time were landscapes onto which late Victorian writers and artists projected their devices and desires and mapped out their fin de siècle anxieties and enthusiasms. But more, when periodical science fiction is read alongside not only the "serious" scientific journals but the more accessible ones, including *Science Gossip* and *Popu-*

lar Science Review, questions are raised about the significance of the distinctions usually drawn between pseudo-, popular, speculative, and "real" science. Links between unseen spirits and physics, for instance, were explored in late-nineteenth-century astronomy texts, magazine science fiction, ladies' journals, and William Crookes's *Quarterly Journal of Science* alike (Wynne 1979; Oppenheim 1985). The more we examine the historiographic boundaries between science fiction and science, the more it becomes apparent that the flow of ideas was not simply one-way, from the sciences to literature. Rather, we are dealing here with a shared context in which distinctions among "scientific," "fictional," "moral," and "social" discourses are blurred. This claim can be supported by reviewing a few now classic tropes — future evolution, the Fourth Dimension, and intelligent Martians — that suggest the interest and value of including science fiction in our contextualized accounts of Victorian scientific culture.

I. Evolution, Entropy, and Time Travel

According to the poet Mathilde Blind in 1886, Darwinian evolution would produce "better, wiser, and more beautiful beings . . . in the ages to come . . ., than we can now have any conception of" (Morton 1984, 53). Many were considerably less sanguine and worried that regression was at least as likely as progression in the natural and social orders (Chamberlin and Gilman 1985; Bowler 1989). Years before Max Nordau's novel *Degeneration* appeared in English translation (1895), Francis Galton's eugenics project began playing to English fears of social and biological degeneration. The "wrong kind" of people were breeding like rabbits. Victorian gentlefolk were facing an invasion of the poor and inferior from below. Even Galton's cousin Charles Darwin was convinced of the baneful effects of our misguided insulation from the bracing and beneficent effects of struggle and natural selection. The devolutionary alarm had been well and loudly sounded by the time Ray Lankester's *Degeneration: A Chapter in Darwinism* appeared in 1880.

Given persistent Victorian hopes for evolutionary ascent, and widespread fears of social descent, and given the search for Darwinian "influences" on literature, it is remarkable that so little of substance has been written on the subject of "Darwin's plots" (Beer 1983) or "biology and the literary imagination" (the subtitle of Morton 1984). The two studies just referred to represent the analytic cream of the crop, along with *Darwin and the Novelists* (Levine 1988). Yet, outside of students of Wells, only Morton seems aware of relevant science fiction literature; historians of science are just as neglectful. Henkin's 1940 study, less refined, was more in touch with what ordinary Victorians were reading.

The evolutionist science fiction underground would win no literary awards, but historians of biology should know it existed, however obscurely. In 1871, the same year Darwin published his *Descent of Man,* there appeared pseudonymously *The Gorilla Origin of Man; or, The Darwin Theory of Development, confirmed from recent travels* (His Royal Highness Mammoth Martinet 1871). In a concave polar world beyond the Arctic Ocean, not only British science, but politics, courts, churches, and the military were satirized. In *Simiocracy* (Brookfield 1884) future left-wing evolutionists succeed in gaining equal rights for orangutans, and English humans end up subjugated. A more serious work is *The Curse of Intellect* ([Constable] 1895), which tells the tragic tale of a monkey with scientifically enhanced intelligence—one of many forgotten stories on the implications of our descent from primate ancestors (e.g., Rickett 1893).

Other stories, well known and still in print, deserve more attention from historians of science. *The Strange Case of Dr. Jekyll and Mr. Hyde* (Stevenson 1886) involves chemically induced devolution, an explicitly biological metaphor for the Fall. Evolutionary explanations for original sin—part of a wider domain of naturalistic theodicies—can be found in Darwin and Huxley as well. Wells, who wrote scientific essays on human evolution, extinction, and degeneration in the early 1890s, subverts progressionist readings of evolution in *The Time Machine* (1895). This paradigmatic story undercuts both communist and liberal bourgeois utopian pretensions in its evocation of cannibalistic Morlocks, dying suns, and Huxley's dark moral vision of nature as a decaying garden in the just-published "Prolegomena" to *Evolution and Ethics* (1894). *The Island of Doctor Moreau* (Wells 1896) can be read as a horrifying anti-Bridgewater Treatise in which the beneficent God of natural theology is replaced by a cruel parody, a vivisectionist who serves as the evil god of evolutionary "uplift." The dark and alien underside of eugenics is depicted in *The First Men in the Moon* (Wells 1901*a*). In *War of the Worlds* (1898) Wells dramatizes the outer limits of Darwinism and forces a rethinking of assumptions about power, progress, and purpose. Victorian evolutionism was shot through with ideologically loaded metaphors of invasion, conquest, colonization, and extermination. If we did not all know this from recent Darwin scholarship, we would know it from reading science fiction.

Visions of dystopia, degeneration, and death did not depend solely on science fiction interpretations of evolution (Pick 1989.) These themes were reinforced by certain readings of the second law of thermodynamics offered by physicists, science popularizers, social prophets, and science fiction writers. The subject of entropy as a physical and social metaphor, with deep connections to biology and literature, is underexamined (but see introductory discussions in Kern 1983; Myers 1985; Beer 1989; and Dale 1989, chap. 9).

Evolution and entropy encouraged science fiction about the future and time travel, which raises the question of the Fourth Dimension. Wells's *Time Machine* hinged on the new idea of time as the Fourth Dimension, and illustrates how science fiction was in dialogue with physics as well as biology. Novelist Israel Zangwill, in his September 1895 review of Wells's novel for *Pall Mall Magazine,* noted that by traveling out into space at a faster-than-light velocity you could watch the light from the past as it caught up with you. Thus, it was possible to look backward at Earth not only in space, but in time. Observing historical events as they happen was an idea presented in the astronomer Camille Flammarion's best-selling novel *Lumen* (1872). Similarities in certain details—for example, looking earthward from a planet orbiting Capella to witness the French Revolution—suggest that English physicist John Poynting may have borrowed this idea in his 1883 scientific paper entitled "Overtaking the Rays of Light" (Poynting 1920). The Fourth Dimension could be seen as spatial as well as temporal, and so this story is even more complicated. Simon Newcomb, the Canadian-born astronomer at the U.S. Naval Institute, editor of the *American Journal of Mathematics,* and sometime science fiction writer, had been lecturing and publishing on the theory of hyperspace and parallel universes from the late 1870s (e.g., Newcomb 1898). On 28 December 1893 Newcomb addressed the New York Mathematical Society on four-dimensional geometry. This was reprinted in *Nature* in February 1894, where it was read by Wells in time to refer to it in the opening pages of *The Time Machine.*

II. The Fourth Dimension

The three individuals who were most responsible for bringing the Fourth Dimension out of the mathematicians' and physicists' academy and into public consciousness were Edwin Abbott Abbott, Charles Howard Hinton, and Herbert George Wells. Wells obtained his B.Sc. in 1890 from London University, with first-class honors in zoology, second-class in geology, after having earlier served for a year as starstruck acolyte to Huxley's high priest at the Normal School of Science in South Kensington. Within three years he had published a two-volume biology textbook, one on physiography, and articles of scientific exposition, popularization, and speculation, and had begun writing science fiction (Haynes 1980; MacKenzie 1987). But the story of the Fourth Dimension began in another place and time.

On 10 June 1854, at the University of Göttingen, a twenty-eight-year-old mathematician, Bernhard Riemann, announced a radical conception of space as curved, based on a new geometry of higher dimensions. His suggestion that hyperspherical space may describe the universe's actual shape led not only to Einstein and scientific cosmology, but to speculations about four or more dimensions in many hundreds of late-nineteenth-century papers (Bork 1964). Riemann imagined two-dimensional creatures whose world was a sheet of paper. What would happen if the flat sheet were crumpled in a third dimension, invisible to the inhabitants? As the creatures tried to move across their world, they would experience a mysterious "force" (a wrinkle in three-dimensional space) that, by pushing against them, prevented straight-ahead movement. Then Riemann imagined our three-dimensional world warped in higher-dimensional space. He concluded that physical forces—gravity, electricity, magnetism—were the result of *geometry,* caused by the curving and crumpling of our three-dimensional world through an unseen fourth dimension. Riemann died at age thirty-nine, before he could complete his work on hyperspatial theory. But his ideas, born at the intersection of imaginative science and science fiction, were taken up by others, including Hermann von Helmholtz, who wrote and lectured extensively on the mathematics of intelligent two-dimensional beings living on spheres (Kaku 1994, 30-43).[3]

One of the people who helped introduce the new geometry into England was the Reverend Dr. Edwin Abbott, New Testament scholar and author of *Through Nature to Christ* (1877), who published *Flatland* in 1884 (Jann 1985; Smith 1994, chap. 6). This was the story of a hierarchical, two-dimensional world in which the women are lines, the (all male) working and middle classes are triangles, professionals, gentry, and nobility are squares, pentagons, and hexagons, all the way up to the ruling priests, who are circles (figure 12.2).

Flatlanders could have no secrets from three-dimensional beings like us. From our higher perspective, we could see into their very bodies as if with x-ray eyes, or enter and leave locked rooms like ghosts (the walls would appear as outlines on the floor to us). By moving them through the third dimension, we could make objects and even people appear and disappear at will. In other words, we would have godlike powers. A three-dimensional object, such as a sphere, passing through Flatland would be perceived only as a series of lines (two-dimensional cross sections seen edge-on) whose length varied with time. Indeed, this is exactly what happens near the novel's end.

Flatland is read today as an original science fiction classic, a clever exploration of geometry, a satiric contribution to the "woman question" debate, and an indictment of the class-obsessed, close-minded status quo. But it is also a theological parable. One day, "the last of the 1999th year of our era"—as the narrator, a square mathematician, recounts—a Stranger arrives, a powerful Presence who ushers in the Third Millennium. This promised Messiah descends into the narrator's living room: a perfect sphere from Spaceland, incarnate in the form of a circle. The square is chosen to be the first apostle of "the Gospel of the Third Dimension." Of course, the priestly

---

3. See Richards 1979 and 1988 on the introduction of non-Euclidean geometry into England. The missing part of her story is the popular social and intellectual career of the new—and imaginative—geometry, in which both science fiction and the visual arts (Henderson 1983) played key roles. Science fiction as a source of ideas and inspiration for professional scientists deserves more study.

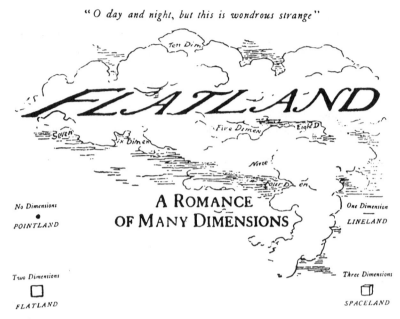

Figure 12.2 Author's frontispiece sketch for *Flatland*, by Edwin A. Abbott (1884).

class seek to stamp out this new heresy and keep control of their world. The story raises a number of questions, including how we could imagine four-dimensional beings and how four-dimensional beings would view us (figure 12.3).

That God's omniscience and omnipotence were explicable if the deity resided in a fourth spatial dimension was one of the theological implications of Charles Hinton's scientific romances, hybrids combining mathematics, religious speculation, fictional narrative, and scientific reflections on physics, causation, and perception. Charles was the son of the extraordinary English surgeon James Hinton, who became the charismatic leader of a sect devoted to polygamy and free love. "Christ was the Saviour of men," he once boasted, "but I am the saviour of women, and I don't envy Him a bit!" (Rucker 1984, 64). This may explain why the younger Hinton, not content with his wedding in 1877 to Mary Everest Boole (daughter of George Boole, the mathematician), also married Maude Weldon (the mother of his twins) in 1885. He lost his job as science master at Uppingham School, was arrested and spent three days in prison when his bigamy was discovered. In 1886, the year Oxford awarded him the M.A. in mathematics, Hinton with his first wife left England for a teaching job in Japan. By 1893 Hinton was in the United States, inventing the automatic pitching machine for the baseball team at Princeton, where he taught mathematics. In 1900, he accepted a position at the U.S. Naval Observatory, and two years later began working

The diminished brightness of your eye indicates incredulity. But now prepare to receive proof positive of the truth of my assertions. You cannot indeed see more than one of my sections, or Circles, at a time; for you have no power to raise your eye out of the plane of Flatland; but you can at least see that, as I rise in Space, so my sections become smaller. See now, I will rise; and the effect upon your eye will be that my Circle will become smaller and smaller till it dwindles to a point and finally vanishes.

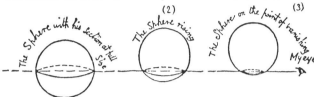

Figure 12.3 The messianic, three-dimensional sphere manifesting itself to Flatlanders as a series of circles. (Author's sketch, Abbott 1884, 73.)

at the U.S. Patent Office in Washington, D.C., where, in the spring of 1907, he dropped dead in the middle of a toast to female philosophers at a reception (Rucker 1980, "Introduction"; 1984, 64–68).

Hinton provides another good example of the tangled web of relations one finds in late Victorian science and science fiction. As an Oxford undergraduate in the 1870s he became interested in the new non-Euclidean geometry being described in the professional journals. His first popular article, "What is the Fourth Dimension?," was an exposition of hyperspace that appeared in the *Dublin University Magazine* in 1880. It received little attention until it was reprinted in the *Cheltenham Ladies' College Magazine* in 1883. The London publisher Swan Sonnenschein issued the essay as a pamphlet in 1884, with the sexy subtitle "Ghosts Explained." It was again reprinted, in volume 1 of Hinton's collected *Scientific Romances*. As Hinton continued breaking new ground in science fiction and mathematics, his work began to be used by others. The mathematician William Rouse Ball (1891) gave priority to Hinton when he used four-dimensional geometry to explain gravitation. Paul Heyl, an American physicist, followed Hinton's ideas in his 1897 doctoral thesis on light. And the religious philosopher Arthur Willink used hyperspace in his 1893 natural theology of the unseen world, believing that God, who dwelt in "Highest Space," was therefore "infinitely near to every point and particle" of each creature. Thus, it was literally, physically true that (as Saint Paul had proclaimed to the Athenians) "in Him we live and move and have our being." What, then, were the boundaries of fiction, knowledge, and belief? This question, with an additional political twist, is even more apposite when it comes to the multivalent, boundary-crossing literature about Mars.

III. Late Victorian Mars

From the 1840s in Britain, there had been great strides made in the accurate figuring and precise polishing of metal mirrors, for reflecting telescopes, and in the pouring and polishing of large glass discs of high optical quality, for refracting telescopes. The first photographs of a celestial object—the moon—also appeared in the 1840s. But it was not until the introduction of more sensitive dry plates after 1878, following years of experiments with wet-plate exposures, that astronomers commonly used cameras. What came to be called the "new astronomy" really began when Gustav Kirchoff and Robert Bunsen showed in 1859 that every element had its own unique absorption pattern, resulting in a signature of dark lines in the spectrum of light passed through a prism. When a prism or diffraction grating was attached to a telescope, a new instrument—the spectroscope—was created. The study of spectral lines therefore could reveal the chemical compositions of stars and planetary atmospheres (Hearnshaw 1993; Crowe 1986, 359–66).

Soon Mars—with its own, albeit thinner, carbon dioxide atmosphere, its rocky, rust-colored terrain, with mountains and desert plains, its polar ice caps, displaying seasonal variations in size and shape, and its daily rotation period and axial tilt almost identical to our own—was seen as more and more earthlike. And if earthlike, and therefore habitable, why not inhabited? And would not intelligent beings everywhere be essentially the same (Pope 1894)?

In 1877, the year a Martian opposition allowed particularly fine telescopic scrutiny, came news that seemed to highlight similarities between Earth and Mars. In Washington, D.C., Asaph Hall announced his discovery that Mars had two hitherto unsuspected moons. From Milan, Giovanni Schiaparelli reported that Mars was covered with *canali*. And at the Greenwich Observatory, Edward Walter Maunder confirmed the presence of water vapor in the Martian atmosphere. Schiaparelli's was the most sensational of these reports, in large part because his *canali* (i.e., grooves or channels) was usually translated as "canals," suggesting they were artifacts, not simply natural features. Using a micrometer attached to an eight-inch refractor, he had carefully studied sixty-two surface features of the Martian landscape. He presented his discoveries in an 1878 treatise that included his first published map of Mars.[4]

Soon after the opposition of 1892, which again afforded exceptionally good views of Mars, many dozens of scientific papers appeared, including

---

4. Maps of Mars were both scientific descriptions and social constructions, as political and professional conflicts over naming features of the Martian landscape reveal. The visual deliverances of astronomical instruments were not purely objective, but mediated, interpreted, and constructed; telescopic images underdetermined perceptions of Mars (Hoyt 1976; Hetherington 1976; Sheehan 1988).

one in 1893 by Schiaparelli on Martian geography that revealed his conversion to the intelligent-alien cause. This paper was translated by U.S. astronomer William Henry Pickering and widely reprinted in professional British and American journals. (It also appeared in 1903 as an appendix to Louis Pope Gratacap's *Certainty of a Future Life in Mars,* more evidence that the membrane separating science and science fiction was semipermeable.) Reports of bright light flashes from the Martian surface—often interpreted as intelligent signals—came from observatories in France and California during 1892, attracting international attention. New sightings of the now infamous "canals" were made. The year also saw the publication of Flammarion's exhaustive survey (1892*b*), the Martian-believer's Bible. The opposition of 1894 was again fine, spawning further spectroscopic studies of the Martian climate and atmosphere. (Fictional depictions of the Martian environment both shaped and reflected revisions to astronomers' descriptions and speculations; see Johnson and Clareson 1964.) In 1895 Percival Lowell published his first Mars book, and in 1896 Francis Galton produced a sixty-page manuscript in which he worked out a language suitable for communicating with alien civilizations (Percival 1895; Galton 1896; see also Flammarion 1892*a;* [Wells] 1896*a*).

This was old hat to science fiction writers like Percy Greg, whose secularist vision of Mars in *Across the Zodiac* (1880) countered Christian and spiritualist renderings of the planet as the abode of higher, older, wiser beings, or of the souls of the dead. As would the anonymous *Politics and Life in Mars* ([Welch?] 1883), Greg's novel included much astronomical detail, with long passages devoted to Martian science, technology, society, and language. The Martian philosophy of life was rationalist and positivist; indeed, it was a crime not to accept the results of science. Alice Jones and Ella Marchant (1893) offered a different ideology in their Martian novel, in which the nameless male narrator discovers a shocking feminist world. Women smoke, drink, run businesses, hold public office, and pick up males for sex with nary a thought of marriage.

But these and many other interplanetary romps represented something more serious than cheap and entertaining ephemera. In Robert Cromie's *A Plunge into Space* (1890), for instance, Mars is a dry and dying world, farther along the evolutionary path that Earth will one day take. The Martians are "at the pinnacle of their perfection," as illustrated by their ability to control nature by creating artificial oases. But the social and scientific wonders encountered by travelers from Earth do not mean that Mars is heaven. For ahead of the Martians "there is no further progress. Their only change must be toward decay" (Cromie [1890] 1891, 103-4).[5] The superintelligent Martians are bored with life and passionless. Their sciences have left them noth-

---

5. Compare Griffith 1901, where the devolution of human-like Martians is set in explicitly Darwinian terms.

ing more to learn. The implication is that this stage of unsurpassable perfection can be followed only by degeneration. Cromie, a Belfast journalist, was holding a sadly skeptical mirror up to Victorian believers in progress.

The tragedy in Cromie's tale was that such a future seems unavoidable. In common with so many evolutionists, he saw the development of life as directional. The relentless workings of the physical and biological laws of nature dictated future progress, followed by inevitable decline and extinction. This was the same cosmic pessimism Huxley expressed in the famous ending of his "Prolegomena" of 1894. This fin de siècle melancholy would reappear with a vengeance in *the* classic Mars novel, Wells's *War of the Worlds* (1898), first serialized during the jubilee year of 1897.

"In the last years of the nineteenth century," Wells wrote in the opening paragraph of his novel, complacent men went about

> their little affairs, serene in their assurance of their empire over matter.... At most [they] fancied there might be other men upon Mars, perhaps ... ready to welcome a missionary enterprise. Yet across the gulf of space, minds that are to our minds as ours are to those of the beasts that perish, intellects vast and cool and unsympathetic, regarded this earth with envious eyes, and slowly and surely drew their plans against us. And early in the twentieth century came the great disillusionment.

That last word was a huge understatement, given the apocalyptic tale that unfolded.

By the novel's end, Earth has been granted a reprieve when the Martians succumb to a terrestrial infection. But in the unsettling epilogue, the narrator anticipates another invasion. "We cannot regard this planet as ... a secure abiding place," he writes. But it may be "that in the larger design of the universe this invasion from Mars is not without" some benefit, for "it has robbed us of that serene confidence in the future" that leads to "decadence." "When the slow cooling of the sun makes this earth uninhabitable, as at last it must do," perhaps humanity will, like the Martians, seek another world to live on. Or maybe not: "To them, and not to us, perhaps, is the future ordained."

In his first chapter, Wells framed his vision in terms of the nebular hypothesis and Darwinism, making clear his political subtext, in which the tables are turned on England's own imperial colonizers. "Life is an incessant struggle for existence," he wrote, "and it would seem that this too is the belief of the minds upon Mars." The Martian's world is crowded, cooling, dying, and "to carry warfare sunward" is their only hope. Before harshly judging them, though, Wells asked his readers to "remember what ruthless and utter destruction our own species has wrought, not only upon animals ... but upon its inferior races." He recalled the Tasmanians' extermination by European immigrants and asked, "Are we such apostles of mercy as to complain if the Martians warred in the same spirit?" After all, had not Darwin himself written in his chapter on natural selection that "in all countries, the natives . . . have allowed foreigners to take firm possession of the land. And as foreigners have thus everywhere beaten some of the natives, we may safely conclude that the natives might have been modified with advantage, so as to have better resisted such intruders"? Had not Darwin averred that natural selection "almost inevitably induces extinction" (Darwin 1859, 83, 433)?

Looking back at his earlier scientific romances, Wells once said that his aim had been to "domesticate the impossible" by situating some extraordinary person or event within the ordinary, comfortable details of daily life (Wells 1934). In *War of the Worlds,* Wells domesticated the alien Martians by locating them among the familiar landmarks of the cozy English countryside. There are over seven hundred English place-names in a book slightly over three hundred pages. At the same time, he defamiliarized the domestic by revealing the alienness of Earth and its human inhabitants. In book 2 of the novel the constantly repeated words "Mars" and "Martians" increase, then overtake the number of English place-names in the text.

In the second chapter of book 2, the Martians are revealed as hideous, sexless vampires, all tentacles and head. Even more horrifying, the narrator figures out that the Martians "may be descended from beings not unlike ourselves, by a gradual development of brain and hands" (figure 12.4). Seeds escape from the Martian spacecraft, and soon red plants begin choking rivers and covering the terrain: "I found about me the landscape, weird and lurid, of another planet" (book 2, chap. 6). A *red* planet. As the earth turns Martian, the Martians are seen as superevolved humans. In the end, for all their terrifyingly advanced weaponry, the Martians are but, like us, the product of Darwinian selection, vulnerable to lowly bacteria to which they have never been exposed.

In his study of fiction about future invasions, Clarke (1992) observes that *War of the Worlds* is about three kinds of war. First, it is a commentary on the war of the Europeans upon the "less civilized" during the imperial expansion of the 1880s and 1890s. This time, however, it is the English who are the Tasmanians, and the Martians who are the colonizers. Second, it is about the biological war taking place everywhere in nature, what Darwin (echoing Malthus) called "the struggle for existence" in "the great battle of life" — the "war of nature" in which (echoing Spencer) "only the fittest survive." Third, it is about the kind of war that might occur if science were devoted to serving military interests by producing advanced weapons of mass destruction.

The novel represents two other kinds of Darwinian warfare as well. Fourth, we witness the invasion of the present by the future, for Mars and

Figure 12.4 Monstrously microbial Martians, flagella waving menacingly, invading Earth in H. G. Wells's *War of the Worlds*. One of their war machines is silhouetted in the background. Warwick Goble illustrated the story when it was first published as a serial in *Pearson's Magazine* (Wells 1897c, vol. 4, no. 23, 559).

its inhabitants stand for the evolutionary destiny of Earth and humankind. And fifth, the invasion and biological salvation of humanity in the face of Martian colonists—the *Origin*'s "foreigners" versus "natives"—is Darwinism on an interplanetary scale.

These last two points hinge on the association of Darwinian and nebular theory. According to popular understanding, the outer planets congealed and cooled before those closer to the Sun. Mars was therefore an older planet than the earth, so presumably life had started there earlier. With a longer evolutionary history, Martian life was now more advanced than ours. But it was also closer to degeneration and eventual death. Though he would inspire a book-length refutation by Alfred Russel Wallace (1907), by the early twentieth century Percival Lowell was arguing that the "canal network" was a last-ditch effort to postpone extinction. For Mars, the once-thriving world, was drying out and canals brought water from the poles to the desiccated equator. An older, more scientifically advanced civilization, faced with the prospect of its inevitable death, might well cast covetous eyes upon the warmer, moist, green Earth.

IV. Conclusions

We have briefly examined some of the forms and functions of science fiction and illustrated the existence of a little-explored mutual traffic between it and late Victorian science. Such an exchange is perhaps not surprising if both activities were situated in a common (although not necessarily simple and coherent) social context. This is not to say that science and science fiction were equal partners, culturally or intellectually, or that paying attention to noncanonical stories putatively about science requires a complete rewriting of the history of science. However, science fiction is worth including in our picture of the Victorian past, for it was both at least and more than a cultural vehicle for popularized and speculative science. In a wider context of other worlds and times, science fiction served as social satire, criticism, and prophecy, reflecting on God, nature, and human nature and on the implications of political, technological, and scientific change.

A recent social history of British anthropology (Kuklick 1992) has shown how the accounts these scientists produced about strange new cultures can be translated into commentaries on British society. Read this way, anthropological texts made appeals for a society based on individual merit and achievement, not inherited status; a social welfare state; and tolerance of cultural diversity. Anthropological literature also informed contemporary public debates on the problems of colonized peoples and the rights of women. The same goes for a lot of late Victorian science fiction.

Victorian science fiction raises many questions of interest to social historians of science, questions that can only be broached in an introductory exploration like this one. Here are some examples.

1. How are scientific beliefs produced and reproduced, received and translated, conscripted and celebrated, resisted and satirized, mediated and marketed?
2. How are publics for science created, and do these publics in turn shape scientific discourses?
3. How are scientific ideas, assumptions, interests, and ideologies expressed and embedded throughout the social landscape and across class, gender, and national lines?
4. How did science fiction both reinforce and subvert secularist, providentialist, or imperialist readings of Victorian science, industry, and social change? More specifically, in its detailed depictions of worlds with alternative histories or futures, with alien social arrangements

and cultural norms, was science fiction a destabilizing commentary on the hegemonic deployment of knowledge and power in late Victorian society?
5. Was science fiction, with its delight in the controversial, unorthodox, and speculative, some kind of attempt to break the monopoly of official, professional science publishing represented by *Philosophical Transactions, Nature,* and other, more specialized journals?
6. Why did Victorian scientists, from William Whewell to William Hudson, from Francis Galton to Hugh MacColl, write science fiction?

In the preface to the 1906 revised edition of his 1887 future utopia, *A Crystal Age,* the naturalist William Henry Hudson wrote: "Romances of the future, however fantastic . . . are born of a very common feeling—a sense of dissatisfaction with the existing order of things combined with a vague faith or hope of a better one to come." That is true, as long as we remember that this literature can celebrate as well as subvert the status quo. Science fiction advanced unsettling end-of-the-century and even end-of-the-world scenarios. It also reassured readers of God's providential care and purpose—or, at least, of some "spiritual" meaning and destiny within or beyond nature. This complex body of writing helped undermine and deflate Victorian self-confidence and complacency. But while it could question scientific triumphalism, it could also serve to reinforce the professionalizing—and somewhat inconsistent—ideology of science as systematized and democratized "common sense," practiced by a new priesthood who alone possessed the hegemonic and legitimating power of universal and authoritative explanation.

The social meanings and uses of late Victorian Mars or the Fourth Dimension are not exhausted by such reflections. There is always more "really going on" in the past than merely what took place. At a mundane level, for example, Mars helped make the late Victorian scientific romance a commercial success. Science fiction helped sell huge numbers of new and often illustrated mass magazines and books (with all their implicit cultural values and politics) through the adaptation and modification of such earlier and popular literary forms as utopias, travel narratives, prophecies, and apocalypses. Transmutations in classical, romantic, and gothic literature resulted in a new species—science fiction—the natural offspring of an age of science, industry, and ambiguous progress.

Bibliographical Note

The social history of science and the critical history of science fiction are virtual strangers but would make fertile partners, as Williams (1990) shows, mining late-nineteenth-century science fiction for visions of imagined subterranean worlds. Bernstein (1967) and Nahin (1993) demonstrate how knowledge of science fiction can add interest and texture to a topic, but such examples are rare. Unawareness of the relevant science fiction literature weakens many otherwise insightful cultural and contextual studies—for example, that of Teich and Porter (1990). Although their focus is elsewhere, Kern (1983), Broks (1990), and Beckson (1992) provide the wider social, scientific, and cultural context for the late Victorian emergence of commercial science fiction. With few exceptions—for example, Morton (1984)—studies of science and literature, analyses of scientific discourses, and work on scientific popularization and periodicals have yet to assimilate science fiction. Scientists' debts to science fiction represents more virgin territory for historians.

So far, the main field for literary discussions of Victorian science and science fiction is the study of Wells; see Bergonzi (1961), Hillegas (1961), Bowen (1976), Vernier (1977), Haynes (1980), McConnell (1981), and MacKenzie and MacKenzie (1987). However, literary scholars' recognition of science fiction's debts to science rarely extend beyond introductory observations—for example, by Welsh (1973), Paul (1979), and Pierce (1987). The critical edition of *War of the Worlds* (1993) is a model of textual scholarship. I am currently studying dozens of late Victorian Martian stories for their relevance to the history of science. Among science fiction critics and historians, Hillegas (1975) is most aware of this subject, but is not in touch with the scientific context. Crowe (1986, pt. 3) treats the professional and popular scientific literature on extraterrestrial life comprehensively but does not discuss science fiction. Guthke (1990, chap. 5), comes closest to integrating the history of Victorian science and science fiction. Two important works appeared too late to be incorporated into my analysis: Dick (1996, esp. chaps. 2, 3, 5) and Sheehan (1996, chaps. 4-8, passim). As did the 1890s, the 1990s are witnessing an explosion of works on Mars, both scientific and fictional.

References

Abbott, Edwin Abbott [A Square, pseud.]. 1884. *Flatland: A Romance of Many Dimensions*. Revised ed. London: Seeley.
Alkon, Paul K. 1994. *Science Fiction before 1900: Imagination Discovers Technology*. New York: Twayne.
Bailey, J. O. 1947. *Pilgrims through Space and Time: Trends and Patterns in Scientific and Utopian Fiction*. New York: Argus.
Ball, Walter William Rouse. 1891. "A Hypothesis relating to the Nature of the Ether and Gravity." *Messenger of Mathematics* 21:20-24.
Beckson, Karl. 1992. *London in the 1890s: A Cultural History*. New York: Norton.
Beer, Gillian. 1983. *Darwin's Plots: Evolutionary Narrative in Darwin, George Eliot, and Nineteenth-Century Fiction*. London: Routledge and Kegan Paul.
———. 1989. "'The Death of the Sun': Victorian Solar Physics and Solar Myth." In *The Sun Is God: Painting, Literature and Mythology in the Nineteenth Century*, edited by J. B. Bullen. Oxford: Clarendon Press, 159-80.

Bergonzi, Bernard. 1961. *The Early H. G. Wells: A Study of the Scientific Romances.* Toronto: University of Toronto Press.

Bernstein, Jeremy. 1967. "Science and Science Fiction." In *A Comprehensive World: On Modern Science and Its Origins.* New York: Random House, 207–69.

Bork, A. M. 1964. "The Fourth Dimension in Nineteenth-Century Physics." *Isis* 55:326–38.

Bowen, Roger. 1976. "Science, Myth and Fiction in H. G. Wells's Island of Dr. Moreau." *Studies in the Novel* 8:318–33.

Bowler, Peter J. 1989. "Holding Your Head Up High: Degeneration and Orthogenesis in Theories of Human Evolution." In *History, Humanity and Evolution,* edited by James R. Moore. Cambridge: Cambridge University Press, 329–53.

Broks, Peter. 1988. "Science and the Popular Press 1890–1914." Ph.D. thesis, University of Lancaster.

———. 1990. "Science, the Press and Empire: 'Pearson's' Publications, 1890–1914." In *Imperialism and the Natural World,* edited by John M. MacKenzie. Manchester: Manchester University Press, 141–63.

Brookfield, Arthur Montagu. 1884. *Simiocracy: A Fragment from Future History.* Edinburgh and London: Blackwood.

Chamberlin, J. Edward, and Sander L. Gilman, eds. 1985. *Degeneration: The Dark Side of Progress.* New York: Columbia University Press.

Clarke, I. F. 1979. *The Pattern of Expectation 1644–2001.* London: Cape.

———. 1992. *Voices Prophesying War: Future Wars 1763–3749.* 2d ed. London: Oxford University Press.

[Constable, Frank Challice]. 1895. *The Curse of Intellect.* Edinburgh and London: Wm. Blackwood and Sons.

Cooter, Roger, and Stephen Pumfrey. 1994. "Separate Spheres and Public Places: Reflections on the History of Science Popularization and Science in Popular Culture." *History of Science* 32:237–67.

Cromie, Robert. [1890] 1891. *A Plunge into Space.* 2d ed. London: Frederick Warne.

Crowe, Michael J. 1986. *The Extraterrestrial Life Debate 1750–1900: The Idea of a Plurality of Worlds from Kant to Lowell.* Cambridge: Cambridge University Press, pt. 3.

Dale, Peter Allan. 1989. *In Pursuit of a Scientific Culture: Science, Art, and Society in the Victorian Age.* Madison: University of Wisconsin Press.

Darwin, Charles. 1859. *On the Origin of Species by means of Natural Selection.* London: John Murray.

———. 1871. *The Descent of Man, and Selection in Relation to Sex.* 2 vols. London: John Murray.

Dick, Steven J. 1996. *The Biological Universe: The Twentieth-Century Extraterrestrial Life Debate and the Limits of Science.* Cambridge: Cambridge University Press.

Evans, Hilary, and Dik Evans, eds. 1976. *Beyond the Gaslight: Science in Popular Fiction, 1895–1905.* London: Frederick Muller.

Flammarion, Camille. [1872] 1897. *Lumen.* Translated by A.A.M. and R.M. London: Heinemann.

———. 1892a. "Inter-Astral Communication." *New Review* 6:106–14.

———. 1892b. *La planète Mars et ses conditions d'habitabilité.* Paris: Gauthier-Villars et Fils.
Frewin, Anthony. 1988. *One Hundred Years of Science Fiction Illustration, 1840–1940.* London: Bloomsbury Books.
Galton, Francis. 1896. "Intelligible Signals between Neighbouring Stars." *Fortnightly Review,* n.s., 60:657–64. (Summarizes sixty-page manuscript preserved in Galton Archives, London University.)
Greg, Percy., "ed." [author]. 1880. *Across the Zodiac: The Story of a Wrecked Record.* 2 vols. London: Trübner.
Griffith, George. 1901. *A Honeymoon in Space.* London: Arthur Pearson.
Guthke, Karl S. 1990. *The Last Frontier: Imaginary Other Worlds from the Copernican Revolution to Modern Science Fiction.* Translated by Helen Atkins. Ithaca: Cornell University Press.
Haynes, Roslynn D. 1980. *H. G. Wells: Discoverer of the Future: The Influence of Science on His Thought.* New York: New York University Press.
———. 1994. *From Faust to Strangelove: Representations of the Scientist in Western Literature.* Baltimore: Johns Hopkins University Press.
Hearnshaw, John B. 1993. "Spectroscopy." In *Cosmology: Historical, Literary, Philosophical, Religious, and Scientific Perspectives.* Edited by Norriss S. Hetherington. New York: Garland, 301–17.
Henderson, Linda Dalrymple. 1983. *The Fourth Dimension and Non-Euclidean Geometry in Modern Art.* Princeton, N.J.: Princeton University Press.
Henkin, Leo J. [1940] 1963. *Darwinism in the English Novel, 1860–1910: The Impact of Evolution on Victorian Fiction.* New York: Russell and Russell.
Hetherington, Norriss S. 1976. "Amateur versus Professional: The British Astronomical Association and the Controversy over Canals on Mars." *British Astronomical Association Journal* 86:303–8.
Heyl, Paul Renno. 1897. "The Theory of Light on the Hypothesis of a Fourth Dimension." Ph.D. thesis, University of Pennsylvania.
Hillegas, Mark R. 1961. "Cosmic Pessimism in H. G. Wells's Scientific Romances." *Papers of the Michigan Academy of Science, Arts, and Letters* 46:655–63.
———. 1975. "Victorian 'Extraterrestrials.'" In *The Worlds of Victorian Fiction,* edited by Jerome H. Buckley. Cambridge, Mass.: Harvard University Press, 391–414.
Hinton, C[harles] H[oward]. 1884–85. *Scientific Romances.* 2 vols. London: Swan Sonnenschein.
———. 1888. *A New Era of Thought.* London: Swan Sonnenschein.
———. 1895. *Stella, and An Unfinished Communication: Studies in the Unseen.* London: Swan Sonnenschein.
His Royal Highness Mammoth Martinet, alias Moho-Yoho-Me-Oo-Oo [pseud.]. 1871. *The Gorilla Origin of Man; or, The Darwin Theory of Development, Confirmed from Recent Travels in the New World called Myu-me-ae-nia.* London: Farrah.
Hooker, J. D. Papers. Library and Archives, Royal Botanic Gardens, Kew.
Hoyt, William Graves. 1976. *Lowell and Mars.* Tucson: University of Arizona Press.
[Hudson, William Henry]. 1887. *A Crystal Age.* London: T. Fisher Unwin.
Huxley, Thomas Henry. 1894. *Evolution and Ethics.* London: Macmillan.
Jann, Rosemary. 1985. "Abbott's Flatland: Scientific Imagination and 'Natural Christianity.'" *Victorian Studies* 28:473–90.

Johnson, William B., and Thomas D. Clareson. 1964. "The Interplay of Science and Fiction: The Canals of Mars." *Extrapolation* 5:37–48.

Jones, Alice, and Ella Marchant. 1893. *Unveiling a Parallel*. Boston: Arena.

Kaku, Michio. 1994. *Hyperspace: A Scientific Odyssey through Parallel Universes, Time Warps, and the Tenth Dimension*. New York: Oxford University Press.

Kern, Stephen. 1983. *The Culture of Space and Time, 1880–1918*. Cambridge, Mass.: Harvard University Press.

Kuklick, Henrika. 1992. *The Savage Within: The Social History of British Anthropology, 1885–1945*. Cambridge: Cambridge University Press.

Lancashire, Julie Ann. 1988. "An Historical Study of the Popularisation of Science in General Science Periodicals in Britain, c. 1890–c. 1939." Ph.D. thesis, University of Kent, Canterbury.

Lankester, E[dwin] Ray. 1880. *Degeneration: A Chapter in Darwinism*. London: Macmillan.

Levine, George. 1988. *Darwin and the Novelists: Patterns of Science in Victorian Fiction*. Cambridge, Mass.: Harvard University Press.

Lowell, Percival. 1895. *Mars*. London: Longmans, Green.

MacKenzie, Norman, and Jeanne MacKenzie. 1987. *The Life of H. G. Wells: The Time Traveller*. Rev. ed. London: Hogarth Press.

McConnell, Frank. 1981. *The Science Fiction of H. G. Wells*. Oxford: Oxford University Press.

Morton, Peter. 1984. *The Vital Science: Biology and the Literary Imagination 1860–1900*. London: George Allen and Unwin.

Moskowitz, Sam, ed. 1968. *Science Fiction by Gaslight: A History and Anthology of Science Fiction in the Popular Magazines, 1891–1911*. Cleveland: World Publishing.

Myers, Greg. 1985. "Nineteenth-Century Popularizers of Thermodynamics and the Rhetoric of Social Prophecy." *Victorian Studies* 29:35–66.

Nahin, Paul J. 1993. *Time Machines: Time Travel in Physics, Metaphysics, and Science Fiction*. New York: American Institute of Physics.

Newcomb, Simon. 1898. "The Philosophy of Hyper-space." *Science* 7 (7 January): 1–7.

Oppenheim, Janet. 1985. *The Other World: Spiritualism and Psychical Research in England, 1850–1914*. Cambridge: Cambridge University Press.

Paul, Terri G. 1979. "Blasted Hopes: A Thematic Survey of Nineteenth-Century British Science Fiction." Ph.D. thesis, Ohio State University.

Pick, Daniel. 1989. *Face of Degeneration: A European Disorder, c. 1848–c. 1918*. Cambridge: Cambridge University Press.

Pierce, John J. 1987. *Foundations of Science Fiction: A Study in Imagination and Evolution*. New York: Greenwood Press.

Pope, Gustavus W. 1894. *Romances of the Planets*. No. 1, *Journey to Mars*. New York: G. W. Dillingham.

Poynting, John H. 1920. *Collected Scientific Papers*. Cambridge: Cambridge University Press.

Richards, Joan L. 1979. "The Reception of a Mathematical Theory: Non-Euclidean Geometry in England, 1868–1883." In *Natural Order: Historical Studies of Scientific Culture*, edited by Barry Barnes and Stephen Shapin. Beverley Hills, Calif.: Sage, 143–66.

———. 1988. *Mathematical Visions: The Pursuit of Geometry in Victorian England.* San Diego: Academic Press.
Rickett, J[oseph] Compton. 1893. *The Quickening of Caliban: A Modern Story of Evolution.* London: Cassell.
Rucker, Rudy, ed. 1980. *Speculations on the Fourth Dimension: Selected Writings of Charles H. Hinton.* New York: Dover.
———. 1984. *The Fourth Dimension: A Guided Tour of the Higher Universes.* Boston: Houghton Mifflin.
Sheehan, William. 1988. *Planets and Perception: Telescopic Views and Interpretations, 1609-1909.* Tucson: University of Arizona Press.
———. 1996. *The Planet Mars: A History of Observation and Discovery.* Tucson: University of Arizona Press.
Sheets-Pyenson, Susan. 1985. "Popular Science Periodicals in Paris and London: The Emergence of a Low Scientific Culture, 1820-1875." *Annals of Science* 42:549-72.
Smale, Fred C. 1900. "The Abduction of Alexandra Seine." *Harmsworth Magazine* 5 (November): 291-98.
Smith, Jonathan. 1994. *Fact and Feeling: Baconian Science and the Nineteenth-Century Literary Imagination.* Madison: University of Wisconsin Press.
Stableford, Brian M. 1985. *The Scientific Romance in Britain 1890-1950.* London: Fourth Estate.
Stevenson, Robert Louis. 1886. *The Strange Case of Dr. Jekyll and Mr. Hyde.* London: Longmans, Green.
Suvin, Darko. 1979. *Metamorphoses of Science Fiction: On the Poetics and History of a Literary Genre.* New Haven, Conn.: Yale University Press.
———. 1983. *Victorian Science Fiction in the U.K.: The Discourses of Knowledge and Power.* Boston: G. K. Hall.
Teich, Mikuláš, and Roy Porter, eds. 1990. *Fin de Siècle and Its Legacy.* Cambridge: Cambridge University Press.
Vernier, J. P. 1977. "Evolution as a Literary Theme in H. G. Wells' Science Fiction." In *H. G. Wells and Modern Science Fiction,* edited by D. Suvin and R. M. Philmus. Lewisburg, Pa.: Bucknell University Press, 70-89.
Wallace, Alfred Russel. 1907. *Is Mars Habitable?* London: Macmillan.
[Welch, Edgar L.?]. 1883. *Politics and Life in Mars: A Story of a Neighbouring Planet.* London: Low, Marston, Searle and Rivington.
Wells, H[erbert] G[eorge]. 1891. "Zoological Retrogression." *Gentleman's Magazine* 271:246-53.
———. 1893a. "On Extinction." *Chambers's Journal* 10 (30 September): 623-24.
———. 1893b. "The Man of the Year Million." *Pall Mall Gazette* 57 (6 November): 3.
———. 1894a. "Popularising Science." *Nature* 50 (26 July): 300-301.
———. 1894b. "The Extinction of Man." *Pall Mall Gazette* 59 (25 September): 3.
———. 1895. *The Time Machine.* London: Heinemann.
[———]. 1896a. "Intelligence on Mars." *The Saturday Review* 81 (4 April): 345-46.
———. 1896b. *The Island of Doctor Moreau: A Possibility.* London: Heinemann.
———. 1897a. "Human Evolution." *Natural Science* 10:242-44.

———. 1897b. *The Invisible Man: A Grotesque Romance.* London: Pearson's.

———. 1897c. *The War of the Worlds. Pearson's Magazine* 3 (April-June): 363-73, 486-96, 598-610; 4 (July-December): 108-19, 221-32, 329-39, 447-56, 558-68, 736-45.

———. 1898. *The War of the Worlds.* London: Heinemann.

———. 1901a. *The First Men in the Moon.* London: Newnes.

———. 1901b. *Anticipations of the Reaction of Mechanical and Scientific Progress upon Human Life and Thought.* London: Chapman and Hall.

———. 1934. "Preface." In *Seven Famous Novels by H. G. Wells.* New York: Alfred A. Knopf, vii-x.

———. 1993. *A Critical Edition of* The War of the Worlds: *H. G. Wells's Scientific Romance.* Introduction and notes by David Y. Hughes and Harry M. Geduld. Bloomington: Indiana University Press.

Welsh, Alexander. 1973. "Theories of Science and Romance, 1870-1920." *Victorian Studies* 17:135-54.

Wendland, Albert. 1985. *Science, Myth and the Fictional Creation of Alien Worlds.* Ann Arbor, Mich.: UMI Research Press.

Williams, Rosalind. 1990. *Notes on the Underground: An Essay on Technology, Society, and the Imagination.* Cambridge, Mass.: MIT Press.

Willink, Arthur. 1893. *The World of the Unseen: An Essay on the Relation of Higher Space to Things Eternal.* New York: Macmillan.

Wynne, Brian. 1979. "Physics and Psychics: Science, Symbolic Action, and Social Control in Late Victorian England." In *Natural Order,* edited by Barry Barnes and Stephen Shapin. Beverley Hills, Calif.: Sage, 167-86.

PART THREE

Practicing Science

13

Practicing Science: An Introduction

FRANK M. TURNER

Many years ago Steven Shapin and Arnold Thackray observed, "The 'scientist' is himself a social construct of the last hundred years or so. And, as usually understood, so are 'science,' 'the scientific community,' and 'the scientific career'" (Shapin and Thackray 1974, 3). That conviction has been one of the chief driving forces in the history of science for the past quarter-century. Historians have directed much attention to the character of the scientist, to the emergence of the various European scientific communities, and to the development of major strands of scientific thought. But historians of science for some time remained hesitant to take a dynamic view of those institutions as they internally organized themselves, as they created the microscientific environments in which scientific ideas were achieved, tested, verified, or rejected and as they interacted with other segments of the society. In other words, historians tended to neglect the manner in which scientific practice is historically problematical and itself in large measure a social construct.

The reasons for that neglect constitute a brief chapter in recent intellectual history. Traditionally, the history of science has been an arena of intellectual history, and as such, science and its development have been seen primarily as a body of abstract thought. This outlook emerged largely as the result of examining first the ideas of the Scientific Revolution, then the ideas associated with evolution, and finally the ideas of modern physics, each of which invited a relatively abstract approach. Within the history of Victorian science, interest in geology and evolution tended and still continues to predominate over concern for chemistry, physics, astronomy, or medicine. Such research concentrated itself on the gentlemen scientists, on scientists who tended to work often more or less alone out of doors, and on scientists for whom personal observation with the naked eye was more important than instrumental research.

Historians understood the problem for their research to be the scientist's interpretation of the evidence rather than instruments or, as Simon Schaffer's and Harriet Ritvo's chapters point out, something both so fundamental and problematical as the systems of metrology and nomenclature, by which evidence was gathered, observed, organized, and then presented. Historians, who themselves often had little hands-on experience with historical scientific instruments or knowledge of their constructions, often simply did not realize that instruments, their location, their operation, and their systems of measurement were historically problematical developments. Here historians of science might well learn much from both historians of music and those art historians who have concerned themselves with material culture. As Jules Prown has written, "An artifact—a made object, whether you call it art or not—is an historical event, something that happened in the past. But unlike other historical events, it continues to exist in the present and can be reexperienced and studied as primary and authentic evidence surviving from the past" (Prown 1995, 2). Such is largely the case with scientific instruments for the history of science, but as the chapters by Ritvo, Jennifer Tucker, and Graeme Gooday suggest, the historically problematical also extends to the naming of species, the verification of observations, and the physical as well as social structure of the laboratory.

While the instruments, the systems of metrology, and the physical settings for the practice of science have long been ignored, the heroic Victorian man of science received the same kind of attention and adulation frequently accorded to the nineteenth-century artistic or musical genius. All were portrayed as minds working in splendid isolation. The heroic scientist—the priest of the new order—exemplified the romantic genius who probed nature more deeply and saw into its depths more clearly than other mortals. The scientists also quite often appeared as men of letters, who, one must recall, constituted one of Carlyle's classes of heroes. Furthermore, just as the stage workers in musical performances were ignored by historians, so were laboratory assistants and both the naval officers and non-Western workers who assisted Victorian scientists with field work in imperial settings.

The emphasis on the genius of the individual scientist with little concern for either physical instruments or human coworkers served a clear ideological purpose during the nineteenth century and after. If science was primarily the work of the gifted, highly trained individual, a clear social line was established between elite science and popular science. That line could distinguish science from quackery, but it could also prevent science from becoming understood as a democratic enterprise or as an enterprise that involved the genuine contributions of artisans or later of salaried laboratory assistants or non-European persons of color. It is now recognized that contrary to the social prejudices of the leadership of the Victorian British Association for the Advancement of Science, various down-to-earth technical skills such as surveying, illustrated in James Moore's analysis of Alfred Wallace, metalcraft, or lens grinding contributed mightily to the Victorian scientific endeavor. But at the time, the official scientific view was that ordinary persons drawn from the lower classes, such as William Whewell or Michael Faraday, might achieve stunning scientific insight, but it was the result of their individual special genius, which distinguished them from the social class of their origin.

Simple prejudice against the history of technology also prevented historians from looking carefully at the practice of science. Here the issue was not unlike a parallel one in military history. Both the history of technology and military history were regarded as narrow pedestrian fields fit for museum curators, weekend amateurs, and history buffs but not for serious academic historians. Military history seemed to many professional historians to be a subject fit for those people who could not deal with real history. It was the realm of re-creators of battles and for people who actually thought ordinary soldiers were important. Military history, however, perhaps ever since the appointment of Sir Michael Howard as Oxford Regius Professor of History, has come to be seen as relating to very sophisticated analysis of strategy, logistics, power relations, economic resources, foreign policy, and the cultural relations of armies to their larger societies. Similarly, the world of science in practice and its technology is now understood to include skilled instrument makers, scientific expeditions, the location and structure of laboratories, the social relations of scientists to their helpers, the relationship of scientists to government agencies, and, as Jane Camerini reminds us, of relationships to military officers as well. We now see that there exists a spectrum of scientific activity extending from theorizing through the construction of instruments and machinery and perhaps beyond.

These new sensitivities return us to a recognition of what Thomas Henry Huxley called "a New Nature created by science" (Huxley 1894, 1:1). The growing technological environment itself became a context for determining scientific problems, calling forth new theory, and manufacturing new instruments. Social and economic historians have given scant attention to technological firms or even the Victorian defense industry. Inventions may be listed, but on the whole their impact on the actual practice of science or government policy has too often been ignored. We have also often not thought very clearly about the manner in which certain technological demands, whether generated by commercial or political needs, could become a driving force in scientific theory and practice. This is the context that becomes so clear in Bruce Hunt's analysis of the relationship of cable technology and electrical physics. Though aware of the connection of commerce and political needs to science and science funding in the present day, we have often assumed it not to have existed in the past. The statement of James Clerk Maxwell quoted by Hunt is quite revealing: "The important applications of electromagnetism to telegraphy have . . . reacted on pure science by giving a commercial value to accurate electrical measurements, and by affording to electricians the use of apparatus on a scale which greatly transcends that of any ordinary laboratory" (Maxwell 1873, 1:vii-viii). This statement is virtually contemporaneous with the assertion of the Devonshire Commission that in the future only the finances of nations could support the needs of science. In point of fact, since the 1870s government support for science around the world has been quite stunning; indeed, no group of intellectuals has been so well supported from public resources. But it is also important for us to remember when considering the practice of science that the corporation, or in the Victorian age the large firm, also could and did provide extensive capital needed for science. Although there exist some studies of industrial research laboratories, academic historians from the late nineteenth century to the present have traditionally been so hostile to private enterprise that they have tended to overlook this corporate support for the practice and finance of science.

To look at the actual practice of Victorian science one must recognize that it lay very much enmeshed in the warp and woof of commercialism, empire, militarism, and capitalism. Indeed, many Victorian scientists took great pride in just those commercial and military associations. Academic historians of this century, however, who have been trained in a tradition of the humanities itself rooted in Victorian anticommercialism or in Marxist approaches, have tended to be uncomfortable with empire, commercialism, and capitalism. They have wanted to see science as being pursued as a good in itself rather than as a field for profit or national aggrandizement. It is paradoxical that the spirit of Newman's *Idea of a University* rather than that of Huxley's educational ideal has informed much academic history of science.

It was the practice of science in its relationship to the commercial and military worlds that brought about the pressure to come to a decision both in resource allocation and in theory. In that regard, it is well to remember that the establishment of standard time zones was the result of the necessities raised by railways spanning continents. Technology as requiring application to commercial and military matters dragged theory kicking and screaming toward decision. It would almost seem that what was good for commerce was good for science and vice versa. That outlook should be compared with the view of academic scientists today, who often spurn commercial funding for the allegedly neutral funding of central governments. In the Victorian era, these commercial concerns easily merged with the various noncommercial and nonscientific values because there had long existed a similar mix of commercial ideas and other nationalistic moral values in the long tradition of English natural theology (Turner 1993, 101-30). It was in part simply one element in the style of rhetoric of English science since the late seventeenth century. In that sense, natural theology itself was a part of the larger commercial context of the Victorian scientific enterprise.

Finally, when scholars originally explored the history of science from the standpoint of social history, they did so largely in terms of macrosocial (often Marxist) theory rather than from the standpoint of the training of the individual practitioner or the actual daily work of the laboratory or the structure and capacity of scientific instruments. Interest in social theory rather than the actual practice of science was the driving intellectual force. For these authors the practice of science was merely the epiphenomenon of larger social or economic forces. There was little or no necessity to get into the details of what scientists actually did. There was a kind of unstated assumption that we knew how scientists spent their days, made their observations, and wrote up their experiments. God was not hidden in the details, but in the big picture.

Although we have explored scientific activity at the margins of respectability in terms of theory, with phrenology being studied most extensively, there has been so little exploration of the practice of science that we shall probably have to rethink where the margins stand. More potential borders will probably appear than we are ready to confront immediately. There will be the border between the scientist who runs the laboratory and the work of his assistants. But there will be another social border between the laboratory, including the scientist and assistants, and the workshops of inventors and makers of instruments. There will also be the world of natural history collectors into which recent Darwin studies have led us. Another frontier, mentioned earlier in this introduction, will be the border between science and technology, across which most historians have been very uncomfortable passing.

One of the most interesting of these borders is that between British scientists and the nonscientific people with whom they interacted in the field. The world of Victorian science in practice within the empire involved a whole host of people about whom we know little. Scientists, who virtually always have come from civilian backgrounds, might suddenly find themselves working alongside naval and military personnel whose views of life were very different from theirs. As Camerini argues, the decisions of the navy often determined where men of science might find their fields of research overseas. Furthermore, scientists working abroad might well form important working relationships with local native colleagues, such as was the case with Wallace. Astronomers who traveled all over the world in search of the precise moments required for their observations depended on the willing cooperation of both local British officials and the local native labor force. At the same time, as Moore suggests, the activities of scientists abroad will also need to be related to the social background and expectations they brought to the field—background and expectations such as Wallace brought to his practice of mapmaking, first in rural Wales, then in South America and the Far East.

Certain domestic cultural contexts will also need to be brought to the fore. These relate to the manner in which scientists presented themselves and their work. Neutrality was the cultural context within which the scientists and their supporters wrapped themselves and their enterprise. It was one of the major achievements of Victorian public science. The Victorian world was sharply riven by religious, political, class, economic, and other divisions. Science was to rise above these divisions. The supposed neutrality of science was itself a powerful cultural force, one that worked against other powerful cultural forces such as individualism, provincialism, and a penchant to cherish eccentricity. Curiously, the scientist himself was to be a gifted individual, but he was to function in theory at least as a neutral observer of nature. Gooday's chapter reveals the manner in which the Victorian scientist labored to portray the laboratory as a realm in which against great odds and difficulties a neutral observation was possible.

The context of moral struggle was also important. The late Victorian scientist, in contrast to the early Victorian scientist, foregrounded the problems and problematics of the laboratory. Indeed, in that sense, again as Gooday's chapter suggests, the comments of certain late Victorian scientists rather remind one of the present-day cultural critics of the laboratory and of its often disorganized character. The late Victorian scientists appear to have believed that their work would gain credibility if they edited in as many of the problems encountered with instruments, laboratory setting, and observation as possible. In part, this tactic may have been simply the use of the rhetoric of overcoming hardships and allowing virtue to triumph. Furthermore, the emphasis on poor conditions may have been intended to achieve better institutional and government funding, but clearly, something else was also at work. One finds that a chief question is the trustworthiness of both the experiment and the experimenter. In this regard, by putting all of the difficulties of the laboratory and the instruments in the foreground, the scientist gained in credibility.

The actual examination of the practice of science—the microcosmic social world of science—may lead eventually to a serious reconceptualization of the history of science. It may lead to a new synthesis in which the old internalist-externalist dichotomy will vanish for good. It will also lead to important and fascinating evidentiary problems. What did men of science actually observe, how did they observe it, and under what hindrances did they observe it. Both Gooday and Tucker also raise the question of how they could persuade their colleagues and the general public of what scientists believed they had observed. Tucker points out that the modes of persuasion often drew upon cultural outlooks and prejudices that had little or nothing to do with science itself. The history of science in practice will be a history that attempts to recapture not simply the theoretical development of the individual scientific mind, but the phenomenology of the production, dispersion, and transmission of what became scientific knowledge. It will be a study of the social relations of scientists within the setting of the laboratory and the working scientific community. In particular, the emphasis on instrumentation opens questions of finance, instrument makers, instrument operation, and lab assistants. All of these changes will consolidate what may be called a nonheroic history of science that may soon produce a new scientific hero, the master of the great laboratory or the developer of important instruments.

In social history the world of the ordinary has in the last three decades produced very important discoveries. The same may very well hold true for the history of science. The core issue, however, will be whether we can produce the right questions and be able to avoid trivial ones. The study of Victorian or twentieth-century scientific fieldwork and laboratories could become the equivalent in the history of science of the village study in social history. Each laboratory or scientific expedition could become the personal possession of individual historians, with no one really ever checking the data after the first investigator has studied it. Also, historians could become so fascinated with their own laboratory that they will lose sight of the larger issues. The study of science as it is practiced will become and remain an important one, but only so long as the fundamental concern is with science as knowledge of nature and not with context or practice for its own sake.

References

Huxley, T. H. 1894. *Collected Essays.* New York: D. Appleton and Company.

Maxwell, James Clerk. 1873. *Treatise on Electricity and Magnetism.* Oxford: Clarendon Press.

Prown, Jules. 1995. "In Pursuit of Culture: The Formal Language of Objects." *American Art* 9, no. 2:2-3.

Shapin, Steven, and Arnold Thackray. 1974. "Prosopography as a Research Tool in the History of Science: The British Scientific Community, 1700-1900." *History of Science* 12:1-28.

Turner, Frank M. 1993. *Contesting Cultural Authority: Essays in Victorian Intellectual Life.* Cambridge: Cambridge University Press.

14

Wallace's Malthusian Moment: The Common Context Revisited

JAMES MOORE

> Naturalists need not be bound by the same rule as politicians, and may be permitted to recognize the just claims of the more ancient inhabitants, and to raise up fallen nationalities. The aborigines and not the invaders must be looked upon as the rightful owners of the soil, and should determine the position of their country in our system of Zoological geography.
>
> A. R. WALLACE (1864, 118-19)

> To draw a boundary is always to make a commitment.
>
> DENIS WOOD (1991, 79)

In the past twenty-five years historical writing on Victorian science has been transformed. Whig-positivist accounts of scientific progress were scouted, externalist-internalist debates found stale—the historiographic equivalent of junk food. Demarcationist dreams of a science so bounded that its phenomena may be ascribed to external or internal "factors" finally gave way, and with them the jerry-built walls between intellectual and social history (Turner 1993, 3-37; Shapin 1992). Scholars began to historicize boundaries, disciplines, and even truth claims, making actors' categories relevant. Controversies were unraveled, black boxes unpacked, and patronage unveiled. Natural knowledge itself came to be seen as a complex cultural product. Today Victorian science no longer stands pristine, beyond social and economic forces, on the edge of time. Theories and practices, instruments and institutions are not fetishized but "contextualized," embedded in their formative social matrix. Historians who offer such interpretations now call themselves "contextualists."

The proximate source of contextualist historiographic discourse was a series of remarkable essays written by Robert M. Young between 1968 and 1972. Young, an expatriate Yank, then discovering Marxism, honed his essays razor sharp in the political struggles of the time. Science was oppressive, society unjust, protest imperative. Brilliantly, he made history his ally by showing that contested values lay at the heart of the nascent, nineteenth-century human sciences. His essays analyzed the Malthusian "common context of biological and social theory," the "fragmentation of a common context" of natural theological debate, and finally the "historiographic and ideological contexts" of that same debate, on "man's place in nature." At this time Young began describing his writings as "relativist" and "contextualist." He shared the view "that ideas do not beget ideas but that people do so in particular historical contexts and that the meaning of those ideas is exquisitely bound to the particularity of those contexts" (Young 1985, 168, 176).

Even so, Young's self-styled "social intellectual history" was primarily about ideas, his "common context" ideological. While apparently canvassing research on social "particularity," he neglected the rich detail of political, economic, demographic, and institutional history, leaving colleagues to supply the defect (Hodge 1993; Bohlin 1991, 1995). A band of rising scholars soon obliged, and their work to date has been impressive. Young's common context is now materialized and differentiated by party, place, and class. Tory Anglican and Whig Malthusian intellectuals no longer enjoy cozy, polite debates. Their clubs and classrooms are besieged by noisy crowds—secularists, spiritualists, phrenologists, mesmerists, radical anatomists, and artisan naturalists—throngs of independent so-and-sos with a glint in their eyes and grit on their hands. Their number can only grow. "Between the Malthusian Whigs and the socialist demagogues lies terra incognita," writes Adrian Desmond, who more than anyone has independently fleshed out Young's work. "It is a territory that should be opened up. In this unexplored terrain all sorts of dissident knowledge flourished: not only varieties of evolution, but a swirling vortex of alternative economic, social, and biological sciences that threatened to wash away the pillars of the establishment edifice" (Desmond 1989, 4).

The same tide is flooding historical scholarship. While much remains to be learned about the sciences of plebeian secularism, the social affinities and secular tendencies of dissident bourgeois science are now indelibly clear. Here biography has proved a powerful contextualizing tool. Major studies of Baden Powell and Robert Chambers place evolution at the center of mid-Victorian cultural contests (Corsi 1988; Secord 1989, 1994). New lives of Charles Darwin and Thomas Huxley go further, showing how even scientists' "most basic ideas" (to use Young's words) were "inside culture, inside society, inside the ideological and socioeconomic forces which shape the rest of the social world" (Young 1987, 213; Moore 1985; Desmond and Moore 1991; Browne 1995-; Desmond 1994).[1] Further contex-

---

1. Compare Young's previously published views with the attack in Young 1994.

tualist biographies will sharpen these points, none more so than a life of Darwin's codiscoverer of natural selection, Alfred Russel Wallace.

I. "Why Do Some Die and Some Live?"

Not that Wallace has been ignored. Four modest biographies have appeared in the last three decades, and a spate of specialist monographs. The most famous also-ran in Victorian science has been studied almost to bits. And this is just the problem. There is no adequate overview of Wallace's life, no full-blooded narrative that treats his fads and foibles and phenomenal achievements as a coherent whole, in their place and time. We have instead Wallace the Welshman, Wallace the geographer, Wallace the group selectionist, Wallace the spiritualist, the land nationalizer, the socialist, Wallace the scourge of vaccination, flat-earth theories, and life on Mars, Wallace the defender of women's rights, and so on. By turns the poor man is made a whipping boy or laughingstock and a genius, seer, or saint. More often he is presented as a foil: Wallace the lesser light, reflecting a greater glory; Wallace otherworldly, orbiting in occult circles; Wallace as "Darwin's Moon."

Or so one biographer calls him (Williams-Ellis 1966). Wallace wrote up the theory of natural selection in 1858, twenty years too late, and his originality was eclipsed. The story is notorious. Darwin read his essay, then scooped the kudos by rushing into print. In his haste he may even have lifted Wallace's concept of evolutionary divergence. The hapless theorist, sweating it out in the Malay Archipelago, remained the perfect gent, but back in Britain a few years later he obstinately went his own way. Within a decade he and Darwin had parted company over spiritualism, sexual selection, spontaneous generation . . . the list goes on. Wallace played the crank to Darwin's political correctness. Only as an afterthought was he asked to bear the great man's coffin in Westminster Abbey (Brackman 1980; Brooks 1984; Beddall 1988; Kottler 1985; Moore 1982, 101).

Such at any rate is the impression left by a random walk through the literature on Wallace and evolution. The subject is rarely broached without dragging Darwin in; the treatment is usually invidious. Wallace's theories apparently lack intrinsic merit, or Darwin's are required to calibrate them. Defenders of Wallace today both exceed and excel his detractors, but few have studied him in his own right, as an independent naturalist.

The contrast with Darwin is striking. His theories have probably been researched more thoroughly than those of any nineteenth-century scientist. His "path" to natural selection has been mapped minutely and the "discovery" itself contextualized. We now know that Darwin could not have "happened to read" the Reverend Thomas Malthus, as he later claimed, let alone simply "for amusement" (C. Darwin 1958*a*, 120; Kohn 1980; Oldroyd 1984; Hodge and Kohn 1985; Erskine 1987, 241–90; Desmond and Moore

1991, 195-298). The *Essay on the Principle of Population* was sacred scripture to Whig poor-law reformers. In London Darwin moved freely in their circles, breathing the oxygen of their publicity. He inhaled deeply on 28 September 1838 and came up with natural selection. Reading Malthus, he grasped that living nature was in effect the workhouse world writ large. Ruthless struggle was everywhere the law, not just among London's starving poor. Adaptation comes through competition. Progress costs lives. He recorded his insight on the spot in a staccato burst of ink.

This was Darwin's famous "Malthusian moment." Its "common context" is today no longer merely ideological, but instead the rich, reformist world of the Whig urban gentry. Within this very social "particularity," three months after Victoria was crowned, an ambitious young gent with a wife in sight and a private fortune in prospect first glimpsed a wider world in which "from death, famine, rapine, and the concealed war of nature . . . the highest good, which we can conceive, the creation of the higher animals has directly come." It now seems practically "inconceivable," as Charles Gillispie once remarked, that such a view could have been expressed "by any Frenchman or German or by an Englishman of any other generation" (C. Darwin 1958*b*, 87; Gillispie 1974).

What then of Wallace's path to natural selection? His theory has been analyzed often enough, almost invariably by comparison with Darwin's, but what were its origins? Where is the common context for this "discovery"—by an Englishman of Darwin's generation?

In 1858 Wallace was working alone, four years and seven thousand miles outside of London. His world was profoundly unlike Darwin's of twenty years before. Indeed, Wallace himself belonged to a different world, socially and economically remote from that of a Cambridge M.A. and country squire. The seventh child of an impoverished solicitor, he had left school at thirteen, trained as a surveyor, and gone abroad as a self-financed specimen collector. (Darwin was actually one of his customers, having spent "a fortune" on carriage for a pair of Balinese birds. See memorandum [December 1855], and letter, 29 Nov. [1856], in Darwin 1985–, 5:510, 6:290.) How then did such a man hit on natural selection in a remote corner of the East Indies? Surely there was no context dependency here. Whatever the role of Darwin's own circumstances, nature itself must have impressed the theory on Wallace. His discovery was a cerebral, not a social event, and purely independent.

Except for this: in February 1858 Wallace himself had a Malthusian moment. It happened, as Young noted, "in the context of [his] ethnological investigations into the origin of human races," and subsequent scholars have agreed (Young 1985, 44; McKinney 1972, 80–96; Brooks 1984, 174–99). Wallace, an evolutionist without a mechanism, was struck by the checks to population growth described in Malthus's *Essay*: the wars, famine, and disease that decimate "savage races" and "barbarous nations." It then occurred to him that these checks must act more severely among animals, which breed many times faster than humans, and this led him to ask, "Why do some die and some live?" His answer would later be dubbed survival of the fittest. Wallace immediately wrote up what for the present he termed his "general principle" and posted the manuscript to Darwin (Wallace 1905, 1:362; Paul 1988; Wallace [1858] 1958, 269).

All of this has long been known and was integral to Young's assertion of a ideological common context for the discovery of natural selection. Wallace's theory, like Darwin's, was indebted to the argument of a political tract. The question now is: Can we go further? Twenty-five years on, can Wallace's Malthusian moment be recontextualized as the product of a social "particularity," as the immanent insight of an actor, like Darwin, whose partisanship is known?

I believe it can. The key to Wallace's discovery lies in the practices by which, as a fledgling naturalist, he made his living in rural Wales.

II. "Somehow My Thoughts Turned"

Let me start by tackling a pair of problems. First, there is no contemporary evidence for Wallace's Malthusian moment, no breathless, dated notebook entry like Darwin's. The earliest known account was written down a decade later, and it is sketchy. Wallace's four fullest recollections (McKinney 1972, 80-81, 160-63) date from around the turn of the century, a further forty years on. By then his story had been well rehearsed, and it is full of retrospect, with an anachronistic "survival of the fittest" dazzling him in a Pauline "flash of insight." A proviso therefore must be entered before contextualizing begins: Wallace's Malthusian moment was malleable. Its original dimensions are not apparent from surviving accounts. There is scope for reconstruction.

This limitation points up the second problem: none of the accounts suggests why Wallace thought of Malthus at a particular place and time. Darwin's reading of his *Essay* in 1838 makes perfect sense: he was in London, the Whig reforms were in place, his brother's companion had been the poor-law apologist Harriet Martineau, and Erasmus himself owned a copy of the *Essay* that he could borrow (Di Gregario and Gill 1990-, 563). Darwin was researching the laws of life, one of which—as every rate-paying Whig and his brother knew—was the Malthusian "principle of population." Wallace's case is wholly different. Though he too was looking for an evolutionary mechanism, what impressed him about the *Essay* was its bearing on racial questions. And he only *remembered* reading it after a lapse of fourteen years. (Which parts struck him he did not say for a further half-century.) Why then the delay? Wallace had long dwelt on ethnological subjects, most fully in his *Travels on the Amazon and Rio Negro*. Already he had spent three years in the Malay Archipelago, observing the struggle for existence, obtaining daily evidence of the checks on native tribes. Yet there was no Malthusian moment until February 1858. Why only then? Why in the Malay Archipelago? Why indeed on the island of Gilolo (now Halmahera)?

Once serendipity is ruled out—the last refuge of the perplexed—and once creative genius ceases to be an explanatory concept (in history as in nature), then contextualizing can begin. Now the task is to explain, not the actual causes, or sufficient conditions, but just the *pre*conditions for Wallace's Malthusian moment, the conditions of its possibility. All or some may be necessary or sufficient, for the event is historically underdetermined and indeterminate. Wallace himself did not settle the question of causality, nor can we.[2] What can and must be done is to describe the various material means—local, epochal, biographical—by which he was prepared to see not merely an evolving world, but a world of ruthless competition, selective survival, and adaptive improvement, a Malthusian world so like the one in which Darwin's own theory had been forged twenty years before.

One ready conduit to this world was the 1845 edition of Darwin's *Journal of Researches,* which contained an added paragraph explaining how Malthusian checks may bring about extinction. Wallace had a copy with him and quoted from it in his journals, though there is no evidence that he read the passage at the time (C. Darwin [1839] 1845, 174-75).[3] A more direct reminder of the struggle for existence was the "intermittent fever"—malaria—from which he suffered in February 1858. It was in fact during one of the "rather severe" cold-and-hot fits, which "lasted together two or three hours," that he recalled the Malthusian checks and then quickly saw their role in natural selection. Among such checks, according to Malthus's *Essay,* were "malignant fevers" (Wallace [1870, 1878] 1891, 20; 1903, as cited in McKinney 1972, 160; Malthus [1798] 1826, vol. 1, chap. 7).[4]

But these are only two of many possible paths to Wallace's Malthusian moment. None of them need be barred as we follow a more familiar route. To repeat, the one context that scholars agree would surely have suggested Malthus was ethnological: Wallace was pondering the origin of human races. Now this view is valid, and helpful as far as it goes. It points up his interest in human development at the very moment he remembered a book

---

2. Wallace thought it a "most interesting coincidence" that he and Darwin were both led to natural selection by Malthus (F. Darwin 1892, 189). Charles Smith asserts (in Wallace 1991, 5) that Wallace's " 'discovery' of natural selection was something of an accident."

3. The passage is well marked in Wallace's copy of the 1873 edition, held at the Linnean Society of London. The words "check" and "geometrical" are underscored.

4. As a child Wallace survived three deathly fevers; his younger brother Herbert, who joined him in Brazil, succumbed to yellow fever there, aged twenty-two (Wallace 1905, 1:46, 282). These episodes suggest how seriously Wallace must have taken his malaria attacks—several had occurred—and add poignancy to his contrast in the famous essay between "the most perfect in health and vigour" and "the weakest and least perfectly organized," who "must always succumb" (Wallace 1958, 272). Thanks to Ralph Colp, Jr., for this point.

that subjected humanity to a biological law, the principle of population. No logical leap was needed, no creative fiat, for him to connect the one with the other. Even so, the questions remain: Why February 1858? Why Gilolo?

Wallace's late recollections give no clue. "Something led me to think" of Malthus's *Essay*, he wrote enigmatically. "Somehow my thoughts turned" (Wallace 1898, 139; 1903, as cited in McKinney 1972, 160). Without excluding multiple causation, I want to propose a contextual solution to this motivational mystery. What provoked Wallace can be found by "pulling focus," by turning our own thoughts from his immediate island environment to its broad historical backdrop, the geographic, economic, and social common context of his earliest practical fieldwork.[5]

III. "There Must Be Some Boundary"

Consider first the most likely unintended clue to the source of Wallace's Malthusian moment. This appears in his 1869 travel narrative, *The Malay Archipelago*. Here eight years' adventures are recounted—the fourteen thousand miles' island hopping, the 125,000 specimens bagged. Not once in a thousand pages does the book mention Malthus or natural selection. Its chief boast is rather Wallace's discovery of the two great zoological regions of the archipelago, with their distinctive faunas. Placental mammals, for instance, are found only in the western, "Indo-Malayan" zone, marsupials only in the "Austro-Malayan" zone to the east. The boundary between these regions is now famous as "Wallace's Line" (Wallace [1869] 1877, xi, 13-16, 590-91; Camerini 1993; Whitmore 1981).

Nor is this the only biogeographic boundary shown on the book's handsome, foldout "physical map," which Wallace himself drew (figure 14.1). There is another, marking the division of "the Malayan and all the Asiatic races, from the Papuans and all that inhabit the Pacific." This ethnological line runs on average a few hundred miles east of the zoological one, and precisely between the Moluccan islands of Ternate and Gilolo. In chapter 22 of *The Malay Archipelago* Wallace first visits Gilolo and quickly perceives how "radically different" the indigenous people are from the Malays on Ternate, just ten miles away. "Here then," he says, "I had discovered the exact boundary line between the Malay and Papuan races, and at a spot where no other writer had expected it" (Wallace 1877, 316-17).[6]

What led up to this bold perception? Let me adjust the focus now and pan quickly over the previous thirteen years.

Wallace arrived in the Malay Archipelago a seasoned evolutionary biogeographer. With typical verve, he had read the scandalous *Vestiges of the*

---

5. For the "ciné theory" of narration, see Moore 1996, 280; and Desmond 1994, xiii-xvii.
6. The chapter is misdated "March and September 1858." The latter text corresponds to the last February entry for Gilolo in Wallace's field journal, 1858, sec. 127 (in Linnean Society, transcribed in Brooks 1984, 61-2, 179).

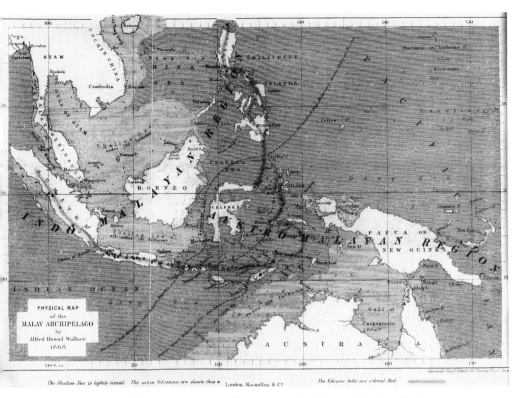

Figure 14.1 Wallace's faunal and racial boundaries in the Malay Archipelago. He first showed the faunal boundary—Wallace's Line—in a map published in 1863; he described the racial boundary a year later: "If we draw a line, commencing on the eastern side of the Philippine Islands, thence along the western of Gilolo, through the island of Bouru, and curving round coast the west end of Flores, then bending back round Sandalwood Island to take in Rotti, we shall divide the archipelago into two portions, the races of which have strongly marked distinctive peculiarities. This line will separate the Malayan and Asiatic from the Papuan and Pacific races, and though along the line of junction intermigration and commixture have taken place, yet the division is on the whole almost as well defined and strongly contrasted as are the corresponding zoological divisions of the archipelago into an Indo-Malayan and Austro-Malayan region" (Wallace 1865, 211). (From Wallace 1877, author's collection.)

Natural History of Creation in 1845, adopted its ingenious development hypothesis, and found support for it in William Lawrence's blasphemous *Lectures* on human anatomy. Three years later he set off on his first collecting expedition, to the Amazon basin, paying his way by selling duplicate specimens. He went expressly to gather facts "with a view to the theory of the origin of species." But there was no rain-forest eureka, no bolt from the Brazilian blue, even if he did hint at life's struggle, and "some . . . principle regulating the infinitely varied forms" of animals. His main achievement was to log the distribution of his specimens, labeling them carefully by locale. "There must be some boundary which determines the range of each

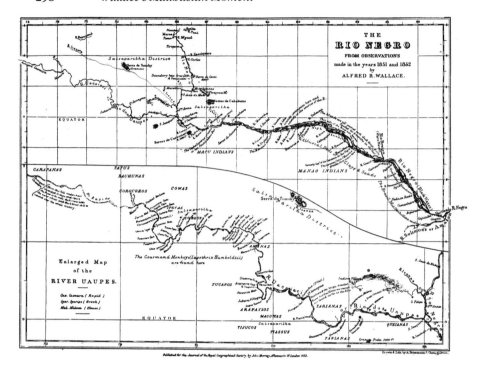

Figure 14.2 Wallace's map of habitats along the rivers Negro and Uaupes, first published in 1853. The original, hand-colored map is in the Royal Geographical Society, London. (From Wallace 1905, vol. 1, facing p. 320, author's collection.)

species," he decided, "some external peculiarity to mark the line which each one does not pass." Rivers were obvious markers, and Wallace prepared a detailed map (figure 14.2) showing, on opposite banks, the habitats of monkeys and native tribes (McKinney 1969, 372; Clodd 1892, xx; Wallace 1853, 83-84, 470; Fichman 1981, 33-34; Wallace 1905, 1:319-21).

Undaunted by losing four years' collections on the homeward voyage—his ship burned and sank at sea—Wallace sailed again in 1854 for the East Indies. There, according to *Vestiges,* he could "expect man to have originated," for the lowest human races, the Malay and the Negro, are found living beside "the highest species of the quadrumana" (Chambers [1844] 1994, 266, 296, 308). He made for Borneo to find out for himself. For fifteen months he studied all primates, Dyaks and orangutans alike. "The more I see of uncivilised people," he wrote home, "the better I think of human nature on the whole, and the essential differences between so-called civilised and savage man seem to disappear" (Wallace 1905, 1:350; Marchant 1916, 1:53, 55). Other essential differences also vanished, the more he saw of orangs. He did take many skins and skeletons—collecting was his job—

but he also adopted a tiny orphan, feeding and bathing the "infant prodigy" for months as if it were his own child.

These great apes intrigued him. Their bodies mocked "the 'human form divine' " and, like tribesmen, they kept in one locale, their range marked by "some boundary line" that they "never pass." He mused on the orang's ancient cousins, "allied species still more gigantic . . . and more or less human in their form and structure." Did the Dyaks descend from these? Such thoughts led Wallace to formulate a grand law of biogeography, which he wrote up on Borneo: "Every species has come into existence coincident both in space and time with a pre-existing closely allied species" (Wallace 1905, 1:343-35; 1856a, 326; 1856b, as cited in Brooks 1984, 110; Wallace 1891, 6).

In 1856 he sailed for the less explored eastern end of the archipelago but was delayed for two months on the islands of Bali and Lombock. While collecting birds—among them the pair for Darwin—he noticed a remarkable change in fauna: "The islands, . . . though of nearly the same size, of the same soil, aspect, elevation and climate, and within sight of each other, yet differ considerably in their productions, and, in fact, belong to two quite distinct zoological provinces" (Wallace 1857b, as cited in Wallace 1991, 233; 1857a, as cited in Brooks 1984, 138). Not just single species, but "genera, families, and whole orders" were absent on one island or the other. Here then, in August 1856, Wallace first drew his famous line.

Months later he was again struck by nature's contrasting "productions." As his native prau approached the island of Ké, off the south coast of New Guinea, he watched in amazement as the dour Malay crew was mobbed by a boatload of indigenous Papuans, "forty black, naked, mop-headed savages, . . . intoxicated with joy and excitement." Comparing the groups "side by side," he realized "in less than five minutes" that they "belonged to two of the most distinct and strongly marked races" on earth. "Had I been blind, I could have been certain that these islanders were not Malays." In the space of a thousand miles, he had crossed into "a new world, inhabited by a strange people." Where, then, was its border? And whence had the Papuans come? (Wallace 1877, 415, 417).[7]

His work instantly acquired fresh zest. The field journal he had begun at Bali now bulged with ethnographic notes. "The human inhabitants of these forests are not less interesting to me than the feathered tribes," he jotted in March 1857. With hummingbirds or humans, life's laws were all the same. Proximity did not entail near ancestry. In the beginning were separate stocks. Mixing was artificial, "transitional" types fictitious. Racial differences ran as deep as the mighty sea that split the archipelago. They were ancient, and their origin was to puzzle Wallace on "all the islands" he vis-

---

7. The text closely follows Wallace's field journal, 1856, sec. 50 (transcribed in Brooks 1984, 163-64).

ited. His new inquisitiveness showed. Having heard from Darwin about his snail's-pace work on species, he jumped the gun in September and asked expectantly whether it would "discuss 'man'" (Wallace cited in Brooks 1984, 137, 164, 168; C. Darwin 1985–, 6:515).

The answer was no—Wallace led the field. He pressed on, sorting the islanders into two, surveying all races alike, birds, beasts, and humans. His mental map now stretched over a million square miles:

> In this Archipelago there are two distinct faunas rigidly circumscribed, which differ as much as those of South America and Africa, and more than those of Europe and North America: yet there is nothing on the map or on the face of the islands to mark their limits. The boundary line often passes between islands closer than others in the same group. I believe the western part to be a separated portion of continental Asia, the eastern the fragmentary prolongation of a former Pacific continent.

From there the Papuans had come. Wallace wrote this passage on Ternate in January 1858 (Marchant 1916, 1:67; Wallace 1905, 1:359; McKinney 1972, 85, 88, 174). He signed and sealed the letter, packed his bags, and made the three-hour crossing to Gilolo. Days later these islands became his ethnological Bali and Lombock. Here too, on identical terrains, within sight of each other, he spotted a great divide. He drew another line between the islands, a racial boundary, but this time he remembered Malthus.

It is the lines Wallace drew, or rather his practice of drawing lines, that point to the source of his Malthusian moment. This is the crucial clue. For just as his ethnology was based on cartography, so his cartography was rooted in economics. Racial distribution was linked to food distribution. His earliest fieldwork was on farms.

IV. "Hospitable Even to the Saxon"

Once more let me pull focus, this time bringing Wallace's first twenty years into frame.

He was born in 1823 at Usk on the Welsh borders, where his father had moved the family from London. Their cottage stood on the west side of the river Usk, the town itself on the east. Wallace never forgot the scenic walk into town, over the old three-arched bridge a quarter-mile away. Crossing it, he would stop and peer upstream to catch a glimpse of "the mountains near Abergavenny, ten miles off." These, he had heard, marked "the beginning of the unknown land of Wales." Most of the locals spoke Welsh, and in town they called him the "little Saxon" for his long blonde hair (Wallace 1905, 1:24, 29). Ethnic differences and boundaries were thus impressed on him from early childhood.

At the age of five Wallace moved with his family back to England. He attended Hertford Grammar School, then in 1837 went to live with his brother John, a builder's apprentice in London. The teenage boys spent evenings at the workingmen's Hall of Science just off Tottenham Court Road. The coffee was free and the lectures stirring, with tirades against private property and religion. Here Wallace picked up the political values that stayed with him more or less for life: human nature is perfectible through education and changed environments; all humans are equal partners in progress. So taught the reformer Robert Owen, whom Wallace once heard lecture. He left town that summer, a budding socialist, to join his big brother William in Bedfordshire as a trainee land surveyor.

It was a boom time for the trade. Just the year before, Parliament had ended the age-old right of farmers to pay tithes in kind. The Tithe Commutation Act substituted a "rent charge" based on the average value of tithable produce and the productive quality of the land. This was apportioned property-by-property or field-by-field, and required an accurate survey. Squads of transit-toting, chain-lugging young men were hired. Their maps had legal status and the format was prescribed. Exact boundaries had to be shown, quantities calculated, and the quality or use of land assessed.[8] Tithe owners pored over every detail, anxious to secure their due. Tenant farmers fumed (Kain and Prince 1985, 5–6, 51–57, 120–21). The rent charge was a tax on gross output, just like the old tithe. The harder they worked, the fatter the squire or the parson became. It was sharecropping gone to seed.

Wallace paced the open fields, reveling in the fresh air and trigonometry. He knew the "well-to-do farmers" but mostly mixed with "labourers" and "mechanics" in pubs. Here poaching songs were sung and grievances aired. His political education progressed. In 1839 the brothers moved to the Welsh borders to make parish maps and survey for the enclosure of commons. This dividing up of open land among landowners was also bitterly resented. Peasants lost their ancient grazing rights and had to pay for them instead. "Legalized robbery of the poor for the aggrandisement of the rich," Wallace would call it, though at the time he simply assumed that, however unpopular, it had "*some* right and reason" (Wallace 1905, 1:109, 114–15, 124–35, 148, 150–51, 158; 1991, 123).

In 1841 the brothers pushed on into the "unknown land of Wales." Wallace now first immersed himself in the culture, lodging in pubs, attending chapels, and admiring "the grand sound of the language." The surveying continued, and late that autumn they arrived at Neath in Glamorganshire to map an enormous parish. They lived and boarded for a year with a "rather rough" hill farmer, himself the bailiff of the four-thousand-acre Duffryn es-

---

8. For surveying practices in the early nineteenth century, see textbooks such as Crocker 1817 and Tate 1848. Instruments are discussed in Bennett 1987. On the ideology and imperial relations of contemporary British mapmaking, see Andrews 1975; and Edney 1993, 1994, and 1997. Livingstone (1995) surveys the "spatial turn" in recent history of science.

tate, owned by the future Lord Aberdare (Wallace 1905, 1:161-67, 179, 186; Howell 1977, 35). Socially and geographically it was a vantage point from which to witness the start of the most violent disturbances in modern Welsh history.

The south Wales farmers were up in arms. Prices had crashed just as cash demands on them soared. Already the new Whig poor law was hated for cutting relief and raising rates. The rent charge was just as loathsome. Calculated from national prices, not depressed local ones, it raised tithes in the region by 7 percent on average and up to 50 percent in places. Payment was due promptly, twice a year and in cash. Peasants on remote farms lacked cash; their payments had always been flexible and in kind. These had been onerous enough, for as Nonconformists they objected to supporting the established church (Jones 1989, 125-35; Howell 1988, 113-25; Evans 1976, 157; Howell 1977, 11, 83-84). On top of other grievances, the new rent charge was intolerable.

Late in 1842 the farmers turned to violence. Tollgates were targeted first, symbols of another hated tax. The rustics swooped at night, breaking and burning; vigilantes in drag, calling themselves Rebecca after their biblical "sister," whose "seed" was to "possess the gate of those which hate them." In the spring full-scale riots broke out across the southwestern counties, in Glamorgan, Cardigan, and Carmarthen. A thousand Rebeccaites stormed the Carmarthen workhouse; troops were sent in and scores arrested. Armed mobs roamed the countryside, avenging every injustice, threatening landlords, tithe owners, and their agents. By the autumn of 1843 attacks on persons and property were running at ten per week. A Celtic conspiracy was suspected. The gentry linked the riots with Irish nationalism, while the press played up another kerfuffle, the "disruption" in the Church of Scotland (Jones 1989, 212 ff., 258, 260, 342-43).

In the chaos, the tithe surveys were halted and Wallace found himself idle for weeks. He seized the opportunity and rambled into the hills, teaching himself botany and geology. In retrospect he saw this as the turning point of his life, the start of his scientific career (Jones 1989, 274; Wallace 1905, 1:188, 191, 196). It was also the moment he became a political journalist, and ever after his science and politics were linked. One of his earliest compositions, "The South-Wales Farmer," dates from the end of 1843. A callow piece, overwrought and overblown, it failed to find a publisher. Yet here Wallace shrewdly exploited his firsthand knowledge of the angry peasants.[9]

The Welsh hill farmer was his actual subject, "a class which, on account

---

9. No one to date has discussed the essay's express purpose as a political commentary: see McKinney 1972, 9; Brooks 1984, 5; Hughes 1989; 1991, 179; and Smith's remarks in Wallace 1991, 9.

of the late Rebecca disturbances, has excited much interest." Wallace began by carefully marking boundaries and borders, excluding English-speaking regions and those where land was "very good and fertile," and agriculture "practised on much better" English principles. In this native Welsh country, he observed, "the system of farming is as poor as the land." Custom reigns supreme, as in the stationary "nations of the East." Nor is there an incentive for improvement, for most farmers are tenants. When asked why he does not do thus-and-so, one will say he "can't afford it" and demand to know he is to get "money to pay people for doing it." Bare survival is hard enough, living "almost entirely on vegetable food," never mind the "turnpike grievances, poor-rates, and tithes."

Wallace's sympathies were socialist. The Welsh hill farmer "lives in a manner which the poorest English labourer would grumble at," yet he is "hospitable even to the Saxon," his "fire, jug of milk, and bread and cheese being always at your service." He works hard and "bears misfortune and injury long before he complains." But the "Rebecca disturbances . . . show that he may be roused, and his ignorance of other effectual measures should be his excuse for the illegal and forcible means he took to obtain redress. . . . It is to be hoped that he will not have again to resort to such outrages as the only way to compel his rulers to do him justice" (Wallace 1905, 1:206-22).

About the time Wallace wrote these lines he left his Celtic neighbors to teach mathematics and technical drawing in Leicester. He remained there for a year, continuing his own education in the town library. Fresh from Wales and scenes of rural distress, he picked up Malthus's *Essay on the Principle of Population* and immediately, in the first twelve chapters, read a harrowing catalog of the "checks to population in the less civilized parts of the world." Native Americans, Nordic shepherds, Asian nomads, African hunters, South Sea islanders—humanity ancient and modern is passed in review, struggling and suffering, maiming and murdering, dying for want of food. Page after page evokes what Wallace had seen and experienced in Wales—the paltry provisions, the filth and squalor, the rude agriculture, the ignorance, the violence. Everywhere the superiority of English customs is assumed (Wallace 1905, 1:232; Malthus 1826, vol. 1, chaps. 3-12). The impact on a young man, now living in green and pleasant Leicestershire, was unforgettable.

V. "Ternate, February 1858"

Wallace crossed into Wales again in 1845 and resumed surveying, this time on a large estate where he was also required to collect the rent charge. The tenants here were very poor; some spoke no English and became confused, others "positively refused to pay." It was wretched work and made him "more than ever disposed to give it all up if I could but get anything else to do" (Wallace 1905, 1:245). About this time he read *Vestiges;* two years later he left for Brazil, never looking back.

One reference point, however, must be marked if Wallace's future path—to the Malthusian moment and beyond—is to be mapped. He embarked on a scientific career less a naturalist than a surveyor, less a biologist than a biogeographer, less an evolutionist than an ethnographer. For seven formative years his job had been prescriptive economic geography. Parish upon parish, field upon field, he had set limits to human livelihoods, marking boundaries, drawing lines. In later years he would become an exemplary naturalist, but always boundaries and borders, habits and habitats, concerned him. Once he even likened the "System of Nature" to a "dissected map," the pieces of which could be assembled in a "mosaic." The picture is of a crowded tithe map (figure 14.3), where field presses on field, niche upon niche, until "all gaps have been filled". Such was a surveyor's view of evolution (Wallace, cited in McKinney 1972, 43; Wallace 1991, 218).

The boundary of which Wallace was uniquely cognizant from early childhood is now called "the Highland Line." It crosses the British Isles from northeast to southwest, dividing the country into two distinctive zones. To the south and east lie rich clay and alluvial soils, well suited for intensive cultivation. To the north and west lies poor irregular terrain above six hundred feet, suitable mainly for grazing. The vast bulk of this highland region falls within the boundaries of Scotland and Wales. Here the Celtic peoples sought refuge from successive invaders, developing their own traditional economies, field systems, and social structures. The peoples who became the English occupied the rich arable lowland and created corresponding but alternative institutions. All the differences noted by Wallace, between Celt and Saxon, chapel and church, tenant and landlord, Welsh farmers and English laborers—all may be mapped along a line drawn from Sunderland in the north, through Sheffield and Bristol, to Exeter and the English Channel: the Highland Line (Hechter 1975, 51-59, 133-37).

It is to the common context—geographic, economic, and social—designated by this line that Wallace's Malthusian moment must be referred. In his swift association, the struggles of early Victorian Wales were transposed into a new provincial setting, along another ethnological line. On either side, within sight of each other, lived different peoples, different races, subsisting in different ways but in mutually dependent social and economic relationships. Malthus had underscored the differences, stressing the conflicts, the checks, as some flourished and others famished. Wallace, making his own comparisons, remembered.[10]

---

10. Brooks (1984) argues that Wallace's Malthusian insight came as he was "pondering the reasons for the 'dying out' of intermediate human tribes as a step leading to the formation of the distinct Papuan race." He needed a "mechanism" to account for the "present distinctness"

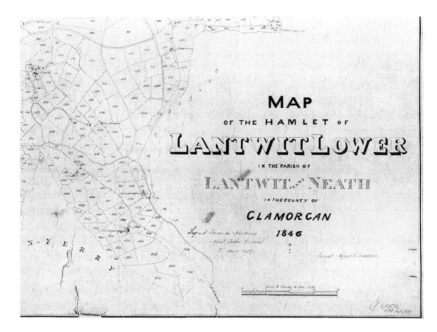

Figure 14.3 Detail from Wallace's last tithe map, showing his signature. The field numbers are keyed to a separate schedule of landowners, occupiers, cultivation, square area, and tithe rent charge. Pressing the poor, Welsh-speaking farmers to pay the rent charge so disgusted Wallace that he quit surveying for natural history. (By permission of Llyfrgell Genedlaethol Cymru/National Library of Wales, Aberystwyth, C2.)

So we return to the historical present, to Wallace's supposed flash of insight, stroke of genius, or ideological brainstorm. Having surveyed its broad backdrop, the geographic, economic, and social common context, we can now focus squarely on the Malthusian moment itself. "Something led me to think" of Malthus's *Essay,* Wallace mused decades later. "Somehow my thoughts turned." Consider, then, the following putative replay of events.

It is February 1858. Wallace has settled on the island of Ternate, with its "ancient town," ample houses, and colonial ambiance. This will be his headquarters for several years. Keen to start collecting, he heads for Gilolo, "rowing and sailing" for three hours. The ferry boat is owned by a Chinese and crewed by slaves, "mostly Papuans." On the island he rents a hut near Dodinga, a small village occupied by Malays—"Ternate men"—and a Dutch government garrison. The village is "completely shut in by low hills"

---

of Malays and Papuans (184, 186). But Wallace was not considering sympatric speciation. For him, the racial differences were primitive. Malays and Papuans hailed from separate continents, like the other fauna in the archipelago. There could be no true "transitional" forms (cf. Wallace 1877, 587–94). Wallace thought of Malthus in a moment of ethnographic, not evolutionary, perplexity.

where ruined forts stand like English castles. Some of the indigenous "semi-Papuans" are "engaged as labourers by the Chinese and Ternate traders"; most have farms in the mountainous northern and eastern parts of the island. From there they bring "rice and sago" to sell in the village, crossing the rugged central isthmus, with its "succession of little abrupt hills and valleys," where "angular masses of limestone rock" stick out, like the Ludlow and Wenlock formations of south Wales (Wallace 1877, 311, 313-14, 316; Secord 1986, 76, 91).

Wallace is on familiar ground. He notes the ethnic differences, the agriculture, the poverty. He thinks of the dominant Ternate men and mentally draws a line, a racial boundary, between the islands. He falls ill. Life is fragile here. He wonders how the locals survive, provisioned from hill farms by primitives.[11] He wonders how *they* survive, these "less civilized" tribes. The Owenite in him thinks of the Welsh farmers; of himself, the Saxon surveyor; of the tithes, the poor rates, the riots. Wars, famine, disease—these cut life off, check it . . . just as Malthus said. Only the fittest to forage remain.

Whereupon, by a deft pulled focus, Wallace arrives at natural selection. His fever passes; he makes notes that evening and returns to Ternate directly, on 1 March, where he writes up the theory on "the two succeeding evenings." His paper is fundamentally about food, where it can be found, how often, in what quantity and with what quality. Among individuals, varieties or "allied species," those that are "best adapted to obtain a regular supply of food, and to defend themselves against the attacks of their enemies and the vicissitudes of the seasons" form the predominant population. The "same laws" apply to birds and mammals—or to Celt and Saxon, he might have said. Wallace inscribes the paper "Ternate, February 1858," the place of completion and the month of conception of his theory. He posts the manuscript to Darwin by mail steamer on the ninth—and the rest is history (Wallace 1905, 1:363; 1958, 271, 272; McKinney 1972, 132).

This is of course only one possible reconstruction of Wallace's Malthusian moment. It is, however, a salutary one, I believe. No longer should it be assumed that Wallace's was an independent discovery of natural selection, or that its common context was merely ideological. Wallace himself left clues enough for us to follow him to that moment when, by an act of contextual transposition, he derived a theory from the same Whig workhouse world that had authorized Darwin's. Twenty years and half the earth away, surrounded by natives on a remote volcanic island, Wallace was reminded of the world that *he* had known, so far from London's gentry. It was an ethnic world, colonized, combative, on the "Celtic periphery" of an island

---

11. In Borneo Wallace reported having "a continual struggle to get enough to eat" because the Dyaks, in debt to Malay traders, would not sell him provisions (Marchant 1916, 1:55).

kingdom (Hechter 1975). To the socialist on Gilolo, in a moment of Malthusian self-consciousness, Wales became a guerilla theater of evolution.

Bibliographical Note

Lest "genius" be thought to explain Wallace's Malthusian insight, see Yeo (1988), Schaffer (1990), and Jackson (1995) on the problem of genius in its cultural and biographical contexts. For the role of "discovery" stories in the history of science, see Schaffer (1986). Next to Wallace's own works, the most important sources for this chapter are the articles by Camerini (1993) and Hughes (1989, 1991), and the monographs by Brooks (1984) and McKinney (1972). Young's original "common context" essay (in Young 1985) is still worth reading, although Bohlin (1991, 1995) has dissected Young's arguments relentlessly. For the Welsh context, Howell (1977), Kain and Prince (1985), and Jones (1989) are essential.

There is no adequate life of Wallace: Fichman (1981) and George (1964) give useful surveys only. The scope for a biography is apparent from the exhaustive bibliography in Wallace (1991). Kottler (1985) and Beddall (1988) clarify Wallace's relations with Darwin. Durant (1979) is the best published study of his socialism, and Kottler (1974) of his spiritualism. Peter Crawford's seventy-five-minute BBC drama, "The Forgotten Voyage" (transmitted 24 December 1982), is the only film biography of Wallace. It makes his Malthusian moment "memorable" (as in *1066 and All That*) with high drama and sumptuous location scenery.

References

Andrews, J. H. 1975. *A Paper Landscape: The Ordnance Survey in Nineteenth-Century Ireland.* Oxford: Clarendon Press.

Beddall, Barbara G. 1988. "Darwin and Divergence: The Wallace Connection." *Journal of the History of Biology* 21:1–68.

Bennett, J. A. 1987. *The Divided Circle: A History of Instruments for Astronomy, Navigation and Surveying.* Oxford: Phaidon-Christie's.

Bohlin, Ingmar. 1991. "Robert M. Young and Darwin Historiography." *Social Studies of Science* 21:597–648.

———. 1995. "Through Malthusian Specs? A Study in the Philosophy of Science Studies, with Special Reference to the Theory and Ideology of Darwin Historiography." Ph.D. diss., University of Göteborg.

Brackman, Arnold C. 1980. *A Delicate Arrangement: The Strange Case of Charles Darwin and Alfred Russel Wallace.* New York: Times Books.

Brooks, John Langdon. 1984. *Just before the "Origin": Alfred Russel Wallace's Theory of Evolution.* New York: Columbia University Press.

Browne, Janet. 1995–. *Charles Darwin.* Vol. 1, *Voyaging.* London: Jonathan Cape.

Camerini, Jane R. 1993. "Evolution, Biogeography, and Maps: An Early History of Wallace's Line." *Isis* 84:700–727.

Chambers, Robert. [1844] 1994. *Vestiges of the Natural History of Creation and Other Evolutionary Writings.* Edited by James A. Secord. Chicago: University of Chicago Press.

Clodd, Edward. 1892. "Memoir." In *The Naturalist on the River Amazons: A Record of Adventures, Habits of Animals, Sketches of Brazilian and Indian Life, and Aspects of Nature under the Equator, during Eleven Years of Travel,* by Henry Walter Bates, new ed. London: John Murray, xvii–lxxxix.

Corsi, Pietro. 1988. *Science and Religion: Baden Powell and the Anglican Debate, 1800–1860.* Cambridge: Cambridge University Press.

Crocker, A. 1817. *The Elements of Land Surveying, Designed Principally for the Use of Schools and Students.* London: printed for Longman, Hurst, Rees, Orme and Brown.

Darwin, Charles. [1839] 1845. *Journal of Researches into the Natural History and Geology of the Countries Visited during the Voyage of H.M.S. "Beagle" round the World, under the Command of Capt. Fitz Roy.* 2d ed. London: John Murray.

———. 1958a. *The Autobiography of Charles Darwin, 1809–1882, with Original Omissions Restored.* Edited by Nora Barlow. London: Collins.

———. 1958b. "Charles Darwin's Sketch of 1842." In *Evolution by Natural Selection,* by Charles Darwin and Alfred Russel Wallace, edited by Gavin de Beer. Cambridge: Cambridge University Press, 41–88.

———. 1985–. *The Correspondence of Charles Darwin.* Edited by Frederick Burkhardt and Sydney Smith. 9 vols. Cambridge: Cambridge University Press.

Darwin, Francis, ed. 1892. *Charles Darwin: His Life Told in an Autobiographical Chapter, and in a Selected Series of His Published Letters.* London: John Murray.

Desmond, Adrian. 1989. *The Politics of Evolution: Morphology, Medicine, and Reform in Radical London.* Chicago: University of Chicago Press.

———. 1994. *Huxley: The Devil's Disciple.* London: Michael Joseph.

Desmond, Adrian, and James Moore. 1991. *Darwin.* London: Michael Joseph.

Di Gregario, Mario A., and N. W. Gill, eds. 1990–. *Charles Darwin's Marginalia.* 2 vols. New York: Garland Publishing.

Durant, John R. 1979. "Scientific Naturalism and Social Reform in the Thought of Alfred Russel Wallace." *British Journal for the History of Science* 12:31–58.

Edney, Matthew H. 1993. "The Patronage of Science and the Creation of Imperial Space: The British Mapping of India, 1799–1843." *Cartographica* 30:61–67.

———. 1994. "Mathematical Cosmography and the Social Ideology of British Cartography, 1780–1820." *Imago Mundi* 46:101–6.

———. 1997. *Mapping an Empire: The Geographical Construction of British India, 1765–1843.* Chicago: University of Chicago Press.

Erskine, Fiona. 1987. "Darwin in Context: The London Years, 1837–1842." Ph.D. thesis, Open University.

Evans, Eric J. 1976. *The Contentious Tithe: The Tithe Problem and English Agriculture, 1750–1850.* London: Routledge and Kegan Paul.

Fichman, Martin. 1981. *Alfred Russel Wallace.* Boston: Twayne.

George, Wilma. 1964. *Biologist Philosopher: A Study of the Life and Writings of Alfred Russel Wallace.* London: Abelard-Schuman.

Gillispie, Charles C. 1974. Comment on "The Evolutionary Theories of Charles Darwin and Herbert Spencer," by Derek Freeman. *Current Anthropology* 15:224.

Hechter, Michael. 1975. *Internal Colonialism: The Celtic Fringe in British National Development, 1536-1966.* London: Routledge and Kegan Paul.

Hodge, Jon. 1993. "One Marxist View of Darwin's Ideas." *Biology and Philosophy* 8:469-76.

Hodge, Jon, and David Kohn. 1985. "The Immediate Origins of Natural Selection." In *The Darwinian Heritage,* edited by David Kohn. Princeton, N.J.: Princeton University Press, 185-206.

Howell, David W. 1977. *Land and People in Nineteenth-Century Wales.* London: Routledge and Kegan Paul.

———. 1988. "The Rebecca Riots." In *People and Protest: Wales, 1815-1880,* edited by Trevor Herbert and Gareth Elwyn Jones. Cardiff: University of Wales Press, 113-38.

Hughes, R. Elwyn. 1989. "Alfred Russel Wallace: Some Notes on the Welsh Connection." *British Journal for the History of Science* 22:401-18.

———. 1991. "Alfred Russel Wallace (1823-1913): The Making of a Scientific Nonconformist." *Proceedings of the Royal Institution* 63:175-83.

Jackson, Myles W. 1995. "Genius and the Stages of Life in Eighteenth-Century Britain and Germany." In *Les âges de la vie en Grande-Bretagne au XVIIIe siècle,* edited by Serge Soupel. Paris: Presses de la Sorbonne Nouvelle, 35-46.

Jones, David J. V. 1989. *Rebecca's Children: A Study of Rural Society, Crime, and Protest.* Oxford: Clarendon Press.

Kain, Roger J. P., and Hugh C. Prince. 1985. *The Tithe Surveys of England and Wales.* Cambridge: Cambridge University Press.

Kohn, David. 1980. "Theories to Work By: Rejected Theories, Reproduction, and Darwin's Path to Natural Selection." *Studies in History of Biology* 4:67-170.

Kottler, Malcolm Jay. 1974. "Alfred Russel Wallace, the Origin of Man, and Spiritualism." *Isis* 65:145-92.

———. 1985. "Charles Darwin and Alfred Russel Wallace: Two Decades of Debate over Natural Selection." In *The Darwinian Heritage,* edited by David Kohn. Princeton, N.J.: Princeton University Press, 367-432.

Livingstone, David N. 1995. "The Spaces of Knowledge: Contributions towards a Historical Geography of Science." *Environment and Planning D: Society and Space* 13:5-34.

Malthus, Thomas Robert. [1798] 1826. *An Essay on the Principle of Population; or, A View of Its Past and Present Effects on Human Happiness, with an Inquiry into Our Prospects Respecting the Future Removal or Mitigation of the Evils Which It Occasions.* 6th ed. 2 vols. London: John Murray.

Marchant, James. 1916. *Alfred Russel Wallace: Letters and Reminiscences.* 2 vols. London: Cassell and Company.

McKinney, H. Lewis. 1969. "Wallace's Earliest Observations on Evolution: 28 December 1845." *Isis* 60:370-73.

———. 1972. *Wallace and Natural Selection.* New Haven, Conn.: Yale University Press.

Moore, James R. 1982. "Charles Darwin Lies in Westminster Abbey." *Biological Journal of the Linnean Society* 17:97-113.

———. 1985. "Darwin of Down: The Evolutionist as Squarson-Naturalist." In *The Darwinian Heritage,* edited by David Kohn. Princeton, N.J.: Princeton University Press, 435-81.

———. 1996. "Metabiographical Reflections on Charles Darwin." In *Telling Lives in Science: Essays on Scientific Biography*. Cambridge: Cambridge University Press, 267-81.

Oldroyd, David R. 1984. "How Did Darwin Arrive at His Theory? The Secondary Literature to 1982." *History of Science* 22:325-74.

Paul, Diane B. 1988. "The Selection of the 'Survival of the Fittest.'" *Journal of the History of Biology* 21:411-24.

Schaffer, Simon. 1986. "Scientific Discoveries and the End of Natural Philosophy." *Social Studies of Science* 16:387-420.

———. 1990. "Genius in Romantic Natural Philosophy." In *Romanticism and the Sciences*, edited by Andrew Cunningham and Nicholas Jardine. Cambridge: Cambridge University Press, 82-98.

Secord, James A. 1986. *Controversy in Victorian Geology: The Cambrian-Silurian Dispute*. Princeton, N.J.: Princeton University Press.

———. 1989. "Behind the Veil: Robert Chambers and 'Vestiges.'" In *History, Humanity and Evolution: Essays for John C. Greene*, edited by James R. Moore. Cambridge: Cambridge University Press, 165-94.

———. 1994. "Introduction." In *Vestiges of the Natural History of Creation and Other Evolutionary Writings*, by Robert Chambers, edited by James A. Secord. Chicago: University of Chicago Press, ix-xlvii.

Shapin, Steven. 1992. "Discipline and Bounding: The History and Sociology of Science as Seen through the Externalism-Internalism Debate." *History of Science* 30:333-69.

Tate, Thomas. 1848. *Principles of Geometry, Mensuration, Trigonometry, Land-Surveying, and Levelling . . .* London: Longman, Brown, Green, and Longmans.

Turner, Frank Miller. 1993. *Contesting Cultural Authority: Essays in Victorian Intellectual Life*. Cambridge: Cambridge University Press.

Wallace, Alfred Russel. 1853. *A Narrative of Travels on the Amazon and Rio Negro, with an Account of the Native Tribes, and Observations on the Climate, Geology, and Natural History of the Amazon Valley*. London: Reeve and Company.

———. 1856a. "A New Kind of Baby." *Chambers's Journal*, 3d ser., 6:325-27.

———. 1856b. "On the Habits of the Orang-utan of Borneo." *Annals and Magazine of Natural History*, 2d ser., 18:26-32.

———. 1857a. Letter dated 21 August 1856. *Zoologist* 15:5414-16.

———. 1857b. "On the Natural History of the Aru Islands." *Annals and Magazine of Natural History*, 2d ser., supplement, 20:473-85.

———. [1858] 1958. "On the Tendency of Varieties to Depart Indefinitely from the Original Type." In *Evolution by Natural Selection*, by Charles Darwin and Alfred Russel Wallace, edited by Gavin de Beer. Cambridge: Cambridge University Press, 268-79.

———. 1864. "On Some Anomalies in Zoological and Botanical Geography." *Natural History Review* 4:111-23.

———. 1865. "On the Varieties of Man in the Malay Archipelago." *Transactions of the Ethnological Society of London*, n.s., 3:196-215.

———. [1869] 1877. *The Malay Archipelago: The Land of the Orang-Utan and the Bird of Paradise; A Narrative of Travel, with Studies of Man and Nature*. 6th ed. London: Macmillan and Company.

———. [1870, 1878] 1891. *Natural Selection and Tropical Nature: Essays on Descriptive and Theoretical Biology.* New ed. London: Macmillan and Company.

———. 1898. *The Wonderful Century: Its Successes and Its Failures.* London: Swan Sonnenschein and Company.

———. 1903. "The Dawn of a Great Discovery (My Relations with Darwin in Reference to the Theory of Natural Selection)." *Black and White* 25:78–79.

———. 1905. *My Life: A Record of Events and Opinions.* 2 vols. London: Chapman and Hall.

———. 1991. *Alfred Russel Wallace: An Anthology of His Shorter Writings.* Edited by Charles H. Smith. Oxford: Oxford University Press.

Whitmore, T. C., ed. 1981. *Wallace's Line and Plate Tectonics.* Oxford: Clarendon Press, 1981.

Williams-Ellis, Anabel. 1966. *Darwin's Moon: A Biography of Alfred Russel Wallace.* London: Blackie.

Wood, Denis. 1991. "Maps are Territories/Review Article." *Cartographica* 28:73–80.

Yeo, Richard. 1988. "Genius, Method and Morality: Images of Newton in Britain, 1760–1860." *Science in Context* 2:257–84.

Young, Robert M. 1985. *Darwin's Metaphor: Nature's Place in Victorian Culture.* Cambridge: Cambridge University Press.

———. 1987. "Darwin and the Genre of Biography." In *One Culture: Essays in Science and Literature,* edited by George Levine. Madison: University of Wisconsin Press, 203–24.

———. 1994. "Desmond and Moore's 'Darwin': A Critique." *Science as Culture* 4:393–424.

15

Doing Science in a Global Empire: Cable Telegraphy and Electrical Physics in Victorian Britain

BRUCE J. HUNT

By the end of the nineteenth century, the British Empire stretched from India, Malaya, Australia, and New Zealand through Canada, much of Africa, and a string of smaller territories and islands. Britain's formal empire took in more than a fifth of the world's land area and a quarter of its people, while its informal empire of trade and investment reached even further and gave London substantial power over much of Latin America, the Middle East, and east Asia as well. This enormous empire affected virtually every aspect of Victorian life; not least, it provided one of the principal contexts for Victorian science.

Recent scholarship has done much to illuminate how the ideology of imperialism was used to motivate and justify various kinds of scientific work in the nineteenth century, as Britain and other European powers moved to appropriate and subordinate wide swaths of the rest of the world. (See Harriet Ritvo's chapter in this volume and the works cited there.) But beyond the effects of this underlying ideology, Victorian science was also shaped simply by the fact that it was carried on within an expanding global empire. Exploration, conquest, and commerce provided materials and posed problems that affected the direction and content of British scientific work in a wide range of disciplines; indeed, much of what was distinctive about Victorian science can be traced to the fact that it was pursued within a global commercial empire. This is perhaps clearest in such fields as geology, botany, zoology, and anthropology, where the exploration of new lands and the cataloging of new plants, animals, and peoples were central components of scientific work. But the imperial context also made itself felt in other sciences, including physics, particularly through the demands and opportunities presented by the technologies used to maintain and advance imperial interests. One of the most important of these was submarine telegraphy; described by one observer at the turn of the century as "the most potent factor in the maintenance of the Empire" (Kappey 1902, 332), it was widely recognized as a major bulwark of British commercial and strategic power in the late Victorian era (Headrick 1991). A quintessential technology of empire, cable telegraphy provided much of the impetus for British work in electrical physics in the second half of the nineteenth century. Through it, the British imperial context shaped Victorian electrical science in deep and distinctive ways.

I. Science and Empire

The influence of the imperial context on Victorian work in the field and observational sciences has been widely recognized in recent years. Historians of science have elucidated the large role "the imperial theme" played in British geology in the nineteenth century, particularly in the work of Roderick Murchison (Secord 1982; Stafford 1989). Similarly, scholars have shown how deeply botanical research at Kew Gardens and elsewhere was shaped by the demands and opportunities of empire (Brockway 1979; Browne 1992; Drayton 1993) and how British work in such fields as navigation, chronometry, and cartography (Howse 1980; Ritchie 1967; Edney 1993), terrestrial magnetism (Cawood 1979), and solar astronomy (Pang 1991, 1993) was also affected by its imperial setting.

In most of these historical studies, the empire is presented mainly as a resource for science, an object of study and a source of data. This roughly corresponds to Phase I of George Basalla's well-known model of the spread of Western science (Basalla 1967). The empire, or more broadly the world beyond the European metropolis, appears here mainly in a passive role, as something on which scientists act and from which they take. Scientists go out into the empire with their tools and theories—as Pang shows, often with a veritable packtrain of investigative equipment—to gather specimens and data. They then return with these to their metropolitan headquarters, there to analyze their significance and use them to bolster, fill out, and perhaps test their theories of the world and its workings. It is a picture that finds a partial parallel in many treatments of technology and empire, which generally focus on the extension of proven technologies into new areas, leading perhaps to some minor adaptations to colonial circumstances but producing little effect on the parent technologies back in Europe (Adas 1989; Headrick 1981, 1988; Bernstein 1960).

This picture of science reaching out into the empire and bringing its prizes back to the metropolitan center is valid as far as it goes, but as a number of scholars have recognized, it needs to be supplemented by other perspectives. The scientific work done in colonial countries must also be examined on its own terms and in relation to the development by such countries of scientific traditions and institutions of their own—a subject that has spawned a literature far too large to be fully discussed here. Attention must also be paid to the active role the imperial context sometimes played in changing the shape and direction of metropolitan science itself. Data and specimens gathered in the empire often posed new puzzles and offered new insights into old ones; the extension of technologies to serve imperial ends often presented engineers and scientists with new phenomena and new problems and drove them to develop new tools and theories with which to address them. The pursuit of science in a global empire thus served both to raise new questions and to cast existing knowledge in a new light, and in so doing sometimes set in train fundamental shifts in scientific belief and practice.

In his fine study of Roderick Murchison, *Scientist of Empire,* Robert Stafford observes that

> the content of science, like the style of an empire, is shaped by its cultural context. British geology and geography, as well as other sciences not treated here, were significantly influenced by Britain's possession of a colonial empire. Imperial concepts, metaphors, data, and career opportunities informed the development of these disciplines. (Stafford 1989, 223)

Stafford illustrates this point by showing how colonial data played a crucial role in shaping geological debates in Britain on theoretical issues ranging from rifting to glaciation (Stafford 1989, 200). Similar examples of the effect of colonial specimens and data on the content of metropolitan science can be enumerated in zoology (Dugan 1987), astronomy (Pang 1993), and other disciplines.

One such example towers above the rest: the role of the imperial context in the origin of Charles Darwin's theory of evolution. That Darwin's voyage around the world on the *Beagle* in the 1830s played a crucial part in the formation of his ideas is a commonplace. It should be remembered that this was an Admiralty surveying voyage undertaken primarily to improve navigational charts of South America, then being drawn more closely into Britain's informal empire of commerce and investment (Desmond and Moore 1992, 105-6). It is very unlikely that Darwin would ever have come up with his theory had he not been exposed while on the *Beagle* to new and exotic species, environments, and ecosystems and particularly to the puzzling facts of biogeographical distribution (Browne 1983). It is a telling fact that both Darwin's codiscoverer of natural selection, Alfred Russel Wallace, and two of his most important early backers, Joseph Dalton Hooker and Thomas Henry Huxley, also went on exploring voyages to distant parts of the world, Hooker and Huxley under British government sponsorship. The wrenching reorientation that biological thought underwent in the mid-Victorian era was the product of many causes, but one of them was surely the confrontation, propelled by the needs of a growing empire, with new facts about the distribution and affinities of living things around the world.

The effects of the imperial context on scientific thinking are perhaps most clearly and strongly seen in the natural history sciences, but they also reached into the "exact sciences," including physics. In a series of books and articles, Lewis Pyenson has provided a wealth of valuable information on the pursuit of the exact sciences in the German, Dutch, and French empires in the nineteenth and early twentieth centuries while treating their role in what he calls the "civilizing mission" of the European powers (Pyenson 1978, 1985, 1989a, 1989b, 1990a, 1990b, 1993a). A key part of Pyenson's larger argument centers on the apparent insulation of the content of work in these exact sciences—particularly astronomy, physics, and geophysics—from overtly imperialist concerns; indeed, he argues that it was precisely the universality and purity of these pursuits that made them useful instruments of European cultural imperialism. He then argues that the independence of the content of work in the exact sciences from local context or imperial circumstances demonstrates the vacuity of a strong social constructivist approach to the history of science: if social and political influences were to show up anywhere in the exact sciences, he says, it should be in the "highly charged" setting of the colonial university or observatory, yet he finds no sign of such influences in the scientific work produced there (Pyenson 1989a, xiv; 1993b, 105). But while Pyenson is no doubt right that much of the scientific work done in colonial outposts was largely indistinguishable from that done in metropolitan centers, this by no means demonstrates that the imperial context did not shape the content of work in the exact sciences, nor does it establish that such sciences were really as pure, universal, and independent of national and imperial differences as he contends (Palladino and Worboys 1993). The degree of consensus on facts, methods, and theories reached in various sciences is indeed a matter of interest and worthy of close study, but it should not blind us to important differences in the science done in different countries at different times or divert us from examination of the sources of those differences.

Electrical physics—surely among the most exact of the exact sciences—presents a remarkable instance in the second half of the nineteenth century of a wide national divergence in theoretical approaches. At a time when action-at-a-distance theories of electromagnetism prevailed in Germany and France, field theory became the favored approach to electromagnetic phenomena in Britain. The roots of this divergence lay, I will argue, in large part in the unique demands and opportunities presented by Britain's global system of submarine telegraph cables and so ultimately in the needs of the empire that system was built to serve. Rather than providing evidence of the insulation of the exact sciences from social and political influences, cable telegraphy presents a striking example of how the imperial context could shape the content of scientific work done not just in colonial outposts, but in the metropolitan centers themselves.

II. The Cable Empire

The first successful undersea cable was laid across the English Channel in 1851 by a group of British entrepreneurs led by Jacob and John Watkins Brett. Within fifteen years cable telegraphy had matured into a reliable ocean-spanning technology, and by the 1870s cables reached nearly around the globe and were changing fundamentally the way the world worked. From its beginnings until at least the 1920s, the world cable industry was completely dominated by British capital and British engineering expertise (Bright 1898; Coates and Finn 1979, 171-72). Fueled by the demands and resources of a growing commercial empire, cable telegraphy emerged as one of Victorian Britain's premier "high-tech" industries. It constituted by far the largest market for advanced electrical knowledge in the third quarter of the nineteenth century, and as such provided much of the impetus for British electrical research during a remarkably productive period. The needs of cable telegraphy directly underlay the British work on electrical units and standards in the 1860s that established substantially the system of ohms, amps, and volts still used today. Cable telegraphy also played an important and previously unrecognized role in the rise and consolidation of field theory, one of the most distinctive and important achievements of Victorian physics.

The reasons for Britain's dominance of the world cable industry are not hard to find. At the time the technology was emerging in the 1850s and 1860s, Britain was by far the leading commercial, industrial, and imperial power in the world, with both the greatest need for rapid transoceanic communications and the greatest resources with which to secure them. Britain built up an early technological lead in cable telegraphy and was able to sustain it for decades, aided by the specialized nature of the technical skills involved and the huge capital investment needed to break into the industry. Britain's imperial and commercial power also reinforced its dominance of the cable industry in quite direct ways. In particular, Britain's control of Malaya and Singapore gave it an effective monopoly on the trade in gutta-percha, a rubber-like substance derived from the sap of certain tropical trees that was the favored material for insulating cables. Hundreds of tons of gutta-percha were needed for a long cable, and since foreign companies were unable to obtain such large quantities, they were effectively shut out of competition (Headrick 1987; Hunt, 1994b; Siemens [1892] 1966, 119, 241).

The early steps in the growth of the cable network had been halting and uncertain. The Brett brothers' first attempt to span the Channel in 1850 had failed after only a few garbled messages were transmitted, but their next try a year later was successful. Indeed, their 1851 Channel cable proved very lucrative—more lucrative than may have been healthy for the budding industry, for the prospect of similar profits soon lured the Bretts and others

Figure 15.1 The first Atlantic cable being readied for shipment at the Glass, Elliot works at East Greenwich on the Thames, March 1857. After two abortive attempts in 1857, this cable was successfully laid from Ireland to Newfoundland in August 1858, but it failed after a few weeks of fitful service. (From the *Illustrated London News* 30 [14 March 1857]: 243.)

into a series of ill-considered attempts to lay cables in the Mediterranean, across the Irish Sea, and elsewhere. Several of these hastily made and laid cables failed completely, raising doubts about the whole enterprise of submarine telegraphy (Bright 1898, 5–22).

These doubts grew stronger after the first Atlantic cable failed in 1858. Led by the American entrepreneur Cyrus Field, a group of British investors made ambitious plans to lay a cable from Ireland to Newfoundland—a much longer distance, and in deeper water, than had ever been attempted before (see figure 15.1). After many reverses, the cable was successfully laid in August 1858. The celebrations had scarcely died away, however, when it became clear that the cable was not working properly; after a few weeks of fitful service it gave out altogether. Recriminations ensued as those involved tried to put the blame on poor design, hasty manufacture, rough handling, or the high-voltage apparatus used by Wildman Whitehouse, the Atlantic Telegraph Company's "electrician-projector" (Bright 1898, 23–55; Dibner 1959; Smith and Wise 1989, 667–78; Hunt 1996).

Meanwhile, another big cable project was going forward in the Red Sea. Worried by slow communications with India during the 1857 mutiny, the British government had backed construction of a cable to run from Suez to Aden and on across the Arabian Sea to Karachi. The cable was laid and tested in sections in 1859 and 1860 but soon failed irreparably. The government, however, had guaranteed the private investors in the cable a 4½ percent return for the next fifty years—whether the cable worked or not. The Red Sea fiasco ended up costing the British Treasury nearly nine hundred thousand pounds (Bright 1898, 57-58; Cell 1970, 225-34, 248-51).

By 1860, deep-sea cable telegraphy was widely viewed as a great *failed* technology. The collapse of the Atlantic and Red Sea cables led to a period of retrenchment and reexamination as investors and cable engineers tried to discover what had gone wrong and how to correct it. The British government and the Atlantic Telegraph Company set up a Joint Committee on Submarine Telegraphs that commissioned studies and took expert testimony from scientists and engineers for more than a year. Its report, issued in 1861, and long a standard reference in the cable industry, drew together the best available information on materials, procedures, and measurement techniques, as well as on the theory and practice of telegraphic signaling. After examining the causes of previous failures, the committee concluded that they were all avoidable, and that if proper care were taken, submarine telegraphy could in the future "prove as successful as it had hitherto been disastrous" (Joint Committee 1861, xxxvi).

Aided by such studies and by a growing body of practical experience, British engineers had mastered the main problems in the manufacture, laying, and operation of cables by the mid-1860s. An important cable in the Persian Gulf, connecting India with an overland telegraph through Mesopotamia and Turkey, was successfully completed in 1864, and confidence was sufficiently restored for the Atlantic project to be relaunched (Harris 1969; Hempstead 1989)(see figure 15.2). A stouter cable was manufactured, and the *Great Eastern,* the only ship large enough to carry all twenty-five hundred miles of it, was chartered to lay it in 1865. Disaster struck again as the cable snapped after about two-thirds had been laid, but yet another effort was made the next year, this time with complete success. Moreover, the engineers managed to grapple the broken 1865 cable from the bottom of the ocean, splice it, and complete it on to Newfoundland—so that by September 1866 there were two working cables across the Atlantic, and the practicability of transoceanic cable telegraphy was clearly established.

The dramatic nature of the Atlantic cable story has understandably made it the focus of most accounts of Victorian submarine telegraphy (Dibner 1959; Coates and Finn 1979). Such accounts often leave the impression that nineteenth-century cable telegraphy was largely an Anglo-American and North Atlantic affair. But the American role even in the first Atlantic cables was quite small; Cyrus Field aside, the principals were all British, and it

Figure 15.2 The shore end of the Persian Gulf cable being landed through the mud at the mouth of the Shatt al Arab in 1864. Sir Charles Bright, who directed the operation, is pictured at left with his arm outstretched. Completion of this cable, along with connecting landlines through Turkey, Mesopotamia, and Pakistan, provided a vital link from Britain to India. (From Bright 1898, 73; based on an engraving originally published in the *Illustrated London News*, 1864.)

should be borne in mind that the cables in fact connected two parts of the British Empire (Ireland and Newfoundland) and were bought, built, laid, and operated almost entirely with British money and by British experts. Nor should the drama of the Atlantic cable story obscure the fact that a far greater length and number of cables were subsequently laid to the east and south, mainly to serve Britain's (formal and informal) empire. In the 1870s, British firms laid cables to Gibraltar, Malta, Egypt, India, Singapore, Hong Kong, Australia, and New Zealand, as well as along the east coast of Africa, the east and west coasts of South America, and throughout the West Indies. In the 1880s and 1890s they added duplicate cables on most of these routes and laid additional ones in the Far East and along the west coast of Africa (Bright 1898, 106–45; Baglehole 1969, 43–46) (see figure 15.3). The total length of the world's working cables jumped from barely 1,100 miles in 1864 to 15,000 by 1870, 86,000 by 1880, and over 210,000 by 1900 (Headrick 1991, 29).

In the late nineteenth and early twentieth centuries Britain's global cable network was often referred to as "the nerves of empire" (Peel 1905); infor-

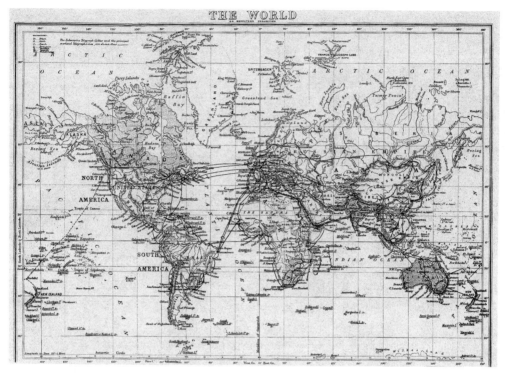

Figure 15.3 The submarine telegraph cables of the world, as of 1900. The main connecting overland lines are also shown. Components of the British Empire are shown shaded or with their names underlined. Most of the cables shown were owned by British companies, and virtually all had been built and laid by British manufacturers. The government-owned British Pacific cable, shown by a dashed line, was completed in 1902. (From *The British Empire Series* 1899–1902, vol. 5, foldout at back of volume.)

mation flowed in along the "mighty electric nerve system" (Kappey 1902, 332), and commands flowed out, binding the empire more closely together and securing Britain's continued political and commercial preeminence. From its position at the center of its web of wires, Britain was able to deploy its naval and military forces with unprecedented efficiency and to exercise more direct control over its far-flung empire than would have been possible in any earlier age. British wire services came to control much of the flow of news around the world, and their dispatches—terse and garbled as they often were—shaped the formation of public opinion in distant lands, often to Britain's advantage. Most important of all, cables provided up-to-the-minute news of foreign markets, oiling the operation of the global trading system that sustained Britain's wealth and power and reinforcing London's position as the banking, investment, and insurance capital of the world (Inglis 1980). As French officials remarked in 1900, while pushing for the construction of their own cable network, "England owes her influence in the world perhaps more to her cable communications than to her navy. She controls the news, and makes it serve her policy and commerce in a marvelous manner" (Kennedy 1971, 748).

Virtually all of the cables that made this possible were manufactured and laid by British firms; indeed, most were controlled by one man, John Pender, a onetime Manchester cotton merchant turned "cable king" (Barty-King, 1979; Bright 1898, 32). After risking and losing a substantial sum on the failed Atlantic cable of 1858, Pender engineered the formation of the mammoth Telegraph Construction and Maintenance Company (TC&M), long the world's largest cable manufacturer, and played a key role in pushing the successful 1866 Atlantic cable through to completion. Pender went on to form the Eastern Telegraph Company and its many affiliates, and from 1870 until his death in 1896 he effectively controlled the manufacture and operation of most of the long cables in the world. By 1892, the Eastern group owned 45.5 percent of the world's cable mileage, other British firms—some also controlled by Pender—an additional 17.5 percent, and British and colonial governments 3.2 percent, for a total of 66 percent in British hands. Even the relatively few cables owned by American, French, or other foreign firms had almost all been manufactured and laid by Pender's TC&M (which accounted for about two-thirds of the world total) or a handful of other British companies headquartered on the banks of the Thames (Headrick 1991, 31, 38; Coates and Finn 1979, 170-72).

Though the great majority of the world's cables were built and operated by private firms like Pender's, most of these had links—sometimes quite close—with the British government. Stung by the 1859 Red Sea fiasco, the Treasury was long reluctant to offer direct financial backing to new projects, but as major customers for cable services, the Foreign and Colonial Offices often helped negotiate deals guaranteeing proposed cables a substantial amount of business. In return, the government was given priority in using the cables—its messages being preceded by the imperious order, "Clear the line, clear the line" (Kennedy 1971, 739; cf. Cain 1971; Jaras 1975; and Cell 1970).

A longtime member of Parliament, Pender worked closely with successive British governments to promote cable interests; "Telegraphs," he liked to say, "know no politics." Pender served the empire by providing secure and reliable communications throughout the world, and the empire served him by providing a ready market for the services he sold (Headrick 1991, 35-37; Coates and Finn 1979, 170-71; Barty-King 1979; *Electrician* 1896). Pender was never very well known to the general public, but leading imperialists recognized the importance of his work; indeed, Cecil Rhodes reportedly once declared that "Pender was 'imperializing the map' while I was just feeling my way" (Wilshaw 1939, 3).

By 1900, the British cable industry represented a capital investment of over thirty million pounds, and its principal firms employed tens of thousands of people in manufacturing, testing, laying, and operating their global network. But as the cable network spread around the world, the expertise needed to run it did not; as Daniel Headrick has noted, "The equipment and knowledge stayed firmly in the possession of a small elite, almost all British and, among the British, almost all of them members of the Eastern group and TC&M" (Headrick 1991, 46). Pender's Eastern group established a special school at Porthcurno, its great cable station near Land's End, to train telegraphers and technicians for service around the world, and several proprietary schools were established in London and elsewhere to prepare young men for careers as cable engineers (Barty-King 1979, 59). Moreover, by the 1870s British universities and technical colleges had begun to set up teaching laboratories that emphasized techniques of electrical measurement modeled on those of cable telegraphy (Gooday 1990).

The growth of the British cable industry inevitably affected British work in electrical science. William Thomson, the leading physicist in Britain through most of the Victorian era, devoted much of his career to advancing the theory and practice of cable telegraphy and turned his pioneering laboratory at Glasgow into a virtual training ground for cable engineers. Knighted in 1867 for his work on the Atlantic cable, he was raised to the peerage as Lord Kelvin in 1891 largely in recognition of his contributions to cable telegraphy and imperial communications (Smith and Wise 1989, 649–83). The other great British electrical physicist of the time, James Clerk Maxwell, never worked as directly on cable telegraphy as Thomson did, but he, too, clearly recognized its important effects on science. In the preface to his *Treatise on Electricity and Magnetism* of 1873 — a work widely regarded as a paradigm of pure and abstract science — Maxwell declared:

> The important applications of electromagnetism to telegraphy have
> . . . reacted on pure science by giving a commercial value to accurate
> electrical measurements, and by affording to electricians the use of
> apparatus on a scale which greatly transcends that of any ordinary laboratory. The consequences of this demand for electrical knowledge,
> and of these experimental opportunities for acquiring it, have already
> been very great, both in stimulating the energies of advanced electricians, and in diffusing among practical men a degree of accurate
> knowledge which is likely to conduce to the general scientific progress of the whole engineering profession. (Maxwell 1873, 1:vii–viii)

Maxwell himself played a major role in supplying the "demand for electrical knowledge" that grew out of the advent of cable telegraphy. His work and that of his followers was deeply shaped, as Thomson's had been, by the demands and opportunities presented by Britain's cable empire.

III. New Phenomena, New Practices, New Ideas

The task of building and operating a global cable network presented British scientists and engineers with a series of novel phenomena and problems, and they responded by developing new knowledge and skills suited to the new circumstances. Cables required more careful manufacture and testing, more precise and sensitive measuring instruments, and much more careful attention to how signals propagated than did overhead landlines. Rather than simply flowing along like water in a pipe, pulses of electricity in cables were found to be affected by complex phenomena in the surrounding dielectric—the electromagnetic field. As resources were poured into cable telegraphy in the third quarter of the nineteenth century, British electrical physics was steered in directions quite different from those followed in Germany, the United States, and other countries with only overland telegraphs.

Cable telegraphy gave an important stimulus to the development of new electrical instruments. Early cables were troubled by the weakening of signals because of leakage through the insulation. Whitehouse had tried to get around the problem on the 1858 Atlantic cable by using such strong initial currents that even after considerable loss in transit enough would reach the far end to operate ordinary receiving instruments. The results, however, were disastrous; the high voltages and heavy currents from his giant induction coils burned out existing weak points in the insulation, hastening the failure of the cable (Hunt 1996). Thomson followed an opposite tack, devising a sensitive "mirror galvanometer" that could detect even very weak and attenuated currents. By cutting the weight of its moving parts to a minimum (for instance, by replacing its "pointer" with a tiny mirror reflecting a beam of light), he was able to give his receiving instrument unprecedented sensitivity and quickness of response (see figure 15.4). Strong and potentially damaging currents like Whitehouse's were no longer needed for effective signaling. Indeed, in 1866 Latimer Clark, a leading telegraph engineer, gave a vivid demonstration of the sensitivity of Thomson's mirror galvanometer: using only a tiny battery made from a silver thimble, he sent signals on a round trip through both Atlantic cables yet was still able to read quite clearly the message spelled out by the moving spot of light (Thompson 1910, 496-97).

Thomson's mirror galvanometer and later siphon recorder (which used a thin jet of ink to mark signals on a moving paper tape) were used on virtually all long cables after 1866. The sale of patent rights brought Thomson and his partners, Fleeming Jenkin and Cromwell Fleetwood Varley, tens of thousands of pounds—a small fraction of the sum the superior performance of their instruments added to the cable companies' profits (Smith and Wise 1989, 700-712). These and other electrical instruments, including electrometers, voltmeters, and ammeters derived from instruments first

Figure 15.4 Signals from the *Great Eastern* being received in the instrument room at Valentia, Ireland, during the unsuccessful 1865 attempt to lay a cable across the Atlantic. Large chemical batteries are shown on the floor, and several of William Thomson's electrical instruments, including two of his sensitive mirror galvanometers, can be seen on the table. Similar instruments soon came into use in British physics laboratories. (From the *Illustrated London News* 47 [5 August 1865]: 117.)

developed for the cable industry, soon came into use in British physics laboratories as well, often supplied by Thomson's manufacturing firm, James White of Glasgow. They played a major part in making precision electrical measurement the characteristic activity of late Victorian laboratory physics.

Cable telegraphy also gave an important impetus to the development of electrical units and standards, particularly of resistance to the flow of current. As Clark noted in 1871, "No very exact measurements are required to be made of overhead lines"; on submarine cables, by contrast, they were crucial, both for the testing and specification of materials and for the accurate location of faults (Clark and Sabine 1871, 7). In the wake of the failure of the first Atlantic and Red Sea cables, British engineers and scientists began to push for the adoption of a uniform standard to which all resistance measurements could be compared. Thomson and Jenkin had called for such a standard in their testimony to the joint committee, and on their initiative (joined by a similar proposal from Clark and his partner, Charles Bright), the British Association for the Advancement of Science established a Committee on Electrical Standards in 1861 (Hunt 1994*a*). Maxwell joined the committee the next year and, together with Jenkin, carried out the main experimental work of calibrating and standardizing the resistance of coils of wire. At Thomson's suggestion, the committee adopted an interconnected system of units for current, potential, and resistance based on Wilhelm Weber's "absolute" system but with the magnitudes adjusted to fit the needs of cable telegraphy.

The British Association Committee focused its main efforts on establishing a standard of resistance, the so-called BA ohm. After years of careful measurement and testing, Jenkin, as secretary of the committee, began to issue approved standards in 1865 (£2 10s., in a box), proudly noting that they went not only to physics laboratories and instrument makers, but also to the Atlantic Telegraph Company and to many British, foreign, and colonial telegraph firms and departments (Smith [1862-1912] 1913, 193-94). By the 1870s the BA ohm was in general use among both scientists and engineers, though some negotiation and correction remained before it was to secure official international adoption (Lagerstrom 1992).

It is significant that the unit adopted around the world was the *British* ohm. It replaced a plethora of more or less rough units previously used in different countries and outflanked the mercury unit devised and promoted by the German electrical industrialist Werner Siemens (Schaffer 1992). That the British unit won out reflected the importance of the British cable industry as the largest and most active market for accurate electrical measurement in the 1860s and 1870s.

In an 1883 lecture entitled "Electrical Units of Measurement" at the Institution of Civil Engineers, Thomson declared that from about 1860 until the early 1870s, British cable engineers had in fact been ahead of all but a few physicists in the practice of electrical measurement. "Resistance coils and ohms," he said, "and standard condensers and microfarads, had been for ten years familiar to the electricians of the submarine-cable factories and testing-stations, before anything that could be called electric measurement had come to be regularly practised in almost any of the scientific laboratories of the world" (Thomson 1891-94, 1:82-83). The development of techniques of electrical measurement, and especially of electrical units and standards, was largely driven in Britain in the 1860s and 1870s by the demands of the cable industry; scientific work in these areas took its tools and many of its concerns from this technological source.

Cable telegraphy shaped not only the tools of British electrical science, but its theories as well. The laws governing electromagnetic phenomena could, in the mid-nineteenth century, be formulated in two distinct but broadly equivalent ways. The action-at-a-distance approach posited direct attractions and repulsions between charged particles and currents; the field approach focused instead on stresses and tensions in the surrounding space and in its strong form regarded charges and currents as little more than epiphenomenal reflections of the state of the field (Buchwald 1985). Both approaches gave the same numerical results for virtually all cases, but each was best suited to the description of particular kinds of phenomena and the solution of particular kinds of problems. Action-at-a-distance theorists took as their paradigm phenomenon the attraction between two particles of opposite charge; field theorists took as theirs the distribution of lines of force around a magnet or a current-carrying coil of wire, as shown by iron filings. Through most of the nineteenth century, action-at-a-distance theories prevailed in Germany and France, while the field approach gained ground rapidly in Britain after the mid-1850s, culminating in the work of Maxwell and his successors between the 1860s and the 1880s.

Why did electrical theory follow such different paths in Britain and on the Continent? No doubt many factors played a part, but an important one was cable telegraphy: the British, with their global cable network, encountered phenomena and problems that their Continental counterparts simply did not face. And some of those phenomena, particularly those involving the propagation of signals, tended to focus their attention on the role of the electromagnetic field. To state the point baldly and a bit too simply, the British did field theory because they had submarine cables, and the Germans did not because they had none.

The peculiarities of signaling through cables were first noticed around 1850 on underground lines and shortly thereafter on the first submarine cables (the two kinds of lines being very similar electrically). Instead of passing along smoothly, crisply, and virtually instantaneously, as they did on overhead lines, pulses of current sent into underground and submarine cables seemed to be sucked up and then only gradually disgorged, so that the signals received at the far end were slightly delayed and substantially stretched out. Such "retardation" posed a serious obstacle to rapid signaling; if one tried to send at too high a rate, the successive pulses jumbled together into an indecipherable blur. After encountering retardation on underground lines between London and Manchester, Clark—thinking it might be of scientific interest and hoping to find a way to lessen its effects—brought the phenomenon to Michael Faraday's attention in October 1853. After carefully examining Clark's experiments on the underground lines and on 110 miles of insulated wire being readied for use in a submarine cable, Faraday hailed their scientific importance; they provided, he declared at the Royal Institution in January 1854, "remarkable illustrations of some fundamental principles of electricity, and strong confirmation of the truthfulness of the view which I put forth sixteen years ago" on the relationship between induction, conduction, and insulation (Faraday 1854, 197; Hunt 1991b).

Faraday had long maintained that the conduction of a current was always preceded by the induction of a state of strain in the surrounding dielectric and the consequent storage of a certain amount of charge. In ordinary cases, including overhead telegraph lines, only a small amount of charge could be stored and the whole process took place too quickly to be noticed.

A long cable, however, with only a thin layer of insulation separating the copper conductor from the surrounding damp earth or seawater, could store an enormous amount of charge even at low potentials; it thus took an appreciable time to set up the inductive state, and that accounted for the retardation and smoothing out of the signals. A property Faraday called the "specific inductive capacity" of the insulating dielectric also came into play, further slowing the process and increasing the amount of charge stored. The induction across the gutta-percha was intimately tied up with the conduction in the copper, Faraday said; neither could be treated in isolation.

When Faraday had first propounded these views in the late 1830s and early 1840s, the idea that the apparently inert dielectric played such an active role in electrical phenomena had been widely dismissed as a needless complication. But the advent of cable telegraphy in the 1850s provided not only new evidence in support of Faraday's ideas, but also a new market for them. Faraday's field approach found a realm of useful application where it seemed to fit quite well. The specific inductive capacity of insulating materials, for instance, was found to have a direct effect on the rate of signaling achievable on a cable. As Clark told the joint committee in 1860, "At the date of Faraday's interesting researches [on specific inductive capacity], it could little be foreseen that such an obscure phenomenon should be destined to become one day, as it has now, a consideration of high national importance, and one which has a direct and most important bearing on the commercial value of all submarine telegraphs" (Joint Committee 1861, 313). Faraday's ideas on the connection between induction and conduction were also of practical use in understanding the cause of retardation and in suggesting ways to lessen its effects—something that was to become a major preoccupation of British electrical scientists and engineers in the 1850s and 1860s.

The retardation problem on underground lines was readily solved: most such lines were simply pulled up in the 1850s and replaced with overhead lines. This option was, of course, not available for submarine cables, and as the British extended their cable network, they had to continue to deal with retardation. Indeed, Thomson soon found theoretically, and Jenkin confirmed experimentally, that the retardation increased with the square of a cable's length, making it an even worse problem on the long cables laid after 1858. Telegraph engineers in Germany, France, and the United States dealt mainly with overhead lines and so did not have to pay much attention to propagation phenomena or to field effects in the space around their wires. British cable engineers and electrical physicists could not ignore such considerations and so were driven to take propagation conditions and field phenomena very seriously.

Thomson played a leading role in this work, both theoretically and experimentally, and in his presidential address to the British Association at Edinburgh in 1871, he drew attention to the influence cable technology had exerted on science. The early investors in the Atlantic cable had been impelled, he said, by a sense of "the grandeur of their enterprise" and of the benefits (and, they hoped, profits) that would flow from its success. "But they little thought," he went on,

> that it was to be immediately, through their work, that the scientific world was to be instructed in a long-neglected and discredited fundamental electric discovery of Faraday's, or that, again, when the assistance of the British Association was invoked to supply their electricians with methods for absolute measurement (which they found necessary to secure the best economical return for their expenditure, and to obviate and detect those faults in their electric material which had led to disaster), they were laying the foundation for accurate electrical measurement in every scientific laboratory in the world, and initiating a train of investigation which now sends up branches into the loftiest regions and subtlest ether of natural philosophy. (Thomson 1891-94, 2:161-62)

Thomson's closing reference was to Maxwell's recent work on the electromagnetic theory of light, which held that light itself was waves in the electromagnetic field, comparable in many ways to the pulses of induction sent skittering along in the gutta-percha insulation of a submarine cable. Thomson himself viewed the phenomenon somewhat differently, and Maxwell never pursued the analogy between light and telegraphic pulses as far as he might have—perhaps out of deference to Thomson's telegraphic "turf."

While Maxwell never did much telegraphic work himself, his theory still had close ties to cable technology: the measurements he used to establish that the speed of his hypothetical electromagnetic waves would be equal to that of light were drawn largely from his work in determining the BA ohm, and his later followers, particularly the former cable engineer Oliver Heaviside, recast his theory into its canonical form as the set of four "Maxwell's equations" in the 1880s largely to make it a better tool with which to treat cable propagation problems. Indeed, most of Heaviside's papers on the subject were first published in the *Electrician*, a weekly trade journal owned by Pender's cable interests. In the late 1880s Heaviside even drew on the theory to devise a way to reduce retardation by "inductive loading" —though this method turned out to be better suited to the new technology of telephone transmission than to the cable telegraphy that had gotten Heaviside started on the problem (Hunt 1991*a*, 48, 122-28, 132-36; see also Yavetz 1995).

By the 1880s British electrical physics was firmly oriented toward field theory; Heaviside and the other British Maxwellians looked on the field as the seat of all electromagnetic actions and regarded charges and currents as little more than surface discontinuities in the distribution of stress and energy in the field. They took it as decisive confirmation of their views when the German physicist Heinrich Hertz detected electromagnetic waves (radio waves) in 1888 and showed them to have just the properties predicted by Maxwell's theory. Only then did field theory come to replace, or at least supplement, the particle-based electrical theories that had so long prevailed on the Continent. But there was a final irony here, for the field theory that had grown to maturity in Britain's cable empire, and drawn so much of its stimulus from problems of cable propagation, had now planted the seeds of a new technology—wireless telegraphy—that, within a few decades, would finally break Britain's monopoly over global communications (Headrick 1991).

IV. "An Imperial Science"

At the end of *Modern Views of Electricity,* his widely-read 1889 book on Maxwellian theory, Oliver Lodge gave a glowing account of Hertz's recent experiments; they had, he declared, "utterly and completely verified" Maxwell's theory and shown conclusively that light was indeed simply electromagnetic waves. Adopting a very up-to-date mode of expression, Lodge declared that "the whole domain of Optics is now annexed to Electricity, which has thus become an imperial science" (Lodge 1889, 307). But electricity was an imperial science in a more direct sense as well. From its ohms and instruments to Maxwellian field theory itself, British electrical science had, by 1889, come to be deeply shaped by the demands and opportunities presented by cable telegraphy, and so by those of the British Empire it served.

The same could be said, to varying degrees, of many other sciences in late Victorian Britain. Building and running a global commercial empire was an enormous task, and its effects spilled over into all areas of Victorian life. The sciences could not help but be affected, and from geology and geophysics to botany and evolutionary theory they were steered in distinctive directions by the imperial context within which they were pursued.

Bibliographical Note

The historical literature on the British Empire is enormous, and that on science and imperialism is large and growing. For background on the important notion of "informal empire," see Gallagher and Robinson (1953); on the role of commerce in the nineteenth-century empire, see Cain and Hopkins (1993). On the development of science in colonial countries, the starting point remains Basalla (1967); for criticism and modifications of his model, see MacLeod (1982) and Inkster (1985). Recent work on related issues is collected in Reingold and Rothenberg (1987) and Petitjean, Jami, and Moulin (1992). For specific studies of Australia, see MacLeod (1988) and Home and Kohlstedt (1991); of Canada, Zeller (1987); and of India, Kumar (1991) and Sangwan (1991). The best case study of how imperial concerns shaped scientific work in Britain itself is Stafford (1989), which combines a detailed examination of Roderick Murchison's work in geology and geography with a useful discussion of broader issues of science and empire.

The best single history of cable telegraphy is still that in Bright (1898). Headrick (1991) gives a clear and accessible account of the role cable telegraphy played in international politics and imperial affairs between 1851 and 1945, while Smith and Wise (1989), Schaffer (1992), and Hunt (1991*b* and 1994*a*) explore the role cable telegraphy played in the development of Victorian electrical physics.

References

Adas, Michael. 1989. *Machines as the Measure of Men: Science, Technology, and Ideologies of Western Dominance.* Ithaca, N.Y.: Cornell University Press.

Baglehole, K. C. 1969. *A Century of Service: A Brief History of Cable and Wireless, 1868-1968.* London: Cable and Wireless.

Barty-King, Hugh. 1979. *Girdle Round the Earth: The Story of Cable and Wireless and Its Predecessors.* London: Heinemann.

Basalla, George. 1967. "The Spread of Western Science." *Science* 156:611-21.

Bernstein, Henry T. 1960. *Steamboats on the Ganges: An Exploration in the History of India's Modernization through Science and Technology.* Bombay: Orient Longman.

Bright, Charles. 1898. *Submarine Telegraphs: Their History, Construction, and Working.* London: Crosby Lockwood.

The British Empire Series. 1899-1902. 5 vols. London: Kegan Paul, Trench, Trübner.

Brockway, Lucile. 1979. *Science and Colonial Expansion: The Role of the British Royal Botanic Gardens.* New York: Academic Press.

Browne, Janet. 1983. *The Secular Ark: Studies in the History of Biogeography.* New Haven, Conn.: Yale University Press.

———. 1992. "A Science of Empire: British Biogeography before Darwin." *Revue d'Histoire des Sciences* 45:453-75.

Buchwald, Jed Z. 1985. *From Maxwell to Microphysics: Aspects of Electromagnetic Theory in the Last Quarter of the Nineteenth Century.* Chicago: University of Chicago Press.

Cain, P. J., and A. G. Hopkins. 1993. *British Imperialism: Innovation and Expansion, 1688-1914.* London: Longman.

Cain, Robert J. 1971. "Telegraph Cables in the British Empire, 1850-1900." Ph.D. diss., Duke University.

Cawood, John. 1979. "The Magnetic Crusade: Science and Politics in Early Victorian Britain." *Isis* 70:493-518.

Cell, John W. 1970. *British Colonial Administration in the Mid-Nineteenth Century: The Policy-Making Process.* New Haven, Conn.: Yale University Press.

Clark, Latimer, and Robert Sabine. 1871. *Electrical Tables and Formulae for the Use of Telegraph Inspectors and Operators.* London: Spon.

Coates, Vary T., and Bernard Finn, eds. 1979. *A Retrospective Technology Assessment: Submarine Telegraphy—The Transatlantic Cable of 1866.* San Francisco: San Francisco Press.

Desmond, Adrian, and James Moore. 1992. *Darwin: The Life of a Tormented Evolutionist.* New York: Warner.
Dibner, Bern. 1959. *The Atlantic Cable.* Norwalk, Conn.: Burndy Library.
Drayton, Richard. 1993. "Imperial Science and a Scientific Empire: Kew Gardens and the Uses of Nature, 1772-1903." Ph.D. diss., Yale University.
Dugan, Kathleen G. 1987. "The Zoological Exploration of the Australian Region and Its Impact on Biological Theory." In *Scientific Colonialism: A Cross-Cultural Comparison,* edited by Nathan Reingold and Marc Rothenberg. Washington, D.C.: Smithsonian Institution Press, 79-100.
Edney, Matthew. 1993. "The Patronage of Science and the Creation of Imperial Space: The British Mapping of India, 1799-1843." *Cartographica* 30:61-67.
Electrician (London). 1896. "Sir John Pender" [obituary], 37 (10 July): 334-35.
Faraday, Michael. 1854. "On Electric Induction—Associated Cases of Current and Static Effects." *Philosophical Magazine* 7:197-208.
Gallagher, John, and Ronald Robinson. 1953. "The Imperialism of Free Trade, 1815-1914." *Economic History Review* 1:1-15.
Gooday, Graeme. 1990. "Precision Measurement and the Genesis of Physics Teaching Laboratories." *British Journal for the History of Science* 23:25-51.
Harris, Christina Phelps. 1969. "The Persian Gulf Submarine Telegraph of 1864." *Geographical Journal* 135:169-90.
Headrick, Daniel R. 1981. *The Tools of Empire: Technology and European Imperialism in the Nineteenth Century.* Oxford: Oxford University Press.
———. 1987. "Gutta-Percha: A Case of Resource Depletion and International Rivalry." *IEEE Technology and Society Magazine,* December, 12-18.
———. 1988. *The Tentacles of Progress: Technology Transfer in the Age of Imperialism, 1850-1940.* Oxford: Oxford University Press.
———. 1991. *The Invisible Weapon: Telecommunications and International Politics, 1851-1945.* Oxford: Oxford University Press.
Hempstead, C. A. 1989. "The Early Years of Oceanic Telegraphy: Technology, Science and Politics." *IEE Proceedings* 136A:297-305.
Home, R. W., and Sally Kohlstedt, eds. 1991. *International Science and National Scientific Identity: Australia between Britain and America.* Dordrecht: Kluwer.
Howse, Derek. 1980. *Greenwich Time and the Discovery of the Longitude.* Oxford: Oxford University Press.
Hunt, Bruce J. 1991a. *The Maxwellians.* Ithaca, N.Y.: Cornell University Press.
———. 1991b. "Michael Faraday, Cable Telegraphy, and the Rise of Field Theory." *History of Technology* 13:1-19.
———. 1994a. "The Ohm Is Where the Art Is: British Telegraph Engineers and the Development of Electrical Standards." *Osiris* 9:48-63.
———. 1994b. "Insulation for an Empire: Gutta-Percha and the Development of Electrical Measurement in Victorian Britain." Paper presented at Princeton University conference "Materials in Science," February 1994.
———. 1996. "Scientists, Engineers, and Wildman Whitehouse: Measurement and Credibility in Early Cable Telegraphy." *British Journal for the History of Science* 29:155-69.
Inglis, K. S. 1980. "The Imperial Connection: Telegraphic Communication between England and Australia, 1872-1902." In *Australia and Britain: Studies in a*

Changing Relationship, edited by A. F. Madden and W. H. Morris-Jones. London: Frank Cass, 21-38.

Inkster, Ian. 1985. "Scientific Enterprise and the Colonial Model: Observations on Australian Experience in Historical Context." *Social Studies of Science* 15:677-706.

Jaras, Thomas F. 1975. "Promoters and Public Servants: The Role of the British Government in the Expansion of Submarine Telegraphy (1860-1870)." Ph.D. diss., Georgetown University.

Joint Committee. 1861. *Report of the Joint Committee on the Construction of Submarine Telegraphs.* British Parliamentary Papers, 1860 [2744] LXII.

Kappey, Ferdinand E. 1902. "Cable and Colonial Telegraphs." In *The British Empire Series.* 5 vols. London: Kegan Paul, Trench, Trübner, 5:332-52.

Kennedy, P. M. 1971. "Imperial Cable Communications and Strategy, 1870-1914." *English Historical Review* 86:728-52.

Kumar, Deepak, ed. 1991. *Science and Empire: Essays in Indian Context, 1700-1947.* Delhi: Anamika Prakashan.

Lagerstrom, Larry Randles. 1992. "Constructing Uniformity: The Standardization of International Electromagnetic Measures, 1860-1912." Ph.D. diss., University of California, Berkeley.

Lodge, Oliver J. 1889. *Modern Views of Electricity.* London: Macmillan.

MacLeod, Roy. 1982. "On Visiting the 'Moving Metropolis': Reflections on the Architecture of Imperial Science." *Historical Records of Australian Science* 5:1-15. Reprinted in *Scientific Colonialism: A Cross-Cultural Comparison,* edited by Nathan Reingold and Marc Rothenberg. Washington, D.C.: Smithsonian Institution Press, 1987, 217-49.

———, ed. 1988. *The Commonwealth of Science: ANZAAS and the Scientific Enterprise in Australia, 1888-1988.* Melbourne: Oxford University Press.

Maxwell, James Clerk. 1873. *Treatise on Electricity and Magnetism.* 2 vols. Oxford: Clarendon Press.

Palladino, Paolo, and Michael Worboys. 1993. "Science and Imperialism." *Isis* 84:91-102.

Pang, Alex Soojung-Kim. 1991. "Spheres of Interest: Imperialism, Culture, and Practice in British Solar Eclipse Expeditions, 1860-1914." Ph.D. diss., University of Pennsylvania.

———. 1993. "The Social Event of the Season: Solar Eclipse Expeditions and Victorian Culture." *Isis* 84:252-77.

Peel, George. 1905. "The Nerves of Empire." In *The Empire and the Century: A Series of Essays on Imperial Problems and Possibilities.* London: Murray, 249-87.

Petitjean, Patrick, Catherine Jami, and Anne Marie Moulin, eds. 1992. *Science and Empires: Historical Studies about Scientific Development and European Expansion.* Dordrecht: Kluwer.

Pyenson, Lewis. 1978. "The Incomplete Transmission of a European Image: Physics at Greater Buenos Aires and Montreal, 1890-1920." *Proceedings of the American Philosophical Society* 122:91-114.

———. 1985. *Cultural Imperialism and Exact Sciences: German Expansion Overseas.* New York: Lang.

———. 1989a. *Empire of Reason: Exact Sciences in Indonesia, 1840-1940.* Leiden: Brill.

———. 1989b. "Pure Learning and Political Economy: Science and European Expansion in the Age of Imperialism." In *New Trends in the History of Science,* ed. R. P. W. Visser, H. J. M. Bos, L. C. Palm, and H. A. M. Snelders. Amsterdam: Rodopi, 209-78. A shorter version appears as "Science and Imperialism" in *Companion to the History of Modern Science,* edited by R. C. Olby, G. N. Cantor, J. R. R. Christie, and M. J. S. Hodge. London: Routledge, 1990, 920-33.

———. 1990a. "Habits of Mind: Geophysics at Shanghai and Algiers, 1920-1940." *Historical Studies in the Physical and Biological Sciences* 21:161-96.

———. 1990b. "Why Science May Serve Political Ends: Cultural Imperialism and the Mission to Civilize." *Berichte zur Wissenschaftsgeschichte* 13:69-81.

———. 1993a. *Civilizing Mission: Exact Sciences and French Overseas Expansion, 1830-1940.* Baltimore: Johns Hopkins University Press.

———. 1993b. "Cultural Imperialism and Exact Sciences Revisited." *Isis* 84:103-8.

Reingold, Nathan, and Marc Rothenberg, eds. 1987. *Scientific Colonialism: A Cross-Cultural Comparison.* Washington, D.C.: Smithsonian Institution Press.

Ritchie, G. S. 1967. *The Admiralty Chart: British Naval Hydrography in the Nineteenth Century.* London: Hollis and Carter.

Sangwan, Satpal. 1991. *Science, Technology and Colonisation: An Indian Experience, 1757-1857.* Delhi: Anamika Prakashan.

Schaffer, Simon. 1992. "Late Victorian Metrology and Its Instrumentation: 'A Manufactory of Ohms.' " In *Invisible Connections: Instruments, Institutions, and Science,* edited by Robert Bud and Susan E. Cozzens. Bellingham, Wash.: SPIE Optical Engineering Press, 23-56.

Secord, James A. 1982. "King of Siluria: Roderick Murchison and the Imperial Theme in Nineteenth Century British Geology." *Victorian Studies* 25:413-42.

Siemens, Werner. [1892] 1966. *Inventor and Entrepreneur: Recollections of Werner von Siemens.* 2d ed. New York: Kelley.

Smith, Crosbie, and M. Norton Wise. 1989. *Energy and Empire: A Biographical Study of Lord Kelvin.* Cambridge: Cambridge University Press.

Smith, F. E., ed. [1862-1912] 1913. *Reports of the Committee on Electrical Standards appointed by the British Association for the Advancement of Science.* Cambridge: Cambridge University Press.

Stafford, Robert. 1989. *Scientist of Empire: Sir Roderick Murchison, Scientific Exploration, and Victorian Imperialism.* Cambridge: Cambridge University Press.

Thompson, Silvanus P. 1910. *The Life of William Thomson, Baron Kelvin of Largs.* 2 vols. London: Macmillan.

Thomson, William. 1891-94. *Popular Lectures and Addresses.* 3 vols. London: Macmillan.

Wilshaw, Edward. 1939. "Lecture to the Country Conference of the Chartered Institute of Secretaries at Manchester, May, 1924." In *The Cable and Wireless Communications of the World: Some Lectures and Papers on the Subject, 1924-1939.* Cambridge: Heffer, for Cable and Wireless, 1-14.

Yavetz, Ido. 1995. *From Obscurity to Enigma: The Work of Oliver Heaviside, 1872-1889.* Basel: Birkhäuser.

Zeller, Suzanne. 1987. *Inventing Canada: Early Victorian Science and the Idea of a Transcontinental Nation.* Toronto: University of Toronto Press.

16

Zoological Nomenclature and the Empire of Victorian Science

HARRIET RITVO

It is by now an old story that the global expansion of European influence in the seventeenth, eighteenth, and nineteenth centuries, whether by means of conquest, politics, or trade, was both mirrored and expedited by the efforts of knowledge workers. Laboring in the wake of men of action, by their sides, and sometimes in their skins, cartographers, geologists, naturalists, and others of their ilk confirmed official and commercial claims by making sense of them. They reduced exotic territories to intellectual order by observing their distinctive natural features and by collecting, preserving, and cataloging their natural productions. As these scientific adventurers shared their basic subject matter with imperial administrators and entrepreneurs, so, mutatis mutandis, did they employ similar methods in their work, exemplifying such Enlightenment values as rational organization, universality of application, and consensus among experts.

This entanglement of the intellectual and the more explicitly pragmatic faces of the imperializing mission—between the conquest of the globe and the conquest of nature—was institutionalized in many forms. For example, the Admiralty cooperated with the Royal Society in sponsoring the voyage of the *Endeavour,* which sailed for the South Seas in 1768 not only to observe the transit of Venus, but also, according to Captain Cook's official instructions, because "the making Discoverys of Countries hitherto unknown and the Attaining a Knowledge of distant Parts which though formerly discover'd have yet been but imperfectly explored, will redound greatly to the Honour of this Nation as a Maritime Power, . . . and may tend greatly to the advancement of the Trade and Navigation thereof" (Beaglehole 1974, 148). Research in natural history continued to figure on the naval agenda well into the nineteenth century; as a result such Victorian

I would like to thank the Museum of Zoology, University of Cambridge, the Edinburgh University Library, Lord Derby, and the Liverpool Record Office for allowing me to quote from manuscript sources.

Figure 16.1 Zoological spoils enjoyed by citizens of the imperial capital: Bullock's Museum, Piccadilly, 1810. The image is from *Ackermann's Repository of Arts,* no. 18 (London, 1810).

naturalists as Charles Darwin, Joseph Hooker, and Thomas Henry Huxley were able to observe the flora and fauna of New Zealand and other remote imperial ports during formative voyages early in their careers. At home, magnates like the thirteenth Earl of Derby, whose interest in natural history was avocational rather than professional, felt free to request the services of imperial officials in the furtherance of their private pastimes. They were not invariably accommodated, but declinations carried no sense that the requests had been inappropriate. Thus, one official in Ceylon thanked the Earl profusely for stating "his wishes relative to the introduction of Singhalese animals into England," but regretted that "my engagements at Columbo place it out of my power to collect" (Templeton 1847). The botanical and zoological spoils of empire were routinely displayed in such institutions of science as the British Museum (Natural History), Kew Gardens, and the Zoological Society of London; reciprocal displays of the triumphs of science formed standard components of such institutions of political and commercial empire as expositions and world's fairs (figure 16.1).

Some of the most significant connections between the intellectual mastery celebrated by Enlightenment naturalists and their successors and the expanding global dominion enjoyed by their nations were not concretely embodied in trophies, but were abstracted in language. In the eighteenth century, the systematic classification of the natural world emerged as one of the quintessential achievements of modern science. Perhaps most overtly, it embodied the replacement of the unstructured ignorance of the past, as exemplified by the efforts not only of learned predecessors but also of unlearned contemporaries, by systematic expertise. At the same time that they bemoaned the undisciplined observations of run-of-the mill explorers and the unreliability of information gleaned from indigenous informants, self-consciously enlightened systematizers castigated the allegedly chaotic works of celebrated Renaissance naturalists like Ulisse Aldrovandi and Konrad Gesner as "insupportably tedious and disgusting" and "so incomplete as scarce to deserve mentioning" (Brookes 1763, I, x-xi). Further, the victory of system over ancient and entrenched error seemed to its advocates the harbinger of broader relevance and extended sway. Scientific classification was hailed as both symbol and agent of a larger intellectual triumph, one that could ultimately reverse the traditional relationship between humans and the natural world. Referring to the nomenclatural wing of his grand taxonomic project, Linnaeus, who was both the standard eponym and the standard synecdoche for Enlightenment systematics, had called himself the second Adam.

More concretely, classification represented European possession of exotic territories, as well as intellectual mastery of their natural history (Pratt 1992, 24-37). Citizens of a prosperous global power like Great Britain easily conflated such metaphorical dominion with more practical or literal modes of appropriation. Thus, naturalists in the mother country automatically claimed the right to classify colonial plants and animals—their subjects in more than one sense. Although this prerogative of possession was more often assumed than articulated, dispossession could provoke an explicit statement. As Thomas Pennant lamented in a preface of 1784, "This Work was designed as a sketch of the Zoology of *North America*. I thought I had a right to the attempt, at a time I had the honor of calling myself a fellow-subject with that respectable part of our former great empire; but when the fatal and humiliating hour arrived, which deprived *Britain* of power, strength, and glory, . . . I could no longer support my clame of entitling myself its humble Zoologist" (Pennant 1784, 3). Since nature knows no boundaries, however, he was able to salvage his research under the title *Arctic Zoology*.

By the Victorian period the classification of animals and plants had long relinquished its position at the cutting edge of science. Even within the enterprise of natural history, itself greatly diminished in glamour, classification had become part of the nuts and bolts. Naturalists continued to debate the merits of rival systems, but, for the most part, their struggles no longer constituted the stuff of headlines. Nevertheless, although relegated to comparative obscurity, systematic taxonomy and the nomenclature devised to express it retained their Enlightenment association with the reduction to order of exotic corners of the globe. After all, they represented a vast web of individual decisions; over a thousand new genera were being named each year by the last decades of the nineteenth century, with the species count correspondingly higher ("Our Book Shelf" 1882, 240; "*The Zoological Record*" 1883, 311). Each of these decisions was potentially contested (although, of course, few actually were) by rival nomenclators who felt they had staked a previous claim. Each of them offered an opportunity for comment on the mission of British natural history—and by association the British nation—in the world.

Nor did the social and political commentary embedded in naming practices refer only to remote colonial territories; botanical and zoological nomenclature also worked to create and reinforce a range of parallel patterns of human hierarchy. After all, the relations between metropolis and periphery, although distinctive, were hardly unique. On the contrary, like the imperial order itself, they mirrored or symbolized complex and often troubled relations within British society, and even within British science, which might also find expression in scientific terminology. In order to consolidate previous conquests as well as to extend its sway, the empire of science required constant and vigorous defense against challengers on many flanks, at home as well as abroad. Such challenges could, of course, simply reflect technical disagreements—or turf wars—among experts, as when Richard Owen, an accomplished skirmisher, appropriated the dinosaurs and the *Archaeopteryx* by definitively christening them (Rupke 1994, 73-74, 133-34). But resentment of hierarchy and struggle against the hegemonic claims implicit in the pursuit of universal knowledge surfaced even within the by then time-honored practices of scientific nomenclature. And although they were widely acknowledged to be as transparent and smoothly functioning as they were time honored, the classification and naming habits of nineteenth-century naturalists left plenty of room for argument. Despite the routine celebration of the Linnaean accomplishment, borrowed more or less verbatim by Victorian scientists from their Enlightenment predecessors, nomenclature remained inconsistent and multiple through the nineteenth century. The conquest of nature proved as protracted, troublesome, and ambiguous in its results as the political enterprises that it paralleled.

In many cases, indeed, the information that nomenclature offered about naturalists may have been more accurate than the information it offered about the animal kingdom. That is, the ability to deploy latinate terminology in accordance with a complex set of rules and conventions may have been a more reliable means of characterizing nineteenth-century naturalists than the names themselves were of species. Despite the long-sustained chorus of praise from naturalists, scientific nomenclature did not work as well as advertised. As with classification in general, there was many a slip between the abstractions of nomenclatural theory and the concretions of nomenclatural practice. Even reduced to a method of indexing, Linnaean nomenclature was far from simple. It required that a complex and ambiguous set of rules be applied to raw material that could be characterized in the same terms, by naturalists who themselves varied widely in culture, disciplinary background, and personal commitment.

The most obvious problem of scientific nomenclature was its failure to eliminate the profusion of synonyms that had long prevented naturalists from confidently identifying specimens and, therefore, definitively establishing the relationships between similar species. Linnaeus himself could falter in this regard. Thus, his admirer George Shaw apologized for the fact that, because of a "confusion and misapplication of synonyms" the variegated baboon (*"Simia mormon"*) had been "confounded with one really different, though very much resembling it," noting that "in so extensive a work as . . . the Systema Naturae" such lapses were "almost unavoidable" (Shaw 1792, 38). Subsequent nomenclators proved no better at avoiding them, and once again designations proliferated, albeit as latinate binomials and trinomials rather than vernacular tags. Thus, one early-nineteenth-century owner of Thomas Pennant's *History of Quadrupeds,* in which every printed entry began with a list, often quite long, of latinate and vernacular synonyma, was repeatedly obliged to pencil additional designations in the margins (Pennant 1793, author's copy). In 1830, the museum of the Zoological Society of London ("We show them first the nat'ral way, / And then our beasts we *stuff,*" as a jingle put it) was criticized for the "barbarous assemblage of names, as if to describe all the mongrels in creation" with which a wild goat was labeled (*Zoological Keepsake* 1830, ix, 152). And in 1896, looking back at more than a century of post-Linnaean primate nomenclature, the naturalist and traveler Henry O. Forbes abjured any attempt to "write a synonymy of the species of Monkeys"—that is, to collect all the scientific names by which each species had been denominated. Not only was the relevant information "scattered over many, often obscure, periodicals," but the consequence of assembling it might be "to introduce a great deal of confusion" (Forbes 1896-97, 1:vii).

Many such gaps between the promise of Linnaean nomenclature and the results it actually delivered reflected technical problems incident to the work of natural history in the eighteenth and nineteenth centuries. To ensure that an apparently new species had not previously been discovered, described, and named by someone else, it was necessary, then as now, to search the literature. Networks of transportation and communication were constantly improving, but not fast enough to guarantee that naturalists would be able to locate and examine all potentially relevant reports—buried as they might be in the proceedings of obscure societies, published in foreign languages. And even if a possible precursor emerged in the printed record, it was often difficult to establish whether the two animals in fact belonged to the same species. A definitive judgment would require physical comparison of the specimens, which might, however, be irrevocably separated by geography or by condition of preservation—even assuming that the two represented the same sex, age, and life phase. Few naturalists had the resources of time, money, and prestige to match the efforts made by Charles Darwin as he worked on his monograph about barnacles; for several years in the 1840s his house was filled with smelly specimens loaned by a global array of scientists, private collectors, and curators (Desmond and Moore 1991, chap. 22).

Even if these formidable difficulties could be overcome, nomenclature might proliferate as a result of legitimate differences of zoological theory or practice. Then as now, taxonomists were divided into "splitters," inclined to recognize new species on the basis of relatively slight differences, and "lumpers," who advocated a higher threshold for separation (Mayr 1982, 240-41). For example, although Linnaeus had established a single genus, *Equus,* to accommodate the horse and all its close relatives, many subsequent naturalists wished to acknowledge the evident subdivisions within this group by creating separate genera for asses (*Asinus*) and for zebras (*Hippotigris*). Legitimate as such disagreements might be on strictly scientific grounds, they played such havoc with the textual apparatus of natural history that the eminent zoologist William Henry Flower, a committed lumper, suggested that nomenclatural or linguistic considerations should take precedence over those of anatomy. He argued that "the great inconvenience of altering the limits of genera is that, as the name of the genus is part of the name by which . . . the animal is designated in scientific works in all languages, every change in the limits of a genus involves some of those endless changes in names which are among the greatest causes of embarrassment in the study of zoology in modern times" (Flower 1891, 71).

Although it seemed attractive on logistical grounds, such an accommodation ultimately proved impracticable. In the case of equines, parties to the nomenclatural controversy were in agreement about the relationships that the species bore to one another; they disagreed only about the proper representation of those relationships. But in cases where affinities themselves were moot or contested, shifts in nomenclature were unavoidable. Thus, when Thomas Hardwicke discovered a small new carnivore in the Himalayas, whose unusual dentition placed it outside any of the recognized genera, he felt special reluctance to devise a name that was not likely to be permanent (Hardwicke 1821, 163-64). And when the balance of zoological opinion shifted to agree with William Jardine that the musk ox, which had been conventionally denominated *Bos moschatus* to signify its taxonomic connection with ordinary cattle, "may perhaps have an appropriate station as an intermediate form connecting this division with the sheep," there was no avoiding the creation of an additional genus, the name of which, *Ovibos,* memorialized its newly liminal position (Jardine 1836, 189; Eisenberg 1981, 205).

Nineteenth-century naturalists were, of course, perfectly aware of these problems, which they regularly lamented at the same time, if not in the same breath, that they celebrated the transformation in their discipline wrought by the introduction of binomial nomenclature (Heppell 1981; Linsley and Usinger 1959; La Vergata 1987). Thus, in 1833 a contributor to the *Field Naturalist* "regretted that . . . the language of zoology and botany is necessarily changing. And what is the consequence? we are overburdened with synonymes, . . . [which] create as much, if not more, confusion than did the provincial terms, in the absence of scientific nomenclature" (Solitarius 1833b, 523). Nor was their reaction to this oddly intractable situation limited to lamentation. The 1840s saw the beginning of a sustained effort at reform on the part of establishment British zoologists. At the 1841 meeting of the British Association for the Advancement of Science, a committee with a small but distinguished membership was charged "to draw up a series of rules with a view to establishing a nomenclature of Zoology on an uniform and permanent basis" (Nomenclature Papers, scrapbook 1). The committee drafted a "Proposed Plan," which was circulated to a long list of British naturalists and a short list of foreigners; a "Proposed Report of the Committee on Zoological Nomenclature," modified in response to their comments, was printed in 1842, and the rules it suggested were adopted by the British Association (*Proposed Plan* 1841; *Proposed Report* 1842; Strickland 1842).

These labors received a good deal of private and public praise. For example, Richard Owen wrote to Hugh Strickland, who chaired the committee, that the proposed code was "ably drawn up," and the popular naturalist Leonard Jenyns characterized it as "extremely good, and likely to confer a great benefit on the science" (Owen 1841; Jenyns 1842). An American reviewer offered the British Association "hearty thanks" for having "undertaken to interpose the weight of its authority in arrest of the growing abuses in nomenclature" (Gould 1843, 1). More tangible results were, however, thinner on the ground, and in 1865, after several disappointing decades, the British Association was moved to readopt the proposal, only slightly modified by the few surviving members of the original committee. Again, the positive impact on zoological practice was difficult to discern. In 1874, Alfred Russel Wallace observed that although "zoologists and botanists universally adopt what is termed the binomial system of nomenclature invented by Linnaeus," nevertheless, "one of the first requisites of a good system of nomenclature—that the same object shall always be known by the same name—has been lost" (Wallace 1874, 258-59). In consequence, as Flower asserted in his 1878 presidential address to the British Association's Department of Zoology and Botany, "All beginners are puzzled and often repelled by the confused state of zoological nomenclature" (Flower [1898] 1972, 167).

From the straightforward perspective of efficiency and utility—the values that were ordinarily invoked in discussions of nomenclature—the manifest reluctance of the Victorian zoological community to accept what was generally acknowledged to be a very sensible set of suggestions might be difficult to explain. After all, the confusing status quo required naturalists to waste valuable time on what were essentially clerical labors. As Alfred Newton, the first professor of zoology at Cambridge, noted in 1879, when he urged his colleagues to adopt the still-orphaned British Association rules, "Nomenclature is so trifling an adjunct to zoology that no true student of the science can fail to grudge the time which he is . . . compelled to bestow upon it, or ought to be ungrateful to those who have expended their toil in preparing some rules for his guidance through the intricate maze of synonyms that . . . enfolds almost every object with which he has to deal" (Newton 1879, 419-20).

Similarly, given the apparently mundane and pragmatic nature of the issues surrounding scientific nomenclature, it could be difficult to account for the tone of anxiety and passion that frequently crept into learned discussions of it. Thus, an early catalog of the Regent's Park zoo began with the following disclaimer: "N. B. It is to be observed, that the Council of the [Zoological] Society do not hold themselves responsible for the Nomenclature used . . . in this publication" (*List of the Animals* 1833, 2). The initial proposal circulated by the British Association committee in 1841 characterized nomenclatural irregularity as an "evil," the result of "neglect and corruption"; it referred to Buffon's practice of labeling new species only in the vernacular and not with latinate binomials as "vicious" (*Proposed Plan* 1841, 2, 3, 9). In his response to the draft proposal, William John Broderip, a successful lawyer as well as a respected amateur naturalist, implicitly acknowledged the political volatility of the topic when he warned against using words like "Parliament" or "legislation," which might give "the appearance of dictation" and thus "excite ridicule" (Broderip 1842). Such language suggested that more was at stake in establishing uniform and consistent zoological nomenclature than the elaboration of a merely technical order.

Indeed, the elite naturalists who drafted the original British Association proposal began by dismissing technical sources of confusion—"those diversities which arise from the various methods of classification adopted by different authors, and which are unavoidable in the present state of our knowledge"—as of secondary concern (*Proposed Plan* 1841, 1). Instead, they focused their attention and their ire on discrepancies that arose from extrascientific causes. Challenges to the scientific authority of elite British naturalists were conflated with challenges mounted on other grounds, more clearly rooted in human nature and therefore more vulnerable to policing. Nomenclature became a medium upon which a variety of frailties and lapses and antagonisms could be inscribed, as well, inevitably, as the representative or symbol of those alternative behaviors and commitments. An energetically enforced standard of nomenclatural propriety would embody and reinforce hierarchical order both inside the zoological community and in the larger society, both national and imperial, to which its members also belonged; at the same time it would identify inappropriate or troublesome colleagues. Consequently, the errors and eccentricities in nomenclature that attracted the most severe and protracted criticism from the British Association committee were those that most clearly associated their perpetrators with groups considered obnoxious for political or cultural or social reasons.

Some of the most provocative challenges were mounted from abroad. In an era of intense international military and political rivalry, scientific claims could be conflated with those of the polity in general; the competitive colonization of soldiers and diplomats had its analogue in the nomenclatural activities of zoologists. Naming constituted a strong, if metaphoric, claim to possession, not only of the newly christened species, but by implication of its native territory; conversely, territorial claims were easier to question in scientific journals than on the battlefield. Thus, Thomas Stamford Raffles, founder of both the city of Singapore and the Zoological Society of London, once found himself in the unhappy position of having to dismiss "two French gentleman who [had] appeared qualified" to help him with the preservation and description of the many specimens he had collected during his colonial service in Southeast Asia lest, because of what he called their "private and national views," "all the result of all my endeavours . . . be carried to a foreign country." What he feared was the integration of his specimens into a Gallicized nomenclature—which he characterized as "speculative and deficient in the kind of information required"—and their consequent loss, not only to himself but to his country (Raffles 1820-21, 239-40). The political edge of the rivalry between British and French science similarly intensified Charles Darwin's "disgust" when, about to commence fossil hunting in South America at his own expense, he discovered that a French collector had already been working his target area for six months, sponsored by the French government (Desmond and Moore 1991, 128; Browne 1995, 267-68).

Ironically, the Linnaean terminology originally designed to serve the supranational scientific community, and for that reason, among others, couched in latinate forms that recalled the universal language of medieval and Renaissance learning, had come to replicate rather than to supersede the separation of rival national cultures. Later in the nineteenth century, such separatism in the clothing of universality drove Edwin Ray Lankester, whose distinguished career featured professorships at Oxford and London, as well the directorship of the British Museum (Natural History), to suggest that British scientists abandon their internationalist aspirations. Instead, he rather quixotically recommended that they content themselves with imposing uniform terminology at home, by the introduction of "a series of terms distinctly English in their etymology, which would be accepted as authoritative and used throughout the country" ("Notes" 1874, 453).

The prominence of political concerns, as well as the fact that, like Raffles, many naturalists also participated in the imperial enterprise as government administrators, military officers, or explorers, meant that the first nomenclatural lapses singled out for criticism by the British Association committee were those committed by foreign naturalists. The published report of 1842 lamented that "the commonwealth of science is becoming daily divided into independent states. . . . If an English zoologist . . . visits the museums and converses with the professors of France, he finds that their *scientific* language is no less foreign . . . than their *vernacular*" (Strickland 1842, 106-7). In making this complaint, the committee followed a trail blazed by earlier British critics, who had identified flaws in the Gallic national character that might account for this willful and uncooperative divergence: the French "rage for innovation" and preference for "forever subdividing where the great aim should be to combine," exacerbated by the inclination of French naturalists to display "their grandeur, or at least their vanity," in coining species names (Lawrence 1807, xvi; Bicheno 1827, 494; Smith 1819, 285).

In any case, the committee's formulation carefully staked out the intellectual high ground, emphasizing, even more than had Raffles, the scientific rather than the territorial dimension of the problem. Nevertheless, it was significant that France, Britain's most serious geopolitical competitor, figured as the primary locus of the linguistic "despair" experienced by traveling British naturalists, with Germany and Russia mentioned only as afterthoughts; perhaps even Buffon's practice would not have seemed so vicious if he had abjured latinate nomenclature for that of some other vernacular. Although nomenclatural disputes between British and American naturalists did not involve a foreign language, they did involve politics. In a continuing effort to shake off a yoke that persisted in cultural matters long after their nation had become independent, American naturalists resisted outside efforts to name and describe, and so claim, species indigenous to their country (Winsor 1991, 89). And foreign policy considerations also shaped the private responses of Strickland's colleagues to the nomenclatural proposals of his committee; for example, the explorer John Richardson prophetically cautioned, "The main difficulty will be in gaining the hearty assent of the European naturalists to the proposed plan" (Richardson 1842). These forebodings were confirmed as the scientific establishments of other nations responded with reciprocal patriotism. The immediate effect of the British Association initiative was to exacerbate the international Babel of nomenclatures, by inspiring naturalists throughout Europe and North America to promulgate their own competing plans (Heppell 1981, 136-37).

But the foreign menace was not the only one with which the British Association committee felt it had to contend. In the view of the establishment naturalists who composed it, the terminological practices of some fellow citizens presented similarly grave challenges to the order and hierarchy that binomial nomenclature had been designed to embody. One of the most disturbing of these, castigated as an "evil" in the committee's initial report, was "the practice of gratifying individual vanity by attempting on the most frivolous pretexts to cancel the terms established by original discoverers, and to substitute new and unauthorized nomenclature in their place" (*Proposed Plan* 1841, 2). The most frequent means of accomplishing this end was by excessive splitting—by carving up previously recognized species, so as to produce new entities that would require new names, or by too carelessly assuming that a new specimen represented a new species rather than combing the literature for likely previous descriptions. In 1878, *Nature* castigated such slack and ignorant colleagues as caring "very little to know what others are doing . . . , and when a few years after, some industrious German or Scandinavian naturalist quietly relegates the name on which they had prided themselves to the limbo of synonyms (perhaps with a mark of admiration which does not mean praise), they accept the rebuff and console themselves with the reflection that 'a fellow can't be expected to know everything'" ("*The Zoological Record*" 1878, 485).

Individual vanity was an issue because the scientific name of an organism was normally cited with the name of the namer appended in parentheses. Therefore, as one critic irascibly put it, "nomenclators who take a *pride* in the manufacturing of names . . . fight among themselves as to which . . . shall have the *honour* of appending his name to them, hoping that by thrusting a Jack Scroggins-of-a-name into notice, it will be handed down to posterity" (Solitarius 1833*a*, 461). Scientific etiquette forbade describers of new species to commemorate themselves in the exercise of their naming prerogative, although this attempt to enforce refined self-effacement was routinely compromised by the tendency of discoverers to name species after their employers or sponsors. For example, Lionel Walter Rothschild, an aggressive collector who sent his proxies to gather the zoological spoils of the British Empire in the late nineteenth and early twentieth centuries, was commemorated in the names of fifty-eight species or subspecies of birds, eighteen of mammals, three of fish, two of reptiles and amphibians, one hundred fifty-three of insects, three of arachnids, one of millipedes, and one of nematodes (Rothschild 1983, 364). (See figure 16.2.) But if ordinary naturalists could not aspire to these heights of immortality, the parenthetical convention provided a strong incitement to those who inappropriately valued glory over truth.

In the view of the committee and its correspondents, these rogue naturalists were not a random sample of the zoological community, distinguished only by intellectual irresponsibility and moral weakness. The terms in which this weakness was denounced by elite naturalists often evoked the getting and spending of the marketplace, thus implying social condescension as well as scientific disapprobation. In his correspondence with Strick-

Figure 16.2 One of the many creatures named to honor the powerful. The *Sciurus rafflesii* offered a welcome "opportunity of paying another tribute to" Sir Stamford Raffles, "a Naturalist to whom science already owes so many obligations." The image and quotation are from Vigors and Horsfield 1828, 112-13.

land, Darwin disparaged those naturalists in need of suppression as mere "species-mongers" who wanted to "have their vanity tickled" and were therefore responsible for a "*vast* amount of bad work"; more harshly, the pseudonymous Solitarius criticized "those whose paltry conceited minds are gratified at the idea of having obtained a little celebrity for themselves, by the shortest and easiest method" (Darwin 1849; Solitarius 1833*b*, 522). As Flower reflected, looking back on three decades in which the "excellent code . . . drawn up in 1842" had languished, its provisions had "undoubtedly been a guide to . . . conscientious workers," but "unfortunately no means exist of enforcing them upon those of a different class" (Flower 1972, 168).

The suggestion of vulgarity latent in these characterizations reemerged in the rules proposed as specific correctives for such egotistical excesses, rules that perhaps surprisingly stressed the importance of philological and grammatical correctness in the creation of zoological nomenclature. Since correctness was understood in terms of a command of classical languages normally acquired only through elite education (if then), it also served to distinguish zoologists with such backgrounds from those whose expertise had been acquired in less genteel academies (Rudwick 1986; Desmond 1990; Delaporte 1987). Thus, it was asserted, "the *best* zoological names are . . . derived from . . . Latin or Greek," and namers were warned against designations that revealed a misunderstanding or half-understanding of classical texts, such as referring to an ancient name for a different animal, or a mythological figure that had no relation to the character of the animal being named (*Proposed Plan* 1841, 11-14). Indeed, as Charles Lyell pointed out in *Principles of Geology*, even terms coined or adapted for the general scientific vocabulary could also betray lack of refinement, if the inclusion of "foreign diphthongs, barbarous terminations, and Latin plurals" signaled insensitivity to the subtleties of English as well as of other tongues (Lyell [1830-33] 1990-91, 3:53). That such strictures reflected a desire to establish a binary taxonomy of Victorian zoologists, rather than any more generalized respect for traditional scholarship, was suggested by the British Association committee's warning to nomenclators inclined to delve too deeply into Aristotle, Pliny, and other figures from the prehistory of zoology. Such propensities might result in "our zoological studies . . . [being] frittered away amid the refinements of classical learning" (*Proposed Plan* 1841, 5).

Whole categories of names were repeatedly banned on the grounds that they revealed lack of taste: for example, nonsense names, names made up of fragments of two different words, hybrid names that combined elements of two languages (the inability or disinclination to distinguish between Greek and Latin roots was often also a telltale indication of extracanonical education), and names that instead of commemorating "persons of eminence as scientific zoologists," celebrated "persons of no scientific reputation, as curiosity dealers . . . , Peruvian priestesses . . . , or Hottentots" (*Proposed Plan* 1841, 11-14). A paleontologist primarily interested in the remains of extinct sea monsters characterized the latter transgression in similar terms as "injurious to the dignity of Science, and the Taste of the Age in which we live" (Hawkins 1840, 9). And the entomologist Francis Pascoe, who would have read the British Association committee's original report as a young naturalist, vociferously restated its complaint in old age, protesting in the introduction to a manual of zoological taxonomy "against the barbarous and other objectionable names (sometimes at variance with good taste and even with decency) that have been introduced into science—such, for example, as Battyghur, Butzkopf, Agamachtschich, Know-nothing, Stuff, Jehovah, Cherubim, or such idiotic names, or rather sounds, as Toi-toi, Sing-sing, Gui, Yama-mai." Having thus exemplified lapses in refinement, he left his readers to imagine still more heinous offenses, concluding that "indecent names need not be further alluded to" (Pascoe 1880, vi). Such critiques conflated fractious British naturalists with the objectionable foreigners, whether from Europe or from farther afield, whose names they insinuated into official scientific discourse.

Nor did nonelite naturalists pose the only domestic challenge to the expert authority represented by a uniform, authoritative, and exclusive code of zoological nomenclature. Interest in the animal kingdom was not restricted to conventionally scientific circles, and the larger audience for zoological curiosities had little interest in the terminological hairsplitting of nomenclatural purists or, for that matter, in any latinate binomials. From this perspective, differences within institutionalized science were less important than the difference between that self-defined elite, which claimed to control a certain sphere of knowledge, and others interested in the same material, who resented both such claims and the confident, even overbearing tone in which they were often made. At home, as well as abroad, the imperium of British science over the natural world was subject to contestation.

Such resentment inevitably crystallized around the distinctive nomenclature used by zoologists to consolidate and symbolize their intellectual dominion. Resistance of this kind had shadowed Linnaean nomenclature since its initial acclamation. Thus, in 1751, as part of an extended satiric attack on the scientific pretensions of the Royal Society, John Hill proposed to commemorate his "Esteem for the very eminent [Royal] Societarian Mr. *Henry Baker*" by giving his name, in the form of "Bakera," to an "Animal which is no Animal," on the grounds that "it is as much an Animal as he is a Philosopher" (Hill 1751, 79-80). A century later, John Ruskin advocated the replacement of scientific terms for birds with "the simplest and most descriptive" English nomenclature, as part of an explicit attempt to reclaim natural history for nonspecialists. He made some attempt to argue in the terms of science, claiming that the need to incorporate new and yet-to-be-discovered information made it "one of the most absurd weaknesses of modern naturalists to imagine that *any* presently invented nomenclature can stand, even were it adopted by the consent of nations, instead of the conceit of individuals." But his strongest objection was more fundamental: "A time must come when English fathers and mothers will wish their children to learn English again, and to speak it for all scholarly purposes; and, if they use, instead, Greek or Latin, to use them only that they may be understood by Greeks or Latins; and not that they may mystify the illiterate many of their own land" (Ruskin [1881] 1906, 21-22).

In addition to providing a target for criticism, scientific nomenclature could itself be subverted to express disapproval and even ridicule. Indeed, even people whose sympathies and talents were usually at the disposal of science might be tempted to poke gentle fun at the pretentiousness implicit in an enterprise that relentlessly reclothed familiar objects in impenetrable new names. For example, the humorist Edward Lear was a distinguished zoological illustrator, but he also produced a "Nonsense Botany," which featured such plants as the *Encoopia Chickabiddia*, the *Tickia Orologica*, the *Washtubbia Circularis*, the *Plumbunnia Nutritiosa*, and the *Manypeeplia Upsidownia* (Lear 1934, 197-219). (See figure 16.3.) With greater hostility, *Punch* occasionally made pseudoscientific nomenclature the metaphor for the love of obfuscation shared by scientists and most other scholarly specialists, referring to the "Clamour-making Cat (*Felis catterwaulans*), which is well known to all Londoners," the "*Felis omnivora*, or Common Lodging-House-Keeper's Cat," the "Learned British Pig (*Porcus Sapiens Britannicus*)," and the "Rum Shrub (*Shrubbus Curiosus*)"("Our Cat Show" 1876, 101; "Notes by a Cockney Naturalist" 1871, 194).

A more common, and perhaps also ultimately a stronger riposte to the appropriative claims implicit in scientific taxonomy was to ignore them completely—to indulge without reservation in what the British Association's Nomenclature Committee referred to as "the vicious taste on the part of the public" for "vernacular appellations" (*Proposed Plan* 1841, 3). Or, as

Figure 16.3 Edward Lear's *Barkia Howlaloudia*, one of the new species allegedly collected by "Professor Bosh . . . in the Valley of Virrikwier" (Lear 1901, 5 [quote], 29 [drawing]).

a writer who signed himself "Bob" more mildly put it in the *Oriental Sporting Magazine*, "I wish that people who write on sporting subjects in the nineteenth century would call the animals . . . by their proper names" (Bob 1833, 411). On the contrary, however, the public insisted on calling a bison a bison, and not a *Bos americanus*. Indeed, it was also disposed to follow its own counsel, rather than that of the scientific establishment, even with regard to vernacular appellations, thus sometimes calling a bison a buffalo or even a bonassus.

And these alternative nomenclatural predilections, however misguided or depraved, could influence the printed record, and even the practice of experts. When financial matters were at stake, zoologists stuck to their terminological guns at their peril. Members of the general public—and even some amateurs of natural history—were apt to regard latinate binomials as "a Torrent of hard Words . . . [a] Parade of indeterminate Sounds," and consequently to avoid linguistic exposure that might prove painful (Roberts 1785-86, 96). Several mid-Victorian zoos failed because their directors refused to pitch the exhibits to the general public rather than to naturalists (Ritvo 1987, 214). In the preface to the 1846 edition of his *Observations in Natural History*, Leonard Jenyns more accommodatingly assured readers that he had "transferred to the notes all the scientific names of . . . animals" (Jenyns 1846, 1:x). With similar shrewdness the author of a midcentury school text on zoology assured his readers that "the repetition of scientific names . . . is, comparatively, of little importance. The great object should be to bring natural-history knowledge home to the personal experience of the pupil" (Patterson 1857). Natural history museums, no matter how learned their primary audience or how august their stature, also attempted to counteract the impression of exclusiveness and inaccessibility produced by latinate labels, by offering a more palatable and familiar alternative. Thus, in the comparative anatomy collection at the University of Cambridge, "for the convenience of casual visitors . . . the trivial as well as the scientific names have been appended to all the large skeletons, and in many cases to the small ones also"; and "all the specimens" in the Zoological Department of the British Museum (Natural History) were "marked with their popular and their systematic names" ("Report to the Vice-Chancellor" 1865; "Returns Relating to the National Collections" 1857, 1).

So, at least as it has been embodied in scientific nomenclature, the old story of global intellectual hegemony requires some revision. In some respects, the claims made by nomenclators were more grandiose—their colonialism was only the leading edge of a much more comprehensive project of domination, encompassing both foreign colleagues and less elite fellow citizens. But these claims were not reliably honored. If in theory classification and nomenclature represented the empire of science over nature, of Britain over its colonies, and of elite British scientists over the possessors of competing expertise, whether at home or abroad, in practice they also exposed gaps in the hierarchical fabric. The structure of nomenclature, although imposing, was not perfect even on its own terms, and it seemed less perfect still to those who subscribed to different standards. And so its very ambitiousness and triumphalism produced opportunities, eagerly grasped, for criticism, dereliction, resistance, and even, occasionally, rebellion.

Bibliographical Note

The notion that eighteenth- and nineteenth-century natural history was imperialistic in both a literal and metaphorical sense—that it at once reflected contemporary political, military, and economic endeavors and also exerted an intellectual dominion of its own—has been explored by scholars in such diverse fields that it can be difficult to view them as participating in a single discussion, or even argument. In *The Order of Things* (1971), Michel Foucault identified scientific classification as one of the main exemplars of the organization of knowledge in the eighteenth century, expressing control through all-encompassing structure. Although Foucault argues that this *epistème* was superseded in the nineteenth century, studies of individual disciplines and subdisciplines have demonstrated the persistent connection between imperial claims in politics and in science (for example, Browne 1983, 1992; Stocking 1987). Sometimes this connection had practical consequences for imperial policies, as demonstrated by MacKenzie (1988), Adas (1989), and Grove (1995).

As with political imperialism, the center of gravity of the imperium of natural history was not in the colonies but in the metropolis. Specimens and trophies poured into Britain from all over the world, where the naturalists who appropriated them by naming practiced only one of several possible modes of possession. The attempt to expand the flora and fauna of a colonizing power by naturalizing—literally taking over—plants and animals belonging to a colony had constituted a pragmatic dimension even of Linnaeus's grandiose systematic enterprise (see Koerner 1994), and acclimatization developed more expansively with the more reliable and convenient transportation of the Victorian period (see Lever 1992; Anderson 1992). In addition, both the productions of colonial nature and the order that had been imposed on them served as concrete demonstrations of imperial power within a variety of institutions devoted to their organized display, including museums, zoos, botanical gardens, and international expositions. Such institutions existed in provincial and colonial administrative centers as well as in the imperial capital. Discussions are included in Brockway (1979), Ritvo (1987), Sheets-Pyenson (1988), and Greenhalgh (1988).

References

Adas, Michael. 1989. *Machines as the Measure of Men: Science, Technology and Ideologies of Western Dominance*. Ithaca, N.Y.: Cornell University Press.

Anderson, Warwick. 1992. "Climates of Opinion: Acclimatization in Nineteenth-Century France and England." *Victorian Studies* 35:135–57.

Beaglehole, J. C. 1974. *The Life of Captain James Cook*. Stanford, Calif.: Stanford University Press.

Bicheno, J. E. 1827. "On Systems and Methods in Natural History." *Linnean Society of London Transactions* 15:479–96.

Bob. 1833. "Animals by their Proper Names." *Oriental Sporting Magazine*, no. 20:411.

Brockway, Lucille. 1979. *Science and Colonial Expansion: The Role of the British Royal Botanical Gardens*. New York: Academic Press.

Broderip, W. J. 1842. Letter of 5 May 1842 to Hugh Strickland. Scrapbook 2. Nomenclature Papers. Hugh E. Strickland Collection. University of Cambridge Museum of Zoology.

Brookes, Richard. 1763. *New and Accurate System of Natural History*. 6 vols. London: J. Newbery.

Browne, Janet. 1983. *The Secular Ark: Studies in the History of Biogeography*. New Haven, Conn.: Yale University Press.

———. 1992. "A Science of Empire: British Biogeography before Darwin." *Review of the History of Science* 45:453–75.

———. 1995. *Charles Darwin: Voyaging*. New York: Alfred A. Knopf.

Darwin, Charles. 1849. Letter of 29 January 1849 to Hugh Strickland. Scrapbook 1. Nomenclature Papers. Hugh E. Strickland Collection. University of Cambridge Museum of Zoology.

Delaporte, Yves. 1987. "*Sublaevigatus* ou *Subloevigatus?* Les usages sociaux de la nomenclature chez les entomologistes." In *Des Animaux et des Hommes,* edited by Jacques Hainard and Roland Kaehr. Neuchâtel, Switzerland: Musée d'Ethnographie, 187-212.
Desmond, Adrian. 1990. *The Politics of Evolution: Morphology, Medicine and Reform in Radical London.* Chicago: University of Chicago Press.
Desmond, Adrian, and James Moore. 1991. *Darwin.* London: Michael Joseph.
Eisenberg, John F. 1981. *The Mammalian Radiations: An Analysis of Trends in Evolution, Adaptation, and Behavior.* Chicago: University of Chicago Press.
Flower, William Henry. 1891. *The Horse: A Study in Natural History.* London: Kegan, Paul, Trench, Trübner.
———. [1898] 1972. *Essays on Museums and Other Subjects Concerned with Natural History.* New York: Books for Libraries Press.
Forbes, Henry O. 1896-97. *A Hand-book to the Primates.* 2 vols. London: Edward Lloyd.
Foucault, Michel. 1971. *The Order of Things: An Archaeology of the Human Sciences.* New York: Random House.
Gould, Augustus A. 1843. "Notice of Some Works, Recently Published, on the Nomenclature of Zoology." *American Journal of Science* 45:1-12.
Greenhalgh, Paul. 1988. *Ephemeral Vistas: The Expositions Universelles, Great Exhibitions and World's Fairs, 1851-1939.* Manchester: Manchester University Press.
Grove, Richard H. 1995. *Green Imperialism: Colonial Expansion, Tropical Island Edens and the Origins of Environmentalism, 1600-1860.* Cambridge: Cambridge University Press.
Hardwicke, Thomas. 1821. "Description of a New Genus of the Class Mammalia, from the Himalaya Chain . . ." *Linnean Society of London Transactions* 15:161-65.
Hawkins, Thomas. 1840. *The Book of the Great Sea Dragons, Ichthyosauri and Plesiosauri.* London: William Pickering.
Heppell, David. 1981. "The Evolution of the Code of Zoological Nomenclature." In *History in the Service of Systematics,* edited by Alwyne Wheeler and James H. Price. London: Society for the Bibliography of Natural History, 134-41.
Hill, John. 1751. *A Review of the Works of the Royal Society of London.* London: R. Griffiths.
Jardine, William. 1836. *The Natural History of the Ruminating Animals, Part II, Containing Goats, Sheep, Wild and Domestic Cattle.* Edinburgh: W. H. Lizars.
———. 1863. "Proposed Reform of Zoological Nomenclature." *Edinburgh New Philosophical Journal,* n.s., 18:260-83.
Jenyns, Leonard. 1842. Letter of 16 March 1842 to Hugh Strickland. Scrapbook 1. Nomenclature Papers. Hugh E. Strickland Collection. University of Cambridge Museum of Zoology.
———. 1846. *Observations in Natural History: With an Introduction on Habits of Observing as Connected with the Study of that Science.* 2 vols. London: John Van Voorst.
Koerner, Lisbeth. 1994. "Linnaeus' Floral Transplants." *Representations,* no. 47:144-69.
La Vergata, Antonello. 1987. "Au Nom de l'Espèce: Classification et Nomenclature

au XIXe Siècle." In *Histoire du Concept de l'Espèce dans les Sciences de la Vie,* edited by Scott Atran et al. Paris: Fondation Singer-Polignac, 193-225.

Lawrence, William. 1807. "Introduction." In *A Short System of Comparative Anatomy,* by J. F. Blumenbach. London: Longman, Hurst, Rees, and Orme.

Edward Lear. 1901. *Nonsense Botany and Nonsense Alphabets.* 7th ed. London: Frederick Warne.

———. 1934. *The Complete Nonsense Book.* Edited by Lady Strachey. New York: Dodd, Mead.

Lever, Christopher. 1992. *They Dined on Eland: The Story of the Acclimatisation Societies.* London: Quiller Press.

Linsley, E. G., and R. L. Usinger. 1959. "Linnaeus and the Development of the International Code of Zoological Nomenclature." *Systematic Zoology* 8:39-47.

List of the Animals in the Gardens of the Zoological Society, with Notices Respecting Them. 1833. London: Richard Taylor.

Lyell, Charles. [1830-33] 1990-91. *Principles of Geology.* 3 vols. Chicago: University of Chicago Press.

MacKenzie, John M. 1988. *The Empire of Nature: Hunting, Conservation and British Imperialism.* Manchester: Manchester University Press.

Mayr, Ernst. 1982. *The Growth of Biological Thought: Diversity, Evolution, and Inheritance.* Cambridge, Mass.: Harvard University Press.

Newton, Alfred. 1879. "More Moot Points in Ornithological Nomenclature." *Annals and Magazine of Natural History,* 5th ser., 4:419-22.

Nomenclature Papers. Hugh E. Strickland Collection. University of Cambridge Museum of Zoology.

"Notes." 1874. *Nature* 10:453.

"Notes by a Cockney Naturalist." 1871. *Punch* 61:104.

"Our Book Shelf." 1882. *Nature* 25:240.

"Our Cat Show." 1876. *Punch* 71:101.

Owen, Richard. 1841. Letter of 25 September 1841 to Hugh Strickland. Scrapbook 1. Nomenclature Papers. Hugh E. Strickland Collection. University of Cambridge Museum of Zoology.

Pascoe, Francis P. 1880. *Zoological Classification: A Handy Book of Reference.* London: John Van Voorst.

Patterson, Robert. 1857. *Introduction to Zoology for the Use of Schools.* London: Simms and M'Intyre.

Pennant, Thomas. 1784. *Arctic Zoology.* London: Henry Hughes.

———. 1793. *History of Quadrupeds.* 2 vols. London: B. and J. White.

Pratt, Mary Louise. 1992. *Imperial Eyes: Travel Writing and Transculturation.* New York: Routledge.

Proposed Plan for Rendering the Nomenclature of Zoology Uniform and Permanent. 1841. London: Richard and John E. Taylor.

Proposed Report of the Committee on Zoological Nomenclature. 1842. London.

Raffles, Thomas Stamford. 1820-21. "Descriptive Catalogue of a Zoological Collection, made on Account of the East India Company, in the Island of Sumatra and its Vicinity." *Linnean Society of London Transactions* 13:239-40.

"Report to the Vice-Chancellor on the Removal of the Collection of Comparative Anatomy." 1865. Professor of Anatomy Collection. University of Cambridge Archives.

"Returns Relating to the National Collections of Works of Art &c." 1857. *Parliamentary Papers* 13:1-7.
Richardson, John. 1842. Letter of 1 March 1842 to Hugh Strickland. Scrapbook 1. Nomenclature Papers. Hugh E. Strickland Collection. University of Cambridge Museum of Zoology.
Ritvo, Harriet. 1987. *The Animal Estate: The English and Other Creatures in the Victorian Age.* Cambridge, Mass.: Harvard University Press.
Roberts, James Watson. 1785-86. "Of the Degeneration of Animals." Papers of the Natural History Society (Special Collections, Edinburgh University Library: MS Da 67 Nat Hist), 4:96-110.
Rothschild, Miriam. 1983. *Dear Lord Rothschild: Birds, Butterflies and History.* Philadelphia: Balaban.
Rudwick, Martin. 1986. *The Great Devonian Controversy.* Chicago: University of Chicago Press.
Rupke, Nicolaas A. 1994. *Richard Owen, Victorian Naturalist.* New Haven, Conn.: Yale University Press.
Ruskin, John. [1881] 1906. *Love's Meinie. Lecture on Greek and English Birds.* Vol. 25 of *The Works of John Ruskin,* edited by E. T. Cook and Alexander Wedderburn. London: George Allen.
Shaw, George. 1792. *Museum Leverianum, Containing Select Specimens from the Museum of the Late Sir Ashton Lever.* London: James Parkinson.
Sheets-Pyenson, Susan. 1988. *Cathedrals of Science: The Development of Colonial Natural History Museums during the Late Nineteenth Century.* Kingston: McGill-Queen's University Press.
Smith, James Edward. 1819. *An Introduction to Physiological and Systematical Botany.* London: Longman, Hurst, Rees, Orme, and Brown.
Solitarius. 1833a. "Faults in Zoological Nomenclature." *Field Naturalist* 1:461-63.
———. 1833b. "Remarks upon Zoological Nomenclature and Systems of Classification." *Field Naturalist* 1:521-28.
Stocking, George W. 1987. *Victorian Anthropology.* New York: Free Press.
Strickland, Hugh. 1842. "Report of a Committee appointed 'to consider of the rules by which the Nomenclature of Zoology may be established on a uniform and permanent basis.'" *Report of the British Association for the Advancement of Science for 1842,* 105-21.
Templeton, R. 1847. Letter of 16 August 1847 to Lord Derby. Stanley Collection. Liverpool Record Office.
Vigors, N. A., and Thomas Horsfield. 1828. "Observations of Some of the Mammalia contained in the Museum of the Zoological Society." *Zoological Journal* 4:105-13.
Wallace, Alfred Russel. 1874. "Zoological Nomenclature." *Nature* 9:258-60.
Winsor, Mary P. 1991. *Reading the Shape of Nature: Comparative Zoology at the Agassiz Museum.* Chicago: University of Chicago Press.
Zoological Keepsake; or Zoology, and the Garden and Museum of the Zoological Society, for the Year 1830. 1830. London: Marsh and Miller.
"The Zoological Record." 1878. *Nature* 18:485-86.
———. 1883. *Nature* 27:310-11.

17

Remains of the Day: Early Victorians in the Field

JANE CAMERINI

> The historian should be able to explain to us how our subject [natural history] in actual fact works: how it is put together, what articulates with what, who keeps the mechanism running. This is not at all an easy matter.
>
> DAVID ALLEN (1976)

> Who're you gonna get to do the dirty work when all the slaves are free?
>
> JONI MITCHELL (1988)

The current generation of scholars in the history of natural history has witnessed a shift in emphasis, from one that privileged ideas and theories to one that incorporates a new appreciation for practice. It is more than a simple shift from ideas to actions; it is a different perspective on what constitutes the activities and products of science. The current emphasis on material and social practices casts a different net over the traditional subjects of natural history research, widening our concern with the collection and analysis of natural objects to embrace a more complex picture of the labor and the people that were involved in natural history pursuits. It seeks to understand natural history practices in the context of social, political, and economic worlds as well as intellectual and scientific ones. What were once considered external factors that might shape the inner core of scien-

---

I would like to express my thanks to Janet Browne, Adrian Desmond, Linda Derksen, Elihu Gerson, Lynn Nyhart, Alan Richardson, the conferees, and two anonymous referees for their helpful comments at various stages of this chapter. I am grateful to the students and faculty of the History of Science Department at the University of Wisconsin, Madison, for their ongoing support. I would also like to thank the University of Chicago Press for permission to use some material from my *Osiris* article, and The Archives, Imperial College of Science, Technology and Medicine, London, for permission to use the sketch by Thomas Huxley, with special thanks to Anne Barrett for her kind assistance in reproducing the sketch.

tific knowledge are now viewed as part of a continuum of conventions, interests, and innovations that shape the very nature of scientific work.

Natural history, including fieldwork, straddles the borders between amateur and professional, sacred and profane, science and emotion. Its common epithets as descriptive science, bug hunting, or stamp collecting reflect problematic underlying dichotomies: natural history as the description of nature versus natural philosophy as the search for laws and causes, and experimental versus historical or observational science. In Victorian times, collecting and studying plants, animals, rocks, and fossils were enormously popular activities, pursued by people as varied as the culture itself. Particular goals of diverse practitioners ranged broadly, and natural history was favored for its constructive and healthful benefits by evangelicals and secularists alike. At the same time, many of the practical activities of collecting were denigrated as amateur pursuits, as inferior to real science, as nontheoretical and merely descriptive, and as unimportant even to the classificatory endeavor (Outram 1984, 61-64; Stevens 1994, 206-7). Attention to the traveling natural scientists permits us to open this ambiguity to analysis and formulate some new questions about how natural history was actually done (Cannon 1978).

This chapter describes the fieldwork of four natural scientists: Charles Darwin; Joseph Dalton Hooker, celebrated plant geographer and taxonomist, friend to Darwin, and director of the Royal Botanical Gardens at Kew, 1865-85; Thomas Henry Huxley; and Alfred Russel Wallace. As established naturalists, these men knew one another as friends and colleagues. Following these four naturalists into the field before they became well known allows us to compare how these men of different social classes engaged in a set of relations to various individuals, social worlds, and institutions in order to pursue natural history research. Through their relationships with other people, a network of opportunities and obligations connected them to two overarching features of the Victorian context: colonialism and industrialization. This broad context is partially delineated below by describing the relationship between natural history and the Royal Navy. We proceed to increasingly specific analyses of the situations and relationships with other people that made fieldwork possible for these four natural scientists. Empire building and an industrial economy defined the purposes and pathways of the voyages on which these naturalists sailed; the specifics of these situations and relationships demonstrate how the broad context shapes field practices. Although the British navy falls into the background in Wallace's fieldwork, the context of empire and industry were no less important for him. His field activities were situated in the colonial network in the East Indian archipelago and in the flourishing trade in natural history specimens in London. We shall see that for each of these naturalists, that which connects the individual to the broader context is relationships with other people.

Historiographically, this essay is an attempt to combine David Allen's social history of natural history with the contextualized approach of recent biographies (Allen 1976, 1994; Desmond and Moore 1991; Desmond 1994; Rupke 1994; Browne 1995). It is informed by writings in the social studies of natural history that treat fieldwork and collaboration in various settings (Rudwick 1985; J. Secord 1985; Griesemer and Gerson 1993; Kohler 1994; A. Secord 1994) and by several recent works that deal directly with field collecting (Larsen 1993; chapters by Outram, Rudwick, Browne, and Larsen in Jardine, Secord, and Spary 1996; and *Osiris,* vol. 11 [1996]). These works point to the importance of fieldwork in the professionalization of geology in early Victorian Britain, to the necessity of cooperation for the formation of natural history museums, and to the social and practical necessities for collecting specimens in the field. Focus on the field as a critical site in the shaping of scientific knowledge reflects widespread interest in the social and situated nature of science as it is practiced (Ophir and Shapin 1991; Shapin 1992).

The purpose of this preliminary overview is to outline the situations and relationships that made fieldwork possible. This might seem to diminish the extraordinary dangers and challenges faced in these voyages, or to overlook the pivotal and symbolic role of voyages in establishing professional credibility, or to ignore the difficult-to-articulate skills that made these individuals excel as scientists. Even worse, this emphasis omits the most romantic and powerful of motives for some naturalists—the love of nature. As central as these issues are, the relationships these naturalists engaged in were a necessary and often overlooked feature of fieldwork. Relationships pervade the practice of fieldwork—the acquisition of specific skills, advice on instruments and books, introductions to local people, hiring servants, and the practical problems of locating, procuring, preserving, and transporting plants, animals, or their proxies back to London and Cambridge. Relationships are the medium that enables each of them to use resources made available by colonial settlement, industrialization, the navy, and the specimen trade to accomplish their fieldwork.

I. An Officer and a Gentleman

The empire-building enterprise, the extension of British rule throughout the world, evolved in conjunction with the social, economic, and political changes wrought by industrialization. At the same time that the colonies provided new sources for markets and raw materials for manufacturing, they became resources for new knowledge of the natural world and thus created new opportunities for individuals to acquire and shape such knowledge (Browne 1992). In the growing industrial economy, expertise and knowledge became more valued; Anglicanism and agriculture, the old standard-bearers of wealth and status, were joined by dissent and manufacturing in the hands of the vocal middle class (Desmond 1989). The redistribution of power and wealth that occurred in Victorian Britain involved tremendous upheaval. Radical reform was in the air, demands for democracy and an end to privilege were sounded among the working and middle classes in the press, in the streets, in the factories, and in the countryside. The upheaval was followed in many cases by both reform and a return to a social order in which scientific expertise was absorbed into existing and new institutions (Morrell and Thackray 1981; Drayton 1993). The ruling elites in the universities and scientific societies needed the scientific expertise of nongentlemen professionals, as the careers of men like Huxley, and many others, demonstrate. In contrast, the careers of men like William Swainson, or Wallace, remind us that, then as now, the cultural currency of scientific institutions has never been based simply on merit (Morrell and Thackray 1981; Desmond 1985).

The culture of early Victorian natural history was shaped by, and in turn shaped, the larger culture. Many of the practices associated with natural history writ large in Britain itself—garden clubs, provincial societies, mechanics institutes, public science lectures, middle-class natural history cabinets, and the growing commercial trade in natural history specimens—can be understood as pursuits made possible by the growing numbers of working- and middle-class people who had some leisure time and access to new railroads, postal services, and cheap presses that permitted participation in natural history collecting and in the exchange of information and specimens. Similarly, long-term changes in social worlds and institutions generated opportunities for Victorian fieldwork abroad through the activities associated with colonialism, especially those of the Royal Navy.

In exploring and navigating the world's oceans, the Royal Navy was in direct competition with the Dutch and the French for trade routes, strategic harbors, and sovereignty over foreign lands and their products. The pathways across oceans, coastlines, and colonies charted the course for Victorian naturalists in their collecting work abroad. Natural history might be accommodated, but the voyages were governed by economic, military, and colonial purposes. The goals of naturalists who sailed these journeys overlapped with those of the navy to varying degrees. This overlap might be thought of as a two-way exchange of material, social, intellectual, and moral practices between naval officers and naturalists.

The tradition that a naval ship's medical officer also served as naturalist while visiting foreign lands goes back to at least the seventeenth century (Keevil, Lloyd, and Coulter 1963, 69). The role of naturalist or surgeon-naturalist was made official in about 1800 through the influence of Sir Joseph Banks, grand patron of science and president of the Royal Society and self-appointed naturalist on James Cook's first voyage (Knight 1974; Allen 1990). Two important, and somewhat conflicting, precedents were explicitly established in the early decades of the nineteenth century by Banks's connections to the Admiralty: the carrying of a naturalist on naval voyages and the relegation of actual collecting work in the field to servant status (Desmond 1985, 174; Drayton 1993, 195). The ambivalence is apparent: natural history is respectable, as long as someone else does the dirty work.

Deep-seated social prejudice inhibited efforts to implement the order for medical officers in the navy to have the status of an officer and a gentleman (Keevil, Lloyd, and Coulter 1963, 11). By and large, medical officers were regarded as an inferior class; promotions were scarce, remuneration dismal, and work conditions generally degrading and demoralizing. During the course of the nineteenth century, repeated efforts to reform the treatment of naval medical personnel were largely unsuccessful until about 1880. The oft-repeated success stories of Huxley, Hooker, and Darwin (who was a guest on a naval ship) were exceptional cases. Many medical officers, artists, and guests on naval vessels pursued natural history, some quite successfully, but that they managed to do so depended on support from their captains and from a small group of scientists within the navy.

While naval exploration proceeded apace during the nineteenth century, with expeditions to the far corners of the globe, support for scientific research was more the result of the commitments of a few stellar individuals than of reform or of an overarching brief. The distinguished career of Sir John Richardson (1787–1865), who began as a surgeon's mate in 1815 and became a knighted archetype of the scientific traveler, provided more than a model for naval scientists. Through his service and influence at Haslar Naval Hospital, promising young medical officers were carefully appointed to appropriate ships, with captains sympathetic to science, and journeys to scientifically interesting locations. Richardson, friend to Darwin, Hooker, and Huxley, was an important—indeed critical—link between the Medical Department of the Navy and the social and intellectual world of natural science.

Some natural history pursuits were more germane to naval interests than others; the study of currents, tides, magnetism and meteorology had obvious relevance to navigation, whereas botany and mineralogy did not. However, uniting these pursuits was "Humboldtianism," a set of concerns with worldwide observations and mappings of a wide variety of natural phenomena (Cannon 1978; Dettelbach 1996). The term is a shorthand for the scientific interests and methods characteristic of traveling scientists in the nineteenth century, and although much refinement of the concept is needed, it emphasizes that the activities of mapping and measurement were not restricted to any one field, such as hydrography. The "Humboldtian" focus on mapping, on geographical distribution, and on using instruments to measure physical phenomena loosely describes a set of goals more or less shared by people involved in both charting the globe and studying its natural phenomena.

One such early Victorian Humboldtian was Admiral Sir Francis Beaufort. Following the end of the Napoleonic war in 1815, Britain increased its attention to hydrography and chart making, and during the tenure of Beaufort as the navy's chief hydrographer from 1829 to 1855, the chart making of the British Admiralty gained an ascendancy befitting the empire. Beaufort commanded one of the most productive and complex periods in the history of hydrographic surveying. Not a professional scientist per se, he played a powerful role in advancing its practice; his writings and accomplishments embody the forces of his day—the struggle against patronage, the obsession with mapmaking and measurement, and the Royal Navy as a vessel of trade and empire. His naval career began at age thirteen, and as a young man he became a member of the Geological, Royal, and Astronomical Societies, and a cofounder of the Royal Geographical Society. It was during his tenure as hydrographer that the department forged links with the scientific communities at Cambridge and London (Ritchie 1967; Friendly 1977; Cannon 1978).

Beaufort was one source of inspiration and pressure for the publication of *A Manual of Scientific Inquiry* (1849), which embodied the wishes of the Admiralty for its naval officers to have practical instructions for making substantial contributions to the various sciences. Edited by John Herschel, and authored by fifteen leading scientists of the day, including George Airy, Edward Sabine, Frederick Beechey, Charles Darwin, William Whewell, Henry De la Beche, James Prichard, Richard Owen, and William Hooker, the manual is a manifesto of Humboldtian science as conceived by members of the Cambridge network, a group of elite academicians who wanted science and its societies controlled by scientists rather than noble lords and physicians (Herschel [1849] 1851; Cannon 1978). The undercurrents of the manual suggest links to large-scale political and social changes, and each of the chapters provides an underused, albeit indirect, window to material and intellectual practices of Victorian science. The obvious link is the manual itself: the Admiralty recognized that expertise was needed and that it would come from the kinds of connections forged by Beaufort, Richardson, and other naval officers with the community of scientists and their expert knowledge. Other, more subtle, links between broad-scale changes and scientific practice are revealed by the emphases of specific chapters.

Owen's chapter on zoology was a revision of his earlier booklet of directions issued in 1835 for both commercial and naval travelers (Rupke 1994, 76–77). As curator of the Hunterian Museum of the Royal College of Physicians and Surgeons, professor of comparative anatomy, and later superintendent of the natural history collection of the British Museum, Owen's interest was in the acquisition of specimens for the museums. Owen's professional trajectory, from modest beginnings to institutional director, demonstrates the increasing value of scientific expertise as political currency and the increasing institutionalization of natural history. The fact that his working life was spent indoors is reflected in the careful instructions for dissecting and preserving specimens and his insensitivity to the practical problems of working on a ship. He seemed unaware that collectors might not have access to reference books he suggested, to adequate conditions of light and humidity for drying specimens, or to preservation materials he recommended. His directions for procuring, anatomizing, skinning, drying, preserving, and packing dominate the sixty-five-page chapter.

Directions for labeling and taking notes on invertebrates, reptiles, and birds are minimal. Somewhat more detailed instructions are given for labeling and recording information about mammals, but the single page of instructions leaves much to the judgment of the collector. Owen emphasized the procurement of specimens, rather than collateral information that might be used for questions about topics other than anatomy. His suggestion that native assistants in the colonies could easily be taught to skin mammals is another indication of the relegation of messy work to others. In contrast, an eight-page section of the chapter written by Darwin and Huxley is noteworthy in its practicality. Their experience enabled them to give clear instructions for how to use a microscope under constrained conditions and what observations to record. Owen's instructions appear more self-serving: bring back lots of specimens for the real science of anatomy and to add to the prestige of British institutions. Even though the manual reflects official support for natural history collecting, Owen's chapter expresses the tension between museum work and field observations, and between science as an activity for experts and the messy work of mere collecting. It demonstrates, par excellence, the colonial sensibility of acquisition.

Sir William Hooker's chapter on botany provided encouragement as well as plain directions. The twenty-two-page chapter is divided into three sections: on collecting living plants for cultivation, on preserving plants for the herbarium, and by far the longest section on collecting interesting vegetable products for the Museum of Economic Botany. Here the traveler is directed to procure not only fruits, seeds, flowers, portions of tree trunks, and specimens of wood used in building, but also gums, resins, dye stuffs, medicines, and other useful products such as tea, paper, and clothing. In its emphasis on products of economic and industrial importance, and on the particular goals of the Museum of Economic Botany (at the Royal Botanical Garden at Kew), the interests that underlay the instructions to travelers are patently institutional and economic as much as they are scientific (Drayton 1993, 300-302; also see Drayton's bibliography for numerous references to this topic).

It may prove difficult to assess the effects of the manual on actual work done by naval officers and others. The Admiralty manual was a relative latecomer in a long line of British guides to observing and collecting, and there is evidence that such manuals, while useful sources of encouragement and information, did not replace teachers or experience in acquiring the complex skills needed to make successful natural history collections (Larsen 1993, 129–36). In traveling to distant shores on naval expeditions, naturalists mastered these skills and overcame practical problems of locating and procuring specimens largely through the help of other people. The relationships that made their fieldwork possible constitute the links, the specific opportunities and obligations, that connected individuals to the empire-building and industrializing world in which they lived.

II. Naturalists at Sea

The navy's evolving interest in science shaped the early careers of many important naturalists, most notably Darwin, Huxley, and Hooker. Their journeys were framed by the colonial enterprise, structuring a natural history practice that helped form early Victorian culture. Social relationships mediated between the individual naturalist and his capacity to take advantage of naval exploration to carry out fieldwork.

It was Beaufort's wish that the *Beagle*, in addition to its main brief to make a geographical and hydrographic survey of southern South America, should take a chain of measurements of longitude at a series of locations around the globe. These were chosen to ensure that the longitudinal intervals between them were approximately equal so as to minimize chronometer rate errors. Thus, Beaufort selected the locations where the *Beagle* made port, determining where Charles Darwin would venture inland to collect.

When HMS *Beagle* arrived in Rio de Janeiro in 1832, four months into its journey, naval surgeon Robert McCormick, the official naturalist assigned to the voyage, stormed off the ship and left his post. He was furious at Captain Robert Fitzroy and at his companion, Charles Darwin. McCormick was a prickly career naturalist, experienced both in overseas commissions as well as in personal disagreements. His natural history was conventional at best. Yet, as head surgeon, he was the official natural historian for the *Beagle* expedition as far as the crown was concerned, by letter and by custom.

Darwin's gentlemanly status, his Cambridge education, and introductions through the elite network gave him a weighted edge over McCormick. Each of the privileges that Fitzroy granted Darwin on the basis of their shared superior status had a direct bearing on enabling the would-be gentleman naturalist to make the most of his opportunities on the voyage. Thus, it was Darwin's microscope, not McCormick's, that sat on the table in the poop cabin, and Darwin's nets and trawls that hung over the side of the ship. These and other favors gave Darwin priority access to the sites for collecting and studying natural history specimens. McCormick and Darwin competed for these privileges because the collection of specimens had so-

Figure 17.1 An inland excursion from a European voyage to South America. Such excursions typically involved several members of the crew as well as local labor. (Illustration from Maximilian of Wied-Neuwied 1820.)

cial, scientific, and financial worth. Darwin strove to use the training, advice, money and privileges he had at his disposal. He knew precisely what was at stake in making a "good collection"—he had to show his teachers, his family, and their upper-class coterie of colleagues and friends that he could use his opportunities for the enhancement of all concerned. Darwin's challenge was to learn how to plan, execute, and record his collecting expeditions. His mentor at Cambridge, Reverend John Henslow, would take care of assiduously printing and circulating Darwin's results to a circle of powerful professors and dons (Browne 1995, 202–10, 335–37).

Darwin accompanied Fitzroy in small landing parties at ports of call, was introduced to local naval officers and gentlemen, and traveled with either his own assistants or locally employed guides (figure 17.1). Early in 1833 Darwin employed one of the sailors, Syms Covington, as his full-time assistant; during his student years, Darwin began the lifelong habit of employing help for the menial aspects of collecting (Browne 1995, 103). From shooting and skinning birds to copying his papers, Covington served Darwin for the rest of the voyage and in England until 1838. They sustained a cordial relationship for many years following the journey (Desmond and Moore 1991). Covington, and many of the officers on the *Beagle,* shared Darwin's love of hunting and avid interest in plants, animals, and fossils, but their potential claims were subsumed by Darwin's dominance and Fitzroy's treatment of him. The social hierarchy on board epitomized the one at home, and the privileged status conferred on Darwin comes as no surprise. It was one of the resources available to him, as were the introductions, instructions, and books given to him before he left.

A major factor shaping Darwin's goal as a gentleman naturalist was his financial security. In addition to the commercial wealth from his mother's family, Darwin's father profited greatly from shrewd investments in real estate, construction, stocks and bonds, and interest on mortgages and other loans to local investors. Darwin's wealth, which enabled him to be a paying guest on the *Beagle* and to confidently hire his own servant while on board, was a direct result of commercial and manufacturing developments in Britain. One specific advantage of his father's influence was Darwin's visit to Richard Corfield, an independent merchant in Valparaiso who owed money to Darwin senior. Darwin benefited from his comfortable stay at Corfield's home and from several introductions to other English businessmen in Chile (Browne 1995, 7-9, 276-77).

Because he was a guest of the *Beagle,* Darwin's collections belonged to him rather than the Admiralty, and even before leaving he thought he would give them to a large and central museum where they would receive the attention he sought. He was anxious about the reaction to his collections and observations, which he had sent to Professor Henslow. After two years of sending materials back (at Admiralty expense), he finally received the approbation he needed from Henslow (Desmond and Moore 1991, 143, 150). Henslow was not only Darwin's beloved mentor, he was a pivotal figure in positioning Darwin in the elite circle of natural history. With Henslow's blessing, he proceeded with renewed confidence and determination to journey inland, shooting, unearthing, and buying specimens, outlining his interpretations and theories along the way.

In addition to surveying and chronometry, Fitzroy's *Beagle* was an agent for political and colonial purposes. Darwin, his microscope, and his journal shared space at the same table in the poop cabin on which Fitzroy and other officers recorded their measurements, wrote their daily logs, and spread the charts on which the British empire was being inscribed. Darwin's position on a naval expedition exemplifies context and practice: he took advantage of Britain's colonial and manufacturing expansion by using relationships with Fitzroy, Henslow, Covington, and many others to carry out his fieldwork.

On another empire-building expedition ten years later, young Thomas Henry Huxley made a name for himself as assistant surgeon-naturalist on HMS *Rattlesnake,* a naval vessel bound for Australia. His achievements were won by availing himself of the scientific opportunities of colonial expansion, and he too could not have done so without key interactions with other people. His impassioned self-discipline drove him from the lower middle class through a stellar career as a student at Sydenham College, one of the newer medical colleges where dissent flourished, to Charing Cross Hospital (for medical training), and from there to a position as assistant surgeon on an exploring expedition to the South Seas. Huxley, exposed to London's voices of rebellion but passionate above all else about anatomy, was seriously in debt. How did he get to be a paid member of a group of naturalists on board an imperial vessel? Having no high-placed patrons, he wrote directly to the physician general of the navy, was interviewed and examined, and then placed under the command of Sir John Richardson, who had been naturalist on the Franklin Arctic expeditions. It was Richardson's devotion to science that led him to look for a good post for the brilliant young microscopist. When Captain Owen Stanley asked Richardson to find a scientifically minded naturalist for his exploring expedition to Australia and New Guinea, Richardson seized the opportunity for Huxley.

Huxley tapped into the scientific network through the good graces of his captain. In the months before their departure in December 1846, Stanley introduced Huxley to Richard Owen, Edward Forbes, a self-made naturalist and geologist, and John Edward Gray, the gifted and industrious keeper at the British Museum. These introductions echo the experiences both of Darwin, some fifteen years earlier, who was introduced to Richardson and Beaufort by the letters of his mentor, Professor Henslow, and of Joseph Hooker, who made the rounds in 1839 through the introductions of his father, Sir William Hooker. While it comes as no surprise that Darwin and Hooker, as young gentlemen, were treated to the best advice their seniors could offer—practical advice about books, instruments, and the identification, notation, preservation, and crating of specimens—it is noteworthy that Huxley, an assistant naval surgeon with no connections, could avail himself of like sources of expertise and knowledge.

Thus armed with advice from top naturalists, Huxley joined the expedition to the southern seas. The brief of the *Rattlesnake* was to make northern Australia and the surrounding seas safe for British settlement and merchantmen. This entailed surveying the passage between northern Australia and New Guinea, charting the southern shores and eastern archipelago, marking channels through the reefs, and assessing possible sites for British colonization. The region was relatively unexplored by naturalists and known to be rich in exotic plants, animals, and peoples. Economic pressures and political interests were the prime motives of the expedition, although Captain Stanley secured several naturalists for the journey in addition to Huxley (Desmond 1994, 42–49).

Aboard the 113-foot, 500-ton ship, inhabited by 180 officers and men, Huxley was pleased to have the privacy of his own alcove off the gun-room, six by seven feet, with a height of four feet, ten inches, a rare privilege for an assistant surgeon. Six feet tall, he had to crouch into the restricted space, which held a cot, clothes, chest, rifle, and desk. For Huxley, much as for Darwin, physical space was a necessity for holding a microscope, for drying plants, or storing bottled animal specimens; space on board was at a premium, and rights to a corner of the table in the poop cabin were indeed a privilege. By the time Huxley sailed on the *Rattlesnake* in 1846, general collections of exotic animals or plants held little promise for his future prospects. In order to distinguish himself in the navy as a scientist, Huxley knew that he had to specialize. He thought he should study what no one else could, the "perishable or rare marine productions," the sort that rarely reached England. He could dissect, draw, and discard jellyfish and their relatives at sea, and this he did (Desmond 1994, 54). His skills, honed in his medical training, in dissecting, observing, and drawing these messy organisms, along with skills to analyze the function and relations of invertebrate forms, resulted in a series of highly regarded and well-published scientific papers (Winsor 1976). Here was no aristocratic gentleman with servants to write his fair copies or perform manual labor; Huxley got his hands dirty, and that, along with acute intelligence and skill, established his authority as a scientist. He embodied the new type of scientist, whose expert knowledge, rather than his social station, was valued by the power elite.

Huxley quickly became friendly with the official naturalist on board, John MacGillivray. They dredged, climbed, and drank together. Captain Stanley not only saw that Huxley had his own cabin and space on the large table in the chart room, he offered him the utmost assistance whenever he could, including sending Huxley's first paper from the voyage to his father, a prominent member of the Linnean Society. Huxley's next paper, a wide-ranging analysis of the anatomical relations among the jellyfish, reached the Royal Society through Stanley's subsequent communications with Richardson and Beaufort. Edward Forbes was another contact to whom Huxley sent subsequent papers for publication. Huxley took full advantage of the opportunities to study animals and fulfilled the expectations of those who supported him by doing good science. For high-ranking men such as Captain Stanley and Admiral Richardson, the pursuit of scientific knowledge was cause for recruiting from outside their rank.

Although he availed himself of Stanley's largesse, Huxley distanced himself somewhat from his skipper's kindnesses. As his biographer, Desmond, paints him, Huxley hated the nepotism of the ruling lords of science and was cynical of a friendship with Stanley, whose family connections were old regime Whigs, although tolerant of dissidents. In a slightly different vein, his stormy moods often kept him isolated for weeks, even months at a time. He spent much of his time alone, either working or condemned to solitude by depression. Nonetheless, Huxley managed the social end of his job more than adequately. His company was eagerly sought by other naturalists for inland explorations (figure 17.2). The social apparatus for getting advice, for attracting the company of like-minded collectors, and for making contact with local people and collecting sites was accessible to him in spite of his modest background and moodiness. He was apparently liked

Figure 17.2 Huxley and other members of the *Rattlesnake* crew in the Australian bush, 1848. (Unsigned pencil sketch, almost certainly by Thomas H. Huxley, from J. Huxley 1935, 134. Photography courtesy of The Archives, Imperial College of Science, Technology and Medicine, London.)

more than well enough to make use of social connections, even with his defiant and black nature.

No such depression befell the twenty-two-year-old Joseph Hooker as he prepared to send home his first shipment of plants as assistant surgeon to HMS *Erebus* and botanist to the Ross expedition (the *Erebus* and the *Terror*) to Antarctica (1839–43). His first collecting venture was on the island of Madeira, followed by Cape Verde and Saint Helena. Although Hooker was raised as a botanist from an early age and trained not only by his father, the distinguished botanist Sir William Hooker, but also by the celebrated botanist and family friend Robert Brown, his first shipment was notably poor. The plants he found were burnt from drought, and the paper he used for pressing them fermented in the damp conditions on board the ship. Both Sir William and Robert Brown were disappointed and made several suggestions for future collecting. He was advised to use brown paper instead of blotting paper and to find an assistant collector to leave him freer from making notes and drawings. Subsequent collections were far more satisfactory (L. Huxley 1918, 1:64-65).

Although there were other naturalists on the expedition, Hooker worked under the wing of the expedition captain, James Clark Ross, distinguished arctic explorer and friend of Richardson. We should keep in mind that Captain Ross was an ardent Humboldtian and a good friend of Joseph's father—Joseph had been presented to Ross at a breakfast hosted by his parents at their home. In his biography of Hooker, Leonard Huxley wrote that he "did not so much learn botany as grow up in it" (L. Huxley 1918, 1:37). More than that, his family life, like Darwin's, prepared him for relationships with the power elite, scientific and otherwise. Hooker was exceedingly grateful for Captain Ross's generosity. Ross directed and instructed the young man; he was kind and attentive and made every effort to support his labors. From space for his work materials, to towing nets for marine organisms, Ross's benevolence bears a similarity to Fitzroy's treatment of Darwin. While Ross may well have been influenced by the social and scientific prestige backing the young Hooker, we have seen in Huxley's case that even a mere assistant surgeon, with no heritage to back him, inspired similar mentorship in Captain Stanley. Not all of the men wielding power in Britain's established institutions were constrained by class boundaries or religious affiliation, particularly those commanding scientific enterprises.

Hooker became a superlative botanist; his social skills were essential to his mastery of the natural world. While Huxley or Wallace would not likely have been at ease traveling with lords and princes, Hooker used his background to great advantage in getting introductions, guidance, and opportunities to botanize in far corners of the world. Some five years after his return from the Antarctic voyage, Hooker embarked on another journey, this time to Sikkim, a part of the Himalayas virtually unexplored by the British. In preparing for the voyage, numerous high-level maneuvers were transacted to secure him a salary of four hundred pounds per annum. Baron Alexander von Humboldt, Lord Aukland (George Eden), and Dr. Hugh Falconer, not to mention Sir William Hooker, all prevailed upon their contacts in the British government and the East India Company to help arrange the mission. The young Hooker traveled to Alexandria on a British navy steamship with Lord James A. B. R. Dalhousie, the new governor general of India. As a result of the friendship that developed between the two, Hooker was invited to join Dalhousie's entourage the rest of the way to Calcutta.

The preface to Hooker's *Himalayan Journals* (1854) is replete with gracious thanks. Here Hooker acknowledged that "the portion of the Himalaya best worth exploring was selected for me" and expresses his sense of obligation to noblemen and officials who housed him, traveled with him, and aided in his collecting activities in various ways (Hooker [1854] 1891).

Noteworthy among these acknowledgments is one to Brian Hodgson, colonial administrator and scholar, in whom Hooker apparently found a kindred spirit. The two worked and lived together for months, Hodgson sharing his extensive natural history library and broad knowledge "like a prince." Hooker wrote of the extraordinary material influence the learned Hodgson had on his studies and travels, adding that the rich content of their exchanges was insensibly incorporated into his understanding of the geography and botany of the Himalaya. A biographer wrote: "If the friendship with Lord Dalhousie provided the key that opened official barriers and made Hooker's journeyings possible, the friendship with Hodgson more than anything else made them a practical success" (L. Huxley 1918, 1:249). Echoing the reliance of Darwin and Huxley on other people, Hooker's relationships with Dalhousie and Hodgson were critical nodes connecting him to other networks of people, knowledge, and skills through which his fieldwork was carried out. These relationships embody the connection between the individual naturalist and the colonial context.

Four mountain expeditions over three years in disputed territory brought their share of political intrigue and physical dangers, including an incident in which he and a companion were held prisoner, but by and large, Hooker maintained his good health. He traveled as gentleman, with plenty of servants and local British support secured through his network of official connections (figure 17.3). During one trip to some mines being worked by

Figure 17.3 Painting based on William Taylor's portrait of Joseph Hooker in Sikkim. (From L. Huxley 1918, 1:286.)

the East India Coal and Coke Company, he experienced his first journey made by elephant, which he described as "traveling in excellent style, the elephants pushing forward the heavy wagons of mining tools with their foreheads" (Hooker 1891, 26). Here then is a picture of science, empire, and industry—the young botanist, awash with botanical curiosity, riding on a local beast of burden, whose weight pushes forth the instruments of so-called economic development.

III. Europe in the Malay Archipelago

Traveling in "excellent style" would hardly do to describe any of Alfred Russel Wallace's voyages. Here we view him as a field-worker by examining three different relationships that connected him to the broader context of colonial development. Wallace's relationship with his agent, with local servants, and with the European community in the East Indian archipelago, enabled him to fulfill his goals as philosophical naturalist and commercial collector (Camerini 1996). With the institutional frames of the navy and the Cambridge network falling into the background, Wallace's experiences highlight different human networks constituting the context of early Victorian fieldwork abroad.

When Wallace set off for the East Indian archipelago in 1854, not knowing he would remain eight years in the region, he had four years of tropical field collecting behind him. In preparing for his first collecting voyage to the Amazon region in 1848, Wallace and his naturalist companion Henry Walter Bates crafted a network for funding their venture through specimen sales. They consulted with Edward Doubleday, curator of insects at the British Museum and assistant to John Edward Gray (who had advised Darwin and Huxley), and with Dr. Thomas Horsfield at the India Museum. They carefully studied existing museum collections and found an agent who would sell their specimens; they practiced shooting and skinning birds. Like Darwin, Huxley, and Hooker, they tapped into existing resources of skills, expertise, books and instruments, albeit on a somewhat lesser scale. They had to rely on their own nascent skills and the vagaries of the specimen trade to survive—no salary or family served as a safety net.

It was Wallace's agent, Samuel Stevens, who had the good business sense to insure the bulk of the Amazonian collection, which was subsequently lost at sea, and who proved to be a critical link to the London natural history community during Wallace's stay in the East Indies. Stevens was an enthusiastic collector of British beetles and butterflies, and brother of Mr. John Crace Stevens, the well-known natural history auctioneer. During Wallace's years in the east (1854–62), Stevens conscientiously used his connections in London's growing specimen trade and his financial acumen to make sales profitable for both of them. Not only did the amounts involved provide for his living expenses, but the excess, invested by Stevens, left

Wallace a modest income for several years (George 1979). Stevens regularly sent Wallace's letters from abroad and advertisements of his collections to natural history journals, and he exhibited the specimens at the Entomological and Zoological Societies. Stevens insured the collections sent from abroad, kept Wallace supplied with cash and supplies as needed, and wrote to him regularly.

By advertising and selling Wallace's collections as he did, Stevens not only maximized their financial worth, but in so doing represented Wallace as a philosophical naturalist collecting scientific novelties. Stevens arranged for Wallace's bird specimens to be described by George R. Gray, brother of John Edward Gray, at the British Museum. The museum agreed to purchase Wallace's bird specimens with the understanding that Gray would describe them all, novel as well as already described species (British Museum 1906, 489; Gunther 1912, 27). Thus, Gray received credit for published descriptions of new species, and the museum purchased one set of bird specimens. The other set, selected by Wallace, was retained by Stevens until Wallace's return. It is significant that the set that Wallace chose for his own collection contained all of the type specimens (i.e., the specimens to which the binomial name is formally tied, and on which the written description is based). This arrangement exemplifies a network of obligation and opportunity: Stevens earned status and money, Gray published scientific descriptions of new species, and Wallace benefited financially, received credit in print for collecting new species, and his bird collection was described by an expert. Wallace's relationship with Stevens highlights how the specimen trade, itself an outgrowth of European expansion, provided a particular set of opportunities and constraints for the natural history collector abroad.

Wallace's interactions with Europeans in the archipelago were enormously significant for his fieldwork. Nearly all of the seventy or more "houses" that Wallace inhabited during eight years of fieldwork were the result of these interactions, and to a large extent determined the locations of his daily collecting sites. Geographically and practically, established European society formed the lattice of this journey. Again and again, the bits of the lattice were pieced together by a brother-in-law here, a letter of introduction there, a Dane lending him a horse, a German giving him a meal, an Englishman providing him with a servant, or land, or materials to help build a temporary abode, and a Dutchman introducing him to a local Rajah (Wallace [1869] 1877; Marchant 1916).

The framework of colonial society that enabled Wallace to carry out his fieldwork in the East Indies had its roots in the labyrinth of European colonialism, especially the intersecting paths of English trade from India to China, and the Dutch settlements in Java, Borneo, and the Moluccas. The steamships that brought Wallace to the archipelago carried not only his mail and specimens to and fro, but also lead shot for his guns, fabric for his clothes, boxes to carry animals, and medicines for his illnesses. He made use of the roads and buildings that enabled agriculturalists, miners, and engineers to tap the resources of these islands. The doctors, missionaries, traders, teachers, and merchants who built lives within the new mixes of cultures served Wallace as nodes, or critical points of departure, for his collecting. It was through their "good offices" that he found lodgings, directions to places for collecting, servants, translators, introductions to local rulers, and letters of introduction to other Europeans.

Just as Wallace's interactions with Europeans formed a lattice that made his fieldwork possible, so too his relationships with his native servants were pivotal for carrying out the daily practice of fieldwork. Servants accompanied Wallace practically everywhere as he met the day-to-day exigencies of living and working in the towns and back hills of the great archipelago.

While finding local help in new locations was fraught with difficulties, one servant whom Wallace employed in 1855 became enormously important for the remainder of this stay. Ali, who referred to himself as Ali Wallace, became Wallace's head servant. He grew highly skilled at locating, shooting, skinning, and pinning insects and birds, and taught Wallace the Malay language. He and Wallace took their meals together, helped one another through numerous illnesses, and over the years developed a finely tuned rapport. Ali trained the other servants who worked for Wallace and was often able to communicate with native residents to help Wallace locate and collect animals in situations beyond the limits of European control. Ali and two other young native men were especially important in Wallace's search for the highly prized birds of paradise, which were found in areas of the archipelago not occupied by colonial peoples (figure 17.4). Describing him as the best native servant he ever had, Wallace knew that his faithful companion was essential for his access to local knowledge of native animals.

The class distinctions among men of various professions—in commerce, the clergy, government, industry, and medicine—were altered in the transference of European society to the colonies. The boundaries between the middle and upper classes were more blurred and forgiving, in a context in which the boundary between European and Other was enhanced. Wallace was able, like many a colonial fortune seeker, to move far more readily through social boundaries in this dislocated European society than he would have been had he remained in England. In this context, even a nonelite person such as Wallace could procure and afford a small staff of servants. The quality of his interactions with Europeans in the East Indies—the domestic openness, the extending of favors, guidance, and privilege, the intellectual exchanges—suggests that the status of gentleman was conferred according to local, colonial conventions. The significance of this is that it allowed Wallace to perform his job. Each favor granted him by a Eu-

Figure 17.4 Native servants on Aru Island hunting birds of paradise for Wallace's collection. (Woodcut, from Wallace 1877, opposite p. 433.)

ropean constituted a piece of his fieldwork. In effect, local European society functioned as an institution, validating and assisting his work, making it possible for him to send animal specimens and scientific articles back to England. Of course Wallace's publications and correspondence with English scientists were pivotal in his professional identity during these fieldwork years. But it was the Europe in the archipelago that allowed Wallace to become a bona fide British naturalist, not merely a collector or working class amateur.

IV. Conclusion

Personal relationships played a key role in all of these voyages. The character of these relationships varied, of course, and they were arguably more critical for Wallace, who lacked any institutional affiliation or source of income, than for the others. The nature and qualities of the relationships reveal as much about the character of their times as about the men themselves: the aristocratic pairing of Darwin and Fitzroy, or Hooker and Dalhousie; the new men of science, Huxley and Richardson, using the navy as a stepping stone or career path; Stevens, the well-connected businessman helping Wallace, the aspiring collector; or Wallace, the self-taught naturalist, befriending colonial landowners. Broad-scale economic and social changes are reflected in relationships that paired old wealth with new expertise: Fitzroy and Beaufort, Stanley and Huxley, Hooker and Hodgson, Darwin and Wallace. For all, the empire-building enterprise conditioned their itineraries and framed the immense opportunities through which they built careers.

These four naturalists, and scores of other less famous ones, traveled to distant countries. Through skills learned in diverse settings, they selected objects and processes of nature to observe, they collected—rocks, plants, animals, fossils, human artifacts—and they preserved and pressed and shipped these objects back to England. They traveled with scientific books at their side and kept in touch with personal and scientific news from home through various sources—families, friends, professors, and agents. They depended on the people with whom they traveled, on contacts made on foreign shores, and on assistants and servants in carrying out their natural history work. This sketch opens a window of topics little developed in existing literature; scores of field notebooks await further inquiry. What information did naturalists record? How did they grapple with practical problems: selecting and identifying what to collect, killing animals and preserving plants for the long journey back to England, and labeling specimens so that they could be later coordinated with their field observations? What was the fate of the specimens; how were they treated as they were disseminated to various museums, experts, or collectors? What were the expectations of those who received specimens, either by sale, donation, or naval order, and how did these expectations shape activities in the field? Questions about the epistemic role of local expertise and servants, and how skills learned through relationships, briefly outlined here, were incorporated into scientific discourse, await further study (Shapin 1994; Camerini 1996).

David Allen, a leading historian of natural history, reminds us that there are dimensions of natural history that are not strictly scientific, but nonetheless integral to the subject (Allen 1976, 510). He speaks of the emotional porosity of natural history, of the slippery realms of aesthetics and taste, and of the interactions between the core pursuit of scientific understanding and the social influences upon it. Here we have seen that the pursuit of scientific understanding is a collective, cultural process. The scientific arenas of natural history—journals and books, networks of specimen and information exchange, museum and university collections, professional societies, amateur clubs, as well as exploration expeditions—are all collective enterprises. At the same time, these activities are carried out by individuals whose motives and achievements vary and whose insights stand at the core of scientific understanding. Rather than analyzing the social influences on science, this chapter articulated the realm between the individual and the social in the practice of fieldwork. Emotions need not be bracketed off because they belong intractably to the individual, nor do political, economic, or social contexts need to be bracketed off because they lie outside the core of science. By focusing on what a few individual naturalists actually did, on the relationships that were necessarily part of the practice of their science, and on the contexts that shaped these practices, it is possible to view simultaneously their individuality and their participation in larger social endeavor.

Bibliographical Note

An important resource for fieldwork is the recent volume of *Osiris* (1996, 2d series, vol. 11) devoted to field sciences, edited by Robert Kohler and Henrika Kuklick. Many essays of direct relevance to fieldwork studies are found in the recent volume edited by Jardine, Secord, and Spary (1996). The family of new biographies of Victorian naturalists is another important resource, in which cultural and institutional contexts are highlighted (Outram 1984; Desmond and Moore 1991; Desmond 1994; Rupke 1994; Browne 1995). The literature on the history of natural history discusses collecting and fieldwork in a larger intellectual, cultural, or institutional context (Allen 1976*a*, 1976*b*, 1977, 1987, 1994; Barber 1980; Browne 1983, 1988; Rehbock 1983; Desmond 1985; J. Secord 1985; Wheeler and Price 1985; Dance 1986; Ritvo 1987; Merrill 1989). Collecting animal specimens is the focus of one recent dissertation (Larsen 1993); collecting plants is discussed in several works by Ewan (e.g., 1992), and animal preservation is the focus of Farber (1977). As this note is by necessity highly selective of the extensive literature in the history of natural history, the reader is referred to the following for their excellent references: Farber 1982; Van Riper 1993; Shortland 1994; Browne 1992; Bowler 1993; Bridson 1994.

The relations of amateur and expert in nineteenth-century science are discussed in the substantial literature on the professionalization of science; further references are found in Knight 1974; Lowe 1976; Morrell and Thackray 1981; Keeney 1992; and A. Secord 1994. Colonialism, imperialism, and the history of exploration provide another set of resources for natural history fieldwork (see Drayton 1993 and Grove 1995 for numerous references). Collecting animals for zoos and museums, and the formation of biological and astronomical field stations, as well as nature reserves, provide other avenues for the study of fieldwork at home and abroad (Winsor 1991; Pang 1993; Kohlstedt 1995).

References

Allen, David. 1976. "Natural History and Social History." *Journal of the Society for the Bibliography of Natural History* 7:509-16.

———. 1977. "Naturalists in Britain: Some Tasks for the Historian." *Journal of the Society for the Bibliography of Natural History* 8:91-107.
———. 1987. "The Natural History Society in Britain through the Years." *Archives of Natural History* 14:243-59.
———. 1990. "Banks in Full Flower (Essay Review)." *Notes and Records of the Royal Society of London* 44:119-24.
———. 1994. *The Naturalist in Britain: A Social History.* Princeton, N.J.: Princeton University Press.
Barber, Lynn. 1980. *The Heyday of Natural History, 1820-1870.* London: Jonathan Cape.
Bowler, Peter. 1993. *The Norton History of the Environmental Sciences.* New York: Norton.
Bridson, Gavin. 1994. *The History of Natural History: An Annotated Bibliography.* New York: Garland.
British Museum. 1906. *The History of the Collections Contained in the Natural History Departments of the British Museum.* Vol. 2. London: British Museum.
Browne, Janet. 1983. *The Secular Ark.* New Haven, Conn.: Yale University Press.
———. 1988. "Passports to Success (Essay Review)." *Journal of the History of Biology* 21:343-49.
———. 1992. "A Science of Empire: British Biogeography before Darwin." *Revue d'histoire des Sciences* 45:453-75.
———. 1995. *Charles Darwin: Voyaging.* New York: Alfred Knopf.
Camerini, Jane R. 1996. "Wallace in the Field." *Osiris*, 2d ser., 11:44-65.
Cannon, Susan F. 1978. *Science in Culture: The Early Victorian Period.* New York: Dawson and Science History Publications.
Dance, S. Peter. 1986. *A History of Shell Collecting.* Leiden: E. J. Brill.
Desmond, Adrian. 1985. "The Making of Institutional Zoology in London: 1822-1836." *History of Science* 23:153-85, 223-349.
———. 1989. *The Politics of Evolution: Morphology, Medicine, and Reform in Radical London.* Chicago: University of Chicago Press.
———. 1994. *Huxley: The Devil's Disciple.* London: Michael Joseph.
Desmond, Adrian, and James Moore. 1991. *Darwin: The Life of a Tormented Evolutionist.* New York: W. W. Norton.
Dettelbach, Michael. 1996. "Humboldtian Science." In *Cultures of Natural History,* edited by Nicholas Jardine, James Secord, and Emma Spary. Cambridge: Cambridge University Press, 287-304.
Drayton, Richard H. 1993. "Imperial Science and a Scientific Empire: Kew Gardens and the Uses of Nature." Ph.D. diss., Yale University.
Ewan, Joseph. 1992. "Through the Jungle of Amazon Travel Narratives of Naturalists." *Archives of Natural History* 19:185-207.
Farber, Paul L. 1977. "The Development of Taxidermy and the History of Ornithology." *Isis* 68:550-66.
———. 1982. *The Emergence of Ornithology as a Scientific Discipline: 1760-1850.* Dordrecht: Reidel.
Friendly, Alfred. 1977. *Beaufort of the Admiralty: The Life of Sir Francis Beaufort.* London: Hutchinson.
George, Wilma. 1979. "Alfred Wallace, the Gentle Trader: Collecting the Amazonia

and the Malay Archipelago 1848-1862." *Journal of the Society for the Bibliography of Natural History* 9:503-14.

Griesemer, James R., and Elihu M. Gerson. 1993. "Collaboration in the Museum of Vertebrate Zoology." *Journal of the History of Biology* 26:185-203.

Grove, Richard H. 1995. *Green Imperialism: Colonial Expansion, Tropical Island Edens and the Origins of Environmentalism, 1600-1860.* Cambridge: Cambridge University Press.

Gunther, Albert. 1912. Appendix to vol. 2 of *The History of the Collections Contained in the Natural History Departments of the British Museum.* London: British Museum.

Herschel, John F. W. [1849] 1851. *A Manual of Scientific Enquiry; Prepared for the Use of Officers in Her Majesty's Navy; and Travellers in General.* 2d ed. London: John Murray.

Hooker, Joseph Dalton. [1854] 1891. *Himalayan Journals; or, Notes of a Naturalist in Bengal, the Sikkim and Nepal Highlands, the Khasia Mountains etc.* London: Ward, Lock, Bowden.

Huxley, Julian, ed. 1935. *T. H. Huxley's Diary of the Voyage of HMS Rattlesnake.* London: Chatto and Windus.

Huxley, Leonard. 1918. *Life and Letters of Sir Joseph Dalton Hooker.* New York: Appleton.

Jardine, Nicholas, James Secord, and Emma Spary, eds. 1996. *Cultures of Natural History.* Cambridge: Cambridge University Press.

Keeney, Elizabeth B. 1992. *The Botanizers: Amateur Scientists in Nineteenth-Century America.* Chapel Hill: University of North Carolina Press.

Keevil, John J., Christopher Lloyd, and Jack L. S. Coulter. 1963. *Medicine and the Navy.* Vol. 4, *1815-1900.* Edinburgh: Livingstone.

Knight, David. 1974. "Science and Professionalism in England." *Fourteenth International Congress on the History of Science Acts* 14:53-67.

Kohler, Robert E. 1994. *Lords of the Fly: "Drosophila" Genetics and the Experimental Life.* Chicago: University of Chicago Press.

Kohlstedt, Sally Gregory. 1995. "Museums: Revisiting Sites in the History of the Natural Sciences (Essay Review)." *Journal of the History of Biology* 28:151-66.

Larsen, Anne L. 1993. "Not since Noah: The English Scientific Zoologists and the Craft of Collecting." Ph.D. diss., Princeton University.

Lowe, P. D. 1976. "Amateurs and Professionals: the Institutional Emergence of British Plant Ecology." *Journal of the Society for the Bibliography of Natural History* 7:517-35.

Marchant, James. 1916. *Alfred Russel Wallace: Letters and Reminiscences.* New York: Harper.

Maximilian of Wied-Neuwied, Prince. 1820. *Travels in Brazil in the years 1815, 1816, 1817.* Pt. 1. London: H. Colborn.

Merrill, Lynn L. 1989. *The Romance of Victorian Natural History.* Oxford: Oxford University Press.

Morrell, Jack, and Arnold Thackray. 1981. *Gentlemen of Science: Early Years of the British Association for the Advancement of Science.* Oxford: Clarendon Press.

Ophir, Adi, and Steven Shapin. 1991. "The Place of Knowledge: A Methodological Survey." *Science in Context* 4:3-22.

Outram, Dorinda. 1984. *Georges Cuvier: Vocations, Science and Authority in Post-Revolutionary France.* Manchester: Manchester University Press.

Pang, Alex Soojung-Kim. 1993. "The Social Event of the Season: Solar Eclipse Expeditions and Victorian Culture." *Isis* 84:252–77.

Rehbock, Philip F. 1983. *The Philosophical Naturalists: Themes in Early Nineteenth-Century British Biology.* Madison: University of Wisconsin Press.

Ritchie, G. S. 1967. *The Admiralty Chart: British Naval Hydrography in the Nineteenth Century.* New York: American Elsevier.

Ritvo, Harriet. 1987. *The Animal Estate: The English and Other Creatures in the Victorian Age.* Cambridge, Mass.: Harvard University Press.

Rudwick, Martin J. S. 1985. *The Great Devonian Controversy: The Shaping of Scientific Knowledge among Gentlemanly Specialists.* Chicago: University of Chicago Press.

Rupke, Nicolaas A. 1994. *Richard Owen: Victorian Naturalist.* New Haven, Conn.: Yale University Press.

Secord, Anne. 1994. "Science in the Pub: Artisan Botanists in Early Nineteenth Century Lancashire." *History of Science* 32:269–315.

Secord, James A. 1985. "Natural History in Depth (Essay Review)." *Social Studies of Science* 15:181–200.

Shapin, Steven. 1992. "Discipline and Bounding: The History and Sociology of Science as Seen through the Externalism-Internalism Debate." *History of Science* 30:333–69.

———. 1994. *A Social History of Truth: Civility and Science in Seventeenth-Century England.* Chicago: University of Chicago Press.

Shortland, Michael. 1994. "Darkness Visible: Underground Culture in the Golden Age of Geology." *History of Science* 32:1–61.

Stevens, Peter. 1994. *The Development of Biological Systematics: Antoine-Laurant de Jussieu, Nature, and the Natural System.* New York: Columbia University Press.

Van Riper, A. Bowdoin. 1993. *Men among the Mammoths: Victorian Science and the Discovery of Human Prehistory.* Chicago: University of Chicago Press.

Wallace, Alfred Russel. [1869] 1877. *The Malay Archipelago: the Land of the Orang-utan and the Bird of Paradise; A Narrative of Travel with Studies of Man and Nature.* 6th ed. London: Macmillan.

———. 1905. *My Life; A Record of Events and Opinions.* 2 vols. New York: Dodd, Mead.

Wheeler, Alwyne, and James H. Price, eds. 1985. *From Linnaeus to Darwin: Commentaries on the History of Biology and Geology.* London: Society for the History of Natural History, British Museum (Natural History).

Winsor, Mary P. 1976. *Starfish, Jellyfish, and the Order of Life.* New Haven, Conn.: Yale University Press.

———. 1991. *Reading the Shape of Nature: Comparative Zoology at the Agassiz Museum.* Chicago: University of Chicago Press.

18

Photography as Witness, Detective, and Impostor: Visual Representation in Victorian Science

JENNIFER TUCKER

> The nineteenth century began by believing that what was reasonable was true and it wound up by believing that what it saw a photograph of was true—from the finish of a horse race to the nebulae in the sky.
> WILLIAM IVINS, *Prints and Visual Communication* (1980, 94)

"Photographs can lie," announced British jurist Ernest Arthur Jelf in "Photography as Evidence," an illustrated essay published in the December 1894 issue of the fashionable London periodical, the *Idler*. While allowing that photography aided scientists as a method for illustrating and proving phenomena too "inaccessible" or "rapid" for ordinary human observation, Jelf disagreed with those, such as former prime minister William Gladstone, who insisted that something "stood upon evidence" just because it had been made the subject of a photograph. A combination print showing Gladstone outside a London pub (figure 18.1) was one of several photographs prepared for the article by trick photographers "to expose the worthlessness of the evidence which photography affords." As Jelf explained, it was absurd to suppose that there was any "foundation in fact" for a photograph

---

This chapter is part of my doctoral dissertation, "Science Illustrated: Photographic Evidence and Social Practice in England, 1870–1920," submitted in September 1996 to the Department of the History of Science, Medicine, and Technology at Johns Hopkins University. I would like to thank the librarians and staff of the British Library, the Museum of the History of Science at Oxford, the Cambridge University Library, the Royal Meteorological Society, the Royal Astronomical Society, and the Wellcome Institute for the History of Medicine for their assistance, Robert Smith, Simon Schaffer, Bernie Lightman, Judith Walkowitz, and the participants of the British History graduate seminar at Johns Hopkins University for their advice and encouragement, and University College London and the Society for Psychical Research for permission to reproduce illustrations in their collections. The research for this project was assisted by grants from the Joint Committee on Western Europe of the American Council of Learned Societies and the Social Science Research Council, the National Science Foundation, and the National Endowment for the Humanities.

Figure 18.1 Prime Minister Gladstone shown standing outside a London pub. Combination print photograph reproduced by Messrs. Boning and Small by making two exposures. (From Jelf 1894, 520).

showing the veteran statesman issuing from a public house in Seven Dials. "Yet there he is in the photograph," mused Jelf: "it is certainly he himself; it is certainly a public-house in Seven Dials, which any of our readers may identify for himself by going to the spot." In conclusion, Jelf warned readers to use judgment when they viewed photographs presented to them as evidence in support of facts. Just as courts of law did not always admit photographs as evidence, photographs outside the legal arena, he urged, "must stand the test of cross-examination." For Jelf, therefore, the conditions for producing photographs demanded explanation: a photographer must "*prove* his photograph" by expounding the manner in which it had been made before the image could be admitted as a matter of fact (Jelf 1894, 520–21, 524).

These remarks by Jelf remind us that photography was a controversial new standard of objective reporting in Britain during the late nineteenth century. Photographic practices were assimilated in a variety of scientific and medical fields in the decades of technological innovation following the invention of the daguerreotype process in 1839. Scholars have detailed the nineteenth century's belief in the power of photographic technology to replicate the act of unmediated seeing, to eliminate human prejudice, and to minimize the errors that allegedly vitiated the objectivity of drawings (Daston and Galison 1992; Tagg 1988; Gernsheim 1969; Ivins 1980; Newhall 1949). From the time of photography's invention, Victorians identified it as a certain type of human: a "witness," a "detective," and a "discoverer." Of the supremacy of photography as an evidential tool at the end of the century, Émile Zola made the famous declaration: "We cannot claim to have really seen anything before having photographed it" (Lemagny and Rouillé 1986, 71).

But what was the power of photography, precisely? Jelf, for example, was not alone in doubting the evidential value of photography at the exact moment of photography's ascendancy as an integral element of British science and culture. Of the "uselessness of the camera as a witness," one observer explained, the "wider knowledge of photography due to its adoption by hundreds of thousands of amateurs as a hobby" in the 1880s and 1890s revealed photography to be "the most elastic of all arts" ("Photographic Lies: Proving the Worthlessness of the Camera as a Witness" 1898, 259). Nineteenth-century debates in Britain over claims made with photographs in a variety of settings, from field outposts to the laboratory to the spiritualist seance, suggest that Victorians did not, in fact, accept photographic evidence as unconditionally true and, indeed, that they interpreted facts based on photographs in a variety of different ways.

To understand the role of photography in late Victorian science, we must look beyond praise for the universality and epistemic unity of photography and begin to unravel how, specifically, photographs were used as evidence in diverse realms of late-nineteenth-century British science and culture. Holly Rothermel aptly writes that even comprehensive histories of photography, which do discuss many of the technical and chemical problems of photography, "rarely venture into the uses of photography in scientific practice, let alone into the issues of representation and authority that might be raised by these usages" (Rothermel 1993, 139). What claims did Victorians make with photographs? How, and where, did they submit photographs as evidence or proof? Were facts presented by means of photographs reinterpreted? Were those claims challenged and, if so, how?

To answer these kinds of questions, we must recover some of the practices that Victorians connected with making, viewing, and extracting meaning from scientific photographs. Discussions about photography in scientific journals, photographic literature, spiritualist reviews, and periodical magazines from roughly 1870 to 1900 reveal that the acceptance of photographs as objective, meaningful representations in Victorian science and culture did not happen automatically; on the contrary, assent to claims supported by the evidence of photographs was contingent upon their meeting several criteria, for example, of production and of use, established in different settings by an emerging pool of experts. Victorians brought into play their understandings of the competence and trustworthiness of witnesses, detectives, and impostors to help to define situations where photographs were presented as evidence and to provide others with information about the guises and disguises that photography might adopt. To understand the place of photography and its associated practices in Victorian science, we must not only reflect on the general meanings invoked by these parallels; we must also investigate their different inflections in distinctive sites of scientific practice. Tracking elaborations of the evidential value of photographs across scientific and cultural fields reveals problems of production, communication, and reception that Victorian praise for photography often simplified.

Each section in this chapter focuses on an area where Victorians deemed photographs to be consequential for knowledge in the 1880s and 1890s. Beginning with photographic practices in late Victorian meteorology, we will see that meteorologists supervised the photography of lightning flashes by creating extended social networks and by connecting amateur photography and science exhibitions to the development of witnessing techniques. Next, we will examine the constellation of setting, narratives of detection, and modes of seeing in late-nineteenth-century discussions of the role of photography in the new bacteriological science. We will then explore the reception of spirit photography in Britain after 1870, focusing especially on practices adopted by spiritualist photographers in the 1880s and 1890s to secure their testimony against accusations of imposture. The chapter concludes with some reflections on photography in nineteenth-century science.

I. Meteorology and Spectatorship: Photography as a Witness

In meteorology, as in other fields, photography's designation as a witness brought into play Victorian understandings of the competence and trustworthiness of expert witnesses to help to define situations where photographs were presented as evidence. In the Royal Meteorological Society, where meteorologists struggled for disciplinary autonomy, promoters of meteorological photography argued that photography would revolutionize meteorological observation: first, by providing mechanical access to the reality of lightning; second, by expanding the network of observers; and third, by offering a basis for classifying lightning into morphological types.

To understand the special burden of representation borne by photography as a witness in meteorology, however, we must reflect not just on the parallels that Victorians invoked between scientific photography and legal witnessing, but also on the impact of transformations in meteorology during the 1880s and 1890s on Victorian uses and perceptions of photographs as meteorological evidence.

Meteorologists who applied photography to the classification of lightning and clouds in the 1880s and 1890s trained others in community traditions of gathering and assimilating evidence. Of photography's importance in meteorology in 1898, Arthur Clayden, a fellow of the Royal Meteorological Society, photographer, and Exeter Technical College principal, boasted: "It would manifestly be impossible to give anything like a full description of what has been done, and is being done at present, in the way of photographing meteorological phenomena in all parts of the world. . . . Some idea of the importance of the camera as a meteorological instrument may easily be gained by counting how many of the papers brought before this Society are more or less dependent upon photographic methods." When used in connection with photographic processes, Clayden announced that "barometers, thermometers, magnetometers, and electrometers can now all be watched by an observer who is always there, who is always looking in the right direction, whose attention does not flag, and who cannot easily make an erroneous record, if only he is properly adjusted and properly wound up at the appointed time." Clayden held that gelatine dry plates and hand cameras, in particular, made it possible to make an objective record of weather phenomena in scattered regions of the world. Photographs of lightning, clouds, rainbows, halos, and frost, he argued, showed "detail which the eye can never hope to perceive" (Clayden 1898, 170, 179). In particular, photography seemed to provide a means for classifying lightning. Stream, sinuous, ramified, meandering, beaded, and ribbon were the names that Victorian meteorologists assigned to different types of lightning, pictured in the figure published in the *Quarterly Journal of the Royal Meteorological Society* in 1888 (figure 18.2).

Applications of the hand camera as a witness of the weather revived long-standing issues about the trustworthiness and authentication of empirical practices in British meteorology. In a young discipline historically associated with astrology and prognostication, the participation of lay observers raised several questions for investigators. These included how to authenticate reports from observers in extended networks and how to assimilate scattered, uncalibrated data. Of meteorologists' need to monitor the quality and handling of meteorological evidence in an age of "weather fallacies," Richard Inwards, the president of the Royal Meteorological Society, argued in an address to the Society in 1895 that many "so-called rules" of atmospheric change had "no kind of foundation in fact," and indeed

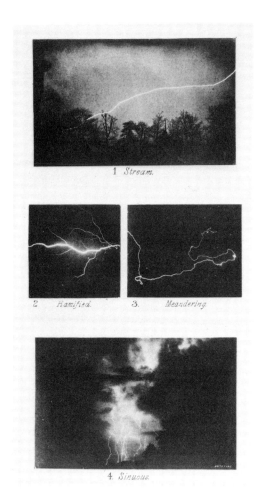

Figure 18.2 "Types of Lightning," selected from specimens in the Royal Meteorological Society, circa 1880. (Reproduced in Abercromby 1888, facing p. 232.)

were based only on a "coincidence of totally independent events" or, worse, imposed on mass credulity by false authorities (Inwards 1895, 50).

Meteorological photography in the 1880s and 1890s built on earlier efforts in the Royal Meteorological Society to collect information about lightning phenomena from scattered lay witnesses. In 1857, the Thunderstorm Committee of the Society, led by George James Symons, founder of a national network of volunteer meteorological observers, asked correspondents throughout Britain to supply observations on the physical appearance, activity, and effects of lightning. As an object of wonder, curiosity, and fear, lightning touched the experiences of Victorians in various ways. It therefore formed a subject that laypersons were not only eager to

Figure 18.3 "A Murderer struck by Lightning" (wood engraving from De Fonvielle 1869, 153). De Fonvielle melodramatized lightning, saying that "works on physical science certainly contain a number of scenes quite worthy of our melodramatic stage" (157).

report, but also well equipped to describe to nonwitnesses through a variety of descriptive resources including personal hearsay, pencil sketches of lightning drawn "direct from nature," instrument readings, and physical objects removed from the sites of storm damage. Symons complained, however, that popular myths about lightning (such as widespread belief in zigzag lightning and thunderbolts) interfered with the gathering of true information about lightning and its effects (figure 18.3). Several people sent Symons specimens of alleged thunderbolts that, they claimed, they found in kitchen grates, on farms, and on suburban streets: for example, residents of Kilburn, London, reported that molten liquid poured down from "infuriated heavens," leaving "fist-sized clinkers" in its wake. Symons exclaimed that while he had looked out for a "real thunderbolt" for over thirty years, alleged thunderbolts were really iron pyrites, fossils, cannonballs, and telegraph wires: "facts twisted" into physical evidence through vaults of imagination. In "The Non-Existence of Thunderbolts," a widely publicized address to the Royal Meteorological Society in March 1888, Symons argued that extinguishing popular belief in thunderbolts was a matter of national and, indeed, masculine pride: "For our credit as Englishmen, we ought to drive the word thunderbolt out of our dictionaries." As a leading member of a research field that placed a premium on collaborative investigation, Symons urged others, when they heard of a thunderstorm having fallen, to "go to the spot, patiently examine the facts, and then explain publicly precisely what has happened" (Symons 1888, 208, 211-22).

Like reconstructing a crime from testimony gathered at the scene, establishing precisely what happened during a sudden thunderstorm was often difficult, however. Even when observations were trusted, they were hard to interpret and compare. Symons's appeals for witnesses of thunderstorms, for example, returned a rich confusion of observations of the color and forms of lightning: "red sheet, and blue curved" from Thwaite; "whitish, not forked" from Uckfield; "rose, zigzag flashes" from Wakefield; "pale lilac, sheet" from London; even "one *vivid* reddish yellow, and one *vivid* purplish pink" from Manchester. To compare the results, Symons reported, he was "obliged to reject or reduce to some of the primary colours the terms employed by some of the observers, such as 'like a sulphur flame,' 'like a burning rope,' 'mauve,' etc." The next time Symons's committee asked for volunteers, investigators restricted the range of information that they requested to four colors (blue, red, white, and yellow) and two forms (sheet and forked) (Symons 1889, 5-6).

Promoters of meteorological photography in the Royal Meteorological Society argued that photography would resolve many of these problems that they associated with the scientific observation of lightning. In June 1887 the Royal Meteorological Society issued two hundred circulars requesting photographic assistance from members of photographic societies in Europe and North America. Clayden, who combined interests in meteorology, geology, and photography, wanted to build "a great army of observers, each capable of taking a creditable photograph" (Clayden 1898, 170). There were only two ways to get the kind of records that meteorologists wanted, Clayden explained, "and they are in the multiplication of people who are both photographers and meteorologists, or in securing a number of local meteorologists who will undertake to try and get some photographing friend of theirs to take suitable views whenever opportunities come" (Clayden 1898, 171). Clayden venerated the photograph for its objectivity and its "clear exposition of argument." With photography, he insisted, there was "no lagging of the record behind the fact" (Clayden 1898, 170). Photographs of phenomena showed details that "the eye can never . . . perceive" (Clayden 1898, 179). "Even apparently poor photographs often contain useful evidence," meteorologist-photographer William Marriott, a fellow of the Royal Meteorological Society, encouraged listeners at a London photographers' meeting (Marriott 1889, 176).

Innovations in the field of photography in the 1880s associated with the large-scale factory production of photographic goods made possible photography of lightning and clouds by nonprofessionals. By 1887, the year that the Royal Meteorological Society launched its appeal for meteorological photographs, the market for amateur photographic equipment was flooded with lightweight cameras of a simple design, new lenses, more sensitive emulsions, and new types of optical glasses. Hand cameras, including the change-box, the magazine, the roll-film, and the reflex, could take a large number of plates in quick succession, offering constant readiness for action. The declining cost of supplies, moreover, made photography more affordable for middle-class hobbyists, with ready-made gelatine plates (the "invaluable ally" of meteorologists) selling in London for one pound per dozen compared to three pounds per dozen during the previous decade (Gernsheim 1969, 323). Gelatine bromide paper, used by amateurs in the mid- to late 1880s, was much more sensitive to light, and thus required shorter exposure time on faint objects.

Meteorologists claimed that photographs enabled them to see lightning with scientific eyes. In an age fascinated by natural marvels, scientific discovery, and machines, meteorologists were right to assume that the chance for Victorians to combine a picturesque outing with opportunities to advance science would send amateur photographers in pursuit of lightning and clouds. Scholars have identified the mid-1880s as the moment when the amateur photography movement gained momentum in Britain, when divisions began to emerge between mass snapshooters and those who sought to advance photography as an art (Henisch and Henisch 1994, 396–430; Gernsheim 1969; see figure 18.4). Identified with travel and with the recording of marvels at home and away, the amateur photographer seemed to many Victorian meteorologists to be an ideal recruit for scientific research. In "Hints on Photographing Clouds," read at the Royal Meteorological Society in 1895, meteorologist-photographer Birt Acres remarked that clouds, whose "forms, movement and measurement" were of immense scientific interest, were also "among the greatest charms of this much abused climate." Recording meteorological phenomena with a camera, moreover, did not require expensive journeys by rail or steamship to faraway countries, for, as Acres pointed out, "the dweller in town has almost equal opportunities with his more fortunate confrére in the open country" (Acres 1895, 160).

Even though they did not gather as many photographs as they had hoped, members of the Royal Meteorological Society came to count meteorological photography among their greatest collective achievements. Photography, far from eliminating concerns about the integrity of meteorological observation, however, raised new concerns about how evidence was gathered. Photography of lightning seemed to be a mechanical question, but, like other methods of picturing nature, it was a human craft. Just as Symons had appealed for the public's assistance in authenticating popular meteorological testimony, Marriott and Clayden delivered illustrated lectures at photographic societies and scientific meetings to discuss the quandaries associated with making meteorological photographs of scien-

Figure 18.4 "The Amateur Photographic Pest," showing amateur photographers (mainly men) running, hiding, climbing trees, and ascending in balloons to photograph others, often women and against their wishes. Note elements portraying surprise, irritation, and outrage at photographers' invasions of privacy. Even the sun, depicted as a white circle in a dark sky, views the photographers' attempts to photograph its spots as an unwanted advance: "Photographing my spots again," it exclaims with a frown. (*Punch*, 4 October 1890, 166.)

tific value (Clayden 1891, 1898; Marriott 1889, 1890). Photographs of lightning, for example, entailed extensive preparation and related skills, including the proper selection of apparatus, adjustment of the camera so that it focused on a distant object before the onset of a storm, and the readying of forms for circumstantial reports. It also required special equipment, including a rapid rectilinear lens with a full aperture and a tripod stand to hold the camera steady. Meteorologists urged observers to give a full account of the circumstances under which each photograph was taken including the time of each flash, the interval in seconds between the lightning and the thunder, and the part of the compass in which the flash appeared (Clayden 1891, 143). Meteorologists called for photographers to stand armed and ready to photograph unusual meteorological formations in the sky (Marriott 1889, 483; Acres 1895, 162).

Thus, meteorological photography was a learned art. Experience taught meteorologists that photography could not redress all the problems that they associated with the collection of testimony from scattered human observers, such as the need to monitor the collection and handling of evidence. While nothing might seem easier than to organize "a great army of observers," Clayden reflected, "how wide a difference there is in getting others to act on their own initiative, and in getting them to fulfill some definite appointed task" (Clayden 1898, 171). When observers missed opportunities for recording unusual storms, "phenomenon after phenomenon passes unregistered," Clayden complained (Clayden 1898, 171). Even when photographs were taken, they were difficult to interpret and compare. Not all photographs had scientific value, meteorologists insisted. The problem with most skyscapes was their "pictorial effect," exclaimed Clayden: "A photographer generally takes his views with an eye to pictorial effect, and does not often care to spoil his plate by inserting anything so ungainly as a yard measure" (Clayden 1898, 171). Meteorologists sometimes blamed lack of skill for photographs of multiple flashes of lightning produced by camera movement.

The authentication of knowledge claims in meteorology did not end with the production of a photograph, moreover. Interpreting photographs of lightning required an experienced eye and a series of judgments: Were unexpected results of lightning photography (narrow ribbon bands, reduplication of the flash, the mysterious phenomenon of the "black" flash) evidence of nature, artifacts of technology, or the result of human error? Had photographs been retouched? Did this worsen or improve the faithfulness of the representation? Debates among meteorologists at the Royal Meteorological Society in the 1880s and 1890s reveal the efforts that scientists exerted to make sense of photographs. Only after numerous discussions and arguments did meteorologists conclude in the early 1890s that photographs demonstrated several things: the reality of narrow ribbon structures of lightning, the existence of multiple, compound discharges in nature, and the nonexistence of artists' zigzag lightning.

Instructing others in the meteorological aesthetic of scientific objectivity became a central project of meteorologists in the last decades of the century, years that witnessed the rise of international and domestic exhibitions featuring scientific wonders. Exhibitions of meteorological photographs and the instruments associated with their production not only informed visitors about how researchers saw lightning and clouds; they were acts of persuasion, reinforcing trust in the empirical practices of scientists at a time when, as we have seen, meteorologists struggled to assert control over the popular interpretation of weather facts. At the Royal Meteorological Society's fourteenth exhibition of meteorological instruments, held at the Institution of Civil Engineers in Westminster in April 1894, visitors saw remarkable photographs of lightning and clouds displayed next to instruments of meteorological investigation. Meteorologists like Clayden expressed hope that exhibiting photographs would stimulate the public's interest in meteorology by "presenting . . . the more picturesque and popular aspects of the Science" rather than the "dry and unintelligible" weather charts and instrumental records constitutive of standard meteorological practice (Clayden 1891, 143). In 1880, following a tour of paintings at the Paris Salon and the Royal Academy, one meteorologist deplored the "meteorological impossibilities" in the skyscapes of modern painters (particularly French artists), such as a "queer sort of sky," "sun too pale," sky "like a cheap wall paper," and, the meteorologist's pet peeve, "an impossible flash of artist's lightning" ("Meteorology and Art" 1880, 50). Among painters, only J. M. W. Turner continued to earn meteorologists' respect: "Long before the time of photography," one meteorologist recollected, Turner saw lightning in its "true form" and "duly noted the same" (Inwards 1895, 56–57).

Scientific photography in meteorology both was impressed by and stood in conflict with the pictorial interests of artists who were working in the tradition of John Constable's "naturalism" and John Ruskin's "truth of observation." Meteorologists appealed to the picturesque allure of meteorological photography to recruit new participants and audiences for meteorology. They remained compelled by "pictorial" values despite the fact that they used photographs to repudiate artistic fictions of the weather, such as zigzag lightning. Descriptions of scientific photography as the fulfillment of an earlier mandate for objective images thus oversimplify the practices that Victorians actually associated with the production and interpretive consumption of photographs in the 1890s. Skilled comparisons between photographs and drawings, the education of photographers, and the reproduction and exhibition of photographs in public settings all played a role in value judgments on meteorological photography. In meteorology as in other scientific contexts the generation of reliable knowledge depended on having a clear concept of trustworthy practice.

II. Seeing Microbes: Photography as a Detective

Unlike meteorological photography, which was connected with knowledge made by amateur meteorologists and nature photographers in the field, photography of bacteria was first produced for a rising class of laboratory experts. The first photographs of bacteria, now viewed as one of the most significant early achievements in the history of scientific and medical imaging, were made on the continent in the 1870s (figure 18.5). At a time when Victorian scientists and others regarded photographs as consequential for proving matters of fact, the camera seemed to offer objective access

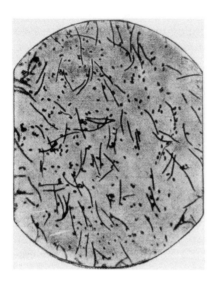

Figure 18.5 Photomicrograph of bacteria (*Proteus vulgaris Bacillus*) pictured as short rods, × 1,000, circa 1891. (From Woodhead 1891, 39.)

to a world of microbial wonders. But photographs of bacteria were neither transparent in meaning nor determined by a single interpretation outside the contexts where visual techniques and viewing practices informed how Victorians saw the war against germs, including magic lantern shows, atlases, medical laboratories, and the illustrated popular press.

Who made photomicrographs during the early years of bacteriology? Where were they circulated? Who saw them? How did viewers respond? During an era of widespread debate about the reality and agency of bacteria, visual representations of microbes made germs tangible. In Britain, factual pictures of microbes were part of a broader set of exchanges over nature, civilization, secrets of the body, and expertise. Many workers in the new field of bacteriology claimed that photographs settled debates about the existence of bacteria. But, as we will see, the introduction of photography to make bacteria recognizable raised new questions about the social and moral values inherent in making and viewing photomicrographs.

The growth of international scientific culture and communication generated demand for photomicrographs of bacteria. Scholars have identified the second half of the nineteenth century as a formative moment in the history of microbiology characterized by new methods for isolating the causative agents of disease, such as techniques for staining and counterstaining bacteria in "colonies." Several factors, including the establishment of protocols and an expert community, influenced how photographs of bacteria were viewed in Britain. These factors shaped the emergence of bacteriology and the meaning of visual images within it.

Our story begins on the continent, where a Prussian physician, Robert Koch, now recognized as one of the founders of bacteriology, made the first photomicrographs of bacteria while working at the Institute of Plant Physiology in Breslau (Brock 1988; Cunningham and Williams 1992). Koch's research on bacteria began while he was working on anthrax as a model for infectious diseases in 1873. The study of infectious diseases was an expanding area of scientific and medical research that depended on many supports from colonial governments in India and Africa. The international nature of scientific debate and exploration is illustrated by Louis Pasteur's work on the microbial basis of putrefaction and Sir Joseph Lister's procedures for antisepsis, both of which generated considerable international medical discussion about infection. The precise role of specific microbes in specific diseases, however, remained a topic of international scientific controversy (Hamlin 1990).

International scientific collaboration helped to provoke innovations in bacteriology. Koch's early experiments with photography coincided with his increased contacts with international scientific researchers, including many in Britain. In August 1876, during a visit to London, Koch met Charles Darwin and John Tyndall. Tyndall popularized Koch's work in newspapers and scientific journals.

Pictorial representation of bacteria both extended and transformed the scientific tradition of ocular demonstrations then associated with microscopic observations. Experts readily acknowledged the benefits derived from the use of the microscope for public health. Bacteriologists now turned to photomicrography to communicate their observations. Koch expressed concern that drawings failed to establish with certainty the status of bacteria. For example, in 1876, Koch criticized drawings of bacteria that failed to show resting spores (Brock 1988, 53).

To communicate his results, Koch overcame some of the difficulties of photographing bacteria, especially those pertaining to staining bacteria. To make better images, he corresponded extensively not only with scientists but also with photographers and optical instrument makers. A special difficulty was the lack of color sensitivity of his photographic emulsions: seeing the bacteria depended on observing color differences between the stained bacteria and the surrounding tissues, but the available emulsions were sensitive only to the blue region of the light spectrum. Koch used a light filter to render the bacteria visible for photography, but because the filter also reduced the intensity of the light that reached the specimen under the microscope, long exposures had to be made. The vibration of the apparatus during these exposures produced fuzzy images, complained Koch (Brock 1988, 77).

While Koch himself doubted that photographs would displace drawings of minute organisms by skilled artists as objects of research and teaching, he stressed the importance of photographs as evidence, urging others to support their discoveries with the "convincing proof afforded by photographic illustrations": "I do not mean to say that photographs should always replace drawings; that will never be the case, and in many cases a drawing alone is possible." But, he added, when photography was applicable, it should be used. If anyone lacked the apparatus, skills, or desire to attempt photography of sections of bacteria, Koch offered to demonstrate the sections by cover-glass preparations in such a way that others could photograph them or have them photographed (quoted in Crookshank 1887*a*, 10).

In 1882, Koch discovered a method for counterstaining preparations of tubercle bacilli with brown dye and photographing them with blue light. Further experiments convinced him that it would be possible by this method alone to distinguish tubercle bacilli from other kinds of bacteria. Koch's paper on the tubercle bacillus, published in Britain in April 1882, prompted vigorous debates over the truth or falsity of his claims. Some failed to find the bacilli and criticized Koch's empirical methods and competence; others traveled to his laboratory, where many corroborated his observations. He himself spent much time preparing drawings, photographs, tables, and models to illustrate his studies of tuberculosis (Brock 1988, 119-20). "I think no one will blame me for only accepting with great reserve drawings of micro-organisms, the accuracy of which I cannot substantiate by examining the original preparation," he announced (quoted in Crookshank 1887*a*, 10).

Over the next several years, bacteriology in England developed not only through contacts between scientific researchers, but through interactions among instrument makers, journalists, and photographers. By 1880, British photographers were publicizing the news of Koch's photomicrographs. One exclaimed, "As these organisms were not visible under the microscope, to photography alone is due their discovery" ("What Photography Does for Science" 1882, 101). Of photography's power to track bacteria beyond human vision, a London correspondent of the *Photographic News* boasted in 1880: "We observe, by means of the micro-camera, objects unseen by the eye" ("With Professor Lister—Photographs of Bacteria" 1880, 410).

When photographers exploited Koch's discovery to promote photography as a utilitarian art, they expanded the audiences of photomicrographs of bacteria. British collaboration with Koch generated demand for his photomicrographs. In 1881, the International Medical Congress invited Koch to demonstrate his microscopical and photographic methods at King's College, London (Brock 1988, 77-79). British apologists for Koch, including the detective fiction writer Arthur Conan Doyle, hoped that Koch's photographs would establish the reality of disease-causing germs and prove the existence of distinct species of bacteria.

As seeing bacteria became a collective activity among specialists, illustrated scientific texts helped form medical understandings of germs. Medical students constituted a major new audience for illustrated atlases of bacteria in the 1880s. From 1886 to 1910, London's major medical teaching schools added laboratory training in bacteriology to instruct students in how to recognize and isolate colonies of bacteria. The vast majority of atlases contained a combination of drawings and photographs. Authors of scientific atlases on bacteria typically either photographed their own preparations or commissioned the work from others.

In the 1880s, laboratory instructors in Britain sometimes had to justify their practice of using photographs to teach students how to see germs. For example, Edgar Crookshank, a professor of bacteriology at King's College and the founder of a distinguished bacteriological laboratory for human and veterinary pathology, defended his practice of teaching students through the medium of scientific atlases illustrated by photographs of culture specimens. So convinced did Crookshank become that photography would advance bacteriological science that he wrote a book devoted solely to the photographing of microscopic organisms, and printed instructions to photographers in the *Photographic News* (Crookshank 1887*a*; 1887*b*, 506).

Crookshank's *Photography of Bacteria,* published in 1887, containing eighty-six photographs reproduced in autotype, was the first text in English devoted solely to the photography of bacteria. Crookshank discussed methods of staining and mounting the specimens, the arrangement of illumination and focus, the operations of the camera shutter, and the need for delicate manipulations. Of the superiority of photographs over original preparations, Crookshank stressed their value in demonstration: "If I gave a microscopical preparation to a person in order to observe a particular part—for example, a lymphatic vessel containing bacteria—I could not in any way be certain that he would look at the right place, or even if he finds it, that he would properly focus and illuminate it" (Crookshank 1887*a*, 6). By contrast, photography reproduced the image for everyone to see under conditions similar to those when it was taken: the same focus, magnification, and illumination. Crookshank explained that one could place a finger on the photograph of a bacterial colony, measure it, and compare it with photographs either of the same colony or of another taken in the same manner (Crookshank 1887*a*, 8).

Thus, Crookshank valued photography as a pedagogical tool. Of the superiority of photomicrographs over drawings for teaching purposes, Crookshank argued that drawings were rarely "true to nature": they were always "prettier" and had sharper lines and more attenuated shadows than the original specimen (Crookshank 1887*a*, 64). Sketches that improved the representation of badly prepared colonies made them less truthful and therefore less instructive, he suggested, for a sharper line or darker shadow could give the figure a different meaning. Crookshank did not reject drawings; in fact, he stated that colored drawings were preferable for double- or triple-stained preparations because they indicated the method of staining. If a colored drawing had to be reproduced by wood engraving, however, thereby losing the information that color conveyed, Crookshank urged that a photograph be made instead (Crookshank 1887a, 60).

Not everyone agreed with Crookshank that photography would revolutionize bacteriological science. Edward Klein, for example, a pioneer of bacteriology and one of the founders of modern histology in Britain, criticized the use of photography to illustrate bacteria in medical textbooks. "The time has not yet come when [photographs] can be said to have supplanted good and accurate drawings," Klein declared in his *Nature* review of Crookshank's *Photography of Bacteria* (Klein 1887, 317). Klein held that photomicrographic methods yielded poorly delineated objects in a small field and allowed some bacteria to "escape reproduction" (Klein 1887, 317).

Scientific photography built on the notion of objectivity, but neither atlases nor photographs escaped judgment. One critic insisted that because atlases reproduced distorting photographs, bacteriologists ought to communicate results in diagrams, not photographs. After comparing a number of photographs, he found that "many so nearly resembled one another that the student would become confused" ("Reviews" 1899, 124). He added: "The teacher who has to rely on . . . photographs cannot know much of his subject" ("Reviews" 1899, 124). As a consequence, photographically illustrated atlases of bacteria were in his view useful neither to novices nor to specialists.

Scientists who wrote illustrated books on scientific topics for diverse (expert and novice) audiences tried to strike an elusive balance between "science" and "show." One critic of George Newman's *Bacteria*, published in 1900, declared that while "any layman of average intellect would be able to derive . . . pleasure from the perusal of it," those "who are taking up bacteriology seriously" would find little value in it ("Reviews" 1901, 262). Journalistic images of the war against germs produced for mass audiences during the 1890s bear witness to the intensifying scientific and popular interest in bacteria and to the readiness of many scientists to exploit military and imperialist iconography and racial stereotypes to show germs as unruly tribes of deadly microorganisms (figure 18.6). The connection between bacteriology and imperialist iconography was more than metaphorical; bacteriology was a science practiced in the outposts of European imperialist landholdings. Newspaper illustrations portrayed dark-skinned "natives" as both hostile warriors and friendly agents of empire just as popular science texts depicted bacteria as secret friends and foes.

In bacteriology as in meteorology, questions about how to accept photographic results crystallized into debates over what practices determined whether a photograph was acceptably "scientific." As international interest in bacteria generated demands for photomicrographs, the forums where Victorians viewed photographs of bacteria expanded, from prints and magic lantern shows to published atlases and the illustrated periodi-

Figure 18.6 Pen and ink sketch by D. E. Wilson for "The Army of the Interior" (Machray and Browne 1899, 264). Victorian representations of indigenous "natives" and military warfare were important sources for journalistic illustrations of germ warfare. The bacteria are depicted in words and pictures in "The Army of the Interior" as a hostile mass of black demons personalized here only with minstrel faces.

cal press. If it was true that broadening participation beyond the traditional circle of photographic inventors eventually confirmed photography as the paragon of objective scientific practice, this broad participation also opened photography to challenges from many different points of view.

III. Can Spirits Be Photographed?

In 1897, the psychical researcher John Godfrey Raupert posed the persisting dilemma of spirit photography in this way: "On the one hand," he remarked, "it is perfectly clear that fraud and trickery have again and again been resorted to with a view of duping the public and to producing many of the now well-known manifestations" (Raupert 1897, 421). Mr. Maskelyne, the famous conjurer at Egyptian Hall, Raupert noted, had succeeded in showing the public how spirit photographs could be faked. There was also, he acknowledged, reason to suspect paid mediums, dark rooms, and cabinets. Phenomena produced under such conditions might well be of a "doubtful and worthless character." But, Raupert pointed out, "We have, on the other hand, the emphatic and unhesitating testimony of eminent scientific men, such as Profs. Crookes, Alfred Russel Wallace, Barrett, and Oliver Lodge—their deliberate assertions, that the phenomena occur, that in very many instances they cannot possibly be due to trickery, and that there is, at present, no known law of science by which they can be reasonably explained and accounted for." The force of this evidence was strengthened by the fact that investigations of spirit photography were carried on according to "strictly scientific principles"; that is, "with no tendency of mind or pre-conceived views, and in many instances, with a strongly marked bias against supernatural or supernormal phenomena of any kind" (Raupert 1897, 423-24).

Debates in Britain over the viability of photographs as evidence of immortality in the last decades of the century dramatize wider Victorian struggles to define objective knowledge. Like meteorologists and bacteriologists, spirit photographers and their supporters employed procedural conventions to portray their photographs as objective. Spirit photography, unlike other fields of photography, never fully escaped imputations of imposture; as in the market for relics, spirit photography had both real and imagined dealers in fraud (Jones 1990, 161). In an age of public speculation about the genuineness or falsity of claims about spirit photographs, spirit photographers were "cross-examined" in a series of "trials." Nineteenth-century trials of spirit photographers, ranging in climate from friendly to hostile, took place in a variety of different locations, including spiritualists' drawing rooms, commercial photographic studios, photographic meetings, newspapers, and police courts. The appellation of "impostor," often raised in connection with nineteenth-century photography, provided Victorians with a way to explain how photography, in which so much trust was invested, could deceive: by pretending to be what it was not.

An early historian of spirit photography has identified two "booms" of interest in spirit photography in Britain: one in the 1870s, following the trial of the American spirit photographer William Henry Mumler in New York City; the second in the 1890s (Coates 1911). Mumler, who gave 5 October 1861 as the date of his first "accidental" spirit photograph, is generally held to have been the first photographer to take "authentic" photographs of spirits. In a case that shocked people on both sides of the Atlantic, an American tabloid journalist accused Mumler of selling him a photograph that Mumler said portrayed one of the writer's relatives, a claim the journalist rejected. Charged with fraud, Mumler appeared before a judge in a New York police court. The testimony of leading citizens, including photographers who tested Mumler, proved crucial to the dismissal of charges against him. Although the judge declared that, personally, he was morally convinced that Mumler had employed trickery and fraud in making his spirit productions, the evidence did not support a conviction (Coates 1911, 4; "Spiritual Photographs" 1869, 285).

Victorian spiritualists and photographers credited the British press coverage of the Mumler case with stimulating interest in spirit photography. On 4 March 1872, three years after the Mumler trial, British portrait photographer Mr. Hudson, who managed a studio in Holloway Road, London, made what is generally understood to be the first spirit photograph taken in Britain (Gunning 1995; Coates 1911). The testimony of leading scientists, including Alfred Russel Wallace, lent credibility to Hudson's results. "The moment I got the proofs the first glance showed me that the third plate contained an unmistakable portrait of my [deceased] mother," Wallace reported (Coates 1911, 37-38). The possibility that photography could record psychic or spiritual forces seemed to some Victorians no more marvelous than other scientific discoveries made with photographic processes in the fields of astronomy, meteorology, physics, and biology (Gunning 1995; Owen 1989; Oppenheim 1985; Turner 1974; Wynne 1979). While few, if any, Victorians believed that all spirit photographs were photographs of spirits, many felt that some photographs at least were authentic and, furthermore, that just one "honestly produced" photograph of a ghost justified the employment of photography to gain access to the reality of a spiritual world. As one subscriber to the *Photographic News* asked: "Who will deny the possibility that spiritual beings can be photographed if they exist?" Spirit photographs might "seem absurd now," argued another, but "so did scientific discoveries ridiculed in other ages" ("Spiritual Photographs" 1869, 285).

In Britain, reflections on the authenticity of spirit photographs compelled viewers to reflect on the quality of the family's testimony. At around the same time that Hudson achieved fame as a spirit photographer, a related cause célèbre was expanding the discourse on photographic evidence, lost kin, imposture, and expert testimony in Britain. From 1871 to 1874, the sensational trial of the "Tichborne Claimant" dominated the London news. At the center of this case was a man who claimed to be the lost son of a dying heiress of the Tichborne estate, south of London. He apparently had died at sea around twelve years earlier. Although the Claimant appeared physically different from the young Roger Tichborne, he presented sufficient personal knowledge of the Tichborne family to convince many people (including Lady Tichborne) that he was, in fact, Roger Tichborne. Over Lady Tichborne's objections, the shocked family prosecuted him as a defrauder. Like spirit photography, the Tichborne trial dramatized witnessing, lost kin, and expert examiners. Witnesses were asked to compare photographs of Tichborne, showing him before he allegedly drowned at sea, with photographs of the Claimant. One doctor, William Matthews, a rival of Francis Galton, attempted to prove their common identity by applying geometric methods; he thought that a person's irises ought to measure the same dimensions over time (Matthews 1876) (figure 18.7).

New technical devices and expanded social audiences for spirit photography transformed the notion of practice in spirit photography in the 1880s and 1890s. Dry-plate photography, invented around 1871, simplified the photographing of spirits, and the growth of the amateur movement in spirit photography produced a wide range of photographs including Martian spirits, allegorical figures, and "extras" of different racial backgrounds. Tales of unexpected manifestations of dead friends, lovers, and relatives became "tiresomely common" in the 1890s, according to Hugh Reginald

Figure 18.7 "The Tichborne Blended Photographs." William Matthews, a physician and vocal advocate of the Tichborne Claimant's cause, used the science of geometry to try to show the common identity of Tichborne, photographed in 1853, and the Claimant, photographed in 1873. The caption reads: "Close scrutiny of these blended photographs, (truthful and authentic beyond all denial) must convince every intelligent person that they represent one and the same man—Sir Roger Tichborne." (Courtesy of University College London [reference: Galton 158/2M].)

Haweis, a British minister and spiritualist. "Only the other day I was told of a young lady who went down to Brighton to an ordinary photographer. She sat as an ordinary sitter, suspecting nothing. The plate came out blurred all over; photographer surprised, and on point of casting plate aside, when sitter begs to see it, and further begs to have it printed off. Result—photo blurred all over, sitter unrecognisable: when subjected to high magnifyer, milky way of blue reveals innumerable faces, but all the same face! Recognised by young lady at once as face of dead lover" (Haweis, quoted in Glendinning 1894, 72).

As in the Victorian antique trade, where identifications of genuine and faked articles were made by skilled workers, discussions about spirit photographs highlighted issues of expertise. John Traill Taylor, as editor of the *British Journal of Photography,* author of several works on photographic chemistry and optics, and Britain's leading witness in photography patent disputes, revived general interest in spirit photography in the 1890s. Taylor declared that his interest in spirit photography was merely that of an experimentalist. Witnesses attended Taylor's famous researches in the presence of the Scottish medium David Duguid at drawing-room seances in the 1890s (figure 18.8). "My conditions were simple," recalled Taylor: "They were, that I for the nonce would assume them all to be tricksters, and, to guard against fraud, should use my own camera and unopened packages of dry plates from dealers of repute." He urged his two witnesses to treat him as under suspicion (Glendinning 1894, 24). In this investigation as in others,

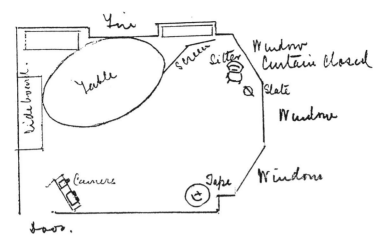

Figure 18.8 Diagram showing the setup of one of John Traill Taylor's spirit photography trials, circa 1892, marking the placement of the camera (*bottom left*) and of the sitter (*top right*). (Cambridge University Library, Society for Psychical Research Papers, Colonel Taylor's notes of sittings with Duguid, David Duguid ["Mediums File"], diagram on p. 4. Courtesy of The Society for Psychical Research. Reproduced by Cambridge University Library.)

Taylor concluded that the plates, some of which showed spirit "extras," had not been tampered with. He refused to state any hypothesis, vowing only to exhibit the results, though they were, he conceded, "fraudulent-looking." Taylor was careful not to state his opinion of how the spirit extras appeared on the plate: "Pictorially they are vile, but how came they there?" he queried (Glendinning 1894, 33).

Like meteorologists, spirit photography investigators such as Taylor compared and analyzed their photographic phantasms at public meetings, where lantern slide projectors were used to project images of ghosts onto the wall. Spirit photographs, as Taylor and his colleague, the Scottish spiritualist Andrew Glendinning, defined them, denoted "photographs of psychic entities who cannot be seen by ordinary persons, but can be photographed by a medium, or with the help of a medium, and with the cooperation of unseen entities" (Glendinning 1894, iv). Their classification of spirit photographs included portraits of psychic entities not seen by normal vision; pictures of objects not seen or thought of by the sitter, the medium, or the photographic operator; pictures that had the appearance of being copied from statues, paintings, or drawings; pictures of wraiths and doubles of persons still in the body; and portraits on plates that developers failed to bring into view but that could be seen and described by clairvoyants and mediums when in trance (figure 18.9).

Skeptics who challenged the authenticity of spirit photography knew little, in fact, about modern photographic techniques and theories; they

therefore could not engage the expert Taylor on a technical level. Instead, they called Duguid and Glendinning "impostors" and "pretenders" to define situations where photographs were presented as evidence of phenomena that they disbelieved. Following a widely publicized presentation by Taylor at the London and Provincial Photographers meeting in March 1893, for example, a "Mr. F. Gass" argued in the *Christian World* that the figures were cut-outs because, when examined by a stereoscope, they appeared flat (quoted in Glendinning 1894, 67). While Taylor had explained that the figures were solid enough to obscure a view of the sitting medium, according to Gass this was not possible. Gass also scorned the possibility of photographing the invisible, of obtaining a portrait without the use of light, and (as Taylor claimed he was able to do) of securing a photographic image without the use of a lens.

Glendinning's defense of Taylor ruled out Gass's authority to deliberate on the question, saying that Gass lacked the necessary knowledge of photography, or of spirits, for that matter. Glendinning waxed sarcastic: "That is, Mr. Gass knows all about solidity, and what could or could not be done by what Mr. Taylor is careful to call a 'psychic entity.'" Of Gass's claim that photography of the invisible was impossible, Glendinning wondered if Mr.

Figure 18.9 "Photograph of a Psychic Lady" (*facing page*) and "Abnormal Portrait of Lady," spirit photographs made during a seance attended by John Traill Taylor, circa 1892 (reproduced in Glendinning 1894, 29 and 35). Taylor, who showed the "Abnormal Portrait" at a London meeting of photographers, deduced that the spirit's image was not formed by the lens because while the two sitters were "stereoscopic," the "psychic figure" was "absolutely flat" (35). He called for scientists to solve the mystery of the "psychic image without a camera."

Gass had "any knowledge of the spectrum, and could he himself 'see' the actinic rays which produce the pictures on the sensitized plate?" If, as Gass had urged, lenses and light were imperative for securing an image, there was an end to the matter, conceded Glendinning, but he insisted that experiments seemed to disprove "the necessity of this 'must.'" Such was the authority of the scientific and technical claims associated with spirit photography that skeptics often were taken to task for their lack of knowledge of the general principles involved (Glendinning 1894, 67–69).

Supporters of spirit photography in Victorian Britain appealed to the honesty, competence, and moral integrity of individuals like Duguid and Taylor who were associated with it. Much as earlier apologists had defended Hudson, those who supported the medium, David Duguid, pointed to his working-class background as evidence that (without spiritual intervention) he could not deceive others so thoroughly. However, as a contributor to the *Practical Photographer* shrewdly pointed out, the "main point of issue" in the experiments was not the medium's morality, but Taylor's competence: "Whether or not the medium was capable of fraud if he got the chance" was not the question, "but whether or not Mr. Taylor is believable when he asserts that the chance for fraud was not given" (Glendinning 1894, 56). Following Taylor's presentation at the London and Provincial Photographers Society, Taylor's colleague Glendinning asked members and their visitors if they believed in Taylor's competence for the investigation. "I do not mean merely, do you consider Mr. Taylor an honest investigator," he explained: "There are many honest men who would not be considered qualified to watch with sufficient care and accuracy experiments in photographic manipulations," who lacked, for example, the necessary knowledge of chemistry and optics and their relation to photography. Others might be honest and skillful, but "too opinionative" to conduct such experiments. Photographers who sought to distance themselves from spirit photography, but who generally respected Taylor as one of its leading investigators, negotiated the politics of disagreeing with Taylor while being professionally courteous. In reply, one photographer stated that as practical photographers, they were "bound to believe Mr. Taylor's statement," but to him — though the pictures shown "*might* be spirit photographs" — they gave the idea of "cut-out prints." Who cut them out he did "not know and did not care," but he did not wish to impute any "complicity or dishonesty" to Glendinning and Taylor (Glendinning 1894, 43).

Like photographic practices in meteorology and bacteriology, spirit photography highlighted issues of witnessing, detecting, and deceiving. While for some, it seemed clear that many spirit photographs were fakes, the testimony of eminent scientists and photographers lent credibility to claims on behalf of spirit photography. For others, truth claims about spirit photographs cast doubt on the validity of knowledge produced by photographic methods in other areas. Spirit photographers never fully succeeded in securing their practices against imputations of fraud: their practices were always suspected and watched. In raising problems associated with the use of photography to represent facts, however, spirit photography was not unusual in Victorian Britain, for, as I have suggested, questions of impartiality, objectivity, and honesty also arose in connection with the use of photography in other research fields, albeit in different ways.

IV. Photographic Evidence and Social Practice

The formation of consensus on the meaning and objectivity of photographs was a complex social process that involved various individuals and institutions. In this chapter I have attempted to show some of the ways that photographs were made and handled as evidence of unseen or illusory phenomena (lightning, bacteria, and spirits) that fascinated Victorian scientists and the general public. Victorians were acutely aware that photography consisted of many distinct steps, any or all of which could potentially determine whether it was "scientific." The role and integrity of photography was interrogated not only in realms such as spiritualism, but in contexts where photographs are not usually thought to have been controversial, such as meteorology and bacteriology. In a variety of fields, technical practices of production and reproduction, the competence of photographers, and the visual education of viewers, including the forms through which individuals learned ways to assess anomalous images, played crucial roles in determining how scientists used photographs to see and to instruct others how to see nature. Far from unproblematically compelling belief, Victorian scientific photographs carried messages whose interpretation varied widely depending on their points of production and reception.

Further research is needed on the place of images and image production in the architecture of Victorian science. To explain how pictures, including photographs, communicated information in the nineteenth century, we must look beyond their points of origin and immediate rhetorical surroundings to their contexts of reproduction and reinterpretation. Assessing the role of photography in Victorian matters of belief compels us to reflect on the impact of a variety of technological, social, and intellectual forces that were at work in the late nineteenth century—forces associated, for example, with scientific discipline formation, the factory production of photographic equipment, the rise of amateur photography, and the growth of "mass" viewing publics created by the new democratic journalism. The study of Victorian scientific illustration provides us with a unique opportunity to explore questions currently being addressed by historians of Victorian science and culture, such as the historical development of scientific disciplines, relations between professionals and amateurs, interconnections between science, technology, and practice, and definitions of science and popular science. It also promises to yield fresh insights into topics in the field of scientific and medical imaging including the changing relations between art and truth claims, texts and pictures, knowledge production and visual practices.

Bibliographical Note

Recent studies of scientific illustration from the late Renaissance to the present day suggest growing interest in the field of visual representation. New and fruitful questions are being asked about the links that connected ideas, images, and knowledge-producing practices in Victorian Britain. Within this area of inquiry, however, more work is needed on issues of representation and authority that images and image production raised in scientific practice. What place did drawings, woodcuts, engravings, photographs, and diagrams have in the Victorian market of ideas? How did the relationship between illustration and investigation evolve in different research settings? What circumstances and conditions surrounded the reception of images as proof or fact?

To answer these types of questions, we must investigate problems lying between the conventional fields of the history of science, art history, and cultural criticism. Studies that explore links between images and knowledge-producing practices in Victorian science and culture include works on natural history, landscape art, and exploration (Blum 1993; Camerini 1993; Klonk 1996; Stafford 1984), on geological illustration (Rudwick 1976, 1992), on astronomical photography (Pang 1993, 1994, 1996; Rothermel 1993; Schaffer 1988, 1995; Lankford 1987), on "mimetic experiment" in meteorology (Galison and Assmus 1988), on anthropology and illustration (Edwards 1992; Cowling 1985), on images of degeneration and madness (Gilman 1985, 1988), on motion pictures (Braun 1992), and on scientific discoveries made with photographs (Darius 1984). Recent research on the history of medical illustration includes Cartwright (1995) and studies focusing on nineteenth-century photography (Maehle 1993; Fox and Lawrence 1988; Sekula 1986; and Gilman 1976) and on European images of the body and sex (Stafford 1991; Laqueur 1990; Jordanova 1989; Schiebinger 1986). For a text that surveys visual representation in science before 1900, see Mazzolini (1993). For information on the technical reproduction of images in Victorian Britain see Gascoigne (1986) and Jussim (1974). Fundamental nineteenth-century theories of art are addressed in Taylor (1987) and Nochlin (1971). Women's contributions to Victorian art are discussed in Orr (1995) and Pollock (1988). On the general state of Victorian photography, see especially Lalvani (1996), Henisch and Henisch (1994), Schaaf (1992), Lemagny and Rouillé (1986), Seiberling (1986), Weaver (1986), Gernsheim (1969), Collins (1990), and Newhall (1949). Problems that photography raised for representation practices in the nineteenth century are treated in Daston and Galison (1992), Crary (1991), Orvell (1989), Tagg (1988), Sekula (1986), Snyder (1980), Benjamin (1979), and Snyder and Allen (1975). For an overview of recent critical perspectives on scientific and medical representation see Jordanova (1990), Porter (1988), and Fyfe and Law (1988). For case studies of scientific image production informed by sociological critiques of scientific practice, see Lynch and Woolgar (1990) and Lynch (1985).

References

Abercromby, Ralph. 1888. "On the Photographs of Lightning Flashes." *Quarterly Journal of the Royal Meteorological Society* 14:226–34.

Acres, Birt. 1895. "Some Hints on Photographing Clouds." *Quarterly Journal of the Royal Meteorological Society* 21:160–65.

Benjamin, Walter. 1979. "The Work of Art in the Age of Mechanical Reproduction." In *Mass Communication and Society,* edited by J. Curran, M. Gurevitch, and J. Woolacott. London: Edward Arnold with the Open University, 384–408.

Blum, Ann Shelby. 1993. *Picturing Nature: American Nineteenth Century Zoological Illustration.* Princeton, N.J.: Princeton University Press.

Braun, Marta. 1992. *Picturing Time: The Work of Étienne-Jules Marey (1830-1904).* Chicago: University of Chicago Press.

Brock, Thomas D. 1988. *Robert Koch: A Life in Medicine and Bacteriology.* Madison, Wis.: Science Tech Publications.

Camerini, Jane R. 1993. "The Physical Atlas of Heinrich Berghaus: Distribution Maps as Scientific Knowledge." In *Non-verbal Communication in Science prior to 1900,* edited by Renato G. Mazzolini. Florence: Leo S. Olschki, 479-512.

Cartwright, Lisa. 1995. *Screening the Body: Tracing Medicine's Visual Culture.* Minneapolis: University of Minnesota Press.

Clayden, Arthur. 1891. "Meteorological Photography." *Quarterly Journal of the Royal Meteorological Society* 17:142-46.

———. 1898. "Photographing Meteorological Phenomena." *Quarterly Journal of the Royal Meteorological Society* 24:169-180.

Coates, James. 1911. *Photography of the Invisible.* London: Fowler.

Collins, Kathleen, ed. 1990. *Shadow and Substance: Essays on the History of Photography in Honor of Heinz K. Henisch.* Bloomfield Hills, Mich.: Amorphous Institute Press.

Cowling, Mary. 1985. *The Artist as Anthropologist: Representation of Human Types in Victorian Art.* Cambridge: Cambridge University Press.

Crary, Jonathan. 1991. *Techniques of the Observer: Vision and Modernity in the Nineteenth Century.* Ithaca, N.Y.: Cornell University Press.

Crookshank, Edgar. 1887a. *Photography of Bacteria.* London: H. K. Lewis.

———. 1887b. "Photo-micrography of Bacteria." *Photographic News* 31:473-506.

Cunningham, Andrew, and Perry Williams, eds. 1992. *The Laboratory Revolution in Medicine.* Cambridge: Cambridge University Press.

Darius, Jon. 1984. *Beyond Vision.* Oxford: Oxford University Press.

Daston, Lorraine, and Peter Galison. 1992. "The Image of Objectivity." *Representations* 40:81-128.

De Fonvielle, W. 1869. *Thunder and Lightning.* Translated from the French and edited by T. L. Phipson. New York: Charles Scribner, 1869.

Edwards, Elizabeth, ed. 1992. *Anthropology and Photography, 1860-1920.* New Haven, Conn.: Yale University Press.

Fox, Daniel, and Christopher Lawrence. 1988. *Photographing Medicine: Images and Power in Britain and America since 1840.* New York: Greenwood Press.

Fyfe, Gordon, and John Law. 1988. *Picturing Power: Visual Depiction and Social Relations.* London: Routledge.

Galison, Peter, and Alexi Assmus. 1988. "Artificial Clouds, Real Particles." In *The Uses of Experiment: Studies in the Natural Sciences,* edited by D. Gooding, T. Pinch, and S. Schaffer. Cambridge: Cambridge University Press, 225-74.

Gascoigne, Bamber. 1986. *How to Identify Prints: A Complete Guide to Manual and Mechanical Processes from Woodcuts to Inkjet.* New York: Thames and Hudson.

Gernsheim, Helmut. 1969. *The History of Photography.* London: Thames and Hudson.

Gilman, Sander, ed. 1976. *The Face of Madness: Hugh Diamond and the Origin of Psychiatric Photography.* Secaucus, N.J.: Citadel Press.

———, ed., with J. Edward Chamberlin. 1985. *Degeneration: The Dark Side of Progress.* New York: Columbia University Press.

———. 1988. *Disease and Representation: Images of Illness from Madness to AIDS.* Ithaca, N.Y.: Cornell University Press.

Glendinning, Andrew, ed. 1894. *The Veil Lifted: Modern Developments of Spirit Photography.* London: Whittaker and Company.

Gooday, Graeme. 1991. " 'Nature' in the Laboratory: Domestication and Discipline with the Microscope in Victorian Life Science." *British Journal for the History of Science* 24:307–41.

Gunning, Tom. 1995. "Phantom Images and Modern Manifestations: Spirit Photography, Magic Theater, Trick Films, and Photography's Uncanny." In *Fugitive Images: From Photography to Video*, edited by Patrice Petro. Bloomington: Indiana University Press, 42–71.

Hamlin, Christopher. 1990. *A Science of Impurity: Water Analysis in Nineteenth Century Britain.* Bristol: Adam Hilger.

Henisch, Heinz, and Bridget Henisch. 1994. *The Photographic Experience 1839–1914: Images and Attitudes.* University Park: Pennsylvania State University Press.

Inwards, Richards. 1895. "Weather Fallacies." *Quarterly Journal of the Royal Meteorological Society* 21:49–62.

Ivins, William M., Jr. 1980. *Prints and Visual Communication.* Cambridge: MIT Press.

Jelf, E. A. 1894. "Photography as Evidence." *Idler* 4:517–25.

Jones, Mark, ed. 1990. *Fake? The Art of Deception.* London: British Museum Publications.

Jordanova, Ludmilla. 1989. *Sexual Visions: Images of Gender in Science and Medicine between the Enlightenment and the Twentieth Centuries.* Madison: University of Wisconsin Press.

———. 1990. "Medicine and Visual Culture." *Social History of Medicine* 3:89–99.

Jussim, Estelle. 1974. *Visual Communication and the Graphic Arts: Photographic Technologies in the Nineteenth Century.* New York: Xerox Publications.

Klein, E. 1887. "Review of Crookshank's *Photography of Bacteria*." *Nature* 36 (4 August): 317.

Klonk, Charlotte. 1996. *Science and the Perception of Nature: British Landscape Art in the Late 18th and Early 19th Century.* New Haven, Conn.: Yale University Press.

Lalvani, Suren. 1996. *Photography, Vision, and the Production of Modern Bodies.* Albany: State University of New York Press.

Lankford, John. 1987. "Photography and the Nineteenth-Century Transits of Venus." *Technology and Culture* 28:648–57.

Laqueur, Thomas. 1990. *Making Sex: Body and Gender from the Greeks to Freud.* Cambridge, Mass.: Harvard University Press.

Lemagny, Jean-Claude, and André Rouillé, eds. 1986. *A History of Photography: Social and Cultural Perspectives.* Cambridge: Cambridge University Press.

Lynch, Michael. 1985. *Art and Artifact in Laboratory Science.* Boston: Routledge and Kegan Paul.

———. and Steve Woolgar, eds. 1990. *Representation in Scientific Practice.* Cambridge: MIT Press.

Machray, Robert, and J. Arthur Browne. 1899. "The Army of the Interior." *Pearson's Magazine,* January-June, 263-67.

Maehle, Andreas-Holger. 1993. "The Search for Objective Communication: Medical Photography in the Nineteenth Century." In *Non-verbal Communication in Science prior to 1900,* edited by Renato G. Mazzolini. Florence: Leo S. Olschki, 563-86.

Marriott, William. 1889. "Instructions for Taking Photographs of Lightning." *Photographic News,* 26 July, 483.

———. 1890. "Application of Photography to Meteorological Phenomena." *Quarterly Journal of the Royal Meteorological Society* 16 (July): 146-51.

Matthews, William. 1876. *Identity Demonstrated Geometrically with Phototype Illustrations.* Bristol: J. Wright and Company.

Mazzolini, Renato G., ed. 1993. *Non-verbal Communication in Science prior to 1900.* Florence: Leo S. Olschki.

"Meteorology and Art." 1880. *Symons's Monthly Meteorological Magazine* 15 (May): 49-51.

Middleton, W. E. Knowles. 1969. *Invention of the Meteorological Instruments.* Baltimore: Johns Hopkins University Press.

Newhall, Beaumont. 1949. *The History of Photography from 1839 to the Present Day.* New York: Museum of Modern Art.

Nochlin, Linda. 1971. *Realism.* London: Penguin Books.

Oppenheim, Janet. 1985. *The Other World: Spiritualism and Psychical Research in England, 1850-1914.* Cambridge: Cambridge University Press.

Orr, Clarissa Campbell, ed. 1995. *Women in the Victorian Art World.* Manchester: Manchester University Press.

Orvell, Miles. 1989. *The Real Thing: Imitation and Authenticity in American Culture, 1880-1940.* Chapel Hill: University of North Carolina Press.

Owen, Alex. 1989. *The Darkened Room: Women, Power and Spiritualism in Late Nineteenth Century England.* London: Virago Press.

Pang, Alex Soojung-Kim. 1993. "The Social Event of the Season: Solar Eclipse Expeditions and Victorian Culture." *Isis* 84:252-277.

———. 1994. "Victorian Observing Practices, Printing Technology and Representations of the Solar Corona." *Journal of the History of Astronomy* 25:249-74.

———. 1996. "The Industrialization of Vision in Victorian Astronomy." In *Cultural Babbage: New Studies in Victorian Science,* edited by Frances Spufford and Jenny Uglow. London: Faber and Faber, 167-92.

"Photographic Lies: Proving the Worthlessness of the Camera as a Witness." 1898. *Harmsworth Magazine* 1:261-64.

"Photography in Court." 1867. *Photographic News,* 31 May, 264.

"Photography in the Witness Box." 1859. *Photographic News,* 29 April, 85.

Pollock, Griselda. 1988. *Vision and Difference: Femininity, Feminism, and the Histories of Art.* London: Routledge.

Porter, Roy. 1988. "Seeing the Past." *Past and Present* 188:186-205.

Raupert, John Godfrey. 1897. "Can Spirits Be Photographed?" *Humanitarian,* December, 421-24.

"Reviews." 1899. *Guy's Hospital Gazette,* 18 March, 124.

"Reviews." 1901. *Guy's Hospital Gazette,* 8 June, 262.

Rothermel, Holly. 1993. "Images of the Sun: De La Rue, Airy and Celestial Photography." *British Journal of the History of Science* 26:137-69.

Rudwick, Martin. 1976. "The Emergence of a Visual Language for Geological Science 1760-1840." *History of Science* 14:149-95.

———. 1992. *Scenes from Deep Time: Early Pictorial Representations of the Prehistoric World.* Chicago: University of Chicago Press.

Schaaf, Larry J. 1992. *Out of the Shadows: Herschel, Talbot and the Invention of Photography.* New Haven, Conn.: Yale University Press.

Schaffer, Simon. 1988. "Astronomers Mark Time: Discipline and the Personal Equation." *Science in Context* 2:115-45.

———. 1995. "Where Experiments End." In *Scientific Practice,* edited by Jed Buchwald. Chicago: University of Chicago Press, 257-99.

Schiebinger, Londa. 1986. "Skeletons in the Closet: The First Illustrations of the Female Skeleton in Eighteenth-Century Anatomy." *Representations* 14:42-82.

Seiberling, Grace, with Carolyn Bloore. 1986. *Amateurs, Photography and the Mid-Victorian Imagination.* Chicago: University of Chicago Press.

Sekula, Allan. 1986. "The Body and the Archive." *October* 39:3-64.

Snyder, Joel. 1980. "Picturing Vision." *Critical Inquiry* 6:499-526.

Snyder, Joel, and Neil Walsh Allen. 1975. "Photography, Vision, and Representation." *Critical Inquiry* 2:143-69.

"Spirit Photographs." 1863. *Photographic News* 7 (13 February): 73.

"Spiritual Photographs." 1869. *Photographic News* 13 (11 June): 285.

Stafford, Barbara. 1991. *Body Criticism.* Cambridge: MIT Press.

———. 1984. *Voyage into Substance: Art, Science, and the Illustrated Travel Account, 1760-1840.* Cambridge: MIT Press.

Symons, George J. 1888. "The Non-Existence of Thunderbolts." *Quarterly Journal of the Royal Meteorological Society* 14:208-12.

———. 1889. "Results of an Investigation of the Phenomenon of English Thunderstorms during the Years 1857-1859." *Quarterly Journal of the Royal Meteorological Society* 15:1-13.

Tagg, John. 1988. *The Burden of Representation.* Amherst: University of Massachusetts Press.

Taylor, Joshua, ed. 1987. *Nineteenth-Century Theories of Art.* Berkeley: University of California Press.

Tucker, Jennifer. 1996. "Science Illustrated: Photographic Evidence and Social Practice in England, 1870-1920." Ph.D. dissertation, Johns Hopkins University.

Turner, Frank. 1974. *Between Science and Religion: The Reaction to Scientific Naturalism in Late Victorian England.* New Haven, Conn.: Yale University Press.

Weaver, Mike. 1986. *The Photographic Art: Pictorial Traditions in Britain and America.* New York: Harper and Row.

"What Photography Does for Science." 1882. *Photographic News,* 3 March, 101.

"With Professor Lister—Photographs of Bacteria." 1880. *Photographic News* 24 (27 August): 409-10.

Woodhead, German Sims. 1891. *Bacteria and Their Products.* London: Walter Scott.

Woodruff, Douglas. 1957. *The Tichborne Claimant: A Victorian Mystery.* London: Hollis and Carter.

19

Instrumentation and Interpretation: Managing and Representing the Working Environments of Victorian Experimental Science

GRAEME J. N. GOODAY

> Any investigation in experimental physics requires a large expenditure of both time and patience; the apparatus seldom, if ever, begins by behaving as it ought; there are times when all the forces of nature, all the properties of matter, seem to be fighting against us; the instruments behave in the most capricious way, and we appreciate Coutts Trotter's saying that the doctrine of the constancy of nature could never have been discovered in a laboratory.
>
> J. J. THOMSON, address to Section A of the British Association for the Advancement of Science, Liverpool, 1896 (J. Thomson 1896, 700)

As a less accomplished experimenter than many at the 1896 meeting of the British Association for the Advancement of Science, Joseph John Thomson was doubtless familiar with the difficulties of interpreting the often disorderly behavior of benchtop instruments.[1] Although said in jest, his utterance to fellow physicists suggests to the late-twentieth-century reader that it was no trivial matter for a Victorian laboratory worker to elicit orderliness from the elusive and refractory "nature." A whole range of chaotic effects could compromise the orderly construction of knowledge from instrumental outcomes. Most obvious among these were the haphazard vibrations caused by passing traffic and pedestrians; more difficult to identify

---

I would like to thank John Christie, Jeff Hughes, Bruce Hunt, Sean Johnston, Bernie Lightman, and Ben Marsden for their advice and criticism of earlier drafts of this chapter, and to acknowledge the constructive support of Jon Agar, Sophie Forgan, Eric Kupferberg, and my colleagues at the University of Leeds.

1. Thomson succeeded Lord Rayleigh as professor of experimental physics at the University of Cambridge in 1884; according to his son, J. J. Thomson was "naturally clumsy with his hands" and thus "more dependent than most physicists of his day on mechanical assistants" (G. Thomson 1964, 73). I am much indebted to Jeff Hughes for drawing my attention to this quotation.

were the subtle forms of interference (thermal or electromagnetic) between adjacent pieces of apparatus; and then there were the mysteriously idiosyncratic performances of laboratory equipment that could even raise doubts about the very orderliness of nature itself.[2] The main aim of this chapter will be to examine how late Victorian practitioners in the physical and life sciences managed and textually represented such day-to-day problems of material ordering in their work; my conclusion will draw out some general points about how closer attention to such activities can enrich our historical accounts of past scientific practice.

Of course, scientific workers in the late nineteenth century had to contend with many diverse problems; these included theological agonies over the age of the earth, the troublesome demarcation of science as a profession, and the defusing of often bitter controversies over priority claims. Against a backdrop of such publicly aired controversies, the everyday, practical problems they faced in handling scientific instruments can be dismissed all too easily as mundane and trite. Historians of the philological persuasion may not be alone in this: historians of the old Edinburgh style, searching for the social "interests" that govern knowledge making, might equally well dismiss such problems as materialistic or internalistic irrelevances. Yet some social and cultural historians find that there is much of interest to be said about the material concerns of Victorian science, especially those who have imbibed from such disciplines as anthropology, archaeology, museum studies, and the history of technology. Whatever the origin of this concern, the growing commitment to it is palpable.[3]

One issue, however, is seldom addressed in this literature: the persistent everyday difficulty experienced by practitioners in getting their devices to behave reliably in the artificial, incompletely controlled, and sometime quite unstable environments of their work. What was at stake here was none other than the stability of the putative knowledge produced with these instruments, as recent studies have shown (Galison 1987; Turnbull 1995). In the early Victorian context, these issues arise most clearly in Sibum's study of James Joule's thermometric practice of the 1840s. Sibum's painstaking practical reconstructions show that the thermometers employed by Joule could produce reliable results only in highly contrived environmental conditions that were unique to his family beer business's underground cellar in Manchester, and only with the temperature-gauging skills unique to his brewery training. Joule had to work with an extraordinarily well controlled ambient temperature ($\pm$ 0.5%) uniquely possible in an environment specifically built for such a purpose; he had to work almost entirely alone, lest body heat disrupt the finely balanced thermal equilibrium; he also required the leisure to work for many hours uninterrupted.

---

2. See the discussion of Knorr-Cetina later in this chapter.
3. See "Bibliographical Note" at the end of this chapter.

Only within these extraordinarily artificial conditions could Joule get anything like a stable "natural" value for the mechanical equivalent of heat. Unsurprisingly, contemporaries of Joule working in less controlled environments had great difficulty in replicating his experiments (Sibum 1995).

This would, of course, be surprising if one uncritically projected Bruno Latour's account of twentieth century instrumental work to the later nineteenth. In *Science in Action,* Latour supposes that instruments can work equally well in any context so long as enough social interests are enrolled to support their credible usage and accurate calibration. For him the materialities of ambient conditions are irrelevant, for modern devices of metrological or graphical representation allegedly act as blackboxes or "immutable mobiles" (Latour 1987, 2, 131, 227). Sibum's study shows that it would be inappropriate to extend this analysis to all instruments in all historical contexts. Joule's extremely delicate thermometer was not after all conspicuously mobile, it was not boxed in black, and perhaps the very last thing that one could say about it was that it was immutable! Hence, we must not take for granted the robustness or reputation of Victorian instruments as if they were transplanted straight from the mass-production hardware racks of the 1990s: they were surely far more dependent on effective techniques of environmental management and adaptation.[4]

Given that this was so, it becomes all the more important to note that Victorian instruments were not habitually used in anything like the disturbance-free convenience of the purpose-built late-twentieth-century research laboratory. In this respect Joule's brewery cellar was a highly atypical place of experiment. As I have suggested elsewhere (Gooday 1990; Gooday and Forgan 1994), the spaces of instrumental work in the late nineteenth century were more generally makeshift and contingently occupied; those working in them had little guarantee that conditions could be made ideal for their purposes or that experimental resources would always be available as needed, or even that the conventions about their usage could be taken for granted. Indeed, the peculiar needs of experimentalists, and the social priority of empirical science undertaken in their spaces were still subject to considerable contestation and compromise.

On this point my account will take a different turn from Sibum's. The problematic environment dependence (and skill dependence) of Joule's practice are almost invisible in his textual representation of experiments. Since they were so very much "backgrounded," Sibum had to infer these features from painstaking practical reconstructions of Joule's labors. By contrast, later nineteenth-century experimenters in experimental physics and biology very much did foreground environmental contingencies in published accounts of their experimental practices. Indeed, the effective

---

4. My thanks to Sean Johnston for pointing out, *pace* Latour, that even physicists in the 1990s do not find their apparatus to be immutable in response to external stimuli.

management and reporting of such contingencies was a major part of establishing that reported phenomena were authentic and not merely artifacts of laboratory disorderliness. Failure to manage and report these problems effectively could provoke considerable skepticism among critical audiences. In 1901, for example, when Harold Pender (a student of Henry Rowland at Johns Hopkins) tried to demonstrate the magnetic effects of convective electric currents, his evidence was dismissed by the Parisian Victor Crémieu as no more than the spurious electromagnetic effect of Baltimore's tramlines—which Pender himself admitted were significantly nearby. To preempt similar assaults on the credibility of his claims, Pender moved his trials far enough into the countryside that they could be two miles from the nearest urban streetcar (Miller 1972, 26).

Yet in the crowded, bustling cities of Victorian Britain, the geographical focus of this chapter, Pender's problem was much more commonplace and his compensating strategy even harder to implement. This was especially so for British practitioners without sufficient funds to support a wholesale translation of their work to the countryside and without ready access to extramural facilities.[5] Instead of remaining silent over the problems of empirical disorder and disturbance, they often adopted the strategy of foregrounding these undeniable environmental contingencies. Textual representations of practice that highlighted the virtues of honesty in the reporting of experimental difficulties, perseverance in identifying sources of possible error, and strenuous labor in overcoming such difficulties were thus deployed in attempts to preempt skeptical accusations of artifactuality.

Before exploring how the Victorians dealt with and represented a recalcitrant and contrary material environment, it is important to reflect on a broader phenomenological issue that must inform any historical study on the daily use of instruments. Half-jokingly, Coutts Trotter and Thomson testified that nature did not customarily appear to act as an orderly agent regulating the behavior of laboratory instruments.[6] The phenomenological problem of accomplishing the "constancy of nature" has recently been addressed by both sociologists of science (Turnbull 1995) and laboratory ethnographers, most notably Karin Knorr-Cetina.[7] She has described the modern laboratory as an "enhanced environment" that in some sense "improves upon" the "natural order as experienced in everyday life." Laboratories rarely work with objects as they are found in nature, she explains, but rather work on their "components, extractions," and "purified" versions.

---

5. Such extramural facilities were procurable in the United States at Johns Hopkins (see Rowland 1878, 281–82) and in Germany at Göttingen (see Kohlrausch 1874, 305).

6. Coutts Trotter (1837–87) was a Fellow of Trinity College, Cambridge, from 1861, and college lecturer in physical science, 1869–84.

7. Phenomenological accomplishment is in contradistinction to the philosophically important metaphysical issues of the "constancy of nature."

Thus the power of the laboratory stems from its "exclusion" of the disorderly forces of "nature as it is independent of laboratories" and in its concomitant "enculturation of natural objects" into the local practices of laboratory life (Knorr-Cetina 1995, 145-46). Knorr-Cetina emphasizes this construction of orderliness in scientists' work because contemporary scientists themselves conspicuously do not,[8] apparently assuming that the robust equipment and architectural convenience of their purpose-built laboratories are somehow sufficient guarantees of a direct access to nature in the laboratory.[9]

By contrast, in the later Victorian accounts, we shall find that it was precisely these practices that were often foregrounded. This was crucial just to prove that there was as little doubt as possible about what it was that instruments were actually registering. This recalls a familiar theme: the tricky business of establishing that instrumental labor has brought some kind of direct unpolluted access to nature. In physics this was canonically achieved through "precision" measurement (Gooday 1990), and in biology one major instrumental technique employed was that of microscopy (Gooday 1991*a*), although it should be stressed that such practices were more problematic than historians have previously acknowledged. It was no easy matter for a Victorian physicist to prove to potentially skeptical audiences that, for example, the mooted electrical resistance of a coil was not the specious result of disturbing extraneous forces acting on a galvanometer. Similarly, it was by no means straightforward for the Victorian naturalist to prove that the fine structure seen on the shell of the algal diatom *Pleurosigma angulatum* was not in fact some artifactual product of the optometrical and illuminative resources that attended his or her use of the microscope. It was crucially important to establish that what was represented in these accounts of experiments was thus indeed the identity of the subject claimed by the experimenter. The security of this identity hinged, as we shall see, on the strategic management of resources in the experimental environment and the effective textual or visual representation of it as part of nature's underlying orderliness.[10]

I. Practical Problems in Physics

The material problems distinctive of experimental physics in the 1860s-1880s stemmed from the less-than-congenial location of its domains of practice, especially those in London, but also in Manchester. Laboratories for physics at Kings College, University College, the Royal College of Sci-

---

8. For an analysis of how scientists typically modalize claims about their experiments to reduce compromisingly circumstantial references see Woolgar 1988, 69-81.

9. See Collins 1985, 79-111, though, for an example of how imputations of external disturbance can be deployed to challenge empirical claims in an experiment of modern cosmical physics.

10. An excellent study of the topic can be found in Lynch and Woolgar 1990.

ence, and the early Owens College were situated in rather busy metropolitan milieux and within bustling institutions of learning. Unlike the laboratories constructed at the Universities of Oxford, Glasgow, and Cambridge (c. 1868–74), the rooms allocated to physics in London were not usually purpose-built for stability, or isolated from proximate disturbance (Gooday 1990). The extent of the problem facing physicists in makeshift laboratories can be seen in the case of Robert Clifton experimenting on the top floor of Owens College, Manchester, in the early 1860s. Although notorious for his later nonresearching career in Oxford (Morrell 1992), in his early career Clifton did attempt some work on spectroscopy. This met with little success, however, as one of his later successors recorded:

> When he (Clifton) looked through his spectroscope he found that every lorry that passed through the street underneath shook the (instrument), so that instead of seeing a point he saw a line vibrating about. He started a research therefore, in order to see whether, by the manner in which it shook, he could distinguish between a lorry of one kind—and a lorry of another kind and an ordinary cab. (Schuster 1924, 30)

Disturbance from extrinsic sources clearly compromised the identity of what physicists could actually determine through laboratory measurement. Without the power to stop metropolitan traffic moving in the vicinity, Clifton could not easily have won credence for claims that the phenomena registered by his instruments were unambiguously spectroscopical in origin.

However, not all cases of civic traffic prohibited measurement experiments among 1860s contemporaries. Instead, the undeniable presence of traffic-induced disturbance was strategically cited to give the fullest criterion for critics to judge the reliability of the experiments. In Kings College, London in 1863, such a tactic was employed by the members of the British Association's committee on electrical standards, Balfour Stewart, Fleeming Jenkin, and James Clerk Maxwell, in attempting an absolute determination of the unit of resistance by spinning a coil of wire within the earth's magnetic field. The Victoria Embankment that lies today between Kings College and the river Thames had not been built in 1863 (Weinreb and Hibbert 1993, 267), so iron steamers often ran within a few dozen yards of their experiments. In the published report of their experiments, they acknowledged this problem in relating the accuracy of their measurements to the baseline of a "background noise" of magnetic disturbances from Thames marine technology:

> The action of [the stabilizing] governor, combined with that of the driving-gear, was such that in many experiments the oscillations in deflection due to a change of speed were not so great as those due to the passage of steamers in the river when all parts of the apparatus were at rest; so that the deflections during twenty minutes could be

> quite as accurately observed as the slightly imperfect zero-point from
> which they are measured. (Jenkin et al. 1863, 120)

Throughout their detailed account, there is a significant emphasis on the resourcefulness, sheer perseverance, courage, and even self-sacrifice with which they conducted their lengthy labors, especially in identifying and solving practical problems of disturbance.[11]

Even so, displays of virtuous conduct were not enough to guarantee the unimpeachable reliability of experiments. Jenkin et al. conceded that the proximity of the iron steamers was nonetheless sufficient to produce somewhat imperfect measurements; this was accompanied by a strategic indication that errors had been minimized by an immense amount of labor and care and yet ironically that a better environment should still be sought to ensure a more reliable measurement:

> The oscillations produced by the passage of steamers on the Thames
> at no great distance from the place of experiments were of very sensible magnitude; and although by carefully observing the limit of every
> oscillation during every experiment the error due to this cause was in
> great part eliminated, it is desirable that any future experiments
> should be conducted in some spot free from all local magnetic disturbances. (Jenkin et al. 1863, app. D ["Description of an Experimental
> Measurement of Electric Resistance, made at King's College"], 173)

German and American critics of these experiments agreed. They claimed to have working conditions effectively free of disturbance in which they could produce measurements to better the British Association committee's results and, a fortiori, cast further doubt on the integrity of its working environment. Writing around 1870, Friedrich Kohlrausch challenged a number of error-inducing features of the British Association's experiment, not least the use of a "strong brass frame" to support the rotating coil: "nowhere" was it explained how Jenkin et al. had convinced themselves of the "unimportance" of the eddy currents induced in this frame by the rotating coil. "No doubt" these would have been "difficult" to detect, Kohlrausch conceded, yet the "neighbourhood of masses of metal" ought to have excited "suspicion" (Kohlrausch 1874, 300–301).[12] By contrast, he trumpeted the unsurpassed "delicacy" and undisturbed "convenience" of the facilities used in his own rival determination, especially in measuring the earth's magnetic field at Weber's Magnetic Observatory in Göttingen. This met his requirements "more completely than any other place," especially as the director of the subsidiary astronomical observatory had had the "goodness" to arrange for local magnetic disturbances to be "avoided" near his suspended instruments (Kohlrausch 1874, 305, 343). From these apparently

---

11. For Victorian debates on the virtues of "character" see Collini 1991, 91–118.

12. The German original was published in Poggendorff's *Annalen,* but all citations are from the later translation (Kohlrausch 1874).

utopian spatial conditions he claimed to have identified an error of nearly 2 percent in the British Association's result for the absolute determination of the ohm (Kohlrausch 1874, 354).

After seeing the English translation of Kohlrausch's work, Henry Rowland challenged it (Rowland 1875) and went on to publish a critique of both British and German results three years later. Very early on in this paper he draws attention to the fact that his experiment to determine absolute resistance was "performed in the back room of a small house near the University." This, he explained, was "reasonably free from magnetic and other physical disturbances," adding significantly that the particular design and execution of his experiment entailed that "it was not necessary to select a region entirely free from [magnetic] disturbance" (Rowland 1878, 282-83). From this spatially privileged viewpoint, Rowland claimed that the British Association's result was subject to an error of ±0.08% and Kohlrausch's of ±0.33%. This stimulated the British to counter with their own redetermination, yet they were not equally able to win access to spaces so easily representable as free of magnetic disturbance. Certainly neither central London nor the backstreets of Cambridge proved ideal for this purpose.

One of the later members of the electrical standards program was George Carey Foster, professor of physics at University College London. Foster had to work in rooms that were inconveniently close to busy corridors and stairwells (Gooday 1990, 39–40; figure 19.1), and on floors so yielding that nowhere could the equilibrium of a delicate balance remain undisturbed if anyone walked around (Foster et al. 1894, 281). Although Foster later gained access to a steady basement room called "the Dungeon," (Gooday 1989, chap. 4, p. 34) there was nonetheless heavy competition for this space from undergraduate teaching. Even when the room was devoted to the instruments of research, such as those assembled by Foster for trials on a new means of determining the British Association resistance standards, the old problems of bustling London life reared up again. His alternative to the spinning coil method depended on the tangent galvanometer and hence required a measurement of the local horizontal component of the earth's magnetic field. Yet, as he reported to Section A of the British Association in summer 1881, this attempt was "made useless by some large mass of iron being brought just outside the laboratory while the experiment was going on" (G. Foster 1881, 427). While Kohlrausch in Göttingen and Rowland in Baltimore could secure assurances of congenial human behavior outside their laboratories, Foster evidently could not. Lacking such institutional power, he pointedly abandoned these experiments, probably in an attempt to embarrass University College into providing him with better facilities.[13] Eventually, in 1894, they did give him a laboratory custom

---

13. A comparison with the difficulties of measuring the gravitational constant G in a London setting can be found in Boys 1890, 265–66; see also and Gooday and Forgan 1994, 176.

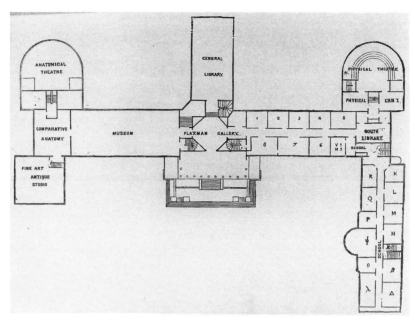

Figure 19.1 Plan of University College London in the 1870s showing the "Physical Laboratory" adjacent to stairways and corridors. (Frontispiece from University College London 1878.)

made with independent stone piers and a superstructure free of troublesome iron (Foster et al. 1894).[14]

Lord Rayleigh's purpose-built Cavendish Laboratory was at least nominally more favorable to delicate measurements than Carey Foster's converted classrooms in London's Gower Street. Even so, in attempting to secure conditions that would render his redetermination of the British Association resistance unit less prone to challenge, Rayleigh and his assistant Arthur Schuster explicitly signaled their careful choice of timing:

> The observations were taken on three different evenings and one afternoon. The evenings (8 h. P.M. to 11 h. P.M.) were chosen on account of the absence of disturbances which, during the usual working hours, are almost unavoidable in a laboratory. (Rayleigh and Schuster 1881, 130)

To present this as a guarantee of the stability of this working environment, Schuster (author of part 2 of the paper) reported figures for the period at

---

14. These were typical of the architectural stratagems embodied in specialist late-nineteenth-century physics laboratories serving to minimize mechanical or magnetic interference in instrumental activity (Forman, Heilbron, and Weart 1975, 90–109).

130   Lord Rayleigh and Dr. A. Schuster.   [May 5,

with the auxiliary magnetometer. A recomparison of resistance with the standard completed the set. The magnet in the centre of the coil should, when no current is passing through the coil, always go through the same changes as the magnet of the auxiliary magnetometer. If this could be insured, the two might be compared once for all, or the comparison might even be omitted altogether, for the difference between the deflections of positive and negative rotations, when corrected for changes in the earth's magnetism, would give the double deflection independently of the actual zero position. Unfortunately, however, and this was our greatest trouble, the comparison between the magnet and the auxiliary magnetometer showed that we had to deal with a disturbing cause, which rendered a frequent comparison between the two instruments necessary. This disturbing cause, which we traced to air currents circulating in the box containing the magnet, will be discussed presently.

The observations were taken on three different evenings and one afternoon. The evenings (8 h. P.M. to 11 h. P.M.) were chosen on account of the absence of disturbances, which, during the usual working hours, are almost unavoidable in a laboratory. We may give, as an example for the regularity with which the magnet vibrated round its position of rest, a set of readings which were taken while the coil revolved about four times in one second, the circuit being closed.

$T = 9^h\ 36^m.$ $t = 13° \cdot 0$ C.

Rotation.	Negative.
374 ·4	362 ·1
373 ·3	362 ·8
372 ·2	362 ·0
373 ·9	361 ·4
372 ·8	362 ·0
372 ·8	362 ·0
372 ·4	363 ·8
371 ·8	364 ·0
371 ·1	364 ·0
370 ·5	
Mean.... 372 ·52	362 ·68

Position of rest, 367 ·60.

$T = 9^h\ 38^m \cdot 5.$ $t = 13° \cdot 0$ C.

The number of readings taken were not always the same, but varied generally between sixteen and twenty.

We used, in the course of our experiments, four different speeds. The method of obtaining and regulating these has been explained by

Figure 19.2 Page from Rayleigh and Schuster's 1881 report of their redetermination of absolute resistance, showing a table purporting to illustrate the lack of disturbance in evening conditions (Rayleigh and Schuster 1881, 130).

which their magnet vibrated round its position of rest.[15] They showed a table (figure 19.2) that nonetheless indicated up to half a percent (!) variation in this period, and no error analysis was given: these figures were clearly somewhat rhetorical in their import. Even with this nighttime guarantee of minimized external interference, intralaboratory disturbance was still very much an issue.

It is revealing to see how explicitly the thorough precautions against circulating air currents and lamp radiation were laid out—and how candid the

15. Little allusion is made, though, to the major corollary of nighttime working: error-inducing fatigue. For an explicit discussion of this in the Cambridge context see E. H. Griffiths's account of his problematic nocturnal efforts to redetermine Joule's constant between 1889 and 1894 (Griffiths 1894).

experimenters were about their limited success in this regard. To give just one of many examples of this cited in Schuster's text,

> Special experiments were now made, and it was found that by placing a lamp about a foot and a half from the magnet box, changes amounting to eighteen divisions of the scale could be observed; greater precautions were taken, in consequence of the experiences thus gained, to secure the box from the radiation of the lamp and gas jets, which could not be dispensed with in the course of the experiments. The magnet box was covered in gold leaf so as to reflect the heat as much as possible. On the last night of work frequent determinations of the [magnet's] position of rest were made, but in spite of all precautions an unknown cause produced a sudden displacement of five scale divisions. The exact time at which this change took place could not be determined, and the [relevant] two spinnings were therefore rejected. During the remainder of the evening the magnet gradually came back to its original position. With the exception of the two spinnings just mentioned we have not rejected any observations. (Rayleigh and Schuster 1881, 133)

From the phenomenological point of view, it is significant just how much familiar laboratory disruption and contingency is cited by the experimenters. By the very articulation of the heating effects of nighttime lighting and of the existence of untraceable sources of discrepancy, Schuster made a cogent bid to maintain his readers' credence in the thoroughness of their experiments and in his honesty in the reporting of them.[16]

Significantly, though, the authors assigned no estimate of error to their conclusion that the ohm "as fixed by the Committee of the British Association" was 0.9893 earth quadrants per second. Unsurprisingly, this figure was contested and reworked on a number of occasions throughout the 1880s: as Schaffer has shown, it was incongruously local to Cambridge when compared with Rowland's value (Schaffer 1995, 162-63). It was indeed to Baltimore, not Cambridge, that the British Association turned for its 1887 determination (Rowland 1901, 239). Only results secured in other experimental environments could adjudicate the credibility of Rayleigh and Schuster's far-from-undisturbed night researches in the Cavendish Laboratory. Evidently not even the most determinedly executed and thoroughly reported campaign against the contingencies of laboratory disorder could be sufficient to win a definitive value for a measurement of electrical resistance.

How, then, can we generalize the day-to-day difficulties of researching a phenomenally inconstant nature in a Victorian physics laboratory? Given that even purpose-built laboratories were not free of disturbance, physi-

---

16. For a valuable comparison with seventeenth-century representational practices see Schaffer and Shapin 1985.

cists drew from a developing repertoire of practical strategies and representational conventions in reporting the readings of instruments in attempts to forestall imputations of artifactuality.[17] Local environmental difficulties in research could be met by various tactics: announcing an intention to carry out future research in a more congenial environment (Jenkin et al.), researching the causes of disturbance instead (Clifton), giving up altogether (Foster), or experimenting at night or deferring to another less disturbed institution (Rayleigh and Schuster). Not that efforts in these directions were generally sufficient for physicists to produce definitive measurements, but there was much more at stake here than getting a "right" answer. However inconclusive, published accounts of their persevering and resourceful labor were crucial to their credibility as physicists. Representing their laboratories as disturbed, even chaotic, environments would have been so familiar a characteristic of the physicists' working day (or night) that certainly specialist audiences would have been rather skeptical had matters been portrayed otherwise. Certainly, candid stories about interference from passing carts, ships, undergraduates, and air currents and from wayward light sources were essential to paint a plausible picture of trustworthy practice that might preempt challenges. These quotidian artifact-inducing problems of traffic and illumination were not unique to experimental physical science but were common to the life sciences also, albeit with a different emphasis.

III. Illuminating Microscopical Biology

> At this time what a mass of thoroughly conflicting evidence there is advanced on almost every question! Three or four views are taught concerning first principles of anatomical and physiological science, each one by an immense amount of what he purports to be [microscopical] observation . . . and yet with what pertinacity are they maintained, and what an amount of work must be done, and what a length of time must elapse before the false facts can be demonstrated to be really false and the true facts proved to be really true!
> LIONEL BEALE, *How to Work with the Microscope* (Beale 1868, 189)

By the 1870s, microscopy was as central to the research practices of British life science as measurement was to contemporary physics (Gooday 1991*a*). The credibility of the microscope as a mediator of nature's finer details was well established among many different audiences (physicians, professional biologists, field naturalists, and domestic hobbyists) and used within a wide range of practical contexts (field, study, museum, laboratory, and even the naturalist's cabin at sea). As the president of the Microscopical Society, Rev-

---

17. For a provocative analysis of similar issues in twentieth-century science see Collins 1985.

erend Joseph Bancroft Reade, declared in 1870: "The character of objects we can neither see nor touch can be revealed by the microscope only" (Reade 1870, 138). True, there were controversies surrounding the chemical contingencies of mounting and staining slide preparation—a point conceded by even traditional historians of microscopy (Bracegirdle 1993). Yet the credence given to the microscope contrasts strikingly with that accorded to other technologies, for example, the clinical thermometer. As one unimpressed physician asserted in 1874, "More than ever now the physician must have knowledge of the soul; must feel, with finer senses, other pulses; and measure heats and chills which no thermometer can gauge" (Lawrence 1985, 515). Why should physicians—and indeed their patients —have regarded the microscope as so much less problematic in clinical usage than the far-from-novel thermometer?

The microscope's comparative ubiquity and accessibility might lead one to describe it as a "boundary object," that is, an object made stable and important in virtue of simultaneously serving the interests of a broad range of different constituencies in appropriate ways (Star and Griesemer 1989). However, on closer inspection, it appears that the microscope was never a malleable single entity—still less was it one of Latour's immutable and universal mobiles (Latour 1987). First, the microscope effectively existed in specific forms for different audiences at different sites: no single universal microscope satisfied the concerns of all users. There were robust, simple microscopes for outdoor fieldwork, including the handy pocket microscopes for the rambler, and compound microscopes for dedicated indoor use. Within the latter category, there were the relatively easy-to-use microscopes of the "third class" for the poorer artisan or autodidact naturalists (e.g., Field's "Society of Arts" prizewinning model); the "second class" of instrument was generally made for the medical student and wealthier home or rambling naturalist (e.g., Hartnack's or Crouch's); the "first-class" microscope, for example, Ross's or Powell and Lealand's (figure 19.3), was the sophisticated specialist model for the laboratory histologist, the clinical pathologist, or the most zealous nonprofessional devotees. It was this audience-specific specialization of form, and of associated practice, that rendered the microscope deployable in diverse domains, enabling biological research among practitioners vastly more heterogeneous and diffuse than those of experimental physics (Gooday 1991*a*).

The other reason (and related to the first) for the microscope's importance to contemporary practitioners in the life sciences was its service as the major experimental tribunal of expert opinion—although by no means stemming controversy, as Lionel Beale's epigraph above shows. Such a role for the microscope emerged in the battle between vitalists such as Beale and neomaterialists such as Thomas Henry Huxley in the 1860s and 1870s over the unresolved structure of cell "protoplasm." Beale maintained that the failure of microscopists to resolve any operative physical structure in

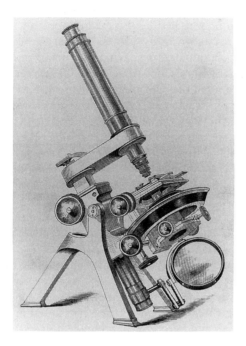

Figure 19.3 Powell and Lealand's "first-class" microscope, circa 1868 (Carpenter 1868, 79).

protoplasm meant that the cell's life-sustaining activity could be explained only by reference to "vital" forces (Beale 1869); Huxley argued against this that the identification of a relevant physicochemical basis for life was merely a matter that awaited future technical progress (Huxley 1870, lxxviii). Voices outside the life sciences, such as John Tyndall, were more circumspect about the jurisdiction of the microscope — as a leading microscopical journal reported of his discourse to the British Association's Liverpool meeting in 1870.[18] Using examples strategically chosen from the physics of crystalline materials and fluids, Tyndall declared that the "world of molecular matter and of motion" was one for which the microscope had "no passport." With a mischievous lack of deference to his friend and auditor, the British Association's president, Huxley, Tyndall concluded that a "little consideration will make it plain to all" that the microscope had "no voice in the real question of germ structure" (Tyndall 1871, 153–57).

On another occasion, Huxley himself was forced to cede ground to his opponents after overinterpreting the structural results of a microscopical enquiry. From slime dredged up in the North Atlantic, Huxley identified in 1868 what he thought was the submarine bearer of pure protoplasm,

---

18. Extracts of Tyndall's discourse appeared in the *Quarterly Journal of the Microscopical Society* (Tyndall 1870).

Bathybius haeckeli. By 1875, though, this identification was chemically deconstructed into something rather less organic: calcium sulphate in an amorphous colloidal state. Rupke suggests that because the scandal was kept quiet, the impact on Huxley's protoplasm theory was not severe (Rupke 1976, 53-62). Yet the Religious Tract Society was not so quiet about Huxley's misjudgment. In 1895 it published Lewis Wright's *Popular Handbook to the Microscope,* in which Wright specifically attacked the recently deceased Huxley for being so "unscientific" as to misuse the microscope to label "primeval slime" as the ultimate bearer of life.

> It was the Microscope that proved such teaching due to sheer ignorance and not to superior knowledge, and, if it could not reveal the Divine mystery of living existence, at least manifested it to us as a greater Mystery than ever. The Microscope, then, has deserved well of the Christian believer. (Wright 1895, 5-6)

Among the brawling biologists and microscopists, whether believers (e.g., Wright) or agnostics (e.g., Huxley), there was one point on which they did seem to agree: that the unequivocal interpretation of microscopical findings lay in the effective management of the physical environment in which microscopy was practiced. When the microscope was claimed by practitioners (pace Tyndall) to have revealed a real structure, guarantees about appropriate stability and lighting, as well as the mounting and staining of specimens, were introduced in order to make that representation of nature less prone to contestation and imputations of artifactuality. In other words, the veracity of the microscope as a mediator of nature did not lie in the instrument itself, but very much in the effective juxtaposition of auxiliary material resources and the persuasive representation thereof to critical audiences. Yet the wide variety of practices among far-flung microscopists—civic, urban, and rural—reveals that there was little consensus upon what constituted the appropriate means of environmental management. This perhaps is the significant counterpart of the physicists' incipient laboratory chaos, the closest parallel lying in microscopists' tendency to foreground the difficulties of establishing unproblematic material stability and lighting.

All such issues were intimately tied up with the material construction of microscopes—as much as the concerns of physics were tied up with the construction of apparatus and its architectural surroundings. Even identifying a reliable microscope was no easy matter, as the experts themselves disagreed quite strongly.[19] Like physicists, microscopists sought physical stability in their working environment, especially at high magnifying powers: at the very limits of optical resolution, the smallest vibration could render accurate focusing and representation highly problematic. For

---

19. In the early 1870s, Huxley's laboratory acolytes and London allies differed sharply over the practical viability of British vis-à-vis Continental models (Lankester 1870; M. Foster 1870; Rutherford 1872).

lower-caliber devices, even the most mundane factors in the local human environment could, according to William Carpenter, easily make them useless.[20] In 1868 he addressed this topic with great candor in *The Microscope and its Revelations:*

> In a badly constructed instrument, even though placed upon a steady table resting upon the firm floor of a well-built house, when high powers are used, the object [under view] is seen to oscillate so rapidly at the slightest tremor—such as that caused by a person walking across the room, or by a carriage rolling-by in the street—as to be frequently almost indistinguishable: whereas in a well-constructed microscope, scarcely any perceptible effect will be produced even by greater disturbances. (Carpenter 1868, 42)

Hence, he recommended to all his readers that they buy well-constructed instruments and test them at the highest powers for their stability. "A Microscope should be unhesitatingly rejected," Carpenter declared, "if the result be unfavourable" (Carpenter 1868, 42–45).

Yet Carpenter soon found that comparable problems could arise even in the most prestigious instruments, and in 1870 he attacked the "first-class" Ross instrument at a meeting of the Royal Microscopical Society. For a microscopist to work effectively in a variable domestic environment, the only way to avoid "tremor," Carpenter claimed, was to make the instrument *"perfectly rigid."* If this were not so, an obtrusive differential vibration could be set up between, for example, the ocular and objective extremities. This, he averred, was the case with the widely used and admired Ross binocular instrument (figure 19.4). Indeed, he said it was hard to "conceive a method of construction" that was more "favourable" to this vibration: its long tubular body, fixed only at the base, was "peculiarly subject" to it, even with the additional use of supportive oblique stays. Countermeasures, such as giving greater solidity to the stem, arm, and body, were to some extent effective, he reported, but made the device cumbersome in its adjustment and nearly impossible to use at the highest powers of magnification—except under the unattainable condition of "perfect stillness" (Carpenter 1870, 183–84).

Carpenter accordingly advised the gathered members of the Microscopical Society to adopt the newly developed type of instrument in preference to the Ross, the better supported—and hence less tremor prone—Jackson (or Lister) model. To bring his point home, Carpenter related his experiences of the two instruments in very different "conditions of stability" during the previous year, namely, on shore in his London laboratory and on a deep-sea exploration mission with Sir Wyville Thomson on HMS *Porcupine*. When sailing under "easy steam," his own Jackson device performed

---

20. For Carpenter's earlier career see the chapter by Alison Winter in this volume.

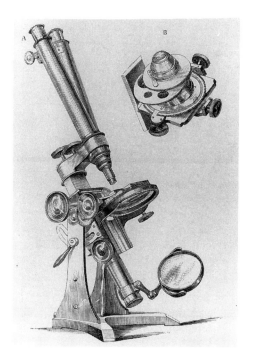

Figure 19.4 Ross's "first-class" binocular microscope, circa 1868 (Carpenter 1868, 76).

well, while Wyville Thomson's heavier, Ross-based device showed tremors even at low magnifications. The difference in performance was most striking, though, in the "peculiarly critical" circumstances of steaming at full power against a "head-sea," when even with a rather large, ⅔-inch objective, Thomson's device produced an image that "danced" very perceptibly —quite unlike Carpenter's own. From this testimony, Carpenter recommended that all microscopes be redesigned to encompass the tremor-resistant construction of the Jackson model, particularly "first-class" research microscpoes in which the primary condition was "*steadiness of the image*" (Carpenter 1870, 185). Carpenter's critique met with considerable dissent from habitual users of the Ross microscope; nonetheless, Carpenter's advice was clearly incorporated into later models such as the Ross-Jackson (figure 19.5) and Ross-Zentmayer (Carpenter and Dallinger 1891, 176-77).[21]

Of all problems regarding the authenticity of the microscope as a device for gaining access to nature, the most important concerned lens construction and illumination (only the latter will be covered here). Illumination is particularly important for the purposes of a comparison between measure-

---

21. For the discussion of Carpenter's opinion of the earlier Ross design, see the discussion summarized in Reeves 1870, 212-13.

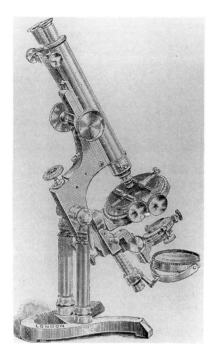

Figure 19.5 Ross-Jackson microscope, circa 1891 (Carpenter and Dallinger 1891, 159).

ment physics and microscopical biology. As we saw in the case of Rayleigh and Schuster's researches, artificial lighting as an essential auxiliary for nighttime physical measurement could become a disturbing factor if ineffectively managed. For microscopy, the issue of illumination—whether by artificial or "natural" light (i.e., sunlight)—was more broadly problematic, since more than being an essential auxiliary, the illumination of the microscope stage was central to the production of the microscopical image. Whether or not a particular kind of lighting could be trusted to produce a faithful representation of nature or served only as a disruptive and misleading source of artifactual results (as with physics) was a matter of debate and anxiety among practitioners of the life sciences. Different authors proposed subtly different parameters for what was an appropriate source of lighting, and even the same author would recommend different sorts of illumination for different contexts and different subjects.

John Queckett, professor of histology to the Royal College of Surgeons, was renowned for his *Practical Treatise on the Use of the Microscope* (1855). In it he argued that a new microscopical object should be viewed under "every description of light," whether "strong or faint, oblique or direct," whether reflected through a mirror or prism, as in the case of a transparent object, or "condensed upon it by a lens" for the opaque. This he maintained was necessary "in order that all the characters of the subject

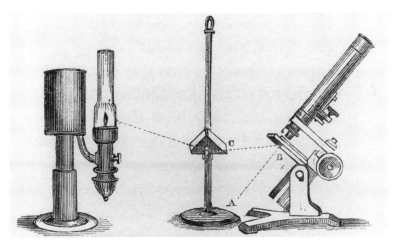

Figure 19.6 Queckett's oblique illumination apparatus with French fountain lamp for viewing the diatom *Navicula* (Queckett 1855, 219).

under investigation may be brought out and rendered perfectly distinct," each distinct form of illumination being somewhat partial in its revelations (Queckett 1855, 210). Thus, for "the perfect definition of the markings of certain Diatomaceae of the genus Navicula" it was necessary to effect oblique illumination through a prism from a French fountain lamp (figure 19.6; Queckett 1855, 212–19). This lamp was particularly valuable in offering a steady flame for such delicate investigations with first-class microscopes, since its construction on the "bird-fountain" principle offered a constant supply of oil to the burner, and it hence gave an extremely steady flame (Queckett 1855, 173–74). Not so, apparently, the "University," or "Cambridge," lamp used for more ordinary pedagogical purposes, which Queckett employed for himself and his students in his lecture theater at the Royal College of Surgeons (figure 19.7; Queckett 1855, 171–72).

Carpenter presented a slightly different emphasis. For the examination of a "greater proportion" of microscopical objects, "*good daylight*" was to be preferred to any other kind of light, but "good lamplight" was nonetheless "preferable to bad daylight." If daylight were employed, then the microscope should be placed near a window whose aspect was as far as possible "*opposite* to the side on which the sun is shining," for the "light of the sun reflected from a bright cloud" was that which the experienced microscopist would "almost always prefer," the rays proceeding from a cloudless blue sky being "by no means so well fitted for his purpose" (Carpenter 1868, 139). Lionel Beale agreed that ordinary daylight or sunlight "reflected from a white cloud" afforded the best possible illumination, adding, "It has been said with truth that microscopical work should be undertaken only by day"; the "most perfect artificial light" that could be obtained was "far infe-

Figure 19.7 Queckett's lecture theater at the Royal College of Surgeons, circa 1855, showing gas lamps on the rear wall and absence of natural daylight (Queckett 1855, frontispiece).

rior to daylight for delicate observation" (Beale 1868, 19). The infelicities of daily life and weather meant that microscopists sometimes had to practice without the luxury of cloud-reflected light. However, a variety of techniques were available to upgrade synthetic lighting or direct sunlight by the use of an "artificial cloud" (figure 19.8): a plane mirror with a surface of "pounded glass" or carbonate of soda, or a disc of plaster of Paris, usually in combination with a substage condenser to provide a sufficient quantity of light (Carpenter 1868, 117).

Lighting was especially important for Carpenter since much of his analy-

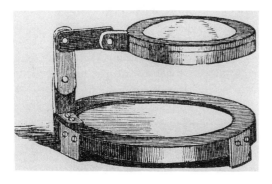

Figure 19.8 An "artificial cloud" or "white-cloud illuminator" for domestic microscopy (Carpenter 1868, 117).

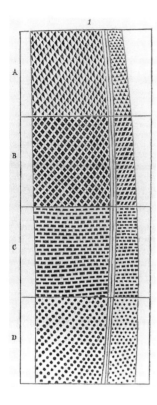

Figure 19.9 Carpenter's illustration of *Pleurosigma formosum*, showing the diverse effects of different angles of illumination (Carpenter 1868, 159).

sis of "errors of interpretation" is taken up with the fallacies arising from the inept management of illumination: diffraction effects from the optical characteristics of the microscope, refraction of light by the structure of the subject itself, or the difficulty of focusing on all of an object at once (Carpenter 1868, 164–70). Especially troublesome in this respect was the effective use of light sources in studying the structure of the algal diatom valves, especially of the genus *Pleurosigma*. While Queckett had recommended that the diatom genus *Navicula* be studied only with oblique lighting, Carpenter illustrated that major problems of structural interpretation arose, for example, in trying to study *Pleurosigma formosum* by this means. According to the angle of obliquity, the surface of this diatom could be represented as a pattern of alternate black and white triangles, as black rhomboids embedded in a white background, or as a black brickwork pattern (figure 19.9). Conversely, if direct central illumination were employed with an achromatic condenser, a pattern of black discs appeared against a white background! (Carpenter 1868, 159–60). Even judgments about coloring and of elevation or depression could be difficult to differentiate from judgments about "correct" focal adjustment. Displaying an enlarged ($\times$ 15,000) photograph of *Pleurosigma angulatum,* Carpenter showed how hexagonal are-

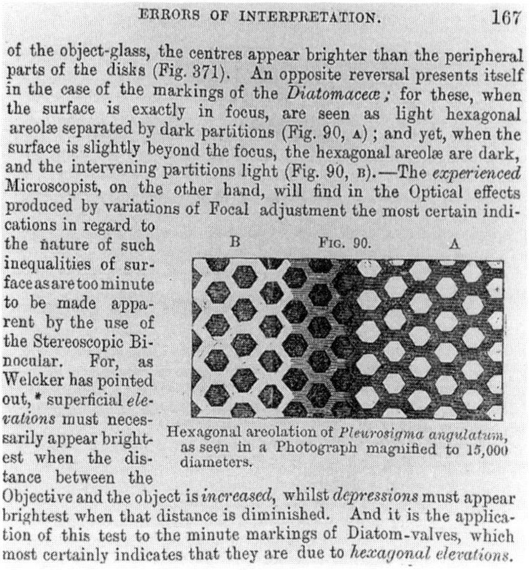

> ERRORS OF INTERPRETATION. 167
>
> of the object-glass, the centres appear brighter than the peripheral parts of the disks (Fig. 371). An opposite reversal presents itself in the case of the markings of the *Diatomaceæ;* for these, when the surface is exactly in focus, are seen as light hexagonal areolæ separated by dark partitions (Fig. 90, A); and yet, when the surface is slightly beyond the focus, the hexagonal areolæ are dark, and the intervening partitions light (Fig. 90, B).—The *experienced* Microscopist, on the other hand, will find in the Optical effects produced by variations of Focal adjustment the most certain indications in regard to the nature of such inequalities of surface as are too minute to be made apparent by the use of the Stereoscopic Binocular. For, as Welcker has pointed out,* superficial *elevations* must necessarily appear brightest when the distance between the Objective and the object is *increased*, whilst *depressions* must appear brightest when that distance is diminished. And it is the application of this test to the minute markings of Diatom-valves, which most certainly indicates that they are due to *hexagonal elevations*.
>
> Hexagonal areolation of *Pleurosigma angulatum*, as seen in a Photograph magnified to 15,000 diameters.

Figure 19.10 Carpenter's illustration of *Pleurosigma angulatum* undergoing inversion of contrast upon moving out of focus (Carpenter 1868, 167).

olations in focus could be seen as white hexagons with black partitions, while this was completely inverted when the pattern was slightly beyond the focus (figure 19.10); he inferred from this that the hexagonal patterns must have been elevations (Carpenter 1868,167).

However, it was on precisely such matters that many microscopists failed to agree—as Beale complained in the epigraphic passage above. With specific reference to the diatoms, most highly beloved among the aesthetic delights of the microscopists, Beale in fact says in *How to Work with the Microscope* that "there is much difference of opinion as to the cause of the markings in many of these," citing Hunt's drawings showing rhomboidal elevations too (Beale 1868, 169, 250–53). Indeed, two years later, Reverend Reade sought to overcome this interpretive chaos. He explained to the readers of the *Popular Science Review* why the structure of *Pleurosigma angulatum* had become the "vexed question of the day."

> How, then, with the aid of the microscope is th[is] diatom-valve described by the host of observers? So far from there being any uniformity of statement, we may almost say, *Quot homines, tot sententiae*.[22] The "Transactions of the Microscopical Society" contain a curious record of the Protean aspects described by different microscopists, and it is amusing to read of the ingenious modes of playing with the illuminating rays, so that the eye, fortified by a little previous theory, may see at will, in one and the same valve, either elevations or depressions, triangular, quadrangular or hexagonal dots, with rhomboids, pyramids or spheres. (Reade 1870, 140)

---

22. Translated "As many men as there are opinions."

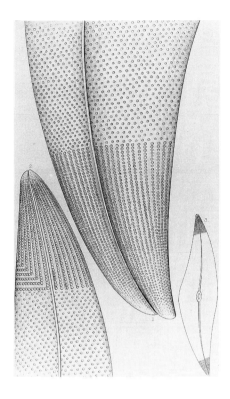

Figure 19.11 Anthony's illustration of the "peculiarities" in the nodular structures of *Pleurosigma angulatum* (Anthony 1870, 121).

Reade's tactic was to privilege the use of the substage condenser, alluded to by Carpenter, as an instrument for giving a "proper" angle to a "suitable" pencil of light, which alone could lead to a consensual view on microscopical structure (Reade 1870, 138). After lengthy study he concluded that previous diversity in the structural interpretations could only have stemmed from cross-beam shadows developed by a promiscuous multiplicity of light beams. A single beam from a condenser revealed to Reade that the "wondrous structure" of the diatom was in fact a "series of beautiful hemispheres"—a view noted briefly by Carpenter but passed over in favor of the hexagonal interpretation (Reade 1870, 142).

Virtually contemporary with Reade's account, though, was a paper read to the Royal Microscopical Society by John Anthony, a Cambridge M.D. He used a quite different illuminative contrivance, namely, a rectangular prism, to describe the structure (figure 19.11). In giving rather more complex detail of the pattern of "nodules" in this structure than Reade had done, Anthony humbly explained his virtuosity:

> Now I do not want anybody to take this peculiarity of structure in our old friend *P. angulatum* for granted . . . it being only fair to say that to make out this structure *well*, will require the very best appliances,

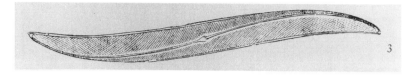

Figure 19.12 Spiers's illustration of the structure of *Pleurosigma angulatum*, circa 1909 (Spiers 1909, 32).

> and no small amount of care in the use of them. I feel sure that the structure is there, that it is not an optical deception, and that a fine instrument, a careful manipulation, and a good sight will not fail to make it out. (Anthony 1870, 122–23)

Most important (and perhaps ironic), Anthony related how "grateful" he was to the minute objects of *Pleurosigma* for the "lessons they have taught me in the use of the microscope"; more particularly, he was grateful for the lessons they had taught him "in the management of light so as to distinguish between true and false" (Anthony 1870, 121). Indeed, the status of *angulatum*'s fine structure was thereafter inverted so as not to be a problem, but rather a test of microscopists' virtuosity, their instruments, and the illuminating qualities of their lighting apparatus.[23] As William Spiers noted in his *Nature Through the Microscope* over thirty years later, *Pleurosigma angulatum* was a "dainty form" now used "often as a test of high-class objectives" (Spiers 1909, 33). Spiers's rendition of this tiny beast was notably different from the representations of either Reade or Anthony, however, even though he was working with the substage achromatic condenser recommended by Reade as the proper means to see diatomic nature through the microscope (figure 19.12).

III. Conclusion

Paying attention to day-to-day problems of material ordering in Victorian microscopy and metrology enrich our historical understanding by locating instrumental practices very specifically in the agonistic contexts of nineteenth century civic life. Experimental work in both fields was prone to major daily environmental perturbations and uncertainties that required hard labor, specialized resources, and much skill to be managed even in an elementary fashion. The major artifact-inducing problem highlighted for physics, that of material disturbance, was paralleled by that of rather "par-

---

23. Debates about the integrity of artificial illumination continued with the development of microphotography, for which see Jennifer Tucker's chapter in this volume and compare Woodward 1870 with Cole 1895, 188.

tial" illumination for biological microscopy. These problems were especially acute in domains of research where disagreement was prevalent about what constituted reliable practices—for example, the appropriate techniques for establishing the numerical values of electrical "constants" or the microscopical structure of simple organisms.

Where such uncertainty existed, writers of scientific papers had to work especially hard to persuade potentially skeptical audiences of their veracity in reporting the results of disturbed or artifact-prone instrumental labor. Despite a palpable interest in playing down the effects of local chaos-inducing conditions, the scientists discussed here actually foregrounded such matters to win the assent of those familiar with such virtually undeniable problems. In the longer term both communities were able to enhance these conditions by reengineering their working environments to achieve greater phenomenological orderliness: physicists by acquiring specially purpose-built laboratories, and microscopists by installing auxiliary devices such as the achromatic substage condenser into the very structure of their instruments.

While this disciplinary comparison may not be satisfactorily sensitive to variations of local context or tradition, it may at least serve as a heuristic for historians' recovery of the material, sociopolitical, and textual practices used by Victorian scientists in presenting their work to each other. A still richer social history of instrumentation could thus emerge from attempts to understand how audiences were, or were not, persuaded to accept such results. Future scholarship might therefore consider how authors chose which practical problems to foreground in their textual accounts, and which to omit; how much citation of such difficulties was necessary to win credence from readers without being so great as to make research appear hopelessly unreliable; how authors chose to situate such citations within their textual accounts for maximum plausibility; how far it was important for an author's credibility to identify all possible causes for any specific disturbance or suspected artifactuality; how important it was to include visual depictions of the working environment to authenticate claims made about its disturbed or undisturbed character; and how relevant strategies of representation may have been subject to contestation and changing convention among networks of authors, editors, referees and their audiences. From such work historians may come to identify better the distinctive cultural embeddedness of Victorian instrumental practice.

Bibliographical Note

Although never entirely eclipsed by the theory-centered historiography of the postwar period, the historical scholarship on instrument and experiment has undergone a major revival during the last decade. An excellent introduction to the politics of experiments is the collection of essays in Gooding, Pinch, and Schaffer (1989), while an interesting insight into the longer-term evolution of technical artifacts is provided by Petroski (1993), with further anecdotal support from Sutton (1986). An informative collection on the makers of Victorian instruments, Clerq (1985), is entertainingly complemented by users' and audiences' perspectives on Victorian material culture in Briggs (1988). Butler, Nuttall, and Brown (1986) provide an excellent introduction to the social history of microscopy, while Smith and Wise (1989), on Lord Kelvin's life and work, is invaluable for any historical study of measurement practices.

The dynamism and increasing sophistication of current research is reflected in the recent devotion of whole issues of major historical journals to studies in the field. For a stimulating overview of approaches to the history of biomedical instrumentation see the issue of *Technology and Culture* (1993, vol. 34, pt. 4) edited by Ruth Schwartz Cowan; particularly pertinent in this volume to the cultural analysis of optical instruments, especially the spectroscope, is Silverman (1993). More recently an annual issue of *Osiris* (1994, vol. 9) edited by Albert Van Helden and Thomas Hankins has been devoted to instruments; in this volume, historians of measurement will find Hunt (1994) particularly rewarding as a study of the social construction and usages of resistance standards.

References

Anthony, John. 1870. "On the Structure of the *Pleurosigma angulatum* and *Pleurosigma quadratum.*" *Monthly Microscopical Journal* 4:121–23.

Beale, Lionel. 1868. *How to Work with the Microscope.* 4th ed. London: Harrison.

———. 1869. "Protoplasm and Living Matter." *Monthly Microscopic Journal* 1:277–88.

Boys, Charles V. 1890. "On the Cavendish Experiment." *Proceedings of the Royal Society* 46:253–68.

Bracegirdle, Brian. 1993. "The Microscopical Tradition." In *The Companion Encyclopedia of the History of Medicine,* edited by W. Bynum and R. Porter. London: Routledge, 1:102–19.

Briggs, Asa. 1988. *Victorian Things.* London: Batsford. Reprinted 1990. Harmondsworth: Penguin.

Butler, Stella, with R. H. Nuttall and Olivia Brown. 1986. *The Social History of the Microscope.* Cambridge: Whipple Museum.

Carpenter, William. 1868. *The Microscope and its Revelations.* 4th ed. London: John Churchill.

———. 1870. "On the Comparative Steadiness of the Ross and Jackson Microscope-stands." *Monthly Microscopical Journal* 3:183–85.

Carpenter, William, and W. H. Dallinger. 1891. *The Microscope and its Revelations.* 7th ed. London: J. and A. Churchill.

Clerq, P. R. de, ed. 1985. *Nineteenth Century Scientific Instruments and Their Makers.* Leiden: Rodopi.

Cole, Arthur. 1895. *Methods of Microscopical Research.* London: Bailliere, Tindall and Cox.

Collini, Stefan. 1991. *The Public Moralists: Political Thought and Intellectual Life in Britain 1850-1930.* Oxford: Clarendon Press.

Collins, Harry. 1985. *Changing Order: Replication and Induction in Scientific Practice.* Thousand Oaks, Calif.: Sage.

Cowan, Ruth Schwartz, ed. 1993. *Technology and Culture* 34, no. 4.

Forman, Paul, John Heilbron, and Spencer Weart. 1975. "Physics c. 1900: Personnel, Funding and Productivity of the Academic Establishments." *Historical Studies in the Physical Sciences* 5:90-109.

Foster, George Carey. 1881. "Account of Preliminary Experiments on the Determination of Absolute Resistance." *British Association for the Advancement of Science Report,* pt. 2:426-31.

Foster, George Carey, T. H. Beare, J. A. Fleming, and T. R. Smith. 1894. "The New Science Laboratories at University College London." *British Architect's Journal* 1:281-308.

Foster, Michael. 1870. "The Microscope." *Nature* 2:255-56.

Galison, Peter. 1987. *How Experiments End.* Chicago: University of Chicago Press.

Gooday, Graeme. 1989. "Precision Measurement and the Genesis of Physics Teaching Laboratories in Victorian Britain." Ph.D. thesis, University of Kent at Canterbury, U.K.

———. 1990. "Precision Measurement and the Genesis of Physics Teaching Laboratories in Victorian Britain." *British Journal for the History of Science* 23:25-51.

———. 1991a. "'Nature' in the Laboratory: Domestication and Discipline with the Microscope in Victorian Life Science." *British Journal for the History of Science* 24:307-41.

———. 1991b. "Teaching Telegraphy and Electrotechnics in the Physics Laboratory: William Ayrton and the Creation of an Academic Space for Electrical Engineering 1873-84." *History of Technology* 13:73-114.

———. 1995. "The Morals of Energy Metering: Constructing and Deconstructing the Precision of the Victorian Electrical Engineer's Ammeter." In *The Values of Precision,* edited by M. Norton Wise. Princeton, N.J.: Princeton University Press, 239-82.

Gooday, Graeme, and Sophie Forgan. 1994. "'A Fungoid Assemblage of Buildings': Diversity and Adversity in the Development of College Architecture and Scientific Education in Nineteenth Century South Kensington." *History of Universities* 13:153-92.

Gooding, David, with Trevor Pinch and Simon Schaffer, eds. 1989. *The Uses of Experiment: Studies in the Natural Sciences.* Cambridge: Cambridge University Press.

Griffiths, E. H. 1894. "The Value of the Mechanical Equivalent of Heat." *Philosophical Transactions of the Royal Society,* ser. A, 184:361-504.

Hunt, Bruce. 1994. "The Ohm Is Where the Art Is." *Osiris,* 2d ser., 9:48-64.

Huxley, Thomas Henry. 1870. "Address." *British Association for the Advancement of Science Report,* pt. 1:lxxiii-lxxxix.

Jenkin, Fleeming, Augustus Matthiessen, James C. Maxwell, and Balfour Stewart. 1863. "Report of the Committee appointed by the British Association on Standards of Electrical Resistance." *British Association for the Advancement of Science Report,* pt. 1: 111-76.

Knorr-Cetina, Karin. 1995. "Laboratory Studies: The Cultural Approach to the Study of Science." In *Handbook of Science and Technology Studies,* edited by Sheila Jasanoff, Gerard Markle, James Petersen, and Trevor Pinch. Thousand Oaks, Calif.: Sage, 140-43.
Kohlrausch, Friedrich. 1874. "Determination of the Absolute Value of the Siemens Mercury Unit of Electrical Resistance." *Philosophical Magazine* 47:294-309, 342-54.
Lankester, Edwin Ray. 1870. "The Microscope." *Nature* 2:213-14.
Latour, Bruno. 1987. *Science in Action.* Milton Keynes: Open University Press.
Lawrence, Chris. 1985. "Incommunicable Knowledge: Science, Technology and the Clinical Art in Britain, 1850-1914." *Journal of Contemporary History* 20:503-20.
Lynch, Michael, and Steve Woolgar, eds. 1990. *Representation in Scientific Practice.* Cambridge, Mass.: MIT Press.
Miller, John D. 1972. "Rowland and the Nature of Electric Currents." *Isis* 63:1-27.
Morrell, Jack. 1992. "Research in Physics at the Clarendon Laboratory, Oxford, 1919-1939." *Historical Studies in the Physical and Biological Sciences* 22:263-307.
Petroski, Henry. 1993. *The Evolution of Useful Things.* London: Pavilion.
Queckett, John. 1855. *A Practical Treatise on the Use of the Microscope.* 3d ed. London: Baillière.
Rayleigh, Right Hon. Lord, and Arthur Schuster. 1881. "On the Determination of the Ohm in Absolute Measure." *Proceedings of the Royal Society* 32:109-81.
Reade, J. B. (Rev.). 1870. "Microscopic Test Objects Under Parallel Light and Corrected Lenses." *Popular Science Review* 9:138-48.
Reeves, Walter. 1870. "Proceedings of Societies: Royal Microscopical Society." *Monthly Microscopical Journal,* 211-15.
Rowland, Henry. 1875. "Note on Kohlrausch's Determination of the Absolute Value of the Siemens Mercury Unit of Electrical Resistance." *Philosophical Magazine* 50:161-63.
———. 1878. "Research on the Absolute Unit of Electrical Resistance." *American Journal of Science* 15:281-91, 325-36, 430-39.
———. 1901. *The Physical Papers of H. A. Rowland.* Baltimore: Johns Hopkins University Press.
Rupke, Nicholaas. 1976. "*Bathybius haeckeli* and the Psychology of Scientific Discovery." *Studies in the History and Philosophy of Science* 7:53-62.
Rutherford, William. 1872. "Notes of a Course of Practical Histology for Medical Students." *Quarterly Journal of the Microscopical Society* 12:1-21.
Schaffer, Simon. 1995. "Accuracy Is an English Science." In *The Values of Precision,* edited by M. Norton Wise. Princeton, N.J.: Princeton University Press, 135-72.
Schaffer, Simon, and Steven Shapin. 1985. *Leviathan and the Air-Pump.* Princeton, N.J.: Princeton University Press.
Schuster, Arthur. 1924. *The Physical Society of London 1874-1924: Proceedings of the Jubilee Meetings.* London: Physical Society. Copy in Institute of Physics Archives, London.
Sibum, Heinz Otto. 1995. "Reworking the Mechanical Value of Heat: Instruments of Precision and Gestures of Accuracy in Early Victorian England." *Studies in the History and Philosophy of Science* 26:73-106.

Silverman, Robert. 1993. "The Stereoscope and Photographic Depiction in the Nineteenth Century." *Technology and Culture* 34:729-56.

Smith, Crosbie, and Norton Wise. 1989. *Energy and Empire: A Biography of Lord Kelvin.* Cambridge: Cambridge University Press.

Spiers, William. 1909. *Nature Through the Microscope.* London: Robert Culley.

Star, Susan L., and J. Griesemer. 1989. "Institutional Ecology, 'Translations' and Boundary Objects: Amateurs and Professionals in Berkeley's Museum of Vertebrate Zoology, 1907-39." *Social Studies of Science* 19:387-420.

Sutton, Alan. 1986. *A Victorian World of Science.* Bristol: Adam Hilger.

Thomson, G. P. 1964. *J. J. Thomson and the Cavendish Laboratory in His Day.* London: Nelson.

Thomson, J. J. 1896. Address to Section A. *British Association for the Advancement of Science Report,* pt. 2: 699-706.

Turnbull, David. 1995. "Rendering Turbulence Orderly." *Social Studies of Science* 25:9-33.

Tyndall, John. 1870. "Limits of the Power of the Microscope." *Quarterly Journal of the Microscopical Society* 10:414-17.

———. 1871. "The Scientific Use of the Imagination." In *Fragments of Science for Unscientific People.* London: Longmans, 125-67.

University College London. 1878. *Calendar.* London: University College London.

Van Helden, Albert, and Thomas Hankins, eds. 1994. Special issue: "Instruments." *Osiris,* vol. 9.

Weinreb, Ben, and Christopher Hibbert, eds. 1993. *The London Encyclopedia.* Rev. ed. London: PaperMac.

Wise, M. Norton, ed. 1995. *The Values of Precision.* Princeton, N.J.: Princeton University Press.

Woodward, J. J. 1870. "On the Magnesium and Electric Light as Applied to Microphotography." *Monthly Microscopical Journal* 3:290-300.

Woolgar, Steve. 1988. *Science: The Very Idea.* Chichester: Ellis Horwood.

Wright, Lewis. 1895. *A Popular Handbook to the Microscope.* London: The Religious Tract Society.

20

Metrology, Metrication, and Victorian Values

SIMON SCHAFFER

> National metrology refers to the comforts, conveniences and most useful employments of the mass of the nation, and especially of the many and the poor. Yet in Great Britain, those who should feel most directly and immediately concerned, do not seem in any way sufficiently awake to the dangerous crisis which is passing.
> CHARLES PIAZZI SMYTH, *Life and Work at the Great Pyramid* (1867)

I. Metrology's Moral Values

At a major South Kensington exhibition of scientific instruments in May 1876, some eminent Victorians gathered to debate their nation's achievements in establishing new sciences and technologies on the basis of accurate standards. "Nearly all the grandest discoveries of science," the entrepreneurial engineer William Siemens explained from the chair, "have been but the rewards of accurate measurement and patient long-continued labour in the minute sifting of numerical results" (figure 20.1). The capacity to succeed overseas also depended on painstaking measures of extreme accuracy. "To resort to a homely illustration," Siemens explained, "let us suppose a traveller in the unknown wilds of the interior of Africa" using such reliable measurements to navigate across the "Dark Continent" (Siemens 1876, 206–7). The very recent exploits of David Livingstone and Henry Morton Stanley may indeed have made this exotic example seem homely to his London audience. Siemens referred to Joseph Whitworth's screw-gauges, George Airy's clocks, and William Thomson's electrometers for ma-

---

I would like to thank the Syndics of Cambridge University Library for allowing me to quote from the Maxwell manuscript of 1877, the Syndics of Cambridge University Library and the Director of the Royal Greenwich Observatory for allowing me to quote from the Airy letter of 1867, and the Glasgow University Library, Department of Special Collections, for allowing me to quote from the Joule letters of 1865.

438

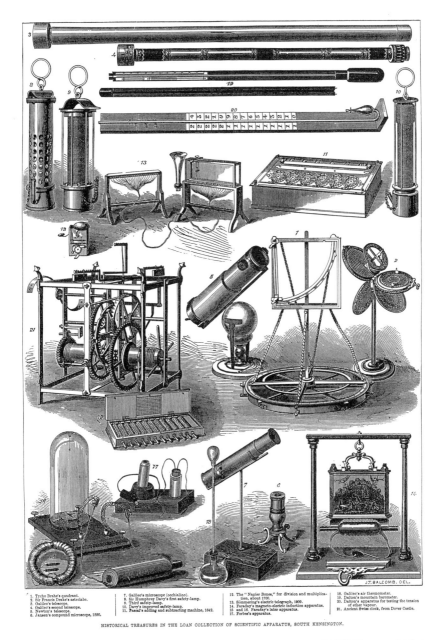

Figure 20.1 Displays from the *Scientific Instruments Exhibition* of 1876. They taught that scientific advance relied on a history of precision technique. (*Illustrated London News* 69 [1876], 269.)

terial examples of commercial vitality and standardized measurement. His fellows set out to depict the campaign for precision through which, as the Manchester manufacturer Whitworth put it, workmen's "vague terms, as a bare sixteenth or full thirty-second," were being systematically replaced by industrially and geographically universal values crucial for economic and military success (Whitworth 1876, 217-18; Rolt 1986, 125-29; Headrick 1981, 98-100).

The development of universal standards was supposed to produce consensual uniformity but was just as likely to breed bitter dispute. In just one evening's debates at South Kensington, there were fights about whether length standards should be taken between two lines engraved in the bar or between its ends, whether the French metric system should be adopted in Britain, whether James Clerk Maxwell's recent electromagnetic theory of light should be adopted on the basis of an identity of light speed and the ratio of electrical units, and whether physicists should be worried by the discrepancy between their best value for the unit of electrical resistance and the results of James Joule's recent work on the mechanical equivalent of heat. Even Whitworth's gauges were notably controversial—he had just emerged from a decade's fight with the army's Ordnance Board about his gun standards (Kilburn 1987, 29-32). None of these were trivial problems. All raised fundamental issues of best practice in the enterprise of precision measurement, which was, after all, supposed to allow science more effortlessly to escape the trammels of interest and judgment by institutionalizing metrology.

The term "metrology" was born with the nineteenth century. Its first recorded English appearance was as the title of an influential 1816 text by the London accountant and mathematician Patrick Kelly, apostle of the science of exchange and founder-member of the Astronomical Society (Ashworth 1994, 422-23). The construction of reliable common standards of measurement held the key to economic, political, and scientific advance. Expansionist free traders urged that the state must guarantee the reliability of values in the new world economy of early-nineteenth-century commercial empires (Hilton 1988, 69). Standards of weight and measure were connected with the state's attempts to impose a uniform system on exchange mechanisms and the fiercely contested contemporary process through which a customary moral economy of the grain market was challenged by the campaign against the Corn Laws. These campaigns spread across the world, whether in statistical surveys of the Scottish Highlands or the ferocious imposition of new property regulations in Bengal (Thompson 1991, 167-75; Bayly 1989, 121-26). The so-called Second Scientific Revolution of the early nineteenth century promoted techniques to further standardization. Inspection and tabulation became marks of scientific control (Kuhn 1977, 60-64, 219-20; Hacking 1990, 47-63). Postwar regulation in the wake of the Weights and Measures Act (1824) depended on a host of interlinked institutions, staffed by public inspectors, statisticians, chemists, and engineers, such as the Statistical Department of the Board of Trade (1832), the Factory Inspectorate (1833), the General Register Office (1836), the observatory of the British Association for the Advancement of Science at Kew (1842), and the Excise Laboratory (1842). Metrologists secured the status of goods as commodities by imposing widespread systems of measures and simultaneously pursued the goal of physical and moral purity, identifying adulterates, verifying gauges, or mapping disease and crime. With the institutionalization of metrology, interested evaluation could be represented as sober valuation (Cullen 1975; Corrigan and Sayer 1985, 123-41).

Metrologists had to define a unit, the numerical measure of some variable, and a standard, the object or technique that embodied the unit. The standard might be represented by some arbitrary artifact or might instead be defined as the result of some carefully defined procedure. The result of this procedure could then be either embodied in a privileged object or else, in principle, reproduced by any worker equipped with the appropriate resources. There were important contrasts between systems in which central authorities engrossed the right to make, distribute, and verify derived standards and those in which users were trusted to produce an embodiment of the unit for themselves (O'Connell 1993, 145). The Victorians inherited a unit of length, the yard, that had traditionally been embodied as a bar held in the vaults of Parliament and the Treasury and a representation derived by distributing copies of this standard throughout the kingdom. Fights about the *standard* raged in the early nineteenth century, when instrument makers, surveyors, and astronomers debated the alternatives of an arbitrary bar or an apparently more natural and procedural standard such as the length of a pendulum beating seconds at London. Fights about the *unit* broke out in the 1860s, when radical free traders actively proposed metrication while patriotic traditionalists urged the supremacy of the imperial yard. Metrologists were the principal designers and users of newly crafted instruments such as pendulums, theodolites, hydrometers, screw gauges, and spectroscopes. These devices relied on the techniques of rather specialist experts who often resisted surveillance of their labor. An 1854 report from the Ordnance Survey insisted that "the draftsman's art is to do justice to nature, and the engraver's to do justice to the draftsman; on neither should rules and methods be imposed, except by superior artistic judgment" (Seymour 1980, 126). It was hard to impose systems of standardized performance. Resistance to such imposition might come from the factory or the agricultural workforce, from the draftsman's studio or the clinic (Porter 1995, 223; Coleman 1973; Lawrence 1985). Victorian metrology therefore uneasily combined the construction of institutions that worked globally to further standard measures with strategies to defend the traditions of corporate rights and moral values. So the history of metrology combines an analysis of metrological institutions' construction with an explication of their apparent autonomy (Barry 1993; Porter 1995, 21-29).

Because of values' practical independence and political and commercial dependence, metrologists constructed their standards in the milieus of conflicting interests. In 1855 George Phillips, chief chemist at the Excise Laboratory, struggled manfully to explain to skeptical members of Parliament who had the best interests of tobacco merchants at heart that "it was an unfair test of chemistry to send us samples of tobacco without our having a standard to fall back on, and then to lay our shortcomings upon the science of chemistry" (Hammond and Egan 1992, 17-18). Sometimes metrology was used to impose moral order: at the General Register Office William Farr designed the post-1851 census classification to use essential occupational types to contest "anarchy, riot, insecurity of life [and] communism" (Higgs 1988, 78; Hacking 1990, 119). Others were markedly less optimistic about the effect of precise definition on moral order. The Oxford chemistry professor Benjamin Brodie told the 1867 Royal Commission on Water Supply that "we can weigh and we can measure, and we can do that with a certain accuracy, and there we stop, but that accuracy is not capable of being multiplied *ad infinitum.*" Brodie could only "say that when you have once put sewage into the water I should be rather reluctant to drink it" (Hamlin 1990, 142). The puzzle often was that traditional wisdom was supposed simultaneously to be challenged and confirmed by experts' measures. One way of solving the puzzle was to make new measures count as traditional. The aim of this chapter is to interpret Victorian metrological innovation as the invention of tradition.

Across Europe, the later nineteenth century witnessed an unparalleled production of invented traditions, cultural forms designed to cope with the dramatic crises of a rapidly changing social order under the guise of immemorial custom and natural right, and the state was often the principal promoter of these inventions (Hobsbawm 1983). Victorian metrology, which had to resolve simultaneous polemical demands for pious morality, capitalist economy, and scientific accountability, is a good example of the traditions invented during the Age of Empire. The careers of three protagonists of metrological work, John Herschel (1792-1871), Charles Piazzi Smyth (1819-1900), and James Clerk Maxwell (1831-79), show how metrological traditions were invented. Herschel, a protagonist of the reconstruction of the imperial yard, identified its status with the quality of the labor involved in producing an artifactual rod rather than the match between this rod's length and some natural quantity, such as the French republican metre. Herschel protested when it seemed as if the logic of unbridled free trade would dictate metrication. His campaigns gave encouragement to conservatives, such as his colleague the Tory astronomer Smyth, who identified national standards with those of divinely validated morality. So from the 1870s a range of metrological positions fought for legitimacy. Some conservatives developed a scriptural metrology based on an original divine warrant for British measures, while reformers argued for what they saw as a rationalized metric system tuned to the commercial dictates of free trade. Interests of imperial power and commercial advantage counted when standards were to be established: in 1871 Maxwell, Cambridge's new experimental physics professor, tried to institutionalize electromagnetic metrology there because of its importance for the boom industry of submarine telegraphy and its use in his new theory of light. A former ally of Smyth, Maxwell nevertheless firmly rejected his extreme scriptural metrology. Yet he had to show how laboratory metrology could shore up orthodox divinity, so he used Herschel's ideas about the morality of standards to reconcile his laboratory program with university culture. Then his arguments drew fierce attacks from eminent scientific naturalists, Thomas Huxley, John Tyndall, and especially the mathematician William Clifford, who denied that any transcendental lessons could be derived from local practical measurements. These controversialists battled for the right to define the proper system of values and invented the scientific traditions that they hoped would secure them.

II. John Herschel: The Yard Drops from the Clouds

A series of commissions that sat in the decade following the end of the Napoleonic War promoted measures to set imperial standards. About one hundred different laws had already been passed on weights and measures, and 230 provincial systems of weights and measures were still in use. While reformers assaulted "the despotic influence of custom with respect to the contents of a certain denomination," Tory journalists in the *Quarterly Review* lauded in Burkean manner the authority of "long usage" and boldly urged that neoteric systems such as the French metre had signally failed (Hoppit 1993, 86-92). The 1824 act attempted to institutionalize new standards on the basis of state-of-the-art London instrumentation already tried in surveys in Scotland and India, including compound pendulums, micrometers, and engraving tools. One of the surveys' protagonists, the astronomer and actuary Francis Baily, sneered at ancient standards fit only for "the mere ordinary purposes of life" (Ashworth 1994, 416-19). Debates focused both on the need to displace custom by accuracy and on the need to identify a truly "natural" standard. Natural length was to be represented through accurate pendulums suspended from carefully engineered supports and timed with high-class clocks. The imperial yard was defined as thirty-six inches, the inch to be derived from the trials carried out for the Ordnance Survey by the army surveyor Henry Kater, who determined that 39.13929 inches would be the length of a pendulum beating seconds at London (Connor 1987, 251-61; Simpson 1993, 179, 183-86).

No sooner was it institutionalized and distributed, than the standard was in trouble. A catastrophic financial crash of 1825-26 prompted intense disputes about the moral and economic meaning of absolute standards of value, whether a gold standard for currency or a terrestrial standard for length. Discussions in pulpits and countinghouses contrasted intrinsic measures of real worth and the conventional basis of current value in social consensus (Hilton 1988, 125-32). Kelly, inventor of the word "metrology," had warned Parliament that "nature seems to refuse invariable standards: for, as science advances, difficulties are found to multiply, or at least, they become more perceptible, and some appear insuperable" (Martineau [1858] 1877, 1:484). The issue of place was crucial. Though the new yards had been deposited at the Exchequer and with Parliament, then sent out through the empire, the effective standards site was a small room in Portland Place where Kater's trials were performed and others came to check his results. During the later 1820s Baily found that Kater's system was unreliable and he got the apparatus back from the House of Commons to retry all the length determinations (Simpson 1993, 185-90).

In 1834 the Houses of Parliament burnt down—Treasury workmen were using the furnaces in the basement of the House of Lords to destroy thousands of wooden tallies, which the government had theretofore used to run its accounts. This dramatically modernizing fire also destroyed the national standards of length and weight. So standards had to be rebuilt. A new Parliament building pointedly Gothic and antique was supposed neatly to reconcile radical demands for a rebuilt constitution and conservative sensibilities tuned to patriotic tradition (Colley 1992, 324-25). New standards of weights and measures had to become part of similarly invented traditions. In the late 1830s the reestablishment of the original standard of imperial length was not made by appeal to pendulum tests, for these had been subjected to criticism in the interim and any "reference to any natural basis" was not thought prudent. The result of eleven years' work until 1845 in the basement of Somerset House by the irascible Cambridge astronomer Richard Sheepshanks was the production of a new imperial standard. His work was debated by the British Association in the mid-1850s and sanctified by legislation in 1855. The imperial yard was ruled by Sheepshanks's allies, the firm of Troughton and Simms, while Airy designed a system of pivoted rollers to carry the new bars and a regime for keeping the end lines fixed. "Parliamentary copies" were carefully listed in order of their authority, one bricked up at Westminster, another at the mint. The imperial yard was now embodied through painstaking labor on an arbitrary piece of metal, not derived from some vulnerable comparison with an external measure. As John Herschel, member of the commission charged with the standards' reconstruction, explained, "The new standard was constructed . . . by an assemblage and most careful comparison of all the scales and standards of any authority which could be got together" (Connor 1987, 257-67; Herschel 1857, 591).

Herschel, preeminent sage and remarkably wealthy amateur of a range of Victorian sciences, held but one paid post in his entire lifetime, as head of the metrologically crucial Royal Mint, and even that briefly (Chapman 1993, 73-76). A handbook of late Victorian piety gave up the task of explaining to its readers the import of Herschel's measures of crystal polarization and of the motions of double stars. "But they will understand that these results could have been attained only by the most assiduous industry and the most unflinching perseverance. And it is on account of this industry and this perseverance that we recommend Herschel as an example to our readers" (*The Story of the Herschels* 1879, 71-72). Herschel used his remarkable status to teach some lessons about the relation between standards and work. "Our yard," he observed, "is a purely individual material object, multiplied and perpetuated by careful copying; and from which all reference to a natural origin is studiously excluded, as much as if it had dropped from the clouds" (Herschel 1867, 432). The imperial yard of 1855, for example, was scientific only to the extent that much high-quality labor had been expended on its production. "Absolutely vague and inadequate" human senses did not provide numbers, so "in this emergency" scientists must work with measuring devices, which required the construction of socially accepted standards (Herschel 1831, 124-25). "Unless we transmit to posterity the units of our measurements, *such as we ourselves have used them,* we, in fact, only half bequeath to them our observations." Herschel, the most famous beneficiary of a singular scientific inheritance from his astronomer-father, proposed the preservation of material standards inside "some great public building," the first British vision of an official standards laboratory (Herschel 1831, 128). His prescription mixed morality with technology. Observers must be trained to be faithful, the division of labor was vital. Herschel even suggested "the circulation of printed skeleton forms" that would "ask distinct and pertinent questions," demand numerical answers, and "call for their transmission to a common centre" (Herschel 1831, 134). This was the program George Airy adopted at Greenwich Observatory and the model for the metrological service led by Herschel's epigones at Kew (Scott 1885, 48-52). The system of standardization, distribution of instructions, division of labor, and rigid hierarchical management seemed to offer the key to Victorian scientific progress.

In his most popular exposition of method, Herschel explained that the successes of the physical sciences "tend of necessity to impress something of the well weighed and progressive character of science on the more complicated conduct of our social and moral relations" (Herschel 1831, 72-73). His key phrase was "well weighed and progressive." Science guaranteed political order because of its use of measurement, and, conversely, measurement revealed the most fundamental fact of natural order—its divine structure. Herschel urged that precision measurement revealed that all atoms of the same kind were exactly alike and repetitive identity was evidence of divine creation. "A line of spinning jennies, or a regiment of soldiers dressed exactly alike, and going through precisely the same evolutions, gives us no idea of independent existence: we must see them act out of concert before we can believe them to have independent wills and properties." Factories and barracks were good signs of control, while disorder manifested willful autonomy. Thus, Herschel concluded, since the atomic world was revealed to be identical everywhere, its atoms possessed "the essential characters, at once, of a *manufactured article,* and a *subordinate agent*" (Herschel 1831, 38). The values of workshop uniformity were the values of precision measurement.

These arguments gained new force in the 1860s. The link between long-range control and standards was evinced in British submarine telegraphy and the contemporary British Association campaign for a new unit of electrical resistance. The absolute resistance unit, eventually baptized the ohm, was to be defined metrically, as 10 million metres, one earth quadrant, per second. Physicists such as William Thomson at Glasgow and Clerk Maxwell in London commissioned ingenious rotating coils that were supposed to embody the new resistance unit, and then set out to link these devices with the techniques of the firms who were making copper wire for the rapidly expanding submarine cable network. Commentators saw a very close connection between the integrity of these laboratory trials and that of the British imperial system (Schaffer 1992; Hunt 1994; see also Hunt's chapter in this volume). These imperial and commercial considerations were used to back more workshops and laboratories in British universities, including Cambridge, where these trials on electromagnetic values would be made by a new, disciplined workforce (Sviedrys 1976; Gooday 1990).

The apostles of international exchange within the British physics community, led by Thomson, never doubted that the units for electrical resistance should be metric. But they noisily attacked the arbitrary *standard* of resistance, derived from the electrical properties of a column of mercury, in vogue among their commercial rivals, the German and French telegraphers (figure 20.2; Jenkin 1862*a*, 126; 1862*b*). By contrast, a Tory such as Joule, who knew that his friend Thomson would "think me a heretic," was "firmly convinced that it will be impossible to make 3/4 of the globe adopt the metre" (Joule 1865*a*). Joule loathed neologisms such as "Ohm," and his perception of a political agenda linked to metrication was confirmed following the extremely controversial Free Trade Treaty with France in 1860. Its negotiator, Richard Cobden, immediately began to promote what he called "free trade in arithmetic," and Cobden's active supporter, the radical member of Parliament William Ewart, who had previously backed the abolition of capital punishment and the introduction of examinations for the

[To face page 340.]

Approximate Relative Values of various Units of Electrical Resistance.

Description.	Name.	Absolute foot second $\times 10^7$.	Thomson's old unit.	Jacobi.	Weber's absolute metre second $\times 10^7$.	Siemens 1864 issue.	Siemens (Berlin).	Siemens (London).	B. A. unit, or Ohmad.	Digney.	Bréguet.	Swiss.	Matthiessen.	Varley.	German Miles.	Observations.
Absolute $\frac{foot}{second} \times 10^7$ electro-magnetic units (new determination)	Absolute $\frac{foot}{second} \times 10^7$	1·000	0·6030	0·4788	0·3316	0·3187	0·3168	0·3131	0·3048	0·3289	0·3123	0·2924	0·2243	0·01190	0·005307	Calculated from the B. A. unit.
Absolute $\frac{foot}{second} \times 10^7$ electro-magnetic units (old determination)	Thomson's unit	1·6585	1·000	0·5029	0·3483	0·3348	0·3328	0·3289	0·3292	0·3455	0·3279	0·3071	0·2357	0·01251	0·005574	From an old determination by Weber.
Twenty-five feet of a certain copper wire, weighing 345 grains	Jacobi	2·068	1·000	1·000	0·6025	0·6055	0·6018	0·6540	0·6367	0·6300	0·6530	0·6106	0·4686	0·02486	0·01108	No measurement made; ratio between Siemens (Berlin) and Jacobi taken from "Weber's Galvanometrie."
Absolute $\frac{metre}{second} \times 10^7$ electro-magnetic units determined by Weber (1862)	Weber's absolute $\frac{metre}{second} \times 10^7$	3·015	1·444	1·000	1·000	0·6007	0·9656	0·9943	0·9191	0·9910	0·9416	0·6817	0·6767	0·03591	0·01655	Measurement taken from a standard sent by Prof. Thomson; does not agree with Weber's own measurement of Siemens's unit, by Weber: 1 Siemens's unit = 1·0215 × 10⁷ metres-second.
One metre of pure mercury, one square millimetre section at 0° C.	Siemens 1864 issue	3·138	2·988	1·503	1·941	1·000	0·9950	0·9829	0·9563	1·033	0·9709	0·9177	0·7047	0·03737	0·01666	Measurement taken from three coils issued by Messrs. Siemens.
One metre of pure mercury, one square millimetre section at 0° C.	Siemens (Berlin)	3·156	3·004	1·511	1·946	1·005	1·000	0·9881	0·9625	1·038	0·9832	0·9227	0·7081	0·03757	0·01675	Measurement taken from coils exhibited in 1862 by Messrs. Siemens, Halske & Co. (well adjusted).
One metre of pure mercury, one square millimetre section at 0° C.	Siemens (London)	3·194	3·040	1·529	1·959	1·017	1·012	1·000	0·9742	1·050	0·9970	0·9337	0·7106	0·03802	0·01695	Measurement taken from coils exhibited in 1862 by Messrs. Siemens, Halske & Co. (well adjusted).
British Association unit	B. A. unit, or Ohmad	3·821	3·123	1·570	1·988	1·0456	1·039	1·026	1·000	1·079	1·024	0·9959	0·7236	0·03905	0·01741	Equal to 10,000,000 $\frac{metres}{second}$ according to experiments of Standard Committee.
One kilometre of iron wire, four millimetres in diameter (temperature not known)	Digney	30·40	28·94	14·56	19·68	9·968	9·634	9·520	9·256	1·000	9·491	8·889	6·822	0·3629	0·1613	From coils exhibited in 1862 (pretty well adjusted).
One kilometre of iron wire, four millimetres in diameter (temperature not known)	Bréguet	32·03	30·50	15·34	19·62	10·20	10·15	10·13	9·760	1·054	1·000	9·365	7·187	0·3814	0·1700	From coils exhibited in 1862 (badly adjusted).
One kilometre of iron wire, four millimetres in diameter (temperature not known)	Swiss	34·21	32·56	16·38	21·34	10·90	10·84	10·71	10·42	1·125	1·068	1·000	7·675	0·4072	0·1815	From coils exhibited in 1862 (badly adjusted).
One English standard mile of pure annealed copper wire 1/16 in. diameter at 15°·5 C.	Matthiessen	44·57	42·43	21·34	14·78	14·10	14·12	13·95	13·59	1·46	1·391	1·303	1·000	0·5306	0·2365	From a coil lent by Dr. Matthiessen (of German-silver wire).
One English standard mile of one special copper wire 1/16 inch in diameter	Varley	84·01	79·96	40·21	27·85	26·75	26·61	26·30	25·61	2·763	2·622	2·456	1·885	1·000	0·4457	From coils lent by Mr. Varley (well adjusted).
One German mile = 8258 yards of iron wire ⅙ inch in diameter (temperature not known)	German mile	188·4	179·4	90·22	62·48	60·03	59·71	59·00	57·44	6·198	5·882	5·509	4·228	2·243	1·000	From coils exhibited in 1862 by Messrs. Siemens, Halske & Co.*

* Messrs. Siemens do not now manufacture coils with this unit, which has been abandoned by them in favour of the mercury unit given above.

Figure 20.2 The British Association's 1863 conversion tables for varying measures of electrical resistance, which revealed the wide differences of unit and standard in use in European telegraphy. (*Report of the 34th Meeting of the British Association for the Advancement of Science (1864)* 1865, 349.)

civil service and the armed forces, soon sought to impose the metric system by law throughout the British Empire. By 1864 a permissive act introducing the foreign system had passed through Parliament. It was argued that "the metric system is the complement and corollary of Free Trade. By adopting it, we shall extend the commerce of England and the commerce of the World" (Morley 1908, 326-27; Munford 1960, 148-50).

In the 1830s Herschel had given some backing to French units, because the earth quadrant, the basis of the metre, allegedly gave a more accurate result for a unit of length than did the length of the seconds pendulum, the quondam basis of the British yard (Herschel 1831, 126-27). But in the interim the British government had severed the link between the yard and the unreliable pendulum. The standard of length now drew its authority from law and labor. So after 1855 Herschel, like Joule, strenuously resisted metrication. Herschel sent a pamphlet to every member of Parliament, in October 1863 he lectured at the Leeds Astronomical Society against the metre, and the following year he published a long paper on celestial metrology in the popular Evangelical magazine *Good Words*. In April 1869 he wrote to the *Times* setting out his opposition to Ewart's bill (Herschel 1867, 176-218, 419-51; Smyth [1864] 1877, 218-19). "So far there is no actual harm done, beyond unsettling opinions and creating uneasiness; but we trust the common sense of the nation will repudiate any attempt to carry out to its designed completion a measure so thoroughly retrograde" (Herschel 1867, 179 n). The grounds of his "common sense" metrological campaign reveal much of the role of tradition in the precision measurements of Victorian culture.

Herschel proposed a tough test for any candidate standard (Herschel 1867, 426). It must be unchanging, reproducible, and generally acceptable. The model was the standardization of divine creation, manifest in the uniformity of all "manufactured" atoms. The metre failed; the imperial inch, suitably modified, passed with flying colors. Herschel reckoned that the friends of metrication displayed "too gratuitous a contempt for our national and time-honoured standards, and too hasty a preference for the apparently more scientifically . . . constructed system of our continental neighbours" (Herschel 1867, 178). He used his attack on French units to teach the right message about the cultural meaning of precision. Exact science could not effortlessly be read in nature or tyrannically imposed on the nation. Here Herschel used Burkean rhetoric against "the political passions" that had moved the republican reformers, a carefree attitude to history and a facile view of scientific labor. "Public opinion" must always treat the national standard with "reverence" (Herschel 1867, 432, 183).

Herschel launched a twofold attack on the metre. He argued it was no better than the imperial unit. Recent heroic British surveys of India were used to give the most reliable value for Earth's dimensions. Herschel described the use of theodolites in the measurement of an arc and the bar standards that the British military surveyors used, designed to be self-compensating under any temperature change and compared by means of a movable microscope focused on the lines carefully engraved along their sides. These measures showed that 10 million metres were not exactly one earth quadrant. So the metre, as currently defined, was a mere approximation (Herschel 1867, 184-87, 440-44). Furthermore, as his colleague Airy had shown, the earth quadrant passing through France was a local and a necessarily inexact measure. Herschel now had to make the yard preferable. He nominated Earth's polar axis as the obvious standard of length, because it was a true universal and, by British geodesy, possible to measure with sufficient accuracy. Then he pointed out that were the imperial units increased in length by 1/1000 of a part, then this axis would be exactly 500 million "geometrical" inches long (Herschel 1867, 446). The increase, he pointed out, was less than the normal inaccuracy of commercial scales. This seemed to Herschel a perfectly satisfactory case of earth commensurability, just as important as the putative earth commensurability of the metre. Herschel proposed a "geometrical cubit" of twenty-five "geometrical inches" as a new standard of length defined this way, by analogy with Isaac Newton's work on the Jewish sacred cubit (Herschel 1867, 191, 450).

Then Herschel played his trump card, appealing graphically to the political and economic facts of life to insist on the value of the imperial yard. British commercial and imperial superiority argued against metrication. "England is beyond all question the nation whose commercial relations, both internal and external, are the greatest in the world. . . . Taking commerce, population and area of soil then into account, there would seem to be far better reason for our continental neighbours to conform to *our* linear unit" (Herschel 1867, 445). Ewart and his radical allies might argue that metrication was the natural accompaniment of free trade; Herschel now riposted that the evidence of imperial trade dictated the retention of imperial units. Commerce and patriotism were inseparable from the choice of standards.

III. Charles Piazzi Smyth: The Great Metrological Monument

Herschel's arguments of the 1860s, which won backing from such authorities as Airy and Joule, provided important resources for late Victorian metrology. The virtues of the nation's metrology were supposed to display the virtues of national life. As several spokesmen argued, French republican values bred a dangerous system of standards. In a significant passage, Herschel also mentioned one of the more fashionable candidates for a theological, and thus certain, basis for such measurements—the Great Pyramid. One protagonist of a pyramidological metrology was the wealthy member of Parliament and army officer Richard William Howard Vyse, who worked in Egypt in 1835-37 and who discussed his results with Herschel in 1839. Another was the London publisher of Herschel's astronomy, John Taylor, who linked the causes of currency reform and metrological conservatism (Smyth 1871a, 204). Herschel noted the claims of the Great Pyramid of Giza as a basis of all civilized standards of measurement. "Of human works, the most permanent, no doubt, and the most imposing as well as generally interesting and respected, are those mighty monumental structures which have been erected as if for the purpose of defying the powers of elementary change" (Herschel 1867, 187, 427).

Taylor and his disciples urged that the dimensions of the Pyramid showed the divine origin of the British units of length. They reckoned that the ratio between the circumference of its base and the Pyramid's height was exactly π. This was a sign that its designers, divinely inspired, had been able to square the circle. Furthermore, Herschel's "geometrical cubit" of twenty-five revised inches was just the length of the Pyramid builders' measuring rods. So these inches were the basis of divine metrology (J. Taylor 1859). Herschel sent Taylor several copies of his own antimetric pamphlets to help with the campaign. But he warned that "the sole means by which we are now enabled to determine the original height [of the Pyramid] consists in a block of the exterior marble casing which will in all probability disappear in the hands of 'the curious' within the next century" (Herschel 1867, 427; Davidson 1938). He was prescient. The curious pyramidologists of late Victorian Britain used Herschel's immense authority, and the opportunities provided by the campaign against French, atheist mensuration, to launch a major campaign for properly divine values.

The spokesman of this new pyramidology, Charles Piazzi Smyth, had excellent credentials. The son of England's leading amateur astronomer, as a young assistant in South Africa in the 1830s Smyth had run a successful geodetic and time service, impressing Herschel, a visitor there, so much that Sir John gave his decisive backing to the young man's "earnestness of character and devotion to the acquisition of knowledge." This backing won Smyth the post of astronomer royal for Scotland in 1845 (Warner 1983, 122). The occupant of one of the few publicly funded posts in astronomy, he completed a massive meridian catalog, and his 1856 Admiralty expedition to Tenerife won plaudits from the leading British astronomers, including Herschel, for its demonstration of the advantages of high-altitude observatories. Then, in January 1864, Smyth read Taylor's remarkable *Battle of the Standards* and during the next few months rapidly completed his own development of this argument, *Our Inheritance in the Great Pyramid.* Much of Smyth's book was devoted to proving the claim that the dimensions of the building were commensurable with those of Earth and the universe. Smyth set out for Egypt in November 1864, completed an exhaustive survey of the Pyramid's dimensions, and reported his results at the Royal Society of Edinburgh in April 1866, publishing them in a bulky, three-volume work, *Life and Work at the Great Pyramid,* in 1867. In the 1880s, Smyth got backing from American railway engineers, who formed the International Institute for Preserving and Perfecting Weights and Measures. Smyth supplied them with ceramic copies of the pyramid cubit as part of a campaign to set up an alternative international metrological system (Brück and Brück 1988, 95-134).

Smyth made no secret of the interests that drove his Egyptian campaign. He claimed that metrication was an invention of the worst of French, atheist republicanism. "Simultaneously with the elevation of the metrical system in Paris, the French nation (as represented there) did for themselves formally abolish Christianity, burn the Bible, declare God to be a nonexistence, a mere invention of the priests, and institute a worship of humanity, or of themselves" (Smyth 1877, 215). The defeat of France by Prussia in 1871 and the behavior of the Commune only confirmed these views. The Communards, he noted, immediately backed metrication and "the abolition of the Christian era" (Smyth 1871*b*). Smyth and his allies saw similar threats closer to home, especially in the wake of the 1867 Reform Act and the subsequent election of a radicalizing administration initially keen on church disestablishment, land reform, and secularized schooling. By 1869 Joule was among those convinced that "no one can say that the British Empire will not be Roman Catholic in five years" (Joule 1869). When he saw Smyth's photographs of the Pyramid's interior Joule immediately told Thomson that he was converted to Protestant pyramidology (Joule 1865*b*). It was not surprising to such conservatives that friends of metrication were radical subversives and the noisy secularists of London science. Smyth identified them as "the merchants and manufacturers of the country, with a section of the scientific men, chiefly of the electrician and chemical camp. . . . The creed that they almost worship consists in . . . making money with the utmost speed" (Smyth 1877, 208). Nor was it odd that the Protestant countries, such as Britain, Denmark, and Prussia, had units of length closest to the pyramid cubit, while those units most distant from the sacred original were found in atheist France and Islamic Turkey (Smyth 1867, 3:595). The Pyramid, a "Metrological monument," was therefore a divine sign of the link between moral values and physical standards, a potent lesson from ancient Egypt to modern Britain (Smyth 1867, 1:xii).

In the wake of the French expedition of 1798, Egypt had been marked by campaigns conducted in the name of allegedly superior European sciences designed to appropriate and displace indigenous cultures (Said 1979, 192-97; Beaucour, Laissus, and Orgogozo 1990, 137-52, 213-23; Godlewska 1994). It became a peculiarly crucial territory for disputes between Gallic and Anglo-Saxon values. In 1841 Britain and France imposed economic control on the khedivate after a punitive military expedition; the British began to build a railway from Alexandria to Suez in the 1850s, and they secured management of the cotton industry during the cotton famine of the Ameri-

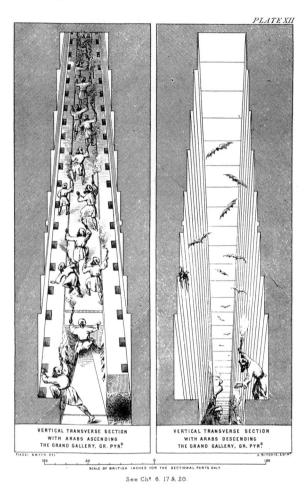

Figure 20.3 Smyth's drawings of Arab workmen climbing the Great Pyramid's tunnels. Smyth included these drawings in his work on divine metrology alongside data on the building's dimensions. (Smyth 1877, plate 11.)

can Civil War. Following a disastrously unsuccessful cable project in 1858–59, the Submarine Telegraph Company, part of John Pender's communications conglomerate, laid a successful line via Egypt to India in 1869, the same year as the completion of the Suez Canal by an Anglo-French consortium. European technology and capital exploited and clashed violently with Egyptian autonomy. By 1882 the country had been absorbed within the British Empire (Headrick 1988, 98–101, 196–204; Mitchell 1988, 15–21; Cole 1993, 23–83). Victorian moralizing taught the appalling and self-serving contrast between the allegedly trustworthy conduct of scientific instruments and assistants and the hostility or incomprehension of indigenous cultures. "The European is a close reasoner," observed Britain's administrator of Egypt from 1882, Lord Cromer. "His statements of fact are devoid of any ambiguity . . . his trained intelligence works like a piece of mechanism." By contrast, according to Cromer, "the mind of the Oriental, like his picturesque streets, is eminently wanting in symmetry. His reasoning is of the most slipshod description" (Said 1979, 38). The supposed universality of European measures meant that other nations could be judged by their inability to share them (Adas 1989; Pang 1993, 274-75).

Smyth represented contemporary Egypt, provisionally occupied by Arabs, as the land of imprecision, and its past, revealed by the British, as the haven of accuracy. Orientalists might enjoy "Cairo and its narrow streets, gaudily dressed population and crowded bazaars," but Smyth found it an obstacle "to a man of moderate means and with a definite task to accomplish within a limited time" (Smyth 1867, 1:20). He drew a pointed contrast between his own mastery of the best British precision instruments and Arab distrust of, and incompetence with, these devices (figure 20.3). European instruments were "proofs in their minds that a European cannot get on at any occupation without some queer and troublesome contrivance to *peep through*" (Smyth 1867, 1:299-300). Smith proudly called his measuring apparatus "the Edinburgh instrumental bow" (Smyth 1867, 1:305-11). It gave him access to a precise past that sacred physics could see, modern industrialism missed, and contemporary Egyptians ignored. "The apparatus proved itself superior in accuracy to the fragments which it had to measure" (Smyth 1867, 2:168). This was a message Smyth was anxious to impress—his instruments were better than the wreckage with which he was forced to deal yet were often matched by the original form of the divine building he was painstakingly reconstructing.

Smyth tried to make the Pyramid into a field station for precision measurement so that the numbers he produced in Egypt would be compelling in Britain. Field stations needed exceptionally secure management to turn them into standards institutions (Pang 1993, 276-77). Airy, for one, found Smyth's Egyptian work "amusing and interesting as an antiquarian speculation" but quickly rejected "speculations on the derivation of modern measures from those of Egypt, or rather the assumption of this as an indubitable truth," and confessed that "I attach less importance" than Smyth to "the accurate determination of the position of the Pyramid" (Airy 1867). Smyth countered with his own fieldwork using an instrument workshop in an abandoned tomb, wittily naming it the Howard Vyse Instrument Tomb in honor of his great predecessor. But he wanted these numbers to look like traditions concealed by more recent Arab detritus, not artifacts of his own instruments. He appealed to the fine engineering of the structure, "so straight that no modern optical instrument-maker could work better straight edges." His flexible mixture of careful antiquarianism and ingenious instrumentation, especially clinometers and theodolites, was de-

Figure 20.4 Smyth's clinometers, which were advertised as the conditions of successful measurement in the Pyramid's interior. (Smyth 1867, vol. 2, plate 1.)

signed to show that precision could be attained even when the techniques of precise surveying might seem impossible (figure 20.4; Smyth 1867, 1:146-56, 11-43).

The director of the Ordnance Survey, Sir Henry James, was Smyth's most dangerous critic. An archetype of British surveying expertise (Secord 1986), in 1866 James used his Southampton headquarters for an epoch-making comparison of triangulation length standards from all major European surveys. Between 1864 and 1869 he dispatched a team to survey biblical sites and, characteristically, to help improve the sewerage of the Holy City. James then summarized their results in *Notes on the Great Pyramid*, in good time to offer a worrisome and authoritative comparison with Smyth's own numbers (Seymour 1980, 154-58). Smyth reckoned he could quash any criticisms. Referring to the scandal of the metrication fight, he celebrated the fact that "although one nation publishes its results to an arithmetical refinement of nine places of figures, it cannot convince any other nation of its correctness beyond the first three places of figures." Smyth pointedly observed the enormous (and wasted) labor and finance expended by the Ordnance Survey. Its project had "less chance than ever of one exact, absolute, and universally admitted conclusion being ever arrived at" because "it is the nature of human science, because it is human and not divine. Human practical science can only go on by approximations, though it work at one and the same simple subject for ages" (Smyth 1877, 45-47).

The Ordnance surveyors found the Pyramid side distinctly shorter than Smyth required (Brück and Brück 1988, 42-43, 50). James threw out all data but his own men's and those of a few reputable railway engineers. Smyth was furious that James had changed his team's numbers "at home and in the closet." He made the plausible allegation that the director of the Ordnance Survey was committed to the traditional classicists' view that the pyramid cubit was identical to that of the Greeks (Smyth 1877, 32-35). Smyth and his allies saw the pragmatic Hellenism of the Ordnance surveyors as one facet of the antiscriptural politicking of a corrupted metrological system: James "went voluntarily to some source on the Continent either pure or impure, picked up there certain anti-British and even heathenish notions and engraved them to the serious misleading of the public on the Ordnance Map of Jerusalem." James answered Smyth in the Edinburgh press: "There are people who cannot see the forest for the trees and who would measure the height and girth of every tree to find the area of the forest" (James 1869, 7; Smyth 1868, 393-96; 1870, appendix, 17).

The worst threat posed by James's men was the suggestion that Smyth was attributing too much precision to the ancient Egyptians (figure 20.5). The most notorious of the pyramidological facts was the π ratio between base circumference and height. James proffered a simpler account of the Pyramid's slope: the stone courses in the Pyramid's sides rose 9 cubits for every 10 cubits of length, a reasonable recipe, the director alleged, to give to ancient Egyptian builders. Then the inclination of the sides would be indistinguishable from the π angle by any of Smyth's instruments. Smyth needed help, and got it from an Evangelical electrical engineer, William Petrie, who had once worked with him in South Africa in the 1830s (Drower 1985, 27-31). Petrie reckoned Egyptian architects had given their workers the 10:9 ratio because it was the best means of building a pyramid whose dimensions were close to the true value of π, and the numbers 10 and 9 were themselves a sign of a higher meaning of the Pyramid's dimensions. The astronomical unit, the mean earth-sun distance, he calculated, was 91,840,000 miles, exactly 10^9 times the Pyramid's height (Smyth 1877, 1-52).

Smyth had good ammunition to defend Petrie's number. In 1854-58 influential astronomers such as Urbain Leverrier had challenged the traditional earth-sun distance of over 95 million miles, reckoning that observations of the moon required a solar distance of around 91 million

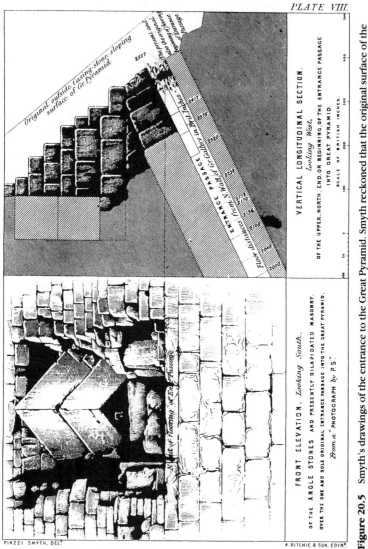

Figure 20.5 Smyth's drawings of the entrance to the Great Pyramid. Smyth reckoned that the original surface of the Pyramid was preternaturally precise, but his critics saw only the rubble that remained. (Smith 1877, plate 8.)

miles, closer to Petrie's pyramidological value. In the same years Léon Foucault showed that light was slower than had been thought, also bringing the astronomical unit much nearer to Petrie's calculation. So in 1864 the Royal Astronomical Society endorsed the reduction of the sun's distance (Clerke 1908, 230-32; Dreyer and Turner 1923, 149, 163-64; Tobin 1993, 278-83). Not only had worldwide consensus brought the astronomical unit much closer to that revealed in the Pyramid, but the change had displayed the instability of any internationally agreed value. The Pyramid "forms therefore in itself, and in all its grand simplicity and antiquity, a single representation of the whole of the numerous, laborious and most costly sun-distance results of all humankind even in the present age" (Smyth 1870, 35-36).

The contrast between divinely validated measures embodied in the metrological monument and the apparent chaos of scientific expertise was persistently exploited. In December 1874 European astronomers were able to observe a transit of Venus across the sun and thus derive a new estimate of the astronomical unit accurate, according to Airy, to within one hundred thousand miles. British and French astronomers launched the first mass invigilation of every observer's personal equation, the mean time delay experienced in registering the passage of an image across their telescopes' crosswires (Schaffer 1988, 125). In such sites as the Egyptian desert, "each observer went out ticketed with his 'personal equation', his senses drilled into a species of martial discipline, his powers absorbed, so far as possible, in the action of a cosmopolitan observing machine" (Clerke 1908, 235) (figure 20.6). Smyth documented the immense mobilization of resources devoted to the transit survey:

> Steam navigation, iron ships, electric telegraphs, exquisite telescopes, both reflecting and refracting, photographic machines of enormous power, refined regulator clocks, and still more refined chronographs, transit instruments, equatorials, spectroscopes, altitude-azimuth circles, all these modern inventions and many others, with all the learning of the universities, and numerous officers and men both of the army and navy, are pressed into the cause. (Smyth 1877, 55-57)

This might seem a characteristic Victorian paean to the dependence of precision measurement on worldwide technical systems. It was not. Smyth correctly, and delightedly, reported that the transit expeditions had failed. Airy was furious, not least because plans for them had been marked by violent debates at the Royal Astronomical Society about his own management of the observers and selection of observing sites. The uncertainty in the astronomical unit remained at about 1,500,000 miles (Dreyer and Turner 1923, 169, 180-82; Tupman 1878; Clerke 1908, 235-37). Smyth pointedly contrasted human failure with divine certainty (Smyth 1877, 57-61). He

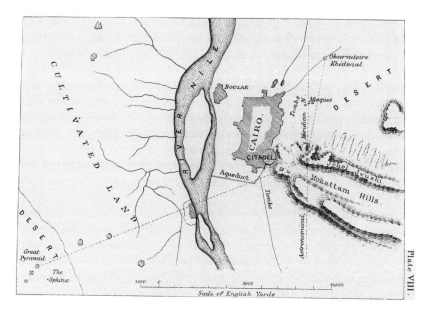

Figure 20.6 Map of the transit of Venus in 1874. British astronomers in Egypt to observe the transit tried to provide reliable data for the earth-sun distance, using the Great Pyramid as a sight line for their observations. (Airy 1881, plate 8.)

won support from Evangelicals and engineers. His antimetric disciples admired Smyth's superb set of photographs, including flash pictures taken from within the Pyramid and sets of lantern slides made from these pictures (Smyth 1870, appendix, 2–4; Brück 1988, 381–82). But much of the force of his precision measurements was removed by the work of Petrie's son, Flinders Petrie, first chair of Egyptology in London, who made himself a national authority on historical metrology. The London professor insisted that no natural standard, and certainly not a pyramidal one, could be determined sufficiently accurately. "A natural standard is therefore only a matter of sentiment" (Brück and Brück 1988, 228–30; Petrie 1888, 478).

In these sentimental debates, Smyth denied that precision was achievable by human convention alone, while his critics denied that it was appropriate for measures of Egyptian rubble (Brück and Brück 1988, 119–22; Ball 1915, 103; Lagrange 1894). Systems of units embodied contrasting systems of social and moral values. Smyth insisted that "there was more of intercommunication in idea and knowledge between the architect of the Great Pyramid and the *origines* of the Anglo-Saxon race, whoever they were, than between the said architect . . . and all the native Egyptian people," thus legitimating prior British rights over Egyptian territory (Smyth 1877, 39). But there were other ways of securing imperial control. The astronomical and military expeditions to Egypt in the 1870s and 1880s relied in large measure on the new rail and telegraphy networks installed by British engineers as part of a world system of intercommunication. The integrity of these networks hinged on the values of electrotechnology being produced in the new physics laboratories of the British universities. These laboratories' managers worked hard to make such values universal, so that their systems of precision measurement could become equally widespread.

IV. James Clerk Maxwell: Uniform Molecules as Manufactured Articles

In the 1860s and 1870s, pyramidologists were not the only group who claimed that modish technologies of precision measurement threatened sacred values. Confronting a newfangled electrotechnological physics advertising its role in communication and trade, many Cambridge dons held that factory science was not proper to their traditional pedagogy. They were not convinced by advocates of laboratory engineering who patiently explained how scientific, commercial, and military links depended on the reliability of the electric standards made in British universities. The introduction into their university of the tools required for such measures seemed subversive of the morally strenuous culture of gentlemanly instruction in the prestigious Mathematics Tripos. Maxwell, head of the new Cavendish Laboratory, himself a product of the Tripos and a devotee of the electromagnetic standards enterprise, understood that these dons would contest the legitimacy of his metrological program. The later success of the Cavendish as a standards institution under the management of his successors Lord Rayleigh and Richard Glazebrook after Maxwell's premature death in 1879 has obscured the initial predicament that the laboratory's sponsors faced (Sviedrys 1970; Schaffer 1992).

Maxwell held that commercial and scientific values must match. In a manuscript essay of the 1870s, he noted that "the want of a unit is felt in buying and selling. . . . It has always been the care of wise governments to provide national standards and to make the use of other standards punishable. . . . The man of business requires these standards for the sake of justice, the man of science requires them for the sake of truth, and it is the business of the state to see that our . . . measures are maintained uniform" (Maxwell 1877). The problem was to connect imperial and commercial values with those of his own university. "In the present day," he conceded in his inaugural lecture of October 1871, "men of science are supposed to be in league with the material spirit of the age, and to form a kind of advanced Radical party among men of learning" (Maxwell 1890, 2:251). He wondered whether laboratory work would foster subversive materialism among Cambridge men, "tainting their mathematical conceptions with material imagery, and sapping their faith in the formulae of the textbooks. . . . Will they not break down altogether?" (Maxwell 1890, 2:247).

University debate about the new physics laboratory between late 1868 and autumn 1870 was rather violent. Norman Lockyer, editor of *Nature* and secretary to a government commission chaired by the Duke of Devonshire on science training, demanded that if Cambridge would not support the laboratory it should be nationalized (Lockyer 1869). University administrators "urged the superior claims of ecclesiastical history and pastoral theology" and complained of "exaggerated statements in favour of physical science as a disparagement of classics and mathematics" (S. Taylor 1870). Only the intervention of Devonshire, a senior wrangler of impeccable credentials, saved the plan and launched the new chair and laboratory. In November 1873 the vice-chancellor, Henry Cookson, described the Devonshire Commission's recommendations, celebrated the completion of the Cavendish Laboratory, but argued that "just in proportion as we give prominence to the study of the sciences which are much connected with what is material and perishing, we ought to foster and encourage studies which have relation to our moral and spiritual nature, and to take care that the latter are not overborne by the former" (Cookson 1873).

Maxwell had to show that precision measurement in the physics laboratory was not solely "connected with what is material and perishing." His public statements of the early 1870s in lectures before the university, before the British Association, and in the *Encyclopaedia Britannica* all addressed this issue (Theerman 1986, 315–16). He conceded that the man of science might "be for a season a calculating machine" (Maxwell 1890, 2:219). Yet by a skillful reinterpretation of the meaning of these calculations, Maxwell was able to make them part of a divine vocation (Maxwell 1890, 2:376). The universe was indeed "material and perishing," but molecular dimensions were not, and so evinced the spiritual power of the deity. "From the ineffaceable characters impressed on them we may learn that those aspirations after accuracy in measurement, truth in statement and justice in action, which we reckon among our noblest attributes as men, are ours because they are essential constituents of the image of Him who in the beginning created, not only the heaven and the earth, but the materials of which heaven and earth consist" (Maxwell 1890, 2:377). Metrological programs might not only demonstrate divine creation; they were, much more importantly for Evangelical physics, a result of that creation.

The "absolute equality" of molecular dimensions was of a different order from any of the earthly candidates for the basis of a system of measures. In 1870 Maxwell reminded the British Association that "you are all aware of the vast amount of scientific work which has been expended . . . in providing weights and measures for commercial and scientific purposes" (Maxwell 1890, 2:225). Unlike those of the Pyramid, "the foundation stones of the material universe remain unbroken and unworn" (Maxwell 1890, 2:377). Thus, molecular dimensions provided the uniquely secure source of values, because they were uniform, and therefore "manufactured," just as Herschel had shown, and their manufacturer was God. "If, then, we wish to obtain standards of length, time and mass which shall be absolutely permanent, we must seek them [in] these imperishable and unalterable and perfectly similar molecules" (Maxwell 1890, 2:225). At the opening of his masterly 1873 *Treatise on Electricity and Magnetism* Maxwell noted that the metre had not been corrected for "new and more accurate measurements of the earth," and he dismissed the use of the earth-sun distance. Instead, he proposed the wavelength of a well-defined sodium line as "the most universal standard of length.... Such a standard would be independent of any changes in the dimensions of the Earth, and should be adopted by those who expect their writings to be more permanent than that body" (Maxwell 1891, 1:2-3).

Maxwell reckoned that the best source of permanence would be the precise determination of the parameters of a universal, continuous ether measured with superb new spectroscopes. Most important was the work of the wealthy London astronomer William Huggins, with whom Maxwell collaborated in 1867 on measures of star spectra (Maxwell 1990-95, 2:306-11). Huggins's work showed that laboratory and stellar hydrogen vibrated at the same frequency (figure 20.7). Maxwell then ingeniously analogized such uniform spectra with the 1869 Ordnance Survey of the Pyramids. James convinced Maxwell that Egyptians and Greeks had a common measure of length; hence, the two civilizations must have been in contact. But, Maxwell now pointed out, there was no way, save through divine will, that the hydrogen standard should be so universally distributed. So recent physics taught that molecules were "manufactured articles" and that they carried "the stamp of a metric system as distinctly as . . . the double royal cubit of the Temple of Karnac" (Maxwell 1890, 2:375). The universal ether and the quality of spectroscopes and spinning coils were the warrants for this standardization. The identity of light speed and the ratio of the electrostatic and electromagnetic units sustained his field theory (Schaffer 1995). Laboratory work on electromagnetic physics thus fitted into University culture.

Maxwell's arguments about manufactured articles did not convince all his audience. In the natural theology popular in Cambridge, divine wisdom had typically been deduced from the adaptation of natural forms to their locally specific purposes. But Maxwell's deity produced uniformity, not diversity. He was challenged on this point by his college friend Cecil James Monro, a classics don already inured to Maxwell's arguments from numerical coincidence between light speed and the units' ratio to the truth of field theory (Maxwell 1990-95, 1:690; Theerman 1986, 313). Monro com-

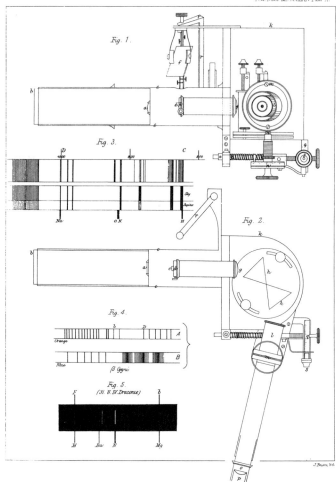

Figure 20.7 Huggins's spectroscopes. His instruments showed that molecules vibrated at the same frequency in the stars and on Earth, whence Maxwell concluded that molecules were divinely manufactured standards. (Huggins and Miller 1864, plate 10.)

plained that a God whose most salient characteristic was tedious uniformity was more like a slum jerry-builder, "a manufacturer who does not care what becomes of his articles the moment he gets them off his hands . . . a manufacturer who cannot solve his own equations except in a grossly approximative fashion." Maxwell had made the existence of God hinge on the accuracy of Huggins's spectroscopes. If these instruments were too imprecise to measure spectral difference, then Maxwell's argument for molecular uniformity would fail (Monro 1874).

It was the uniformity of universal accuracy, not that of factory production, that Maxwell wished to highlight, so Monro's sly reference to the demeaning role of the divine jerry-builder was inappropriate. In an 1875 article entitled "Atoms" Maxwell distinguished several senses of the term "uniformity of manufacture." The uniformity of manufactured objects might make them cheap, like army shoes, or useful, like the Whitworth screws used in Maxwell's own laboratory, or accurate, like the national standardization of weights and measures, and this last feature was found in molecular creation (Maxwell 1890, 2:483-84). Then, in late 1876, the bishop of Bristol, a veteran of the Mathematics Tripos and author of a useful mechanics textbook, asked for further clarification of these differences, since he wished to use the argument in a sermon. Maxwell told him that he did not intend "uniformity in the process of formation," evident in Whitworth's works, but sought, instead, to reconcile his claims with the morality of adaptation. The uniformity of God was equally evident in universal standards and in "the special utility of each individual thing." Maxwell worked hard to get his arguments about the difference between factory standards and divine values into clerical currency (Campbell and Garnett 1884, 300-301).

It was not enough to link the Cavendish Laboratory's ohm with Cambridge's God. Maxwell also had to distinguish his moralized measures from the pyramidologists' wholesale rejection of commercial values. The task was made easier in autumn 1873, when Smyth seized on Maxwell's references to the metrication of molecules and to James's work on the Egyptian cubit. Smyth had known Maxwell for many years. In 1859 he supported Maxwell's unsuccessful application for the Edinburgh natural philosophy chair, while Maxwell regarded Smyth as an expert in spectroscopy and instrument design (Maxwell 1990-95, 1:497, 502, 630; Brück and Brück 1988, 226-27). Smyth responded immediately to Maxwell's endorsement of James in September 1873. Whence had the Cambridge professor got his facts? Did he not understand the crucial difference between a pyramid side of 9,120 inches, as James now calculated, and the true length Smyth had found at Giza, 9,140 inches? With such carefree attitudes to precision, what could the comparison with manufactured articles mean? Maxwell answered that he had used James's reports to provide evidence of "a desire of accuracy and just dealing" in ancient as in modern metrology, and thence to urge that hydrogen molecules would necessarily provide a better standard of length than any terrestrial phenomenon. Maxwell was prepared to concede pyramidologists' claims that earth commensurability had "practical value" but not the innate superiority of ancient Egyptian values (Brück and Brück 1988, 175-76). Smyth was not mollified. James's work was obviously convincing men of science that the Great Pyramid was an imperfect human attempt to match natural standards. Smyth submitted a paper to the Royal Society correcting "the justly celebrated Professor Maxwell" and attacking

James. The Society's secretary, George Gabriel Stokes, Cambridge's distinguished mathematics professor, refused to publish it, and Smyth promptly resigned his fellowship. Expert colleagues criticized his decision, while the pyramidological community, including William Petrie, rallied in support (Brück and Brück 1988, 176-80; Smyth 1877, 304-7). The fight was damaging for Smyth and useful for Maxwell, who saved press cuttings on the matter, for it decisively showed the public the difference between his precision measures and scriptural physics (Maxwell 1873).

Maxwell did not merely seek to show that his ether physics was safe, but had to show that it was effective. An excellent way was to contest the molecular physics of the principal enemies of the Cambridge divines, the scientific naturalists such as Thomas Huxley, John Tyndall, and William Clifford (Turner 1974, 17-30). For these secularist metropolitan spokesmen, donnish moralities of the spiritual and imperishable were risible and sinister. From their newly secured platforms in the press and the lecture halls they easily satirized conservative pyramidology to undermine the appeal of Maxwell's molecular metrology (Jacyna 1980, 34-41; Desmond 1982, 122-23, 158-64; Block 1989, 219-24; Turner 1993, 171-200). In 1868 Tyndall told the British Association that only human consciousness seemed to lie beyond the boundary of self-moving matter, but that this barrier might soon be breached. He reminded his audience that "the human mind is as little disposed to look without questioning at pyramidal salt-crystals, as to look at the pyramids of Egypt, without enquiring whence they came." It was just as silly to deny the naturalistic origin of any material or mental form as it more obviously was to trace the Pyramids to an original divine fiat (Tyndall 1879, 2:80-81). Similarly, in an essay commissioned for Victoria's golden jubilee, Huxley compared "the idea that atoms are absolutely ingenerable and immutable manufactured articles" with the notion of the fixity of species his hero Charles Darwin had now destroyed. "The supposed constancy of the elementary atoms," Huxley sneered in a significant phrase, was as baseless as antievolutionist appeals to "the constancy of species in Egypt since the days of Rameses or Cheops." Time was on the side of scientific naturalists, not sacerdotal conservatives (Huxley 1893, 79).

Much of Maxwell's public work of the late 1860s and early 1870s was designed to counter the materialist implications of Tyndall's molecular physics and Huxley's evolutionism. The "demon" of thermodynamic reversibility, an intelligence that first appeared as a humble "pointsman on a railway" in Maxwell's letters to the Edinburgh professor Peter Guthrie Tait in 1867 and the young physicist Lord Rayleigh in 1870, stood alongside molecular uniformity as a sign of divine purpose and free will (Maxwell 1990-95, 2:332, 582; Porter 1981, 102-7; Smith and Wise 1989, 617-33). Maxwell presented his argument for molecular uniformity as a response to Tyndall's venture into "that sanctuary of minuteness and of power where molecules obey the laws of their existence, clash together in fierce collision, or grapple in yet more fierce embrace, building up in secret the forms of visible things." The Cambridge professor then composed a "Tyndallic Ode" against the derivation of intelligence from molecular motions alone (Maxwell 1890, 2:216; Campbell and Garnett 1884, 412-14). In autumn 1874 Tyndall gave the notorious presidential address before 1,800 people at the British Association meeting at Belfast and interrupted his derivation of mind from active matter and his assault on the rights of theologians to criticize what he judged Maxwell's ludicrous attempt to extract a transcendental moral from spectroscopy (Tyndall 1879, 2:162-63; Jacyna 1980, 74-76; Barton 1987, 119). Maxwell again satirized Tyndall in verse: "Let us honour the atom, so lively, so wise and so small," he sang (Campbell and Garnett 1884, 284, 415-17). Maxwell tried to differentiate the precise reliability of molecular measurements from the unreliable contingency of molecular evolution. Maxwell's new natural theology of precision measurement allowed him to balance the authority of quantitative measurement with the humility of human knowledge before divine will.

It was hard to combine the apparent arrogance of physicists' claims to precision with theological submissiveness. William Clifford, second wrangler in 1867 and mathematics professor at University College London, exploited the confusion in lectures of August 1872 and April 1874 (Lightman 1987, 168-72; Richards 1988, 109-13). Clifford and Maxwell were well matched in the field of precision technique. Maxwell wrote a laudatory reference for Clifford's application for the London chair (Clifford 1879, 1:14), and they jointly contributed authoritative notes to the catalog of the 1876 instruments exhibition at South Kensington (Maxwell 1890, 2:505-27; Clifford 1879, 2:3-30). Clifford was a foe of the natural theologies of the 1870s, especially those of Tait and his allies, who sought to derive Trinitarian religion from ether physics (Heimann 1972; Smith 1979, 69-70; Jacyna 1980, 77-79; Myers 1989, 327-30). Clifford denied any primitive Anglo-Saxon metrological religion in ancient Egypt and instead claimed that orthodox religion was a debased form of polytheism, "retained and refurbished by the bishops of Alexandria out of the wreckage of Egyptian superstition. . . . If you choose to find one thing in the chain of ethers, we may quite lawfully find another" (Clifford 1879, 1:251-52). What he found was evolutionary naturalism. Clifford followed what he called "the true Christ, humanity," and lectured that "the subject of science is the human universe, that is to say, everything that is, or has been or may be related to man" (Clifford 1879, 1:49, 126; Jacyna 1980, 88-89). Uniformity could not be derived from physicists' measures, but was assumed in order to make these measures meaningful. His audience might suppose, with Maxwell, that there was a principled difference between commercial exactness, which referred to the mundane reliability of tradesmen's scales, and mathematicians' exactness, which claimed a measure better than any possible scale. But non-Euclidean geometries, espoused by Clifford and contested by conservative Oxbridge mathematicians, showed that spatial rectilinearity was only a useful approximation. "We assume this universality, and we find that it pays us to assume it. But a law would be theoretically universal if it were true of all cases whatever, and this is what we do not know of any law at all" (Clifford 1879, 1:138; Richards 1988, 110-11).

This argument was potentially devastating for Maxwell's molecular theology and metrology. Maxwell himself judged that Clifford's views wanted "trouncing" (Campbell and Garnett 1884, 325). Clifford pointed to the role that social status played in the public's attitude to scientific arguments. He used this against Maxwell's claim about molecules as manufactured articles. "We may see from this example," he lectured in spring 1874, "how great is the influence of authority in matters of science." Clifford reckoned that "if anyone not possessing [Maxwell's] great authority had put forward an argument" that derived claims about eternity and exactitude from scientific measures, "we should say 'past eternity, absolute exactness, this won't do', and pass on to another book" (Clifford 1879, 1:202-4). So Clifford straightforwardly denied Maxwellian ambitions to perfect exactitude and transcendental morality. "We never get at conclusions which we have a right to say are absolutely exact," and confidence must be limited "by our present modes of measurement" (Clifford 1879, 1:204, 212; Lightman 1987, 171).

In view of the intimate connection between late Victorian metrology and imperialism, Clifford chose a telling illustration of his claim that inference depended on convention. At the *Scientific Instruments Exhibition* in South Kensington, William Siemens might use the news of Stanley's rescue of Livingstone to exemplify how precision measurement allowed Victorians to triumph over alien climes and cultures. At the 1872 British Association meeting in Brighton Clifford used exactly the same episode to hammer home the argument that this triumph relied on local practical judgments:

> When a telegram arrived stating that Dr Livingstone had been found by Mr Stanley, what was the process by which you inferred the finding of Dr Livingstone from the appearance of the telegram? You assumed over and over again the existence of uniformity in nature. That the newspapers behaved as they generally do in regard to telegraphic messages; that the clerks had followed the known laws of the action of clerks; that electricity behaved in the cable exactly as it behaves in the laboratory; that the actions of Mr Stanley were related to his motives by the same uniformities that affect the actions of other men; that Dr Livingstone's handwriting conformed to the curious rule by which an ordinary man's handwriting may be recognized as having persistent characteristics even at different periods of his life. (Clifford 1879, 1:141-42)

Apostle of pragmatic humanism, Clifford reckoned that these practical assumptions were the sources, not the results, of nature's projected uniformity (Hume 1888, 196). Spokesman of devout natural philosophy, Maxwell claimed that the reliability of long-range systems showed that his local values were truly universal throughout creation.

Victorian metrologists proved capable of combining these lessons by fusing the powers of traditional cultures and radically transformative techniques and interests. In the face of newly powerful challenges, whether from working-class protest or from increased international competition, agnostic naturalists and conservative moralists of the British scientific establishment eventually found a common cause in linking natural standards with the social values of nationalism, racism, and managerialism (Jacyna 1980, 304-9). Delegates of this establishment forged an effective front at successive international conferences and in powerful lobbies in their development of scientific and industrial standards. At the century's end, in direct response to German industrial might and the need for closer links between state, science, and industry, a National Physical Laboratory was founded in London under the sponsorship of Maxwell's immediate successors, Rayleigh and Glazebrook, to establish and police scientific metrology (Moseley 1978; Alter 1987, 138-49). These measures were backed by a classical and aristocratic culture whose authority survived within the academy and the polity. Rayleigh's kinship with the Tory leaders Lord Salisbury and Arthur Balfour was decisive in the establishment of the new standards institution (Barrell 1964). The new laboratory was designed to meet the challenge of the enterprise of mass production, networks of light, power, and transportation, and electrotechnology. The values of practical science were then explicitly mobilized in the face of economic warfare and colonial and industrial strife. This combination of aristocratic tradition, contemporary industrial science, and imminent conflict was indicative. Late Victorian intellectuals typically responded with traditional resources that exploited new practices to cope with these dangerous threats (Anderson 1984, 105). Especially crucial was the emergence of a set of cultural traditions, rich codes of practice that soon came to specify the type of authority that nation-states might wield and to which loyalty should be devoted (Hobsbawm 1983, 265). Victorian national standards of measurement forged through public controversy and painstaking labor were notable examples of these newly minted traditions.

Bibliographical Note

Metrology's history is charted in Kula (1986), a European survey of the virtues of standard measures. Scaff (1989) and Megill (1994) contain important discussions of metrology's social meanings. The idiosyncratic British experience is described in Watson (1910), Connor (1987), and Hoppit (1993). The argument of Latour (1987) that metrology is how scientific technique works outside its initial setting is usefully elaborated in O'Connell (1993) and summarized in Barry (1993). Porter (1995) is the best account of nineteenth-century metrology and accountancy, emphasizing the political meaning of standardized judgment. For Victorian standardization of human performances, see Montgomery (1965), Cullen (1975), Macleod (1982), and Hacking (1990). Some important Victorian metrological institutions, especially those backed by the newly expansive state, are documented in Scott (1885), Seymour (1980), Macleod (1988), Hamlin (1990), and Hammond and Egan (1992). Electromagnetic metrology was a key factor in the institutionalization of Victorian physics: see Sviedrys (1976), Smith and Wise (1989), Gooday (1990), and Hunt (1994). The imperial role of Victorian metrology is a major theme in Headrick (1988) and Hunt's chapter in this book. Metrologists' tools have not been systematically studied, but important sources are Rolt (1986), Brooks (1988), and Simpson (1993).

References

Adas, Michael. 1989. *Machines as the Measure of Men.* Ithaca, N.Y.: Cornell University Press.

Airy, George. 1836. "Address on Presenting the Honorary Medal to Sir J. F. W. Herschel." *Memoirs of the Royal Astronomical Society* 9:303–12.

———. 1867. Letter to E. J. Stone, 15 June. Royal Greenwich Observatory. MS 6/240, folios 330–31.

———, ed. 1881. *Account of Observations of the Transit of Venus, 1874, December 8.* London: Her Majesty's Stationery Office.

Alder, Ken. 1995. "A Revolution to Measure: The Political Economy of the Metric System in France." In *Values of Precision,* edited by M. Norton Wise. Princeton, N.J.: Princeton University Press, 39–71.

Alter, Peter. 1987. *The Reluctant Patron: Science and the State in Britain 1850–1920.* Oxford: Berg.

Anderson, Perry. 1984. "Modernity and Revolution." *New Left Review* 144:96–113.

Ashworth, William J. 1994. "The Calculating Eye." *British Journal for the History of Science* 27:409–41.

Ball, W. Valentine, ed. 1915. *Reminiscences of Sir Robert Ball.* London: Cassell.

Barrell, H. 1964. "The Rayleighs and the National Physical Laboratory." *Applied Optics* 3:1125–28.

Barry, Andrew. 1993. "The History of Measurement and the Engineers of Space." *British Journal for the History of Science* 26:459–68.

Barton, Ruth. 1987. "John Tyndall, Pantheist." *Osiris* 3:111–34.

Bayly, C. A. 1989. *Imperial Meridian: The British Empire and the World 1780–1830.* London: Longmans.

Beaucour, Fernand, Yves Laissus, and Chantal Orgogozo. 1990. *The Discovery of Egypt.* Paris: Flammarion.

Block, Ed. 1989. "T. H. Huxley's Rhetoric and the Popularization of Victorian Scientific Ideas, 1854–1874." In *Victorian Science and Victorian Values: Literary Perspectives,* edited by James Paradis and Thomas Postlewait. New Brunswick, N.J.: Rutgers University Press, 205–28.

Bourguet, Marie-Noelle. 1988. *Déchiffrer la France*. Paris: Archives Contemporaines.
Brantlinger, Patrick, ed. 1989. *Energy and Entropy: Science and Culture in Victorian Britain*. Bloomington: Indiana University Press.
Brooks, Randall. 1988. "Standard Screw Threads for Scientific Instruments." *History and Technology* 5:59-76.
Brück, H. A., and M. T. Brück. 1988. *The Peripatetic Astronomer: The Life of Charles Piazzi Smyth*. Bristol: Adam Hilger.
Brück, Mary T. 1988. "The Piazzi Smyth Collection." *Vistas in Astronomy* 32:371-408.
Campbell, Lewis, and William Garnett. 1884. *Life of James Clerk Maxwell*. London: Macmillan.
Cannon, Walter F. 1961. "John Herschel and the Idea of Science." *Journal of the History of Ideas* 22:215-39.
Cardwell, Donald S. L. 1989. *James Joule: A Biography*. Manchester: Manchester University Press.
Chapman, Allan. 1993. "An Occupation for an Independent Gentleman: Astronomy in the Life of John Herschel." *Vistas in Astronomy* 36:71-116.
Clerke, Agnes. 1908. *A Popular History of Astronomy during the Nineteenth Century*. London: Adam and Charles Black.
Clifford, William Kingdon. 1879. *Lectures and Essays*. Edited by Leslie Stephen and Frederick Pollock. 2 vols. London: Macmillan and Company.
Cole, Juan. 1993. *Colonialism and Revolution in the Middle East*. Princeton, N.J.: Princeton University Press.
Coleman, D. C. 1973. "Gentlemen and Players." *Economic History Review* 26:94-116.
Colley, Linda. 1992. *Britons: Forging the Nation 1707-1837*. London: Pimlico.
Connor, R. D. 1987. *The Weights and Measures of England*. London: Her Majesty's Stationery Office.
Cookson, Henry. 1873. "Speech." *Cambridge University Reporter*, 4 November, 68-69.
Corrigan, Philip, and Derek Sayer. 1985. *The Great Arch: English State Formation as Cultural Revolution*. Oxford: Blackwell.
Cullen, Michael. 1975. *The Statistical Movement in Early Victorian Britain*. New York: Barnes and Noble.
Davidson, David. 1938. *Herschel's Geometrical Inch*. Leeds: Davidson.
Desmond, Adrian. 1982. *Archetypes and Ancestors: Palaeontology in Victorian London 1850-1875*. London: Blond and Briggs.
Dreyer, J. L. E., and H. H. Turner. 1923. *History of the Royal Astronomical Society 1820-1920*. London: Wheldon and Wesley.
Drower, Margaret S. 1985. *Flinders Petrie: A Life in Archaeology*. London: Gollancz.
Gerth, H. H., and C. Wright Mills. 1948. *From Max Weber*. London: Routledge.
Godlewska, Anne. 1994. "Napoleon's Geographers." In *Geography and Empire*, edited by Anne Godlewska and Neil Smith. Oxford: Blackwell, 31-53.
Gooday, Graeme. 1990. "Precision Measurement and the Genesis of Physics Teaching Laboratories." *British Journal for the History of Science* 23:25-51.

Graham, Thomas. 1863. "On the Molecular Mobility of Gases." *Philosophical Transactions of the Royal Society* 153:385-405.

Hacking, Ian. 1990. *The Taming of Chance.* Cambridge: Cambridge University Press.

Hamlin, Christopher. 1990. *A Science of Impurity: Water Analysis in Nineteenth Century Britain.* Bristol: Adam Hilger.

Hammond, P. W., and Harold Egan. 1992. *Weighed in the Balance: A History of the Laboratory of the Government Chemist.* London: Her Majesty's Stationary Office.

Headrick, Daniel R. 1981. *The Tools of Empire: Technology and European Imperialism in the Nineteenth Century.* Oxford: Oxford University Press.

———. 1988. *The Tentacles of Progress: Technology Transfer in the Age of Imperialism 1850-1940.* Oxford: Oxford University Press.

Heimann, P. M. 1972. "The Unseen Universe: Physics and the Philosophy of Nature in Victorian Britain." *British Journal for the History of Science* 6:73-79.

Herschel, John. 1831. *Preliminary Discourse on the Study of Natural Philosophy.* London: Longman.

———. 1857. *Essays from the Edinburgh and Quarterly Reviews with Addresses and other Pieces.* London: Longman.

———. 1867. *Familiar Lectures on Scientific Subject*s. London: D. Strahan.

Higgs, Edward. 1988. "The Struggle for the Occupational Census 1841-1911." In *Government and Expertise: Specialists, Administrators and Professionals 1860-1919,* edited by Roy Macleod. Cambridge: Cambridge University Press, 73-86.

Hilton, Boyd. 1977. *Corn, Cash and Commerce: The Economic Policies of the Tory Governments 1815-1830.* Oxford: Oxford University Press.

———. 1988. *The Age of Atonement: The Influence of Evangelicalism on Social and Economic Thought, 1795-1865.* Oxford: Clarendon.

Hobsbawm, Eric. 1983. "Mass-Producing Traditions: Europe 1870-1914." In *The Invention of Tradition,* edited by Eric Hobsbawm and Terence Ranger. Cambridge: Cambridge University Press, 263-307.

Hobsbawm, Eric, and Terence Ranger, eds. 1983. *The Invention of Tradition.* Cambridge: Cambridge University Press.

Hoppit, Julian. 1993. "Reforming Britain's Weights and Measures." *English Historical Review* 108:82-104.

Huggins, William, and W. A. Miller. 1864. "On the Spectra of Some of the Fixed Stars." *Philosophical Transactions of the Royal Society* 154:413-36.

Hume, David. 1888. *A Treatise of Human Nature.* Edited by L. A. Selby-Bigge. Oxford: Clarendon Press.

Hunt, Bruce. 1994. "The Ohm Is Where the Art Is." *Osiris* 9:48-63.

Huxley, Thomas H. 1893. *Method and Results: Essays.* London: Macmillan.

Jacyna, L. S. 1980. "Scientific Naturalism in Victorian Britain: An Essay in the Social History of Ideas." Ph.D. thesis, University of Edinburgh.

James, Henry. 1869. *Notes on the Great Pyramid.* Southampton: Ordnance Survey.

Jenkin, Fleeming. 1862a. "Provisional Report of the Committee Appointed by the British Association on Standards of Electrical Resistance." *British Association Reports,* 125-35.

———. 1862b. "Circular Addressed to Foreign Men of Science." *British Association Reports,* 156-59.
Joule, James. 1865a. Letter to William Thomson, 6 February. Kelvin Manuscripts. Glasgow University Library. MS J 176.
———. 1865b. Letter to William Thomson, 30 December. Kelvin Manuscripts. Glasgow University Library. MS J 180.
———. 1869. Letter to William Thomson, 16 April. Kelvin Manuscripts. Glasgow University Library. MS J 204.
Kilburn, Terence. 1987. *Joseph Whitworth, Toolmaker.* Cromford: Scarthin Books.
Kinglake, Alexander. [1844] 1982. *Eothen, or Traces of Travel brought Home from the East.* Oxford: Oxford University Press.
Kuhn, Thomas S. 1977. *The Essential Tension: Selected Studies in Scientific Tradition and Change.* Chicago: University of Chicago Press.
Kula, Witold. 1986. *Measures and Men.* Princeton, N.J.: Princeton University Press.
Lagrange, Charles-Henri. 1894. *The Great Pyramid.* London: Charles Burnet.
Latour, Bruno. 1987. *Science in Action.* Milton Keynes: Open University Press.
Lawrence, Christopher J. 1985. "Incommunicable Knowledge: Science, Technology and the Clinical Art in Britain, 1850-1914." *Journal of Contemporary History* 20:503-20.
Lightman, Bernard. 1987. *The Origins of Agnosticism: Victorian Unbelief and the Limits of Knowledge.* Baltimore: Johns Hopkins University Press.
Lockyer, J. Norman. 1869. Editorial. *Nature* 1:587.
Macleod, Roy, ed. 1982. *Days of Judgement: Science, Examinations and the Organization of Knowledge in Late Victorian England.* Driffield: Nafferton.
———, ed. 1988. *Government and Expertise: Specialists, Administrators and Professionals 1860-1919.* Cambridge: Cambridge University Press.
Martineau, Harriet. [1858] 1877. *History of the Thirty Years' Peace.* 4 vols. London: Bell.
Maxwell, James Clerk. 1873. Press cuttings on pyramidology. Cambridge University Library. MS ADD 7655 / V (I) 12.
———. [1873] 1891. *Treatise on Electricity and Magnetism.* 3d ed. 2 vols. Oxford: Clarendon Press.
———. 1877. "Dimensions of Physical Quantities." Cambridge University Library. MS ADD 7655 / V (h) 4.
———. 1890. *Scientific Papers.* Edited by W. D. Niven. 2 vols. Cambridge: Cambridge University Press.
———. 1990-95. *Scientific Letters and Papers.* Edited by P. M. Harman. 2 vols to date. Cambridge: Cambridge University Press.
Megill, Allan, ed. 1994. *Rethinking Objectivity.* Durham: Duke University Press.
Mitchell, Timothy. 1988. *Colonizing Egypt.* Berkeley: University of California Press.
Monro, C. J. 1874. "Manufactured Articles." *Nature* 10:481.
Montgomery, Robert J. 1965. *Examinations: An Account of Their Evolution as Administrative Devices in England.* London: Longmans.
Morley, John. 1908. *Life of Richard Cobden.* 2 vols. London: Macmillan.
Moseley, Russell. 1978. "The Origins and the Early Years of the National Physical Laboratory." *Minerva* 16:222-50.
Munford, W. A. 1960. *William Ewart: Portrait of a Radical.* London: Grafton.

Myers, Greg. 1989. "Nineteenth Century Popularizations of Thermodynamics and the Rhetoric of Social Prophecy." In *Energy and Entropy: Science and Culture in Victorian Britain*, edited by Patrick Brantlinger. Bloomington: Indiana University Press, 307-38.

O'Connell, Joseph. 1993. "Metrology: The Creation of Universality by the Circulation of Particulars." *Social Studies of Science* 23:129-73.

Pang, Alex Soojung-Kim. 1993. "The Social Event of the Season: Solar Eclipse Expeditions and Victorian Culture." *Isis* 84:252-77.

Paradis, James, and Thomas Postlewait, eds. 1985. *Victorian Science and Victorian Values: Literary Perspectives*. New Brunswick, N.J.: Rutgers University Press.

Petrie, W. Flinders. 1888. "Weights and Measures." In *Encyclopaedia Britannica*. 9th ed. 24 vols. Edinburgh: Adam and Charles Black, 24:478-91.

Porter, Theodore M. 1981. "A Statistical Survey of Gases: Maxwell's Social Physics." *Historical Studies in Physical Sciences* 12:77-116.

———. 1994. "Objectivity as Standardization." In *Rethinking Objectivity*, edited by Allan Megill. Durham, N.C.: Duke University Press, 197-238.

———. 1995. *Trust in Numbers: The Pursuit of Objectivity in Science and Public Life*. Princeton, N.J.: Princeton University Press.

Report of the 34th Meeting of the British Association for the Advancement of Science (1864). 1865. London: British Association for the Advancement of Science.

Richards, Joan L. 1988. *Mathematical Visions: The Pursuit of Geometry in Victorian England*. San Diego, Calif.: Academic Press.

Rolt, L. T. C. 1986. *Tools for the Job: A History of Machine Tools to 1950*. London: Her Majesty's Stationery Office.

Said, Edward. 1979. *Orientalism*. New York: Vintage Books.

Scaff, Lawrence. 1989. *Fleeing the Iron Cage: Culture, Politics and Modernity in the Thought of Max Weber*. Berkeley: University of California Press.

Schaffer, Simon. 1988. "Astronomers Mark Time: Discipline and the Personal Equation." *Science in Context* 2:115-45.

———. 1992. "Late Victorian Metrology and Its Instrumentation: A Manufactory of Ohms." In *Invisible Connexions: Instruments, Institutions and Science*, edited by Robert Bud and Susan Cozzens. Bellingham, Wash.: SPIE Press, 23-56.

———. 1995. "Accurate Measurement Is an English Science." In *Values of Precision*, edited by M. Norton Wise. Princeton, N.J.: Princeton University Press, 135-72.

Schweber, S. S. 1985. "Scientists as Intellectuals: The Early Victorians." In *Victorian Science and Victorian Values: Literary Perspectives*, edited by James Paradis and Thomas Postlewait. New Brunswick, N.J.: Rutgers University Press, 1-37.

Scott, Robert Henry. 1885, "The History of the Kew Observatory." *Proceedings of the Royal Society of London* 39:37-86.

Secord, James A. 1986. "The Geological Survey of Great Britain as a Research School, 1839-1855." *History of Science* 24:223-75.

Seymour, W. A., ed. 1980. *History of the Ordnance Survey*. London: Dawson.

Siemens, Charles William. 1876. "Opening Address." In *Conferences Held in Connection with the Special Loan Collection of Scientific Apparatus 1876: Physics and Mechanics*. London: Chapman and Hall, 204-16.

Simpson, A. D. C. 1993. "The Pendulum as the British Length Standard: A 19th Cen-

tury Legal Aberration." In *Making Instruments Count,* edited by R. G. W. Anderson, J. A. Bennett, and W. F. Ryan. London: Variorum, 174-90.
Smith, Crosbie. 1979. "From Design to Dissolution: Thomas Chalmers' Debt to John Robison." *British Journal for the History of Science* 12:59-70.
Smith, Crosbie, and M. Norton Wise. 1989. *Energy and Empire: A Biographical Study of Lord Kelvin.* Cambridge: Cambridge University Press.
Smyth, Charles Piazzi. [1864] 1877. *Our Inheritance in the Great Pyramid.* 3d ed. London: Daldy, Isbister.
———. 1867. *Life and Work at the Great Pyramid.* 3 vols. Edinburgh: Edmonston and Douglas.
———. 1868. *On the Antiquity of Intellectual Man.* Edinburgh: Edmonston and Douglas.
———. 1870. *A Poor Man's Photography at the Great Pyramid.* London: Henry Greenwood.
———. 1871*a*. "The Great Pyramid of Egypt from a Modern Scientific Point of View." *Quarterly Journal of Science* 1:16-35, 177-214.
———. 1871*b*. "The Paris Observatory." *Nature* 4:120.
The Story of the Herschels: a Family of Astronomers. 1879. London: Thomas Nelson.
Sviedrys, Romualdas. 1970. "The Rise of Physical Science at Victorian Cambridge." *Historical Studies in the Physical Sciences* 2:127-45.
———. 1976. "The Rise of Physics Laboratories in Britain." *Historical Studies in the Physical Sciences* 7:405-36.
Taylor, John. 1859. *The Great Pyramid: Why was it Built? and Who Built it?* London: Longman.
Taylor, Sedley. 1870. "Physical Science at Cambridge." *Nature* 2:28.
Theerman, Paul. 1986. "James Clerk Maxwell and Religion." *American Journal of Physics* 54:312-17.
Thompson, E. P. 1991. *Customs in Common.* London: Merlin.
Tobin, William. 1993. "Toothed Wheels and Rotating Mirrors: Parisian Astronomy and Mid-Nineteenth Century Experimental Measurements of the Speed of Light." *Vistas in Astronomy* 36:253-94.
Tupman, G. L. 1878. "On the Mean Solar Parallax as Derived from the Observations of the Transit of Venus." *Monthly Notices of the Royal Astronomical Society* 38:429-57.
Turner, Frank Miller. 1974. *Between Science and Religion: The Reaction to Scientific Naturalism in Late Victorian England.* New Haven, Conn.: Yale University Press.
———. 1993. *Contesting Cultural Authority: Essays in Victorian Intellectual Life.* Cambridge: Cambridge University Press.
Tyndall, John. 1879. *Fragments of Science.* 6th ed. 2 vols. London: Longman.
Warner, Brian. 1983. *Charles Piazzi Smyth, Astronomer-Artist: His Cape Years.* Cape Town: Balkema.
Watson, C. M. 1910. *British Weights and Measures.* London: John Murray.
Whitworth, Joseph. 1876. "On Linear Measurement." In *Conferences Held in Connection with the Special Loan Collection of Scientific Apparatus 1876: Physics and Mechanics.* London: Chapman and Hall, 216-27.

Wise, M. Norton, ed. 1995. *Values of Precision.* Princeton, N.J.: Princeton University Press.

Yeo, Richard. 1989. "Science and Intellectual Authority in Mid-Nineteenth Century Britain." In *Energy and Entropy: Science and Culture in Victorian Britain,* edited by Patrick Brantlinger. Bloomington: Indiana University Press, 1–27.

Contributors

Jane Camerini
36 Bagley Court
Madison, Wisconsin 53705
United States of America

Paul Fayter
Departments of Natural Sciences and
  Multidisciplinary Studies
Glendon College
York University
2275 Bayview Avenue
North York, Ontario M4N 3M6
Canada

Martin Fichman
Department of Multidisciplinary
  Studies
Glendon College
York University
2275 Bayview Avenue
North York, Ontario M4N 3M6
Canada

Barbara T. Gates
Department of English
University of Delaware
Newark, Delaware 19716
United States of America

Graeme J. N. Gooday
Division of History and Philosophy of
  Science
Department of Philosophy
University of Leeds
Leeds LS2 9JT
United Kingdom

Bruce J. Hunt
Department of History
Garrison Hall
University of Texas
Austin, Texas 78712
United States of America

George Levine
Center for the Critical Analysis of
  Contemporary Culture
Rutgers University
8 Bishop Place
New Brunswick, New Jersey 08903
United States of America

Bernard Lightman
Office of the Dean of Arts
S922 Ross Building
York University
4700 Keele Street
North York, Ontario M3J 1P3
Canada

Contributors

Douglas A. Lorimer
History Department
Wilfrid Laurier University
Waterloo, Ontario N2L 3C5
Canada

James Moore
Department of History of Science and
  Technology
Open University
Milton Keynes MK7 6AA
United Kingdom

James G. Paradis
Program in Writing and Humanistic
  Studies
Massachusetts Institute of Technology
Cambridge, Massachusetts 02139
United States of America

Evelleen Richards
School of Science and Technology
  Studies
University of New South Wales
Sydney, New South Wales 2052
Australia

Joan L. Richards
History Department
Brown University
Providence, Rhode Island 02912
United States of America

Harriet Ritvo
History Department
Massachusetts Institute of Technology
Cambridge, Massachusetts 02139
United States of America

Margaret Schabas
Department of Philosophy
S428 Ross Building
York University
4700 Keele Street
North York, Ontario M3J 1P3
Canada

Simon Schaffer
Department of History and Philosophy
  of Science
Free School Lane
University of Cambridge
Cambridge CB2 3RH
United Kingdom

Ann B. Shteir
Graduate Programme in Women's
  Studies
S717 Ross Building
York University
4700 Keele Street
North York, Ontario M3J 1P3
Canada

Jennifer Tucker
Division of Humanities
California Institute of Technology
1200 East California Boulevard
Pasadena, California 91125
United States of America

Frank M. Turner
Department of History
Yale University
New Haven, Connecticut 06520
United States of America

Alison Winter
Division of Humanities and Social
  Sciences
California Institute of Technology
Pasadena, California 91125
United States of America

Index

à Beckett, Gilbert, 148, 149
Abbott, Edwin Abbott, 264–66
Abernethy, John, 27
Aborigines Protection and Anti-Slavery Society, 214, 217, 228
Acres, Birt, 386
agnosticism, 101, 103, 129, 167, 222, 467
Airy, George Biddel, 65, 66, 359, 438, 444, 445, 449, 457
Alison, William Pulteney, 40
Allen, David, 181, 356
alternative sciences, 28–31, 44. *See also* mesmerism; phrenology
Analytic Society, 52, 53
Anatomical and Ethnological Museum, 184
anatomy: comparative, 102, 150, 213, 216, 219, 220, 222, 224, 226, 228, 349, 359; human, 297; human brain, 157; and Huxley, 364; and lower-class evolutionists, 7; and nomenclature, 339; and Robert Grant, 26, 37; and Robert Owen, 360; of women, 121
Anglican Church, 3, 5, 40, 55, 56, 79, 105
animal magnetism, 30, 34, 37
Anthony, John, 431, 432
Anthropological Institute of Great Britain and Ireland, 102, 215, 216, 217, 225

Anthropological Society of London, 102, 126, 215, 216, 217
anthropology, 183, 212–30, 273, 404
antivivisection, 130–35
Appleman, Phillip, 9
Arbuthnot, John, 146
aristocratic science, 3, 7, 24, 27, 237
Arnold, Matthew, 147, 162, 170
Arnold, Thomas, 161
artificial selection, 106
artisan botanists, 190
Association for the Advancement of Medicine by Research, 135
Astronomical Society, 440
astronomy: and Clerke, 203–4, 245; and Clodd, 222; and earth-sun distance, 455; and history of Victorian science, 283; and imperialism, 313, 314, 315; and measurement, 441, 443, 445; and photography, 397; and political economy, 81; and probability theory, 53; and Proctor, 199–200; and science fiction, 268–73
atheism, 57
Atlantic cable, 317, 318, 321
Atlantic Telegraph Company, 317, 318, 325
Aukland, Lord, 367
Aveling, Edward, 103, 137
Ayton, Emily, 241

478 Index

Babbage, Charles, 39, 42, 53, 54, 57, 59, 60, 62, 75, 76, 88, 146
bacteriology, 389-95, 403
Bailey, Francis, 443
Bain, Alexander, 80, 82
Balfour, Arthur, 2, 467
Ball, William Rouse, 267
Banks, Sir Joseph, 357
Bates, Henry Walter, 369
Bathybius haeckeli, 423
Beagle, HMS, 361-63
Beale, Lionel, 421, 427, 430
Beaufort, Sir Francis, 359, 361, 365
Becker, Lydia, 249, 250, 251
Bedford College, 250
Beechey, Frederick, 359
Beer, Gillian, 8, 9
Bellamy, Edward, 108-11
Bentham, George, 241
Benthamite. *See* utilitarian
Berg, Maxine, 75
Bergson, Henri, 170
Besant, Annie, 103, 137
biology: and economic ideas, 87; and economics, 89; and evolution, 15, 85, 86; and experiment, 411; and literature, 262, 263; marine, 192; and measurement, 413; microscopical, 426; and photography, 397; and politics, 22, 94-114; and science fiction, 264; and society, 7; and specialized disciplines, 10; and Wells, 264
Blackwell, Antoinette Brown, 137
Blackwell, Elizabeth, 131
Blind, Mathilde, 262
Bodington, Alice, 208
Bohlin, Ingemar, 6
Boole, George, 266
Boole, Mary Everest, 266
botany: and artisans, 190; chair of, at Dublin, 195; and imperialism, 312, 313; and J. D. Hooker, 368; and the navy, 360; and Wallace, 302; and women, 237-43, 245, 246, 248, 250, 251
Boyle, Robert, 5, 145
brain, 157, 168. *See also* mind

Brett, Jacob, 316
Brett, John Watkins, 316
Brewster, David, 100
Bridgewater Treatises, 55
Bright, Charles, 324
Brightwen, Eliza, 183, 195-97, 208
British Association for the Advancement of Science: and electrical resistance, 417, 419, 446, 447; and electrical standards, 324, 414, 415, 416; and imperial standard, 444; and Maxwell, 460; and measurement, 441; and nomenclature committee, 340-44, 346, 347; and political economy, 75; and prejudice against technical skills, 284; and *Punch,* 154; and Tory patrons, 105
British Economic Association, 73
British Empire. *See* imperialism
British Journal of Photography, 398
British Medical Journal, 183
British Museum, 335, 342, 359, 364, 369, 370
Broderip, William John, 341
Brodie, Benjamin, 442
Brown, Robert, 183, 220-22, 366
Bryan, Margaret, 208, 238
Buckland, William, 105, 143
Buckley, Arabella (Mrs. Fisher), 197-99, 207, 208
Bullock's Museum, 335
Bunsen, Robert, 268
Butler, Samuel, 3

Cairnes, John Elliott, 77, 80, 81, 88
Cambridge, University of, 75, 250, 409, 414, 459, 460, 461. *See also* Cavendish Laboratory
Cambridge network, 359
Candolle, Augustin-Pyramus de, 242, 245, 250
Cannon, Walter (Susan), 5
capitalism, 75, 108, 110, 111, 112, 286
Carlile, Richard, 34
Carlyle, Thomas, 16-22, 147, 161, 179, 284

Carpenter, William Benjamin, 22, 24, 35–44, 80, 250, 424, 425, 427–31
cartography, 313
Catlow, Agnes, 244
Catlow, Maria, 244
Cavendish Laboratory, 417, 419, 459, 460, 463
Chalmers, Thomas, 86
Chambers, Robert, 7, 28, 45, 124, 143, 145, 207, 291, 297, 304
Charing Cross Hospital, 364
Chartism, 28, 123
chemistry, 75, 79, 149, 240, 251, 283, 398, 402, 442
Christianity, 59, 166, 212, 213, 222, 423, 451
chronometry, 313
Clark, Latimer, 323, 324, 327
Clark, William, 40
class: biases and scientific popularization, 183; boundaries and science, 367; distinctions and colonies, 371; divisions of Victorian society, 33, 137; and popularization of science, 206; and science fiction, 257; and Twining, 246. *See also* aristocratic science; artisan botanists; middle class; radical science
Clayden, Arthur, 382, 385, 386, 388
Clerke, Agnes Mary, 203–4, 208, 245
Clifford, William Kingdon, 5, 465, 466
Clifton, Robert, 414
clinometer, 454
clocks, 443
Clodd, Edward, 5, 222–24
Cobbe, Frances Power, 122, 129–36, 139
Cobden, Richard, 446
Comic Almanack, 147, 148, 150–52, 154
comic press, 147–60
Committee on Electrical Standards, 324
Comte, Auguste, 101
Conrad, Joseph, 9
Constable, John, 389
contextualism, 3, 4, 7, 9, 16, 190, 290
Conybeare, William Daniel, 40
Cookson, Henry, 460

Cooter, Roger, 189, 190, 239
Corn Laws, 440
correspondence networks, 33
Cournot, Antoine Augustin, 74
Covington, Syms, 362
Coyne, Joseph, 149
Craft, William, 218
Crémieu, Victor, 412
Cromie, Robert, 269, 270
Crookes, William, 262
Crookshank, Edgar, 393, 394
Crosse, Andrew, 31
Cruikshank, George, 148, 151, 152
Cubitt, Sir William, 143
Culler, Dwight, 19

Dalhousie, Lord James A. B. R., 367
Darwin, Charles: and agnosticism, 103; and antivivisection movement, 130, 134, 135; and barnacles, 339; and the *Beagle,* 361, 362; biography of, 291; and degeneration, 262; and feminism, 139; and fieldwork, 361–63; and French science, 342; and Henry George, 109; and human progress, 106; and imperialism, 271, 314; and London, 1; and Malthus, 88, 89, 292, 293, 294; and moral sensibility, 86; and natural historians, 182; and natural selection, 97; and naval manual, 359, 360; and neutrality of science, 95; and nomenclature, 345; and political economy, 87, 88; and psychology, 85; and racial and sexual stereotypes, 121; and satire, 170; and Victorian fiction, 171; views on women, 8, 119, 129, 132, 133, 137; and Wallace, 307
Darwinian, 3, 7, 87, 120, 121, 122, 127, 128, 136, 137, 138, 139, 212, 215, 221, 223, 262, 269, 271, 272
Darwinism, 95–97, 119–25, 129, 137, 138, 139, 263, 270, 272. *See also* evolution; Social Darwinism
degeneration, 262, 263
De la Beche, Henry, 359

De Morgan, Augustus, 21, 53, 54, 60, 62, 64, 66, 68
design, 20, 55, 59, 87, 199, 200, 204, 270. *See also* natural theology
Desmond, Adrian, 7
determinism, 20, 32, 59, 60, 212, 217, 218
Devonshire Commission, 286, 460
Dickens, Charles, 3, 9, 148
Disraeli, Benjamin, 2, 155
Dissent, 7, 40, 55, 364
Dos Santos, Jean Battista, 183, 184
Doubleday, Edward, 369
Doyle, Arthur Conan, 261, 392
dualism, 5, 6, 7, 80, 158, 169
Duguid, David, 398, 400, 401
du Maurier, George, 151

Eastern Telegraph Company, 321
Economic Journal, 75
economics, 79-88. *See also* political economy
Eden, George, 367
Edgeworth, Francis Ysidro, 76, 81, 82, 88
Edgeworth, Maria, 208
education: expansion of, 228; and girls' schools, 246; home-based, 244; and Huxley, 256; medical, 150, 154; and popularization of science, 205; reform of, 103, 227; and science writing for women, 239; women's, 249, 250, 251
electrical standards, 324, 414, 416
electrical theory, 326
Electrician, 328
electricity, 30, 265, 323, 326, 329, 466
electromagnetic theory of light, 440
Eliot, George, 2, 8, 171, 181
Elliotson, John, 30, 34, 35, 36, 38, 41
Ellis, Robert Leslie, 21, 54, 63, 64-68
embryology, 97
English Mechanic, 205
Enlightenment, 68, 74, 78, 80, 86-88, 97, 237, 243, 251, 334-37
entomology, 238, 240, 244
entropy, 263
Erebus, HMS, 366

ether, 464
Ethnological Society of London, 102, 112, 125-28, 214, 215, 217, 220, 228
eugenics, 99, 100, 106, 111, 262
evangelicalism, 196, 213, 239, 448, 455, 460
evolution: and Buckley, 198, 199; and Cobbe, 130; economic component of, 87; and *International Scientific Series*, 205; and Kingsley, 162; and politics, 94-114; and race, 214, 215, 216, 224; and satire, 150, 151; and science fiction, 262, 263; social, 97; and Wallace, 108-13; and women, 8, 119-22, 127, 128, 137, 138. *See also* Darwinism
Ewart, William, 446
Excise Laboratory, 441, 442
experiment, 31, 288, 411, 415, 416, 433
externalism, 4, 5, 6, 288, 290
extraterrestrial life, 199, 275

Fabians, 88, 111
Factory Inspectorate, 441
Falconer, Dr. Hugh, 367
Faraday, Michael, 30, 34, 285, 326, 327, 328
Farr, William, 442
Fawcett, Henry, 88
Fechner, Gustav Theodore, 80, 82
feminism, 110, 122-24, 129, 135, 139, 269
Field, Cyrus, 317
field theory, 9, 315, 316, 326, 328, 329, 461
fieldwork, 354-74
Figaro, 147, 148
Fitzroy, Captain Robert, 361
Flammarion, Camille, 264, 269
Fletcher, James, 39
Flower, William Henry, 339, 340, 345
Forbes, Edward, 364, 365
Forbes, Henry O., 338
Forbes, James D., 65
Foster, George Carey, 416, 417

Foucault, Léon, 457
Foucault, Michel, 349
Foxwell, Herbert Somerton, 76, 84
Francis, George William, 239, 241
Frye, Northrop, 144, 145

Galton, Francis, 95, 99-101, 183, 217, 262, 269, 397
galvanometer, 413
Gamble, Eliza Burt, 110, 137
Gass, F., 400
Gatty, Margaret, 192-95, 207, 241
gender: and attack on Clerke, 204; and contextualism, 8; and Darwinism, 121, 139; and Huxley, 126; and narrative form of science, 237, 248-52; and professional science, 241; and science fiction, 273; and science writing, 239; and Victorian authorship, 244. *See also* women
General Register Office, 441, 442
genetics, 99, 229
genius, 307
Geoffroy Saint-Hilaire, Étienne, 36
geography, 314, 330
Geological Society, 34
geology: and divine action, 40; and electricity, 30; and fieldwork, 356; and genesis, 2; and history of Victorian science, 283; and imperialism, 312, 313, 314, 329, 330; and mining, 75; and satire, 153; status of, 75; and Wallace, 302; and H. G. Wells, 264; and women, 237, 238, 240
geometry, 264, 265, 267, 465
George, Henry, 104, 109
Gifford, Isabella, 244
Gilman, Charlotte Perkins, 110
Gladstone, William Ewart, 2, 155, 378, 379
Glasgow, University of, 414
Glazebrook, Richard, 459
Glendinning, Andrew, 399-402
Gobineau, Joseph, 229
Good Words, 448
Grant, Robert, 26, 36, 38
Gratacap, Louis Pope, 269

Gray, George R., 370
Gray, John Edward, 364, 369, 370
Great Eastern, 324
Greene, John, 5
Greenwich Observatory, 268, 445
Greg, Percy, 269
Gregory, Richard A., 204
Grove, William Robert, 35
gutta-percha, 316, 327

Hall, Asaph, 268
Hall, Marshall, 36, 37
Hall, Spencer Timothy, 28
Hamilton, William, 18
Hardwicke's Science Gossip, 240, 252
Hardwicke, Thomas, 339
Hardy, Thomas, 2, 8, 9
Hartley, David, 80
Harvey, William Henry, 195
Haslar Naval Hospital, 358
Haweis, Hugh Reginald, 398
Haywood, Eliza, 237
Heaviside, Oliver, 328
Helmholtz, Hermann von, 82, 265
Henslow, John Stevens, 40, 41
Henslow, Reverend Robert, 362, 363, 364
heredity, 100
Herschel, John: and Analytic Society, 53; and Babbage, 54; and borders of knowledge, 51; and Carpenter, 40, 41; and design, 20; and direct personal knowledge, 58; and Ellis, 65; and gentlemen of science, 24; and measurement, 442, 444-50; and nature of science, 100, 146; and naval manual, 359; and political economy, 75; and probability theory, 62, 67, 68; and religion, 52
Hertz, Heinrich, 328, 329
Heyl, Paul, 267
Hibberd, Shirley, 241
Hinton, Charles Howard, 256, 264, 266, 267
histology, 426
historiography, 139, 214, 283, 289, 290, 433. *See also* contextualism; externalism; internalism

Hobhouse, Leonard Trelawney, 228
Hodgkin, Dr. Thomas, 214
Hodgson, Brian, 368
Holland, Henry, 40, 41
Hooke, Robert, 145
Hooker, Joseph Dalton, 101, 314, 355, 364, 366–69
Hooker, Sir William, 359, 360, 364, 366, 367
Horsfield, Dr. Thomas, 369
Howard, Sir Michael, 285
Hudson, William Henry, 274
Huggins, William, 461, 462
Humboldt, Baron Alexander von, 367
Humboldtianism, 358, 359, 367
Hume, David, 59, 76, 78, 87, 88
Hunt, Dr. James, 215, 217, 218
Hunterian Museum, 359
Huxley, Thomas Henry: and agnosticism, 167; and Anthropological Institute, 215; and Anthropological Society, 127; and antivivisection movement, 130, 134; and atoms, 464; biography of, 291; and degeneration, 263; and education, 183; and feminism, 139; and fieldwork, 363–66; and imperialism, 314; and irony, 147, 161; and microscope, 421; and naval manual, 360; and neutrality of science, 95, 101, 102–6; and popularization of science, 187, 207, 256; and protoplasm, 423; and *Rattlesnake,* 363–66; and satire, 164–68, 169, 171; satire of, 147, 157; and scientific naturalism, 3, 5; views on women, 8, 120–21, 125–28, 138
hydrography, 359
hydrometer, 441

Idler, 261, 378
Illustrated London News, 259
illustrations, scientific, 403, 404
imperialism, 8, 100, 228, 229, 270, 271, 312–15, 329, 334–50, 355–74, 394, 468
India Museum, 369

industrialization, 3, 75, 99, 257, 355, 356
Inkster, Ian, 9
Institute of Plant Physiology, 391
Institution of Civil Engineers, 389
instruments, scientific, 284, 289, 323, 409–34, 438, 441, 443, 454, 448. *See also* microscope; spectroscope; telescope
internalism, 4, 5, 6, 190, 288, 290
International Institute for Preserving and Perfecting Weights and Measures, 451
International Scientific Series, 205, 207
Inwards, Richard, 382
Irish Royal College of Science, 250

James, Sir Henry, 454, 455, 463
Jardine, William, 339
Jelf, Ernest Arthur, 378, 379
Jenkin, Fleeming, 88, 323, 324, 325, 327, 414, 415
Jennings, Richard, 80, 88
Jenyns, Leonard, 340, 348
Jerrold, Douglas, 148, 149
Jevons, William Stanley, 73, 74, 76, 77, 81, 82, 85, 88
Johns, Charles Alexander, 239
Johnstone, George, 195
Jones, Alice, 269
Joule, James, 410, 411, 440, 446, 448, 451
Jussieu, Bernard de, 245

Kater, Henry, 443, 444
Keane, Augustus Henry, 224–28
Kelland, Philip, 66
Kelly, Patrick, 440
Kelvin, Lord, 8, 207, 322–25, 327, 438, 446
Kew Gardens, 313, 335, 355, 360
Kew observatory, 441
Keynes, John Maynard, 77, 83
Keynes, John Neville, 77
Kierkegaard, Søren, 144, 169
Kings College, 413
Kingsford, Anna, 131

Kingsley, Charles, 86, 147, 161–64, 169, 181, 192, 208, 251, 252
Kingsley, Mary, 185
Kirchoff, Gustav, 268
Klein, Edward, 394
Knoepflmacher, U. C., 10
Knox, Dr. Robert, 215, 229
Koch, Robert, 390–92
Kohlrausch, Friedrich, 415, 416
Krasner, James, 182

laboratory, 250, 284, 287–99, 322, 324, 389, 393, 409–14, 420–24, 441–46, 467. *See also* Cavendish laboratory
Ladies' Companion at Home and Abroad, 240
laissez-faire, 78, 84, 104
Lamarck, Jean-Baptiste, 26, 97
Lamarckism, 96, 98
Lancet, 183
Lane-Fox, Colonel Augustus, 216
Lankester, Edwin Ray, 5, 262, 342
Lankester, Phebe, 245
Laplace, Pierre-Simon de, 57, 58
Lardner, Dionysius, 34
Latham, Robert Gordon, 220, 225
Latour, Bruno, 411, 421, 467
Lawrence, William, 27, 34, 297
Lear, Edward, 347, 348
Lee, Sarah Bowdich, 244
Leech, John, 149, 150, 158
Leigh, Percival, 149, 150
Lemon, Mark, 148, 149, 151
Lennox, Charlotte, 238
Leverrier, Urbain, 455
Levine, George, 9
Lewes, George Henry, 2, 5
libel laws, 42
Lindley, John, 242–44, 251
Linnaean system, 239, 242, 245, 337, 342, 347
Linnaeus, Carolus, 336, 338, 339, 350
Linnean Society, 365
Linton, Eliza Lynn, 122–28, 135–39
Linton, William, 123, 137
Lister, Sir Joseph, 391
Liston, Robert, 34

literature and science, field of, 171
Lockyer, Norman, 204, 460
Lodge, Oliver, 86, 329
London, University of, 242, 250, 413, 414
London School of Economics, 75
Loudon, Jane, 240, 245, 248
Lovejoy, A. O. J., 4
Lowell, Percival, 269, 272
Lubbock, John, 5, 101
Lyell, Charles, 24, 42, 87, 98, 197, 345

MacGillivray, John, 365
Madden, William A., 9
Malthus, Thomas Robert, 6, 26, 74, 79, 87, 88, 257, 292, 303, 304
Marcet, Jane, 208, 244, 247
Marchant, Ella, 269
Marriott, William, 385, 386
Mars, 268–73, 275
Marshall, Alfred, 72, 73, 74, 77, 78, 81, 83, 85, 88, 89
Martin, Benjamin, 243
Martineau, Harriet, 294
Marx, Karl, 76
Marxism, 290. *See also* socialism
materialism: and Carpenter, 38, 39, 41; and Cobbe, 130, 134, 138; and conservative thinkers, 27; and early Victorian life sciences, 26, 27; and Elliotson, 35; French, 24; and French Revolution, 21; and Huxley, 166, 167; and intentions of scientist, 32; and Lawrence, 34; and Linton, 124, 125; and Maxwell, 459; and popularization of science, 182; and radicalism, 29; and Spencer, 99
mathematics, 31, 51–69, 464. *See also* geometry; probability theory
matter, 27, 36, 38, 39, 80, 157, 162, 409, 464, 465
Maudsley, Henry, 5, 80
Maunder, Edward Walter, 268
Maxwell, James Clerk: and cable telegraphy, 322; and electrical standards, 324, 414; and electromagnetic theory, 8, 328, 440; and

Maxwell, James Clerk (*continued*)
field theory, 326, 329; and measurement, 442, 443, 459-67; and popularization of science, 207; and resistance unit, 446; on science and commerce, 285
Mayhew, Henry, 148, 149
McClure's Magazine, 261
McCormick, Robert, 361
McCulloch, John Ramsay, 77
measurement, 284, 325, 413, 434, 438-68
medicine, 30, 75, 133, 134, 135, 150, 154, 179, 283. *See also* anatomy; physiology
mesmerism, 25, 28, 29, 31, 44, 45
Metaphysical Society, 167, 168
metaphysics, 150. *See also* dualism; idealism; materialism
meteorology, 31, 381-89, 397, 403, 404
metrication, 441, 446, 449, 451
metric system, 440, 448
metrology. *See* measurement
microbiology, 390
microscope, 182, 201-3, 244, 360, 364, 391, 392, 413, 420-34, 449
Microscopical Society, 420
middle class: complacency, 158; and Darwinism, 120; families and science writing, 247; feminists, 126; and Lindley, 251; and popularization of science, 188; values, 111; and Victorian intellectuals, 99; and Wood, 219. *See also* class
Mill, James, 80
Mill, John Stuart, 17, 20, 21, 72, 73, 75, 76, 77, 79, 85, 86, 88, 100, 126, 129
Millar, Frederick, 112
mind: and the economy, 77-80, 83-87; evolution of, 130; and knowledge, 67; and machines, 53; materiality of, 29; and matter, 38, 465; and organic processes, 27; and physical forces, 35; and probabilities, 58, 68; sciences of, 31. *See also* brain
Mivart, St. George, 101

monogenesis, 215, 216, 220, 225
Monro, Cecil James, 461
Moore, James, 10
morality, 17, 130, 132, 134, 154, 205, 222, 401, 442, 443, 445, 463, 466
Morley, John, 101
Morrell, Jack, 7, 9
Mumler, William Henry, 396
Murchison, Roderick, 8, 156, 313, 314, 330
Murry, Ann, 238
Museum of Economic Botany, 360
museums, 184, 185, 238, 285, 335, 338, 359, 360, 363, 369, 373, 410, 420. *See also* British Museum
Myers, Greg, 247
mysticism, 20, 38

narrative form of science, 192, 207, 242, 247, 252
nationalism, 302, 467
National Physical Laboratory, 467
Natural History and Antiquities of Selborne, 151
natural history: aesthetics of, 373; and British culture, 357; and Darwin, 363; and fieldwork, 354, 355; and imperialism, 336, 349, 350; and landscape art and exploration, 404; and London, 369; and narrative form, 171; and natural theology, 247; and navy, 334; and nomenclature, 337, 339; and periodicals, 239; and professionalization of science, 182; scholarship on, 374; and schools, 205; social studies of, 356; and Victorians, 181; and J. G. Wood, 218
naturalists, 181, 339, 341, 343, 344, 346, 358, 361, 374
natural law, 19, 24, 26, 29-31, 33, 36, 38-41, 43, 77, 79, 120, 199, 207
natural order, 180
natural philosophy, 24, 355, 463, 466
natural selection, 96, 97, 106, 108, 111, 120, 271, 293, 306
natural theology: and Anglicanism, 3, 6, 55; and Brightwen, 195; and Buckley, 199; and Clerke, 204;

and Clifford, 465; and common intellectual context, 189; and Darwinism, 120; and Gatty, 193, 194; historiography of, 69; and Kingsley, 161; and Maxwell, 465; narratives of, 247; and personal knowing, 63; and popularization of science, 188, 206; and probability theory, 54, 62; and Proctor, 200; and University of Cambridge, 461; and women, 239; and J. G. Wood, 200. *See also* design
Nature, 274, 460
navigation, 313
Navy, British 355, 357, 358, 359, 361, 363, 367
nebular hypothesis, 270, 272
Newcomb, Simon, 264
Newman, George, 394
Newman, John Henry, 21, 54, 56, 57, 59
Newnes, George, 261
Newton, Alfred, 341
Newton, Isaac, 5
nomenclature, 284, 335-50
Nonconformity. *See* Dissent
Nordau, Max, 262
North, Marianne, 185

ohm, 325, 328, 416, 419, 446, 463
Ordnance Survey, 441, 443, 454, 455
Ordnance Survey of the Pyramids, 461
ornithology, 237
Owen, Richard, 101, 102, 143, 149, 157, 337, 340, 359, 360, 364
Owen, Robert, 301
Owenites, 88
Owens College, 414
Oxenford, John, 153
Oxford, University of, 414

Paley, William, 6, 247
Pall Mall Magazine, 261
Paradis, James, 10
Pascoe, Francis, 346
Pasteur, Louis, 391
Pearson, Karl, 99, 103, 111
Pearson's Magazine, 261, 272

Pearson's Weekly, 256, 261
Pender, Harold, 412
Pender, John, 321, 322, 452
pendulums, 441, 443, 448
Pennant, Thomas, 336, 338
Petrie, Flinders, 458
Petrie, William, 455, 457, 464
Phillips, George, 442
Philosophical Transactions, 274
Photographic News, 393
photography, 226, 268, 378-404
phrenology, 25, 28, 31, 32, 44, 103
physics: and cable telegraphy, 316, 323, 324, 330; and economic theory, 86; Evangelical, 460; and experiment, 411, 413, 421, 432; and field theory, 328; and imperialism, 312, 313, 315; institution of, 468; and instruments, 423; laboratories of, 414, 416, 417, 419, 460; and materialism, 464; and measurement, 325, 413, 426, 446, 453, 459, 460; and photography, 397; and political economy, 77, 88; and probability theory, 58; and science fiction, 264, 266; as specialized discipline, 10; and university culture, 461; and unseen spirits, 262; and women, 251
Physiocrats, 78, 79
physiology: and Bain, 80; and Brightwen, 195; and Carpenter, 35-39; and Darwin, 130; and definition of disciplines, 31; and gender difference, 121; and Huxley, 134, 167; and William Lawrence, 34; and J. S. Mill, 85; of plant tissue, 246; and political economy, 79; and psychology, 80; and Roget, 39
Pickering, William Henry, 269
Playfair, Lyon, 101
political economy, 72-89
political science, 88
polygenesis, 215, 216, 224, 226, 229
Pope, Alexander, 146
popularization of science, 182, 183, 187-208, 252, 256, 257

popularization of scientific racism, 218-30
Popular Science Review, 430
Porcupine, HMS, 424
Postlewait, Thomas, 10
Powell, Baden, 40, 291
Poynting, John, 264
Pratt, Anne, 244
Prichard, James Cowles, 39, 40, 225, 359
probability theory, 51-69
Proctor, Richard Anthony, 199-200, 208, 222
professionalization of science: and amateurs, 181, 194, 374; and evolutionary biology, 94, 102; and exclusion of nonprofessionals, 205; and gender, 241; and neutrality, 95; and popularization of science, 182, 188, 189, 190, 191, 206, 207; and scientific naturalism, 101; and secularization, 228
protoplasm, 421, 422, 423
pseudosciences. *See* alternative sciences
psychology, 8, 16, 74, 77, 79, 80, 82-87, 98, 213, 227
publishing, 261
publishing technology, 259, 260
Pumfrey, Stephen, 189, 190, 239
Punch, 143-45, 148-60, 170, 347
Pyenson, Lewis, 315
pyramidology, 450-58, 463, 464

Quarterly Journal of Science, 262
Queckett, John, 426-29
Quesnay, François, 78
Quetelet, Adolph, 68

race, 99, 100, 108, 109, 183, 212-30
radical science, 7, 26-28
Raffles, Thomas Stamford, 342, 343
Rationalist Press Association, 222
Rattlesnake, HMS, 363-66
Raupert, John Godfrey, 395
Rayleigh, Lord, 409, 417, 418, 419, 459, 464, 467
Reade, Reverend Joseph Bancroft, 421, 430, 431, 432

reading audience, Victorian, 190, 191, 205-7, 236-52, 257, 259, 261
Reeve, Lovell, 244
religion: and Darwinian revolution, 179; and inseparability from science, 52; and interaction with science, 69; and knowing, 67; and lessons of nature, 191; and orthodoxy in science, 40; and probability theory, 59, 60; reconciled with science, 256
religious knowledge, 52, 57
Religious Tract Society, 423
resistance, electrical, 324, 325, 413, 414, 417, 419, 440, 446, 447. *See also* ohm
resistance standards, 416, 434
Ricardo, David, 73, 75, 76, 78, 85-87
Richard, David, 72
Richards, Evelleen, 8
Richardson, Sir John, 343, 358, 364, 365
Riemann, Bernhard, 264, 265
Roberts, Mary, 244
Roget, Peter Mark, 34, 39, 40
Roman Catholic Church, 166
Rorty, Richard, 167
Ross, James Clark, 367
Rothschild, Lionel Walter, 344
Rowland, Henry, 412, 416
Royal Academy, 389
Royal Astronomical Society, 199, 457
Royal Botanical Gardens at Kew. *See* Kew Gardens
Royal College of Science, 413
Royal College of Surgeons, 426, 427, 428
Royal Commission on Water Supply, 442
Royal Economic Society, 73
Royal Geographical Society, 213, 359
Royal Meteorological Society, 381-86, 388
Royal Microscopical Society, 424, 431
Royal Society, 145, 146, 347, 357, 365
Royal Society for Prevention of Cruelty to Animals, 130
Royal Society of Edinburgh, 450

Rudwick, Martin, 7
Ruskin, John, 3, 192, 208, 249, 347, 389
Russett, Cynthia, 8

Sabine, Edward, 359
Saint-Simon, Claude Henri, 101
Salisbury, Lord, 467
Schiaparelli, Giovanni, 268, 269
Schools Inquiry Commission, 250
Schuster, Arthur, 417-19
Science Defence Association, 135
science fiction, 183, 256-75
Science Gossip and Popular Science Review, 261, 262
science journals, 183, 191, 205, 208, 240, 252, 261, 262, 274, 430, 460
scientific naturalism: and agnosticism, 103; and Brightwen, 196; and Buckley, 198; and Cambridge divines, 464; and change, 214, 217; cultural authority of, 3, 218; and Galton's eugenics, 100; and Gatty, 193; ideological dimensions of, 114; and irony, 161, 166; and Kingsley, 162; and Linton, 124; and middle-class professionals, 5; and neutrality of science, 95, 106; and optimism, 114; and popularization of science, 188, 206; and race, 216, 229; and secularization, 101, 191; social objectives of, 170; and Tory-Anglican establishment, 8
Scientific Revolution, 94, 96, 180, 283, 440
screw gauge, 438, 441
Scrope, George Poulett, 88
secularists, 451
secularization, 213, 215, 228
Sedgwick, Adam, 24
Senior, Nassau, 75, 76, 77
sexual selection, 106, 108-14
Shadwell, Thomas, 145
Sharpey, William, 34
Shaw, George, 338
Sheepshanks, Richard, 444
Shore, Emily, 185

Siemens, Werner, 325
Siemens, William, 438, 466
Smith, Adam, 79, 86, 87
Smith, Albert, 149, 150, 154
Smith, Crosbie, 8
Smith, James Edward, 245
Smith, John Pye, 40
Smith, Southwood, 36
Smyth, Charles Piazzi, 442, 450-58, 463, 464
Social Darwinism, 7, 88, 121, 179. *See also* Darwinism
socialism, 84, 106, 108, 109, 110, 112, 290, 301, 303, 307
social science, 8, 53, 87, 94, 96, 99, 182, 227
Somerville, Mary, 207, 236
spectroscope, 198, 200, 204, 268, 414, 434, 441, 461-63, 465
Spencer, Herbert, 5, 80, 85, 88, 95, 98-99, 101, 121, 124
Spiers, William, 432
spiritualism, 10, 28, 35, 45, 86, 110, 112, 131, 166, 292, 307, 395-403
Stafford, Robert, 8, 314
Stanley, Captain Owen, 364, 365
Statistical Department of the Board of Trade, 441
Stead, William Thomas, 188
Stephen, Leslie, 101
Stevens, John Crace, 369
Stevens, Samuel, 369, 370
Stevenson, Robert Louis, 263
Stewart, Balfour, 86, 207, 414
Stokes, George Gabriel, 86, 464
Strand, 256, 261
Strickland, Hugh, 343, 345
Sturgeon, William, 30
Sully, James, 80, 82
surveying, 301, 449
Swainson, William, 357
Swift, Jonathan, 146
Sydenham College, 363
Symons, George James, 383-86

Tait, Peter Guthrie, 86, 464
Taylor, John, 450
Taylor, John Traill, 398-402

Telegraph Construction and Maintenance Company, 321
telegraphy, 312–30, 443, 446, 458
telescope, 182, 200, 268
Tenniel, John, 156, 170
Tennyson, Alfred, 2
Tennyson, G. G., 10
terrestrial magnetism, 313
Terrot, George, 66
Thackeray, William, 148
Thackray, Arnold, 7
theism, 129, 138
theodicy, 7
theodolites, 441, 448
theology, 167, 460, 466. *See also* natural theology
thermodynamics, 263
thermometer, 410, 411, 421
Thirlwall, Connop, 160, 161
Thomson, Joseph John, 409, 412
Thomson, Sir Wyville, 424, 425
Thomson, William. *See* Kelvin, Lord
Tichborne trial, 397, 398
Tomlinson, Sarah, 248, 249
Tory, 5, 8, 29, 102, 146, 242, 291, 443, 467
Trollope, Anthony, 9
Trotter, Coutts, 412
Turner, Frank, 5, 21
Turner, J. M. W., 389
Twining, Elizabeth, 246, 247
Tylor, Edward Burnett, 5, 219, 222
Tyndall, John: and Belfast Address, 105, 465; and Carlyle, 21; and Koch, 391; and microscope, 422; and popularization of science, 187, 207; satire of, 156, 157; and scientific naturalism, 3, 5, 95, 101, 464

Unitarian, 39, 40, 83
University College, 30, 34, 413, 416, 417
utilitarian, 3, 20, 27, 28, 29, 86, 130, 191

Varley, Cromwell Fleetwood, 323
Veblen, Thorstein, 77

vivisection. *See* antivivisection
Vyse, Richard William Howard, 449

Wade, Emily, 123
Wakefield, Priscilla, 244, 247
Wallace, Alfred Russel: biographies of, 292, 307; and Darwin, 108, 292, 307; early life of, 300–301; and ethnography, 299; and ethnology, 295; and fieldwork, 369, 371; and "Hall of Science," 185; and imperial context, 314; and Keane, 225; and Malthus, 293–96, 303, 304, 305; and mapmaking, 288; and Mars, 272; and natural selection, 306; and neutrality strategy, 95, 101; and nomenclature, 340; and primates, 298, 299; and sexual selection, 108–13; and socialism, 103, 104, 107–14, 137, 301, 303, 307; and spirit photography, 396; and spiritualism, 307; and surveying, 301; and *Vestiges,* 304; and "Wallace's Line," 296; and Welsh farmers, 302, 305
Walras, Leon, 74, 76
Ward, Mary, 244
Webb, Sidney, 76
Weber, Wilhelm, 325
Weights and Measures Act, 441
Weismann, August, 97
Weldon, Maude, 266
Wells, Herbert George, 183, 256–58, 261, 263, 264, 270, 272, 275
Wheatstone, Charles, 34
Whewell, William: as apologist for science, 18; and atheism, 57; and Babbage, 60; and Cambridge, 53; caricature of, 149; class origins of, 285; and cultural authority of science, 100, 105; and deductive science, 59; and design, 20, 55; and gentlemen of science, 24; and German Romantic philosophy, 21; and inductive science, 56; and mathematics, 68, 69; and nature of knowledge, 64, 146; and naval manual, 359; and political econ-

omy, 77; and process of discovery, 63; and word "scientist," 16
Whig, 29, 290, 291, 293, 294, 302, 306, 365
White, Gilbert, 151
White, James, 324
Whitehouse, Wildman, 317, 323
Whitworth, Joseph, 438, 440
Wicksteed, Philip Henry, 76, 81, 83
Wilberforce, Bishop Samuel, 164
Willink, Arthur, 267
Wise, M. Norton, 8
Wolff, Michael, 9
women: and capitalism, 110; and evolutionary theory, 119-39; and professions, 113; and science, 8, 10, 21, 112, 181, 183, 188, 205, 236-52; and sexual selection, 110, 111; and spiritualism, 45. *See also* gender

Wood, Reverend John George, 182, 183, 185, 200-203, 208, 218, 219, 221
Working Men's College, 246
Wright, Lewis, 423
Wundt, Wilhelm, 80

X Club, 34, 103, 170

Young, Robert M., 6-7, 11, 87, 189, 290, 291

Zangwill, Israel, 264
Zoological Society of London, 335, 338, 342
zoology, 244, 264, 312, 314, 339, 341, 346, 348, 359